Fuel Systems and Emission Controls

Second Edition

**By Chek-Chart Publications,
a Division of
H. M. Gousha**

Roger Fennema, *Editor*
Gordon Clark, *Contributing Editor*
Victoria Easterday, *Contributing Editor*

HarperCollins*Publishers*

Acknowledgments

In producing this series of textbooks for automobile mechanics and technicians, Chek-Chart has drawn extensively on the technical and editorial knowledge of the nation's carmakers and suppliers. Automotive design is a technical, fast-changing field, and we gratefully acknowledge the help of the following companies in allowing us to present the most up-to-date information and illustrations possible:

Allen Testproducts
American Motors Corporation
Borg-Warner Corporation
Caldo Automotive Supply
Champion Spark Plug Company
Chrysler Motors Corporation
Coats Diagnostic (Hennessy Industries)
Ford Motor Company
Fram Corporation, A Bendix Company
General Motors Corporation
   AC-Delco Division
   Delco-Remy Division
   Rochester Products Division
   Saginaw Steering Gear Division
   Buick Motor Division
   Cadillac Motor Car Division
   Chevrolet Motor Division
   Oldsmobile Division
   Pontiac Division
Jaguar Cars, Inc.
Marquette Mfg. Co. (Bear Mfg. Co.)
Mazda Motor Corporation
Nissan Motors
The Prestolite Company, An Eltra Company
Robert Bosch Corporation
Sun Electric Corporation
Toyota Motor Company
Volkswagen of America

The authors have made every effort to ensure that the material in this book is as accurate and up-to-date as possible. However, neither Chek-Chart nor HarperCollins nor any related companies can be held responsible for mistakes or omissions, or for changes in procedures or specifications made by the carmakers or suppliers.

The comments, suggestions, and assistance of the following contributors were invaluable:

Les Clark, Russ Suzuki, Angel Santiago, Al Bauer, and Bryan Wilson; General Motors Training Center, Burbank, Calif.

Pete Egus; AC-Delco, Los Angeles, Calif.

Robert Baier, Dan Rupp, Nick Backer, and Bill Takayama; Chrysler Training Center, Ontario, Calif.

Robert Van Antwerp, Jim Milum, and Ed Moreland; Ford Training Center, La Mirada, Calif.

Bob Kruze; Merry Oldsmobile, San Jose, Calif.

At Chek-Chart, Ray Lyons managed the production of this book. Original art and photographs were produced by Gordon Agur, John Badenhop, Jim Geddes, C. J. Hepworth, Janet Jamieson, Kalton C. Lahue, and F. J. Zienty. The project is under the direction of Roger L. Fennema.

FUEL SYSTEMS AND EMISSION CONTROLS, Second Edition, Classroom Manual and Shop Manual Copyright © 1988 by Chek-Chart, Simon & Schuster Inc.

All rights reserved. Printed in the United States of America. No part of this publication may be reproduced, stored in a retrieval system, or duplicated in any manner without the prior written consent of Chek-Chart, Simon & Schuster Inc., Box 49006, San Jose CA 95161-9006.

Library of Congress Cataloging and Publication Data:
Chek-Chart, 1988
   Fuel Systems and Emission Controls
   (HarperCollins/Chek-Chart Automotive Series)
v. 1. Classroom Manual. v. 2. Shop Manual.

ISBN: 0-06-454016-2 (set)
Library of Congress Catalog Card No.: 87-33433

Contents

On the Cover:
Front — A cutaway of a port-fuel-injected V-6 showing the airflow through the engine, courtesy Chevrolet Division of the General Motors Corporation.
Rear — Schematic drawings of a typical EGR valve and diverter valve, courtesy AC-Delco Division of the General Motors Corporation.

Contents

Introduction to Fuel Systems and Emission Controls

Fuel Systems and Emission Controls is part of the Harper & Row/Chek-Chart Automotive Series. The package for each course has two volumes, a *Classroom Manual* and a *Shop Manual*.

Other titles in this series include:
- Automatic Transmissions and Transaxles
- Automotive Brake Systems
- Automotive Electrical and Electronic Systems
- Automotive Engine Repair and Rebuilding
- Engine Performance Diagnosis and Tune-Up

Each book is written to help the instructor teach students to become excellent professional automotive mechanics. The 2-manual texts are the core of a complete learning system that leads a student from basic theories to actual hands-on experience.

The entire series is job-oriented, especially designed for students who intend to work in the car service profession. A student will be able to use the knowledge gained from these books and from the instructor to get and keep a job. Learning the material and techniques in these volumes is a giant leap toward a satisfying, rewarding career.

The books are divided into *Classroom Manuals* and *Shop Manuals* for an improved presentation of the descriptive information and study lessons, along with the practical testing, repair, and overhaul procedures. The manuals are to be used together: the descriptive chapters in the *Classroom Manual* correspond to the application chapters in the *Shop Manual*.

Each book is divided into several parts, and each of these parts is complete by itself. Instructors will find the chapters to be complete, readable, and well thought-out. Students will benefit from the many learning aids included, as well as from the thoroughness of the presentation.

The series was researched and written by the editorial staff of Chek-Chart, and was produced by Harper & Row Publishers. For over 59 years, Chek-Chart has provided car and equipment manufacturer's service specifications to the automotive service field. Chek-Chart's complete, up-to-date automotive data bank was used extensively to prepare this textbook series.

Because of the comprehensive material, the hundreds of high-quality illustrations, and the inclusion of the latest automotive technology, instructors and students alike will find that these books will keep their value over the years. In fact, they will form the core of the master mechanic's professional library.

How To Use This Book

Why Are There Two Manuals?

This two-volume text — **Fuel Systems and Emission Controls** — is not like any other textbook you've ever used before. It is actually two books, the *Classroom Manual* and the *Shop Manual*. They should be used together.

The *Classroom Manual* will teach you what you need to know about basic electricity and the electrical systems in a car. The *Shop Manual* will show you how to fix and adjust those systems, and how to repair the electrical parts of a car.

The *Classroom Manual* will be valuable in class and at home, for study and for reference. It has text and pictures that you can use for years to refresh your memory about the basics of automotive electrical systems.

In the *Shop Manual*, you will learn about test procedures, troubleshooting, and overhauling the systems and parts you are studying in the *Classroom Manual*. Use the two manuals together to fully understand how the parts work, and how to fix them when they don't work.

What's In These Manuals?

There are several aids in the *Classroom Manual* that will help you learn more:

1. The text is broken into short bits for easier understanding and review.

2. Each chapter is fully illustrated with drawings and photographs.

3. Key words in the text are printed in **boldface type** and are defined on the same page and in a glossary at the end of the manual.

4. Review questions are included for most chapters. Use these to test your knowledge.

5. A brief summary at the end of most chapters will help you to review for exams.

6. Every few pages you will find short blocks of "nice to know" information, in addition to the main text.

7. At the back of the *Classroom Manual* there is a sample test, similar to those given for National Institute for Automotive Service Excellence (NIASE) certification. Use it to help you study and to prepare yourself when you are ready to be certified as an expert in one of several areas of automobile mechanics.

The *Shop Manual* has detailed instructions on overhaul, test, and service procedures. These are easy to understand, and may have step-by-step, photo-illustrated explanations that guide you through the procedures. This is what you'll find in the *Shop Manual*:

1. Helpful information tells you how to use and maintain shop tools and test equipment.

2. Safety precautions are detailed.
3. System diagrams help you locate trouble-spots while you learn to read the diagrams.
4. Tips the professionals use are presented clearly and accurately.
5. A full index will help you quickly find what you need.
6. Test procedures and troubleshooting hints will help you work better and faster.

Where Should I Begin?

If you already know something about a car's basic electrical system and how to repair it, you may find that parts of this book are a helpful review. If you are just starting in car repair, then the subjects covered in these manuals may be all new to you.

Your instructor will design a course to take advantage of what you already know, and what facilities and equipment are available to work with. You may be asked to take certain chapters of these manuals out of order. That's fine. The important thing is to really understand each subject before you move on to the next.

Study the vocabulary words in boldface type. Use the review questions to help you understand the material. While reading in the

Classroom Manual, refer to your *Shop Manual* to relate the descriptive text to the service procedures. And when you are working on actual car systems and electrical parts, look back to the *Classroom Manual* to keep the basic information fresh in your mind. Working on such a complicated piece of equipment as a modern car isn't always easy. Use the information in the *Classroom Manual,* the procedures of the *Shop Manual,* and the knowledge of your instructor to help you.

The *Shop Manual* is a good book for work, not just a good workbook. Keep it on hand while you're working on equipment. It folds flat on the workbench and under the car, and can withstand quite a bit of rough handling.

When you do test procedures and overhaul equipment, you will also need a accurate source of manufacturers' specifications. Most auto shops have either the carmaker's annual shop service manuals, which lists these specifications, or an independent guide, such as the **Chek-Chart Car Care Guide**. This unique book, which is updated every year, gives you the complete service instructions, electronic ignition troubleshooting tips, and tune-up information that you need to work on specific cars.

PART ONE

Fuel System and Emission Control Fundamentals

1

Introduction to Fuel Systems and Emission Controls

This is a book about automotive gasoline engine fuel systems. It is also a book about automotive emission controls. The internal combustion engine produces power by burning fuel and changing the chemical energy of that fuel into thermal (heat) energy. The thermal energy is then converted into mechanical power.

Not all of the chemical energy of the gasoline is converted into useful power. Much of it is wasted. The engine's combustion process also produces some harmful byproducts. These are discharged from the engine and become air pollutants. Emission control systems are necessary to minimize the formation and discharge of these pollutants. Therefore, an engine's fuel system and its emission controls are closely interrelated.

The fuel system has three important jobs. It must:
1. Store liquid fuel
2. Deliver liquid fuel to the carburetor or fuel injectors for mixing with air
3. Distribute the air-fuel mixture uniformly to each cylinder of the engine.

Emission control requirements are important considerations in the design and operation of all parts of the fuel system. The ignition system, which provides the spark to begin combustion, plays an equally important role in emission control.

Because emission controls are now a vital part of the design of a modern automotive gasoline engine and its fuel and ignition systems, we will begin our study of fuel systems and emission controls by examining the harmful byproducts of combustion as a major cause of air pollution.

AIR POLLUTION — A PERSPECTIVE

Air pollution usually is defined as the introduction of contamination into the atmosphere in an amount large enough to injure human, animal, or plant life. There are many types and causes of air pollution, but they all fall into two general groups: natural and manmade. Natural pollution is caused by such things as the organic plant life cycle, forest fires, volcanic eruptions, and dust storms. While pollution from such sources is often beyond our control, manmade pollution from industrial plants and automobiles *can* be controlled to a certain degree.

Most urban and large industrial areas around the world suffer periodic air pollution. During the late 1940's, a unique form of air pollution was identified in the Los Angeles area. This combination of smoke and fog, which forms

Figure 1-1. Have a nice day.

irritating chemical compounds, is called **photo-chemical smog**, figure 1-1. As this phenom-enon increased both in intensity and frequency, it posed more of a problem. California took the lead in combating it by becoming the first state to place controls on motor vehicles. As smog gradually began to appear in other parts of the country, the Federal government moved into the area of regulation. To understand why, we must look at the elements produced by the automobile which form air pollution and smog.

MAJOR POLLUTANTS

An internal combustion engine emits three major gaseous pollutants into the air: **hydro-carbons (HC), carbon monoxide (CO)** and **oxides of nitrogen (NO$_x$),** figure 1-2. In addi-tion, an automobile engine gives off many small liquid or solid particles, such as lead, carbon, sulfur and other **particulates**, which contribute to pollution. By themselves, all these emissions are not smog, but simply air pollutants.

Photochemical Smog: A combination of pol-lutants which, when acted upon by sunlight, forms chemical compounds that are harmful to human, animal, and plant life.

Hydrocarbon (HC): A chemical compound made up of hydrogen and carbon. A major pollutant given off by an internal combustion engine. Gasoline, itself, is a hydrocarbon compound.

Carbon Monoxide (CO): An odorless, colorless, tasteless poisonous gas. A major pollutant given off by an internal combustion engine.

Oxides of Nitrogen (NO$_x$): Chemical com-pounds of nitrogen given off by an internal combustion engine. They combine with hydro-carbons to produce smog.

Particulates: Liquid or solid particles such as lead and carbon that are given off by an inter-nal combustion engine as pollution.

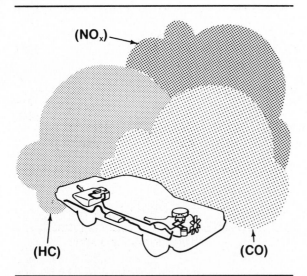

Figure 1-2. Hydrocarbons (HC), carbon monoxide (CO) and oxides of nitrogen (NO$_x$) are the three major pollutants emitted by an automobile.

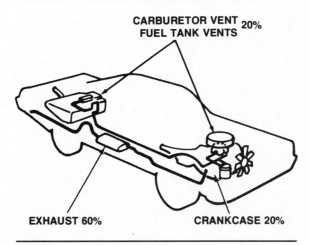

Figure 1-3. Sources of hydrocarbon emissions.

Hydrocarbons (HC)

Gasoline is a hydrocarbon compound. Unburned hydrocarbons given off by an automobile are largely unburned portions of fuel. Over 200 different varieties of hydrocarbon pollutants come from automotive sources. While most come from the fuel system and the engine exhaust, others are oil and gasoline fumes from the crankcase. Even a car's tires, paint, and upholstery emit tiny amounts of hydrocarbons. Figure 1-3 shows the three major sources of hydrocarbon emissions from an automobile:

1. Fuel system evaporation — 20 percent
2. Crankcase vapors — 20 percent
3. Engine exhaust — 60 percent.

Hydrocarbons are the only major automotive air pollutant that come from sources *other than* the engine's exhaust.

Hydrocarbons of all types are destroyed by combustion. If an automobile engine burned gasoline completely, there would be no hydrocarbons in the exhaust, only water and carbon dioxide. But when the vaporized and compressed air-fuel mixture is ignited, combustion occurs so rapidly that gasoline near the sides of the combustion chamber may not get burned. This unburned fuel then passes out with the exhaust gases. The problem is worse with engines that misfire or are not properly tuned.

Carbon Monoxide (CO)

Although not part of photochemical smog, carbon monoxide is also found in automobile exhaust in large amounts. A deadly poison, carbon monoxide is both odorless and colorless. Carbon monoxide is absorbed by the red corpuscles in the body, displacing the oxygen. In a small quantity, it causes headaches and vision difficulties. In larger quantities, it is fatal.

Because it is a product of incomplete combustion, the amount of carbon monoxide produced depends on the way in which hydrocarbons burn. When the air-fuel mixture burns, its hydrocarbons combine with oxygen. If the air-fuel mixture contains too much fuel, there is not enough oxygen to complete this process, so carbon monoxide is formed. To make combustion more complete, an air-fuel mixture with less fuel is used. This increases the ratio of oxygen, which reduces the formation of CO by producing harmless carbon dioxide (CO_2) instead.

Oxides of Nitrogen (NO$_x$)

Air is about 78 percent nitrogen, 21 percent oxygen, and 1 percent other gases. When the combustion chamber temperature reaches 2,500°F (1,370°C) or greater, the nitrogen and oxygen in the air-fuel mixture combine to form large quantities of oxides of nitrogen (NO$_x$). NO$_x$ also is formed at lower temperatures, but in far smaller amounts. By itself, NO$_x$ is of no particular concern. But when the amount of hydrocarbons in the air reaches a certain level and the ratio of NO$_x$ to HC is correct, the two pollutants will combine chemically to form smog.

Lowering the temperature of combustion in the engine reduces the amount of NO$_x$ formed. However, it also results in less efficient burning of the air-fuel mixture. This, in turn, increases hydrocarbons and carbon monoxide, both of which are formed in large quantities at lower combustion chamber temperatures. To combat this problem, automakers have used various

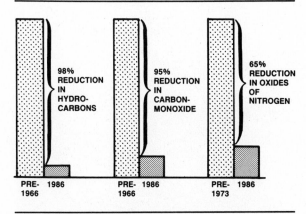

Figure 1-4. Automotive emission reductions.

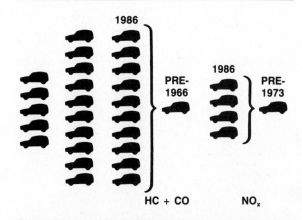

Figure 1-5. Vehicle emission comparisons.

emission control systems which we will study in later chapters.

Particulates

Particulates are microscopic solid particles, such as dust and soot. These fragments remain in the atmosphere for a long time. Because of this, particulates are prime causes of secondary pollution. For example, particulates such as lead and carbon tend to collect in the atmosphere. These are all harmful substances, large amounts of which can injure our health.

Particulates produced by automobiles are a small percentage of the total particulates in the atmosphere. Most come from fixed sources, such as factories. While automobiles *do* produce particulates, the amount can be reduced considerably. This is accomplished by eliminating certain additives such as lead from gasoline, and by changing other characteristics of the fuel. As a result, the amount of additives used in gasoline has been reduced and the types of additives are now carefully controlled.

Sulfur Oxides

Sulfur in gasoline and other fossil fuels (coal and oil) enters the atmosphere in the form of **sulfur oxides**. As these oxides break down, they combine with water in the air to form corrosive sulfuric acid, which is a secondary pollutant. In the past decade, there has been more publicity about this type of pollution, especially in the northeastern United States.

POLLUTION AND THE AUTOMOBILE

While it is true that a single car gives off only a microscopic amount of these pollutants, remember that there are more than 150 million automobiles in use in the United States. Multiply each car's contribution toward air pollution by that figure and you have the potential for a staggering amount of pollution. Without emission controls, automobiles would create almost as much air pollution as all other sources combined.

Great progress has been made in reducing — almost eliminating — automobile air pollution since 1966. Those HC emissions that come from the engine crankcase and fuel system (40 percent of the total) have been almost totally eliminated. The other 60 percent of HC emission, from the exhaust, has been lowered considerably. Total hydrocarbon emissions have been reduced by about 98 percent. Total carbon monoxide emissions have been reduced by a similar amount, about 95 percent, since 1966. Between 1973 and 1986, total NO_x emissions have been reduced about 65 percent. These accomplishments are shown in figure 1-4. Figure 1-5 shows us that twenty-five 1986 cars emit less HC and CO than one pre-1966 model, and four 1986 cars emit less NO_x than one pre-1973 model.

SMOG — CLIMATIC REACTION WITH AIR POLLUTANTS

Smog and air pollution are not the same thing: smog is a form of air pollution, but air pollution is not necessarily smog. Although all are by-products of combustion, each of the three major pollutants is created in different ways.

Sulfur Oxides: Chemical compounds given off by processing and burning gasoline and other fossil fuels. As they decompose, they combine with water to form sulfuric acid.

Figure 1-6. Smog engulfs the Los Angeles Civic Center in the 1960's. When the base of the temperature inversion is 1500 feet (457 meters) above ground, the inversion layer — a layer of warm air above a layer of cool air — prevents the natural dispersion of air contaminants into the upper atmosphere. (LA County)

Hydrocarbons come mostly from unburned fuel; carbon monoxide results from air-fuel mixtures that contain too much fuel; and oxides of nitrogen are created by high combustion chamber temperatures. HC and NO_x are the two principal materials that combine in the atmosphere to form smog.

The Photochemical Reaction

Although smog can be created in a laboratory experiment, scientists still do not completely understand the nature of smog. They do know, however, that three things must be present for smog to form in the atmosphere: sunlight, relatively still air, and a high concentration of hydrocarbons and oxides of nitrogen. When these three elements exist at the same time, sunlight causes a chemical reaction between the HC and NO_x. The result is smog.

Temperature Inversion

Normally, air temperature decreases at higher altitudes. Warm air near the ground rises and becomes cooler by contact with the cooler air

above it. When nature follows this normal pattern, smog and other pollutants are carried away. But some areas experience a natural weather pattern called a **temperature inversion**. When this occurs, a layer of warm air prevents the upward movement of cooler air near the ground. This inversion acts as a "lid" over the stagnant air. Since the air cannot rise, smog and pollution collect.

When the inversion layer is several thousand feet high, the smog will rise enough to provide reasonable visibility. But when the inversion layer is within a thousand feet (300 meters) of the ground, it traps the smog. This reduces visibility, making the distant landscape impossible to see, figure 1-6. Many people experience eye irritation, headaches and difficulty in breathing as a result. Temperature inversion was first noted in the Los Angeles area, which provides a classic example of this phenomenon. Surrounding mountains in that area form a natural basin in which temperature inversion is present to some extent for more than 300 days a year.

YEAR	REGULATIONS	HC	CO	NO$_x$
Before Controls		850 ppm (16.8 g/mi)*	3.4% (125.0 g/mi)*	1000 ppm (4.0 g/mi)*
1966-67	Calif.	275 ppm	1.5%	none
	U.S. Federal	none	none	none
1968-69	Calif.	275 ppm	1.5%	none
	U.S. Federal	275	1.5%	none
1970	Calif. & U.S. Federal	4.6 g/mi	46.0 g/mi	none
1971	Calif.	4.6 g/mi	46.0 g/mi	4.0 g/mi
	U.S. Federal	4.6 g/mi	46.0 g/mi	none
	Canadian	2.2 g/mi**	23.0 g/mi**	none
1972	Calif.	3.2 g/mi	39.0 g/mi	3.2 g/mi
	U.S. Federal	3.4 g/mi	39.0 g/mi	none
	Canadian	3.4 g/mi	39.0 g/mi	none
1973	Calif.	3.2 g/mi	39.0 g/mi	3.0 g/mi
	U.S. Federal	3.4 g/mi	39.0 g/mi	3.0 g/mi
	Canadian	3.4 g/mi	39.0 g/mi	3.0 g/mi
1974	Calif.	3.2 g/mi	39.0 g/mi	2.0 g/mi
	U.S. Federal	3.4 g/mi	39.0 g/mi	3.0 g/mi
	Canadian	3.4 g/mi	39.0 g/mi	3.0 g/mi
1975-76	Calif.	0.9 g/mi	9.0 g/mi	2.0 g/mi
	U.S. Federal	1.5 g/mi	15.0 g/mi	3.1 g/mi
	Canadian	2.0 g/mi	25.0 g/mi	3.0 g/mi
1977-78	Calif.	0.41 g/mi	9.0 g/mi	1.5 g/mi
	U.S. Federal	1.5 g/mi	15.0 g/mi	2.0 g/mi
	Canadian	2.0 g/mi	25.0 g/mi	3.0 g/mi
1979	Calif.	0.41 g/mi	9.0 g/mi	1.5 g/mi
	U.S. Federal	0.41 g/mi	15.0 g/mi	2.0 g/mi
	Canadian	2.0 g/mi	25.0 g/mi	3.0 g/mi
1980	Calif.	0.39 g/mi	9.0 g/mi	1.0 g/mi
	U.S. Federal	0.41 g/mi	7.0 g/mi	2.0 g/mi
	Canadian	2.0 g/mi	25.0 g/mi	3.0 g/mi
1981-87	Calif.	0.39 g/mi	7.0 g/mi	0.07 g/mi
	U.S. Federal	0.41 g/mi	3.4 g/mi	1.0 g/mi
	Canadian	2.0 g/mi	25.0 g/mi	3.1 g/mi

ppm = parts per million
g/mi = grams per mile
* approximate estimations of uncontrolled levels
** theoretical flow rate

Figure 1-7. Exhaust emission limits for new cars by model years.

AIR POLLUTION LEGISLATION AND REGULATORY AGENCIES

Once the problem of air pollution was recognized, it became the subject of intense research and investigation. By the early 1950's, it was believed that smog in Los Angeles was caused by the photochemical process. California, as we noted earlier, became the first state to enact legislation designed to limit automotive emissions. Standards established by California one year often become U.S. Federal standards the next year.

Emission Control Legislation

Beginning with the 1961 new cars, California required control over crankcase emissions. This became standard for the rest of the United States with new 1963 cars. That year, domestic automakers voluntarily equipped their new models with a blowby device which virtually eliminated crankcase emissions on all cars.

California followed by requiring that 1966 and later new cars sold within its boundaries have exhaust emission controls. The use of exhaust emission control systems was extended nationwide during the 1967-1968 model years.

The First Federal air pollution research program began in 1955. In 1963, Congress passed the Clean Air Act, providing the states with money to develop air pollution control programs. This law was amended in 1965 to give the Federal government authority to set emission standards for new cars, and was amended again in 1977-1978. Under this law, emission standards were first applied nationwide to 1968 models, figure 1-7.

In addition to the Clean Air Act, a new approach to air pollution was enacted in 1967. The Air Quality Act and its major amendments of 1970, 1974, and 1977 instituted changes that were designed to turn piecemeal programs into a unified attack on pollution of all kinds. Canada attacked its own smog problem with vehicle emission requirements established by the Ministry of Transport that took effect with 1971 models.

Regulatory Agencies

The Environmental Protection Agency (EPA) is the U.S. Federal agency responsible for enforcing the Air Quality Act. It was formed as part of the Department of Health, Education, and Welfare. The EPA first established standards which required that HC and CO levels for 1975 cars be reduced 90 percent from 1970 levels, with a 90-percent reduction in NO$_x$ by 1976. These standards were later amended to those shown in figure 1-7.

In addition, the EPA has established standards in other automotive areas, such as fuel consumption and fuel additives. California established its own Air Resources Board (ARB), whose authority roughly parallels that of the Federal EPA. The California ARB operates under permission from the EPA, but its authority extends only to those vehicles sold in or brought into California. Canadian vehicle emission standards are established by the national Ministry of Transport.

Temperature Inversion: A weather pattern in which a layer or "lid" of warm air keeps the cooler air beneath it from rising.

PASSENGER CAR		LIGHT DUTY TRUCKS *			
MODEL YEAR	MPG	4 × 2	4 × 4	COMBINED MPG	GVWR
1978	18	none	none	—	—
1979	19	17.2	15.8	—	0-6000
1980	20	16.0	14.0	—	0-8500
1981	22	16.7	15.0	—	0-8500
1982	24	18.0	16.0	17.5	0-8599
1983	26	19.5	17.5	19.0	0-8500
1984	27	20.3	18.5	20.0	0-8500
1985	27.5	19.7	18.9	19.5	0-8500
1986	26.0	20.5	19.5	20.0	0-8500
1987	27.5	21.0	19.5	20.5	0-8500

* Manufacturer can choose to meet individual or combined standard for 1983-85.

Figure 1-8. Corporate Average Fuel Economy (CAFE) standards imposed by Congress.

Corporate Average Fuel Economy (CAFE) Standards

Long gasoline lines and near-rationing of fuel during the energy crisis of 1973 (and again in 1979) resulted in focusing national attention on fuel economy. This resulted in the establishment of Corporate Average Fuel Economy (CAFE) standards, figure 1-8, as a part of the Federal Energy Act of 1975. The EPA and Department of Transportation (DOT) are responsible for administering the CAFE standards.

This gave automotive engineers two conflicting goals. They had to continue reducing emissions while improving fuel economy. If the CAFE standards were not reached each year, automakers were required by law to pay a penalty on each vehicle sold — the so-called "gas guzzler" tax. However, automakers who *exceeded* the yearly average on a corporate basis were granted a credit which could be applied to later years.

By downsizing vehicles, using smaller engines, and paying particular attention to reducing weight and improving aerodynamic efficiency, the domestic automotive industry transformed itself in the early 1980's to meet both goals. During this period, there was a concerted national effort to conserve fuel because of high prices at the pump and a desire to reduce the nation's dependence on foreign oil.

This national effort worked so well that oil prices, which had stabilized by 1983, collapsed in 1985-1986. At the same time, car owners indicated a desire to return to larger cars by ignoring the fuel efficient small cars and purchasing the less efficient cars with bigger engines and more room. This led General Motors and Ford to petition the EPA for a "rollback" of CAFE standards to avoid massive fines resulting from meeting customer demand. Over the objections of Chrysler and other small automakers who had met the standards on time, a temporary return to the 1983 level was enacted for 1986, with the 1987 standard returning to that of 1985, figure 1-8. As of this writing, the future of CAFE standards and what will happen to them is clouded in a national debate as to whether they should be retained, modified, or abandoned.

AUTOMOTIVE EMISSION CONTROLS

Early researchers dealing with automotive pollution and smog began work with the idea that *all* pollutants were carried into the atmosphere by the car's exhaust pipe. But auto manufacturers doing their own research soon discovered that pollutants were also given off from the fuel tank and the engine crankcase. The total automotive emission system, figure 1-9, contains three different types of controls. This picture illustrates that the emission controls on a modern automobile are not a separate system, but part of an engine's fuel, ignition, and exhaust systems.

In order to service a car's fuel system and emission controls, you must have a basic understanding of the internal combustion engine and how it works. Engine operating principles and air-fuel requirements are explained in Chapters 2 and 3. Emission controls are explained through the remainder of this book as they relate to different parts of the fuel system. Chapters 17 through 21 of this book cover those major emission controls that can be studied separately from the fuel system.

Automotive emission controls can be grouped into major families, as follows:

1. *Crankcase emission controls* — Positive crankcase ventilation (PCV) systems control HC emissions from the engine crankcase. These are described in Chapter 17.
2. *Evaporative emission controls* — Evaporative emission control (EEC or EVAP) systems control the evaporation of HC vapors from the fuel tank, pump, and carburetor or fuel injection system. These are described in Chapter 4.
3. *Exhaust emission controls* — Various systems and devices are used to control HC, CO, and NO_x emissions from the engine exhaust. These controls can be subdivided into the following general groups:

   a. *Air injection systems* — These systems provide additional air to the exhaust system to

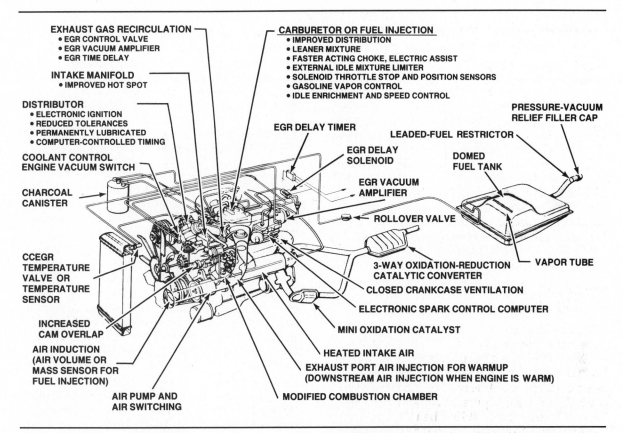

EXHAUST GAS RECIRCULATION
- EGR CONTROL VALVE
- EGR VACUUM AMPLIFIER
- EGR TIME DELAY

INTAKE MANIFOLD
- IMPROVED HOT SPOT

DISTRIBUTOR
- ELECTRONIC IGNITION
- REDUCED TOLERANCES
- PERMANENTLY LUBRICATED
- COMPUTER-CONTROLLED TIMING

**COOLANT CONTROL
ENGINE VACUUM SWITCH**

**CHARCOAL
CANISTER**

**CCEGR
TEMPERATURE
VALVE OR
TEMPERATURE
SENSOR**

**INCREASED
CAM OVERLAP**

**AIR INDUCTION
(AIR VOLUME OR
MASS SENSOR FOR
FUEL INJECTION)**

**AIR PUMP AND
AIR SWITCHING**

CARBURETOR OR FUEL INJECTION
- IMPROVED DISTRIBUTION
- LEANER MIXTURE
- FASTER ACTING CHOKE, ELECTRIC ASSIST
- EXTERNAL IDLE MIXTURE LIMITER
- SOLENOID THROTTLE STOP AND POSITION SENSORS
- GASOLINE VAPOR CONTROL
- IDLE ENRICHMENT AND SPEED CONTROL

EGR DELAY TIMER

**EGR DELAY
SOLENOID**

**EGR VACUUM
AMPLIFIER**

ROLLOVER VALVE

**3-WAY OXIDATION-REDUCTION
CATALYTIC CONVERTER**

CLOSED CRANKCASE VENTILATION

ELECTRONIC SPARK CONTROL COMPUTER

MINI OXIDATION CATALYST

HEATED INTAKE AIR

**EXHAUST PORT AIR INJECTION FOR WARMUP
(DOWNSTREAM AIR INJECTION WHEN ENGINE IS WARM)**

MODIFIED COMBUSTION CHAMBER

**PRESSURE-VACUUM
RELIEF FILLER CAP**

LEADED-FUEL RESTRICTOR

**DOMED
FUEL TANK**

VAPOR TUBE

Figure 1-9. The complex emission controls on a modern automobile engine are an integral part of the fuel, ignition, and exhaust systems. (Chrysler)

■ How New Cars are Emission-Certified

Whether or not you can buy a new car each year depends upon the success of the automakers in completing the emission certification process. The constant volume sampling test is performed by the Federal Environmental Protection Agency (EPA), although California's Air Resources Board tests some vehicles.

As each new model year approaches, automakers build prototype or emission data cars for EPA use. The automakers are responsible for conducting a 50,000-mile (80 000-kilometer) durability test of their emission control systems. The test cars are driven for 4,000 miles (6500 kilometers) to stabilize the emission systems before testing.

The first step is to precondition the car. Then it stands for 12 hours at an air temperature of 73°F (22°C) to simulate a cold start. The actual test is done on a chassis dynamometer, using a driving cycle which represents urban driving conditions. The car's exhaust is mixed with air to a constant volume and analyzed for harmful pollutants.

The entire test requires about 41 minutes. The first 23 minutes are a cold-start driving test. The next 10 minutes are a waiting, or hot-soak period. The final eight minutes are a hot-start test, representing a short trip in which the car is stopped and started several times while hot. If the emissions test results for all data cars are equal to, or lower than, all of the HC, CO, and NO$_x$ standards, the manufacturer receives certification of the engine "family" and the car can be offered for sale to the public.

This explains why some engines have disappeared — they were too "dirty" and could not be "cleaned" up. It also explains why some powertrain combinations may not be available in California (which has different requirements) when they can be purchased in the other 49 states. Automakers are now facing the prospects of certifying their vehicles for 100,000 miles (160 000 kilometers) or more, a development which may be necessary in the near future.

help burn up HC and CO in the exhaust and to aid catalytic conversion. They are described in Chapter 18.

b. Engine modifications — Various changes have been made in the design of engines and in the operation of fuel and ignition system components to help eliminate all three major pollutants. They are identified in appropriate sections throughout this book.

c. Spark timing controls — Automakers have used various systems to delay or retard ignition spark timing to control HC and NO_x emissions. Most of these systems modify the distributor vacuum advance; however, late-model cars with electronic engine control systems have eliminated the need for mechanical or vacuum timing devices. Ignition spark timing and its role in controlling emissions are discussed in Chapter 19. Electronic engine control systems are discussed in Chapter 14.

d. Exhaust gas recirculation — An effective way to control NO_x emissions is to recirculate a small amount of exhaust gas back to the intake manifold to dilute the incoming air-fuel mixture. Exhaust gas recirculation (EGR) systems are described in Chapter 20.

e. Catalytic converters — The first catalytic converters installed in the exhaust systems of 1975-76 cars helped the chemical oxidation or burning of HC and CO in the exhaust.

Later catalytic converters, which began to appear on 1977-78 cars, also promoted the chemical reduction of NO_x emissions. Catalytic converters are discussed in Chapter 21.

SUMMARY

The automobile is a major source of air pollution, resulting from the gasoline burned in the engine and the vapors escaping from the crankcase, the fuel tank, and the rest of the fuel system. The major pollutants are unburned hydrocarbons (HC), carbon monoxide (CO), and oxides of nitrogen (NO_x). The use of emission controls in the past two decades has reduced automotive pollutants from 65 to 98 percent.

Emission controls began as separate "add-on" components and systems, but are now completely integrated into engine and vehicle design. The major emission control systems used are PCV systems, evaporative control systems, air injection, spark timing controls, exhaust gas recirculation, catalytic converters, and electronic engine control systems.

Increasingly stringent emission standards combined with the CAFE regulations imposed during the late 1970's have reshaped the domestic automotive industry. The end result has been smaller, more fuel efficient vehicles that produce much less pollution than those of two decades ago.

Review Questions

Choose the single most correct answer.
Compare your answers with the correct answers on page 414.

1. Smog:
   a. Is a natural pollutant
   b. Cannot be controlled
   c. Is created by a photochemical reaction
   d. Was first identified in New York City

2. The three *major* pollutants in automobile exhaust are:
   a. Sulfates, particulates, carbon dioxide
   b. Sulfates, carbon monoxide, nitrous oxide
   c. Carbon monoxide, oxides of nitrogen, hydrocarbons
   d. Hydrocarbons, carbon dioxide, nitrous oxide

3. Fuel evaporation accounts for what percentage of total HC emissions?
   a. 10%
   b. 60%
   c. 33%
   d. 20%

4. Carbon monoxide is a result of:
   a. Incomplete combustion
   b. A lean mixture
   c. Excess oxygen
   d. Impurities in the fuel

5. Which of the following is *not* true?
   a. High temperatures increase NO_x emissions
   b. High temperatures reduce HC and CO emissions
   c. Low temperatures reduce HC and CO emissions
   d. Low temperatures reduce NO_x emissions

6. Particulates are:
   a. Microscopic particles suspended in the atmosphere
   b. Produced in emissions by additives such as lead
   c. Harmful to our lungs
   d. All of the above

7. Sulfur oxides are harmful because:
   a. They damage catalytic converters
   b. They combine with water in the air to form sulfuric acid
   c. They are primary pollutants
   d. They are visible

8. Since 1966, total HC emissions have been reduced by about:
   a. 65%
   b. 47%
   c. 82%
   d. 98%

9. Smog is created by a combination of sunlight, still air, and:
   a. High levels of CO and HC
   b. High levels of CO and NO_x
   c. High levels of HC and NO_x
   d. High levels of HC and sulfur oxides

10. A temperature inversion increases air pollution by:
    a. Pushing cool air up
    b. Forming a "lid" over stagnant air
    c. Pushing warm air down
    d. None of the above

11. U.S. Federal emission limits are established by the:
    a. Air Resources board
    b. Ministry of Transport
    c. Environmental Protection Agency
    d. Department of Transportation

12. Corporate Average Fuel Economy (CAFE) standards were first applied to vehicles for which model year?
    a. 1976
    b. 1978
    c. 1979
    d. 1981

13. Sulfur byproducts of combustion can combine with water to form:
    a. Sulfates
    b. Particulates
    c. Oxides of sulfur
    d. Sulfuric acid

14. The only major type of pollutant that comes from a vehicle source *other than the exhaust* is:
    a. HC
    b. CO
    c. CO_2
    d. NO_x

2

Engine Operating Principles

Engines used in automobiles are called internal combustion engines because fuel burns inside the engine. The several types of internal combustion engines used in cars are covered in this chapter. You will also learn the basic principles of engine operation that relate to fuel systems and emission control.

AIR PRESSURE — HIGH AND LOW

You can think of an internal combustion engine as a big air pump. As the pistons move up and down in the cylinders, they pump in air and fuel for combustion and pump out exhaust gases. They do this by creating a difference in air pressure. The air outside an engine has weight and exerts pressure. So does the air inside an engine.

As a piston moves down on an intake stroke with the intake valve open, it creates a larger area inside the cylinder for the air to fill. This lowers the air pressure within the engine. Because the pressure inside the engine is lower than the pressure outside, air flows into the engine to fill the low-pressure area and equalize the pressure.

The low pressure within the engine is called **vacuum**. You can think of the vacuum as sucking air into the engine, but it is really the higher pressure on the outside that forces air into the low-pressure area inside. The difference in pressure between the two areas is called a **pressure differential**. The pressure differential principle has many applications in automotive fuel and emission systems.

An engine pumps exhaust out of its cylinders by creating pressure as a piston moves upward on the exhaust stroke. This creates high pressure in the cylinder, which forces the exhaust toward the lower pressure area outside the engine.

Pressure differential can be applied to liquids as well as to air. Fuel pumps work on this same principle. The pump creates a low-pressure area in the fuel system that allows the higher pressure of the air and fuel in the tank to force the fuel through the lines to the carburetor or the injection system.

Again, you can think of vacuum, or suction, drawing fuel through the lines, but that is simply another way of looking at the pressure differential. We will study more about air-flow and air pressure in the next chapter, and again in Chapter 8 as we study carburetors in detail.

TOP OF BLOCK

WATERJACKET HOLE FOR CIRCULATING
WATER AROUND CYLINDERS

CYLINDERS

BOTTOM OF BLOCK

CRANKSHAFT BEARING SUPPORTS

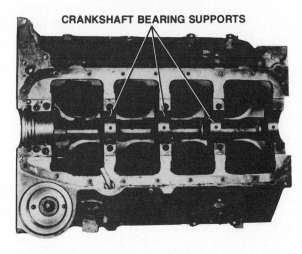

Figure 2-1. The engine block contains the cylinders
and water jackets for cooling. The crankshaft turns in
main bearings at the bottom of the block.

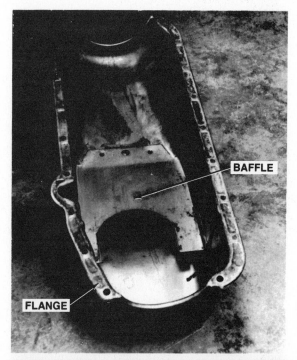

BAFFLE

FLANGE

DRAIN PLUG

FLANGE

Figure 2-2. The oil pan is fastened to the bottom of the
block.

RECIPROCATING ENGINE COMPONENTS

Reciprocating means "up and down" or "back and forth". It is the up-and-down action of a piston in the cylinder that produces power in a **reciprocating engine**. Almost all engines of this type are built upon a cylinder block, or engine block, figure 2-1. The block is an iron or aluminum casting that contains engine cylinders and passages called water jackets for coolant circulation. The top of the block is covered with the cylinder head, which contains additional water jackets and which forms the combustion chamber. The bottom of the block is covered with an oil pan, or oil sump, figure 2-2. A major exception to this type of engine construction is

the air-cooled Volkswagen engine, figure 2-3. It is representative of the horizontally opposed air-cooled engines used by Porsche, Chevrolet (Corvair), and some other automobile manufacturers in years past.

Vacuum: A pressure less than atmospheric pressure.

Pressure Differential: A difference in pressure between two points.

Reciprocating Engine: Also called piston engine. An engine in which the pistons move up and down or back and forth, as a result of combustion of an air-fuel mixture in the top of the piston cylinder.

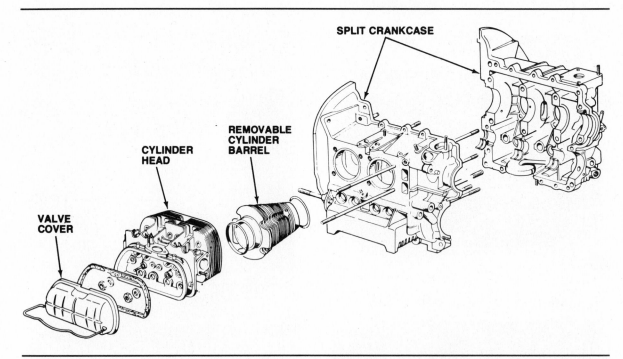

Figure 2-3. The Volkswagen horizontally opposed engine is built with a split crankcase. Individual cylinder castings bolt to the split crankcase, and cylinder heads attach to each pair of cylinders.

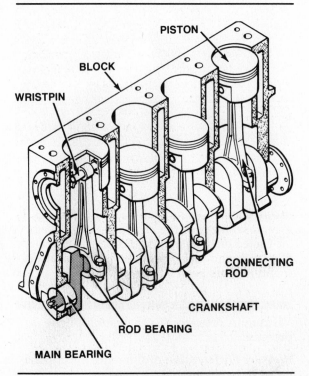

Figure 2-4. The crankshaft changes the reciprocating motion of the pistons to rotating motion.

The diameter of the cylinder is called the engine's **bore**. The distance that the piston travels from its top position to its bottom position in the cylinder is called the engine's **stroke**.

Power is produced by the inline motion of a piston in a cylinder. However, this *linear* motion must be changed to rotating motion to turn the wheels of a car or truck. The piston is attached to the top of a connecting rod by a pin, called a piston pin or wristpin, figure 2-4. The bottom of the connecting rod is attached to the crankshaft. The connecting rod transmits the up-and-down motion of the piston to the crankshaft, which changes it to rotating motion. The connecting rod is mounted on the crankshaft with large bearings called rod bearings. Similar bearings, called main bearings, are used to mount the crankshaft in the block, figure 2-4.

The combustible mixture of gasoline and air enters the cylinders through valves. Automotive engines use **poppet valves**, figure 2-5. The valves can be in the cylinder head or in the block. The opening and closing of the valves is controlled by a camshaft. **Lobes** on the camshaft push the valves open as the camshaft rotates. A spring closes each valve when the lobe is not holding it open. The most common arrangements of engine cylinders and valves are discussed later.

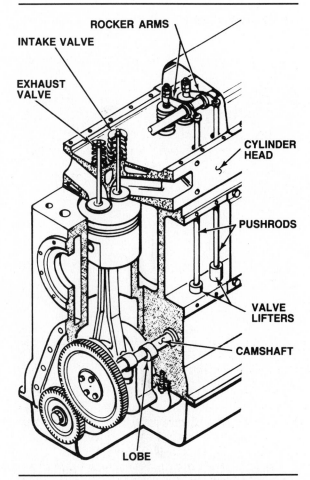

INTAKE VALVE
ROCKER ARMS
EXHAUST VALVE
CYLINDER HEAD
PUSHRODS
VALVE LIFTERS
CAMSHAFT
LOBE

Figure 2-5. The valves in this engine are located in the cylinder head. They are operated by the camshaft lobes, valve lifters, pushrods, and rocker arms.

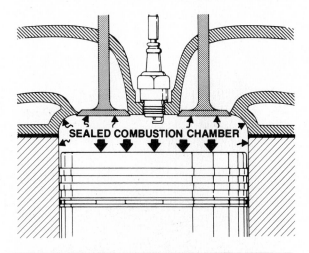

SEALED COMBUSTION CHAMBER

Figure 2-6. For combustion to produce power in an engine, the combustion chamber must be sealed.

In addition to being called internal combustion, reciprocating engines, most automotive engines are classified as:
- 4-stroke or 2-stroke engines
- Spark-ignition (gasoline) or compression-ignition (diesel) engines.

Because the most common automotive engine is the 4-stroke gasoline engine, we will begin our study of operating cycles with that design.

THE 4-STROKE CYCLE

Gasoline by itself will not burn; it must be mixed with oxygen (air). This burning is called combustion and is a way of releasing the energy stored in the air-fuel mixture. To do any useful work in an engine, the air-fuel mixture must be compressed and burned in a sealed chamber, figure 2-6. Here the combustion energy can work on the movable piston to produce mechanical energy. The combustion

chamber must be sealed as tightly as possible for efficient engine operation. Any leakage from the combustion chamber allows part of the combustion energy to dissipate without adding to the mechanical energy developed by the piston movement.

The term "stroke" is used to describe the *movement* of the piston within the cylinder, as well as the *distance* of piston travel. Depending on the type of engine, the operating cycle may require either two or four strokes to complete. The **4-stroke engine** is also called the Otto cycle engine, in honor of the German engineer, Dr. Nikolaus Otto, who first applied the principle in 1876.

In the 4-stroke engine, four strokes of the piston in the cylinder are required to complete one full operating cycle. Each stroke is named

Bore: The diameter of an engine cylinder.

Stroke: One complete top-to-bottom or bottom-to-top movement of an engine piston.

Poppet Valve: A valve that plugs and unplugs its opening by axial motion.

Lobes: The rounded protrusions on a camshaft that force, and govern, the opening of the intake and exhaust valves.

4-Stroke Engine: The Otto cycle engine. An engine in which a piston must complete four strokes to make up one operating cycle. The strokes are: intake, compression, power, and exhaust.

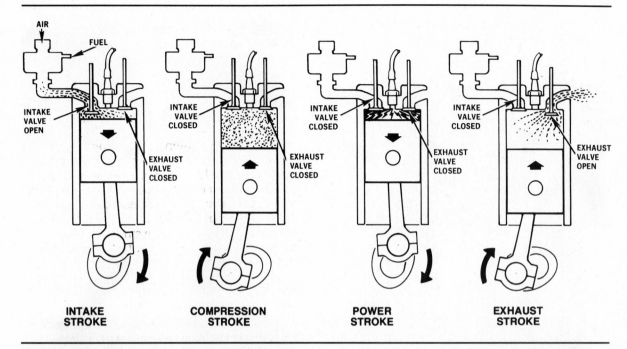

Figure 2-7. The downward movement of the piston draws the air-fuel mixture into the cylinder through the intake valve on the intake stroke. On the compression stroke, the mixture is compressed by the upward movement of the piston with both valves closed. Ignition occurs at the beginning of the power stroke, and combustion drives the piston downward to produce power. On the exhaust stroke, the upward-moving piston forces the burned gases out the open exhaust valve.

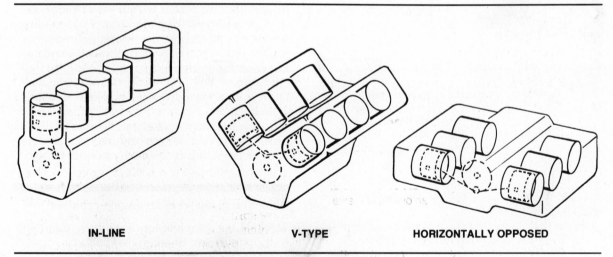

Figure 2-8. Common cylinder arrangements for automotive engines.

after the action it performs — intake, compression, power, and exhaust — in that order, figure 2-7.

1. *Intake stroke* — As the piston moves down, the vaporized mixture of fuel and air enters the cylinder past the open intake valve.

2. *Compression stroke* — The piston returns up, the intake valve closes, the mixture is compressed within the combustion chamber, and ignited by a spark.

3. *Power Stroke* — The expanding gases of com-bustion force the piston down in the cylinder. The exhaust valve opens near the bottom of the stroke.

4. *Exhaust stroke* — The piston moves back up with the exhaust valve open, and the burned gases are pushed out to prepare for the next intake stroke. The intake valve usually opens just before the top of the exhaust stroke.

This 4-stroke cycle is continuously repeated in every cylinder as long as the engine remains running.

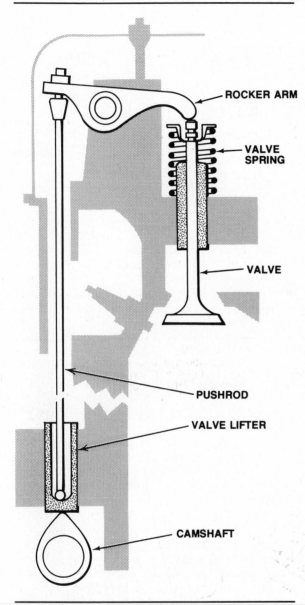

Figure 2-9. Valve mechanism for an overhead-valve engine.

A "2-stroke" engine also goes through four actions to complete one operating cycle. However, the intake and compression actions are combined in one stroke, and the power and exhaust actions are combined in the other stroke. The term 2-stroke cycle or 2-stroke is preferred to the term 2-cycle, which is really not accurate.

CYLINDER AND PISTON ARRANGEMENT

Up to this point, we have been talking about a single piston in a single cylinder. While single-cylinder engines are common in motorcycles, outboard motors, and small agricultural implements, automotive engines have more than one cylinder. Most cars engines have 4, 6, or 8 cylinders, although engines with 5 and 12 cylinders are being built, and we are beginning to see a new generation of 3-cylinder engines in commuter cars.

In automobile engines, all pistons are attached to a single crankshaft. The more cylinders an engine has, the more power strokes are produced for each revolution. This means that an 8-cylinder engine produces a power stroke twice as often as a 4-cylinder engine. The 8-cylinder engine runs more smoothly because the power strokes are closer together in time and in degrees of engine rotation.

The cylinders of multi-cylinder automotive engines are arranged in one of three ways, figure 2-8.

1. Inline engines use a single back of cylinders. Most 4-cylinder and many 6-cylinder engines are of this design. The cylinders do not have to be vertical, as shown in figure 2-8. They can be inclined to either side.
2. V-type engines use two equal banks of cylinders, usually inclined 60 degrees or 90 degrees from each other. Most V-type engines have 6- or 8-cylinders, although V-4 and V-12 engines have been built.
3. Horizontally opposed or "pancake" engines have two equal banks of cylinders 180 degrees apart. These space-saving engine designs are often air-cooled, and are found in the Chevrolet Corvair, Porsches, Subarus, and Volkswagens. Subaru's design is liquid cooled. Late-model Volkswagen vans use a liquid-cooled version of the air-cooled VW horizontally opposed engine.

As described earlier, each cylinder contains intake and exhaust poppet valves which are opened by lobes on the camshaft and closed by valve springs, figure 2-9. To coordinate this opening and closing action with piston movement, the camshaft rotation must be synchronized, or timed, with that of the crankshaft.

■ Sixteen Cylinders

The chances are remote that you'll ever be called on to service a 16-cylinder 1935-37 Cadillac, but in case you do, this is the firing order: 1-8-9-14-3-6-11-2-15-10-7-4-13-12-5-16. The firing order on the 1938-40 V-16 models is: 1-4-9-12-3-16-11-8-15-14-7-6-13-2-5-10. Odd-numbered cylinders are on the left bank, even-numbered cylinders are on the right.

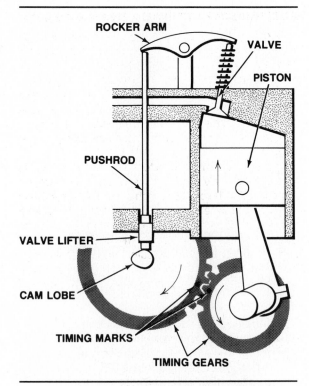

Figure 2-10. Timing marks on the timing gears synchronize valve action with piston movement.

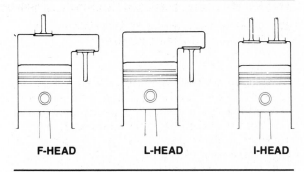

Figure 2-11. Common valve arrangements for 4-stroke automotive engines.

This synchronization is accomplished in one of three ways: by gears, by a chain and sprockets, or by a timing belt and sprockets. Timing marks on the gears or sprockets are used to synchronize the two shafts with each other, figure 2-10.

On most engines, camshaft gear meshes with a driven gear on the ignition distributor to synchronize or "time" spark plug firing with piston and valve positions at the beginning of the piston stroke. The order in which the air-fuel mixture is ignited within the cylinders is called the **firing order**, and varies with different engine designs. Only one power stroke takes place at any instant, regardless of the number of cylinders in the engine. These power strokes follow each other in a rapid sequence to produce a smooth flow of power.

VALVE ARRANGEMENT

Intake and exhaust valves on modern engines are located in the cylinder head. Because the valves are "in the head", this basic arrangement is called an I-head design. In the past, valves have been arranged in three different ways, figure 2-11.

1. The L-head design locates both valves side-by-side in the engine block. Because the cylinder head is rather flat and contains only the combustion chamber, water jacket, and spark

plugs, L-head engines also are called "flat-heads". Once very common, this valve arrangement has not been used in a domestic engine since the early 1960's.

2. The F-head design locates the intake valve in the cylinder head and the exhaust valve in the engine block. A compromise between the L-head and I-head designs, the F-head was last used in the 1971 Jeep.

3. The I-head or overhead valve design, also known as an overhead valve or overhead cam design, locates both the intake and the exhaust valves in the cylinder head. All modern engines use this design.

In the overhead valve engine, the camshaft is in the engine block and the valves are opened by valve lifters, pushrods, and rocker arms, figure 2-12. In the overhead cam engine, the camshaft is mounted in the head, either above or to one side of the valves, figure 2-12. This improves valve action at higher engine speeds. The valves may open directly by means of valve lifters or cam followers, or through rocker arms. The double overhead cam engine has two camshafts, one on each side of the valves. One camshaft operates the intake valves; the other operates the exhaust valves.

Number of Valves

As you saw in figures 2-5 and 2-7, most automobile engines have one intake and one exhaust valve per cylinder. This means that a 4-cylinder engine has 8 valves; a 6-cylinder engine has 12 valves; and a V-8 has 16 valves.

Many engines have been built, however, with more than two valves per cylinder. Engines with four valves per cylinder have been

Firing Order: The order in which combustion occurs in the cylinders of an engine.

Plus-One Performance (If Two Valves Are Good, More Must Be Better)

Four-valve heads have been around a long time in motorcycles and automobile racing engines. Every Miller, Offenhauser, and Cosworth engine that has ever won at Indianapolis has had four valves per cylinder. Although the 4-valve design has been refined in racing applications, it has only appeared in production engines in the past decade.

Everything else being equal, a 2-valve head with a high compression ratio and large enough valves to flow enough air to produce power at high rpm would be ideal — if the combustion chamber could be a perfect sphere, with the spark plug in the center. Unfortunately, while it sounds easy, designing the perfect combustion chamber isn't that simple.

The 4-valve head, with two intake valves on one side and two exhaust valves on the other, allows the designer to use a flatter piston and a centrally located spark plug. Four small valves will have greater total circumference than two valves of equal area, which means greater air flow. The smaller valves are also lighter and allow the valve train to run faster and last longer.

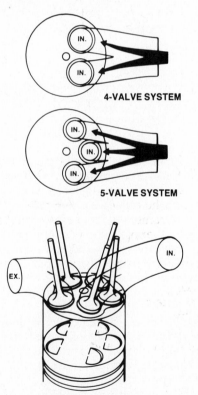

4-VALVE SYSTEM

5-VALVE SYSTEM

INNOVATIVE 5-VALVE CONFIGURATION

Yamaha's 5-valve motorcycle engine uses individual intake and exhaust cam lobes to operate each valve through inverted bucket-type cam followers.

Dropping the center exhaust valve solved the cooling problem, and opening each valve with its own cam lobe and bucket-type tappet let the cam make greater surface contact with the lifter resulting in higher lift. While the total valve area of the 5-valve head is less than one with four valves, the effective intake area around the three intake valves is about 14 percent greater. This increased air intake means higher power output.

Yamaha has built and extensively tested a 7-valve head; three intake and three exhaust valves around the outside of the chamber, with a fourth intake valve in the center between two spark plugs. While it ran like a lion, it was too complex and expensive to manufacture and sell, so the design was downgraded to six valves and one spark plug. Power output and fuel consumption were still good, but valve train complexity remained, and cooling the center exhaust valve proved difficult.

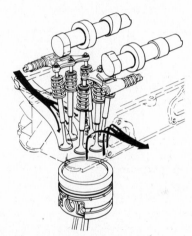

Maserati's 6-valve head uses a single intake and exhaust cam lobe to operate each set of three valves through a finger-type cam follower, or rocker arm. (Maserati)

Maserati has applied multiple valve technology to its 2.0-liter Biturbo V-6 for an amazing 43 percent gain in power — output increased from 180 horsepower (134 kW) at 6,000 rpm to 243 horsepower (181 kW) at 7,200 rpm. This high-performance powerplant was a bit unusual to begin with, because it used one large exhaust and two different-sized intake valves. The redesigned version, designated the 6.36 (for 6 cylinders and 36 valves), is equally unusual because it contains three intake and three exhaust valves arranged concentrically around the bore and inclined at different angles for greater mixture swirl. Each valve trio is operated by the same cam lobe through a wide follower.

The 6.36 is also a complex and expensive engine to manufacture, but Maserati says the performance makes it ideal for limited production use in a 2-seat sports car.

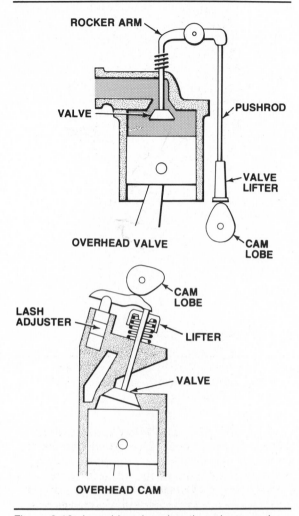

Figure 2-12. In an I-head engine, the valves may be operated by a camshaft in the block or in the head.

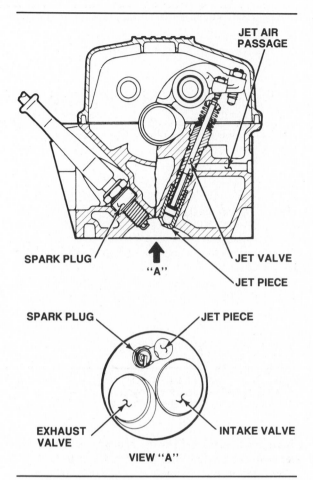

Figure 2-13. The Mitsubishi MCA engine uses a small auxiliary valve to admit a high-speed jet of air on the intake stroke. (Mitsubishi).

used in race cars since the earliest days of motoring. Racing motorcycle engines with five and even six valves per cylinder have been built in recent years. In spite of performance advantages, the higher costs and greater complexity of engines with more than two valves per cylinder have kept these designs from being used in production engines until recently.

To understand the principles of 3-valve, 4-valve, and other similar designs, it is best to distinguish between auxiliary valves and main intake and exhaust valves. Honda's Compound Vortex Controlled Combustion (CVCC) engine is a type of stratified charge engine that has a small precombustion chamber above the main chamber. The precombustion chamber has its own small intake valve that admits a rich air-fuel mixture to begin combustion. The Honda CVCC engine is explained and illustrated in detail at the end of this chapter.

Mitsubishi engines with the Mitsubishi Clean Air (MCA) combustion chamber design have a third small valve in each chamber, figure 2-13. Mitsubishi calls this a "jet" valve because it admits a high-speed jet of air to create a swirl effect that promotes complete combustion. This MCA jet valve design has been used in the engines of many Mitsubishi and Chrysler imports since the mid-1970's. In the mid-1980's Mitsubishi redesigned the head and renamed it the "cyclone" combustion chamber.

Toyota builds a 3-valve engine with a smaller, secondary intake valve that opens later than the primary intake valve. The two intake valves, combined with a unique port design, promote mixture turbulence and complete combustion of lean mixtures. The Honda, Mitsubishi, and Toyota 3-valve designs are examples of auxiliary-valve engines.

Since the mid-1980's several Japanese and European carmakers have been building engines with four main valves per cylinder, figure 2-14. The triple goals of obtaining best economy,

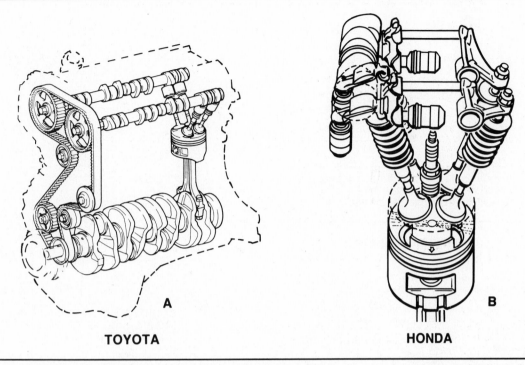

A

TOYOTA

B

HONDA

Figure 2-14. All 4-valve production engines use overhead cams. The engine may have separate intake and exhaust camshafts as in the Toyota example (A), or a single cam, such as the Honda Acura V-6 design (B) that operates the exhaust valves through short pushrods. (Toyota; Honda)

emission control, and performance, plus the competitive nature of the auto market in the 1980's, led manufacturers to adopt these designs for production vehicles. The benefits of 4-valve engines involve a lot of complex engineering considerations that are beyond the scope of this text. The following points, however, summarize the advantages of multiple valves:

● Two small valves in place of one large one can provide a larger total inlet or exhaust area, which improves volumetric efficiency.
● Smaller individual valves, springs, and retainers reduce the weight of each valve assembly. This reduces valve train inertia and allows higher maximum engine speeds.
● Smaller valves allow the engineer more flexibility in combustion chamber design. Valve angles and ports can be designed for improved air-fuel turbulence and combustion. Also, valve installations can be designed for lower overall engine height under the hoods of late-model cars.
● Separated intake and exhaust ports can be designed, or "tuned", for optimum intake and exhaust velocity. Individual, smaller diameter ports of larger *total* area can provide increased intake and exhaust volume at higher speed.
● Electronically controlled fuel injection combined with two intake valves allows engineers

to design induction systems with separate air-fuel flow for different engine speeds and loads. On some engines, the two intake valves open and close at different times to fine-tune the air-fuel metering for different operating conditions. Chapter 7 explains divided induction systems and split-level manifolds in detail.

Modern Combustion Chamber Design

Engineers have worked with combustion chamber design since the automobile was first invented. In the years before exhaust emission controls, much of the experimentation and design work was done with racing engines in an effort to make them go ever faster. The need to reduce exhaust emissions, however, refocused attention on the combustion chamber. Efforts were made to promote rapid, uniform burning of the air-fuel charge to control emissions and improve fuel economy.

Combustion of the air-fuel charge in a cylinder is not an instantaneous explosion, but rather, a controlled burning of the charge by the spark from the spark plug. When the spark ignites the air-fuel mixture, a flame front spreads out across the combustion chamber to consume the mixture. Movement of the flame front is called burn time and requires about 3 milliseconds.

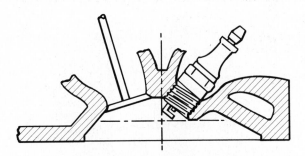

Figure 2-15. Central location of the spark plug in the combustion chamber reduces the distance the flame front must travel to reach the cylinder walls. (Ford)

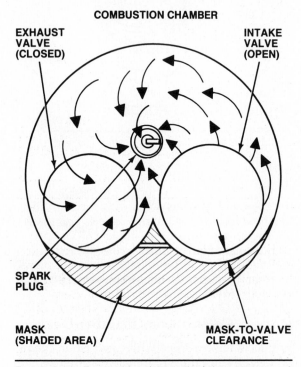

COMBUSTION CHAMBER

EXHAUST VALVE (CLOSED)

INTAKE VALVE (OPEN)

SPARK PLUG

MASK (SHADED AREA)

MASK-TO-VALVE CLEARANCE

Figure 2-16. By masking (shrouding) the area around the intake valve with additional metal, the air-fuel charge is directed into the combustion chamber with a swirling motion to promote more even distribution. (Ford)

However, combustion chamber design, temperature, pressure, and gasoline quality can combine to cause an unwanted explosion or detonation of the air-fuel charge before the flame front reaches it. In the ideal combustion chamber design, the entire air-fuel charge would burn completely, leaving no unburned areas to be exhausted and eliminating the possibility of detonation. In actual practice, however, there is always some part of the mixture that does not completely burn.

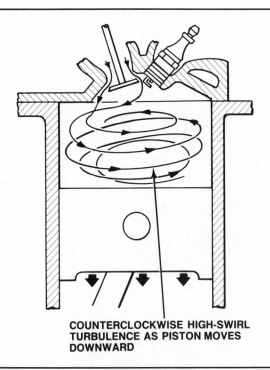

COUNTERCLOCKWISE HIGH-SWIRL TURBULENCE AS PISTON MOVES DOWNWARD

Figure 2-17. Downward piston movement in cylinders with shrouded valves causes a high-swirl turbulence of the air-fuel charge. (Ford)

Current combustion chamber design favors the **fast-burn** or **high-swirl combustion chamber** in which the combustion process is completed in a shorter period of time. This design usually incorporates the following features:

● *Compact combustion chamber* — By providing a smaller amount of surface area for a given chamber volume, the flame front is reduced and the time required for combustion is shortened.

● *Centralized spark plug location* — Positioning the spark plug electrode closer to the center of the combustion chamber, figure 2-15, reduces the distance the flame must travel to the edges of the chamber. This also shortens the combustion period.

● *Masked or shrouded intake port* — By masking or shrouding the intake valve area in the combustion chamber, figure 2-16, the air-fuel mixture is directed in a concentrated stream as it is drawn in through the valve and subjected to turbulence or swirl as the piston moves downward, figure 2-17.

● *Higher compression ratio* — Positioning the spark plug deeper in the combustion chamber reduces the chamber volume and increases the compression ratio. In some designs, this is advantageous. In other designs, the piston crown is dished to offset plug positioning and maintain the compression ratio at a point where no

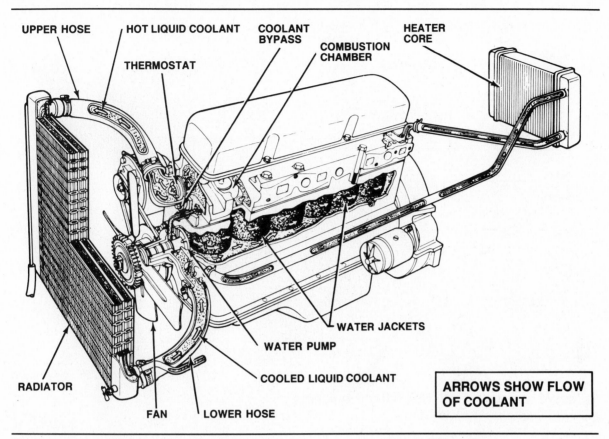

UPPER HOSE   HOT LIQUID COOLANT   COOLANT BYPASS   COMBUSTION CHAMBER   HEATER CORE

THERMOSTAT

WATER JACKETS

WATER PUMP

COOLED LIQUID COOLANT

RADIATOR

FAN   LOWER HOSE

ARROWS SHOW FLOW OF COOLANT

Figure 2-18. Components and operation of a typical automotive cooling system. (Chrysler)

decrease in spark advance is required. The desired end result of using a higher compression ratio is to obtain a more densely compressed and fully atomized air-fuel charge for more complete combustion in less time. Fast-burn combustion chambers with four valves per cylinder have become common in late-model engine design.

ENGINE COOLING SYSTEM

We have already mentioned that automotive engine blocks and cylinder heads contain passages, called a **water jacket**, for coolant circulation. Since the average combustion chamber temperature is approximately 1,500°F (800°C) and peak combustion chamber temperature can reach 6,000°F (3,300°C), it is easy to see why this water jacket is necessary. The engine changes about one-third of the heat created into actual energy; the remaining two-thirds must be removed, or dissipated. In some engines, the heat is dissipated by radiation into the air. On most engine designs, however, heat is removed by the liquid cooling and exhaust systems. Figure 2-18 shows the components and operation of a typical liquid cooling system.

A mixture of ethylene glycol antifreeze and water is used as a coolant and circulated through the engine by the cooling system water pump to absorb the unwanted heat. The coolant passes out of the engine water jacket and into a radiator where air circulation through the radiator absorbs the heat carried by the coolant. The coolant then returns to the engine to absorb more heat in a continuous process.

Fast-Burn Combustion Chamber: A compact combustion chamber with a centrally located spark plug. The chamber is designed to shorten the combustion period by reducing the distance of flame front travel.

High-Swirl Combustion Chamber: A combustion chamber in which the intake valve is shrouded or masked to direct the incoming air-fuel charge and create turbulence that will circulate the mixture more evenly and rapidly.

Water Jackets: Passages in the head and block that allow coolant to circulate throughout the engine.

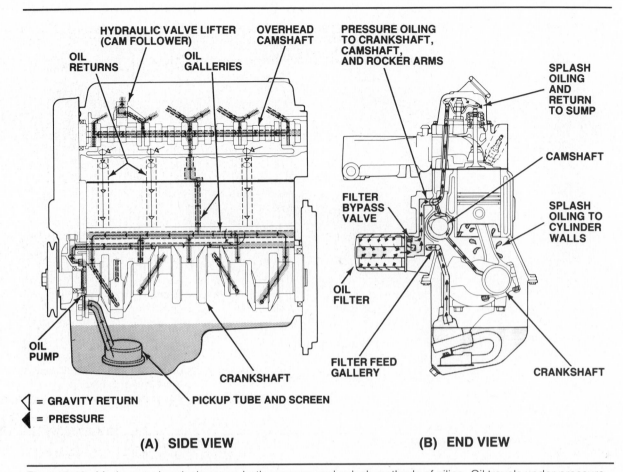

Figure 2-19. Modern engine design uses both pressure and splash methods of oiling. Oil travels under pressure through galleries to reach the top end of the engine (A); gravity flow or splash oiling lubricates many parts. A bypass valve is used to prevent oil starvation if the filter plugs or clogs (B). (Ford)

The coolant passages in the block and cylinder head must be designed so that the coolant will flow uniformly through the engine without collecting in pockets. If coolant flow is not uniform, transfer of heat from the engine to the coolant will be uneven. This means that some parts of the engine may be undercooled while others are overcooled.

If the engine runs too cold, the fuel will not vaporize properly. If liquid fuel reaches the cylinders, it will reduce lubrication by washing the oil from the cylinder walls and diluting the engine oil. This causes a loss of performance, an increase in HC emissions, and premature engine wear.

An engine that runs too hot may suffer from **pre-ignition**, where the air-fuel charge is ignited prematurely from excessive combustion chamber temperature. Viscosity of the oil circulating in an overheated engine is reduced. Hot oil also forms varnish and carbon deposits and may be drawn into the combustion chamber where it

increases HC emissions. This also causes poor performance, premature wear, and may even result in engine damage.

The engine cooling system actually is a temperature-regulation system. For a late-model engine, the cooling system must maintain a temperature that is high enough for efficient combustion but not so high that the engine will be damaged. The two jobs of the cooling system are to carry excess heat away from the engine and maintain uniform temperature throughout the engine. These requirements are very critical for an engine with electronic controls that must maintain precise air-fuel ratios for economy and emission control.

ENGINE LUBRICATION SYSTEM

An automotive engine cannot run without proper lubrication. Engine components work under conditions of extreme heat and must

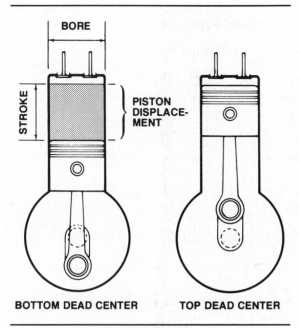

Figure 2-20. Basic engine dimensions.

maintain close tolerances. Engine oil performs a number of important tasks:

- Reduces friction
- Removes heat from engine parts
- Helps keep the engine clean
- Seals the combustion chambers
- Cushions engine parts
- Prevents corrosion and rust.

To do these jobs, engine oil must be formulated to flow easily while retaining its primary characteristic of a lubricating film at all times.

The lubrication system of an automotive engine consists of a reservoir (called the oil pan, sump, or crankcase) to hold the oil supply, a pump to develop pressure, a filter for cleaning, and valves which control flow and pressure.

Two methods are used to circulate oil through an engine: pressure and splash. In modern engine designs, these two methods are often combined, figure 2-19. Pressure is developed by the oil pump, which delivers oil to the filter for cleaning before it is sent to the camshaft and valve train components at the top of the engine. Other components in figure 2-19 are lubricated by splashing oil and by a network of passages, or galleries.

Oil circulates through the passages drilled or cast in the block, head, and crankshaft. These passages are called **oil galleries**. They serve as a network to carry the oil where it is needed. All engines must have at least one main gallery, but larger V-type engines may have two.

Smaller galleries lead away from the main gallery to carry oil to the crankshaft and camshaft bearings. Engines with an overhead camshaft, figure 2-19, will have a vertical gallery to send the oil directly to the cylinder head where it is distributed to the camshaft bearings. Holes, grooves, nozzles, and orifices also are used to ensure that proper lubrication reaches all parts of the engine at all times. A bypass valve is used to prevent oil starvation if the filter should clog or plug.

As the oil carries out its primary job of lubrication, it absorbs engine heat and carries it back to the crankcase, or sump, where the heat is dissipated. Hot oil cannot do its job properly and will quickly lose the qualities required for proper lubrication. To prevent this, an oil cooler is often used, especially with diesel, turbocharged, and air-cooled engines. Heat is removed from oil passing through the cooler by airflow or coolant circulation, depending upon the cooler design.

ENGINE DISPLACEMENT AND COMPRESSION RATIO

Two frequently used engine specifications are **engine displacement** and **compression ratio**. Displacement and compression ratio are related to each other, as we will learn in the following paragraphs.

Engine Displacement

Commonly used to indicate engine size, this specification is really a measurement of cylinder volume. The number of cylinders is a factor in determining displacement, but the arrangement of the cylinders or valves is not. Engine displacement is calculated by multiplying the

Pre-Ignition: An engine condition in which the air-fuel mixture ignites prematurely due to excessive combustion chamber temperature.

Oil Galleries: Passages in the block and head that carry oil under pressure to various parts of the engine.

Displacement: A measurement of the volume of air displaced by a piston as it moves from bottom to top of its stroke. Engine displacement is the piston displacement multiplied by the number of pistons in an engine.

Compression Ratio: The total volume of an engine cylinder divided by its clearance volume.

CUBIC INCH DISPLACEMENT CHART

Bore	(3½) 3.50	(3⁹/₁₆) 3.56	(3⅝) 3.63	(3¹¹/₁₆) 3.69	(3¾) 3.75	(3¹³/₁₆) 3.81	(3⅞) 3.88	(3¹⁵/₁₆) 3.94	4.00	(4¹/₁₆) 4.06	(4⅛) 4.13	(4³/₁₆) 4.19	(4¼) 4.25	(4⁵/₁₆) 4.31	(4⅜) 4.38
2.50	24.05	24.91	25.80	26.69	27.61	28.53	29.48	30.44	31.41	32.40	33.41	34.43	35.46	36.51	37.58
2.60	25.01	25.90	26.83	27.75	28.71	29.67	30.66	31.65	32.67	33.70	34.74	35.80	36.88	37.97	39.08
2.70	25.97	26.90	27.86	28.82	29.82	30.81	31.84	32.87	33.92	34.99	36.08	37.18	38.30	39.43	40.58
2.80	26.93	27.90	28.89	29.89	30.92	31.95	33.02	34.09	35.18	36.29	37.41	38.56	39.72	40.89	42.09
2.87	27.61	28.59	29.62	30.64	31.69	32.75	33.84	34.94	36.06	37.20	38.35	39.52	40.71	41.92	43.14
2.94	28.28	29.29	30.34	31.38	32.47	33.55	34.67	35.79	36.94	38.10	39.29	40.48	41.70	42.94	44.19
3.00	28.86	29.89	30.96	32.03	33.13	34.23	35.37	36.53	37.69	38.88	40.09	41.31	42.55	43.81	45.09
3.10	29.82	30.89	31.99	33.09	34.23	35.38	36.55	37.74	38.95	40.18	41.42	42.69	43.97	45.28	46.60
3.20	30.78	31.89	33.02	34.16	35.34	36.52	37.73	38.96	40.21	41.47	42.76	44.07	45.39	46.74	48.10
3.25	31.26	32.38	33.54	34.69	35.89	37.09	38.32	39.57	40.84	42.12	43.43	44.75	46.10	47.47	48.85
3.30	31.74	32.88	34.05	35.23	36.44	37.66	38.91	40.18	41.46	42.77	44.40	45.44	46.81	48.20	49.60
3.38	32.51	33.68	34.88	36.08	37.33	38.57	39.86	41.15	42.47	43.81	45.17	46.54	47.94	49.39	50.81
3.40	32.71	33.88	35.09	36.30	37.55	38.80	40.09	41.40	42.72	44.07	45.43	46.82	48.23	49.66	51.11
3.44	33.09	34.27	35.50	36.72	37.99	39.26	40.56	41.88	43.22	44.58	45.97	47.37	48.80	50.24	51.71
3.50	33.67	34.87	36.12	37.36	38.65	39.94	41.27	42.61	43.98	45.36	46.77	48.20	49.65	51.12	52.61
3.56	34.25	35.47	36.74	38.00	39.31	40.62	41.98	43.34	44.73	46.14	47.57	49.02	50.50	51.99	53.51
3.60	34.63	35.87	37.15	38.43	39.76	41.08	42.45	43.83	45.23	46.66	48.11	49.57	51.07	52.58	54.11
3.62	34.82	36.07	37.36	38.64	39.98	41.31	42.69	44.07	45.49	46.92	48.37	49.85	51.35	52.87	54.41
3.64	35.02	36.27	37.56	38.86	40.20	41.54	42.92	44.32	45.74	47.18	48.64	50.13	51.63	53.16	54.72
3.66	35.21	36.47	37.77	39.07	40.42	41.77	43.16	44.56	45.99	47.44	48.91	50.40	51.92	53.46	55.02
3.68	35.40	36.67	37.97	39.29	40.64	41.99	43.39	44.81	46.24	47.70	49.17	50.68	52.20	53.75	55.32
3.69	35.50	36.77	38.08	39.39	40.75	42.11	43.51	44.93	46.37	47.83	49.31	50.81	52.34	53.89	55.47
3.70	35.59	36.87	38.18	39.50	40.86	42.22	43.63	45.05	46.49	47.95	49.44	50.95	52.48	54.04	55.62
3.75	36.07	37.36	38.70	40.03	41.41	42.79	44.22	45.66	47.12	48.60	50.11	51.64	53.19	54.77	56.37
3.78	36.36	37.66	39.01	40.35	41.74	43.14	44.57	46.02	47.50	48.99	50.51	52.05	53.62	55.21	56.82
3.80	36.56	37.86	39.21	40.57	41.96	43.36	44.81	46.27	47.75	49.25	50.78	52.33	53.90	55.50	57.12
3.87	37.23	38.56	39.94	41.31	42.74	44.16	45.63	47.12	48.63	50.16	51.71	53.29	54.90	56.52	58.17
3.90	37.52	38.86	40.25	41.63	43.07	44.51	45.99	47.48	49.00	50.55	52.11	53.71	55.32	56.96	58.62
3.94	37.90	39.26	40.66	42.06	43.51	44.96	46.46	47.97	49.51	51.07	52.65	54.26	55.89	54.54	59.23
4.00	38.48	39.86	41.28	42.70	44.17	45.65	47.17	48.70	50.26	51.84	53.45	55.08	56.74	58.42	60.13

Figure 2-21. To use this cubic inch displacement chart, find the number where the bore and the stroke dimensions of your engine intersect. Then multiply by the number of cylinders.

number of cylinders in the engine by the piston displacement of one cylinder. The total engine displacement is the volume displaced by all the pistons.

The displacement of one cylinder is the space through which the piston's top surface moves as it travels from the bottom of its stroke (**bottom dead center**) to the top of its stroke (**top dead center**), figure 2-20. It is the volume displaced by the cylinder by one piston stroke. Piston displacement can be calculated as follows:

1. Divide the bore (cylinder diameter) by two. This gives you the radius of the bore.
2. Square the radius (multiplying it by itself).
3. Multiply the square of the radius by 3.1416 (pi or π) to find the area of the cylinder cross section.
4. Multiply the area of the cylinder cross section by the length of the stroke.

You now know the piston displacement for one cylinder. Multiply this by the number of cylinders to determine the total engine displacement. The formula for the complete procedure reads:

$R^2 \times \pi \times$ stroke $\times$ No. of cylinders = displacement

For example, to find the displacement of a V-6 engine with a 3.80-inch bore and a 3.40-inch stroke:

radius $= \dfrac{3.80}{2} = 1.9$ in.

radius squared $= 1.9$ in. $\times 1.9$ in. $= 3.61$ in.^{2}

cross section $= 3.61$ in.$^2 \times 3.1416 = 11.3412$ in.^{2}

displacement of one cylinder $= 11.3412$ in.^{2} $\times 3.40$ in. $= 38.56$ in.^{3}

total displacement $= 6 \times 38.56$ in.$^3 =$ 231.36 in.^{3}

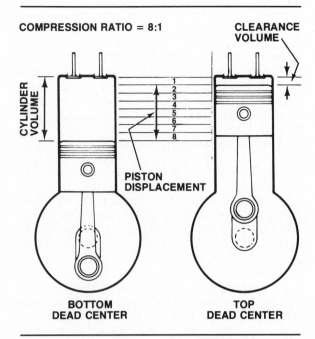

COMPRESSION RATIO = 8:1

CLEARANCE VOLUME

CYLINDER VOLUME

PISTON DISPLACEMENT

BOTTOM DEAD CENTER

TOP DEAD CENTER

Figure 2-22. Compression ratio is the ratio of the total cylinder volume to the clearance volume.

The engine displacement is 231 cubic inches. Fractions of an inch are usually not included.

This procedure can be greatly simplified by using a cubic inch displacement chart, figure 2-21. You simply locate that point on the chart where the bore and stroke specifications for a given engine intersect, then multiply that figure by the total number of cylinders in the engine to find its displacement.

The greater the engine displacement, the more air-fuel mixture the cylinders can accept, and so the greater the power output (assuming all other factors remain equal).

Metric displacement specifications

When stated in English values, displacement is given in cubic inches; the engine's cubic inch displacement is abbreviated as "cu. in." or "cid". When stated in metric values, displacement is given in cubic centimeters (cc) or in liters (one liter equals approximately 1,000 cc). To convert engine displacement specifications from one value to another, use the following formulas:

● To change cubic centimeters to cubic inches, multiply by 0.061 (cc × 0.061 = cid).
● To change cubic inches to cubic centimeters, multiply by 16.39 (cid × 16.39 = cc).
● To change liters to cubic inches, multiply by 61.02 (liters × 61.02 = cid).

Our 231-cid engine from the previous example is also a 3,792-cc engine (231.36 × 16.39 = 3,792). When expressed in liters, this figure

would be rounded off to 3.8 liters. Metric displacement can also be calculated directly with the displacement formula, using centimeter measurements instead of inches.

Compression Ratio

This specification compares the total cylinder volume to the volume of only the combustion chamber, figure 2-22. Total cylinder volume may seem to be the same as piston displacement, but it is not. Total cylinder volume is the piston displacement plus the combustion chamber volume. The combustion chamber volume

Bottom Dead Center: The exact bottom of a piston stroke. Abbreviated: bdc.

Top Dead Center: The exact top of a piston's stroke. Also a specification used when tuning an engine. Abbreviated: tdc.

■ **"I Wouldn't Touch Metric Conversion With A 3.049-Meter Pole"**

The attempt to convert the United States system to metrics has taken many years, cost enormous amounts of money, and has not yet been completely successful, although many measurements are now in metric units.

For instance, cameras and film generally use millimeters (mm) for measuring the focal length of lenses or the size of the film. In electronics, we use the metric system of seconds, volts, watts, amperes, and hertz (cycles per second). The drug industry changed over to metrics more than two decades ago. Domestic automakers are not only using some imported engines with metric designations (such as the 2.3-liter engine), but have built many of their own engines and cars with metric measurements.

Despite the fact that metric measurement is still not universally used, technicians need to know and understand the metric system. Domestic tire manufacturers use kilopascals (kPa) as well as pounds per square inch (psi) to indicate inflation pressure. The fasteners used on many automotive components may be either U.S. or metric, which means that you need two sets of some tools. And tightening torque valves given in metric units are useless if you only have a foot-pound torque wrench and can't convert the specification.

Yes, the grand dream of completely converting the automotive industry to metrics is still alive; it's just taking longer than anyone thought.

Figure 2-23. The uneven firing intervals of Buick's early V-6 engine. (Buick)

with the piston at top dead center is often called the **clearance volume**.

Compression ratio is the total volume of a cylinder divided by its clearance volume. If the clearance volume is ⅛ of the total cylinder volume, the compression ratio is 8 (8 to 1). The formula is as follows:

$$\frac{\text{Total volume}}{\text{Clearance volume}} = \text{Compression ratio}$$

To determine the compression ratio of an engine in which each piston displaces 31.12 cu. in. and which has a clearance volume of 4.15 cu. in.:

$$31.12 + 4.15 = 35.27 \text{ (total cylinder volume)}$$

$$\frac{35.27}{4.15} = 8.498$$

The compression ratio is 8.498; this would be rounded and expressed as a compression ratio of 8.5. This can also be written 8.5:1.

In theory, the higher the compression ratio, the greater the efficiency of the engine, and the more power an engine will develop from a given quantity of fuel. The reason for this is that combustion takes place faster because the fuel molecules are more tightly packed and the flame of combustion travels more rapidly.

But there are practical limits to how high a compression ratio can be. Because of the unavailability of high octane fuel, most gasoline-burning engines are restricted to a compression ratio no greater than 11.5 to 1. Ratios this high, however, create high combustion chamber temperatures. This in turn creates oxides of nitrogen (NO_x), a primary air pollutant. In the early

1970's, compression ratios were lowered to around 8 to permit the use of lower octane, low-lead or unleaded fuel, and to reduce NO_x formation. Advances in electronic engine controls in the 1980's have allowed engineers to raise compression ratios to the 9 and 10 to 1 range for optimum performance and economy.

IGNITION INTERVAL

Every two strokes of a piston cause the crankshaft to rotate 360 degrees, as we have seen. Therefore, every four strokes of a piston causes the crankshaft to rotate 720 degrees (360 + 360 = 720). Because four strokes of a piston equal one engine operating cycle, one engine operating cycle equals 720 degrees of crankshaft rotation.

During the four strokes of the operating cycle, the spark plug fires only once, at the beginning of the power stroke. In a 1-cylinder engine, there would be only one ignition spark every 720 degrees of crankshaft rotation. These 720 degrees are called the **ignition interval**, or **firing interval**, of the engine. It is the number of degrees of crankshaft rotation that occur between ignition sparks.

The more cylinders an engine has, the more power strokes are produced per engine revolution. A 4-cylinder engine produces a power stroke four times as often as a 1-cylinder engine. The 4-cylinder engine has power strokes that are closer together in terms of degrees of crankshaft rotation.

Common Ignition Intervals

Since a 4-cylinder engine has four power strokes during 720 degrees of crankshaft rotation, one power stroke must occur every 180 degrees (720 ÷ 4 = 180). The ignition system must produce a spark for every power stroke, so it produces a spark every 180 degrees of crankshaft rotation. This means that a 4-cylinder engine has an ignition interval of 180 degrees.

An inline 6-cylinder engine has six power strokes during every 720 degrees of crankshaft rotation, for an ignition interval of 120 degrees (720 ÷ 6 = 120). An 8-cylinder engine has an ignition interval of 90 degrees (720 ÷ 8 = 90).

Unusual Ignition Intervals

Most automotive engines have 4, 6, or 8 cylinders, but other engines are in use today. Some companies, such as Jaguar, Ferrari, and BMW, produce 12-cylinder engines, with a 60-degree firing interval. Audi and Mercedes have 5-cylinder engines with a 144-degree firing interval; Suzuki produces a 3-cylinder engine with a

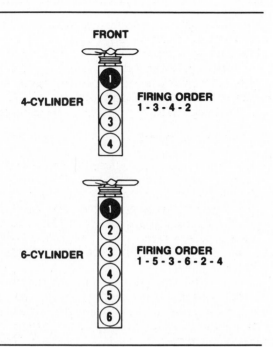

Figure 2-24. Cylinder numbering of an inline engine.

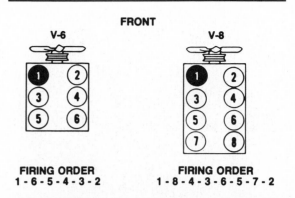

Figure 2-25. American Motors, Chrysler Motors, and most General Motors V-type engines are numbered in this way.

240-degree firing interval used in Chevrolet's Sprint, as does Subaru.

Other unusual firing intervals result from other engine designs. General Motors has produced two different 6-cylinder engines from 8-cylinder engine blocks. The Buick engine developed in the 1960's has alternating 90- and 150-degree firing intervals, figure 2-23. The uneven firing intervals resulted from building a V-6 with a 90-degree crankshaft and block. This engine was modified in mid-1977 by redesigning the crankshaft to provide uniform 120-degree firing intervals, as in an inline six. (The Honda Acura V-6 uses the same block and crankshaft arrangement). In 1978, Chevrolet introduced a 90-degree V-6 engine that fires at alternating 108- and 132-degree intervals. Chrysler's 3.9-liter V-6, derived from its 90-degree 318-cid (5.2-liter) V-8, has split crank pins offset by 22 degrees instead of the 30 degrees necessary to produce uniform firing. This gives it an alternating firing interval of 112 and 128 degrees.

Spark Frequency

In a spark-ignition engine, each power stroke is caused by a spark igniting the air-fuel mixture. Each power stroke needs an individual spark. An 8-cylinder engine, for example, requires four sparks per engine revolution (remember that there are two 360-degree engine revolutions in each 720-degree operating cycle). When the engine is running at about 1,000 rpm, the

ignition system must deliver 4,000 sparks per minute. At high speed (about 4,000 rpm), the ignition system must deliver 16,000 sparks per minute. Precise ignition system performance is needed to meet these demands.

FIRING ORDER

To cause each cylinder in an engine to fire once within 720 degrees of crankshaft rotation and at regular intervals, the pistons and connecting rods are arranged on the crankshaft in a specific order. This is called the firing order, and it varies with different engine designs. Firing orders are designed to reduce the vibration and imbalance created by the power strokes of the pistons.

Engine cylinders are numbered for easy identification. However, the cylinders do not fire in the order in which they are numbered. Straight or inline engines are numbered from front to rear, figure 2-24. A typical 4-cylinder engine firing order is 1-3-4-2. That is, the number 1 cylinder power stroke is followed by the number 3 cylinder power stroke, then the number 4 power stroke, and finally the number 2 power stroke. The cycle then repeats itself. A few 4-cylinder engines have different firing orders. For example, the English Ford and Pinto 1,600-cc engines fire 1-2-4-3.

Clearance Volume: The volume of a combustion chamber when the piston is at top dead center.

Ignition Interval (Firing Interval): The number of degrees of crankshaft rotation between ignition sparks.

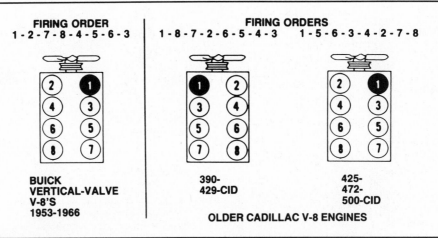

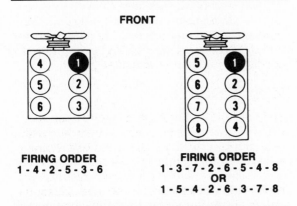

Figure 2-26. Cadillac V-8's and older Buick V-8's have unusual firing orders.

Figure 2-27. Ford numbers its V-type engines in this way.

The cylinders of an inline 6-cylinder engine also are numbered from front to rear, but do not fire in that order. The firing order for all inline 6-cylinder engines is 1-5-3-6-2-4, figure 2-24.

Except for Ford vehicles and a few GM engines, the cylinders of all domestic V-type engines are numbered the same, figure 2-25. The front cylinder on the left (driver's) side is number 1. The front cylinder on the right (passenger's) side is number 2. Behind number 1 is number 3; behind number 2 is number 4, and so on. The firing order for a V-6 numbered this way is 1-6-5-4-3-2. For a V-8, the firing order is 1-8-4-3-6-5-7-2.

Besides Fords, the exceptions to this rule for V-type engine cylinder numbering are:
1. Vertical-valve Buick V-8's built before 1967
2. Older Cadillac V-8's
3. The Chevrolet 173-cid (2.8-liter) 60-degree V-6.

The older Buick V-8's have the number 1 cylinder at the front of the right side, figure 2-26, rather than the left side. The firing order is 1-2-7-8-4-5-6-3.

Cadillac 425-, 472-, and 500-cid engines are numbered like the older Buicks, figure 2-26, but the firing order is 1-5-6-3-4-2-7-8. Cadillac 390- and 429-cid engines are numbered in the more conventional manner, with the number 1 cylinder at the front left, figure 2-26. However, the firing order for these engines is 1-8-7-2-6-5-4-3.

The Chevrolet 60-degree 173-cid (2.8-liter) V-6 engine has the cylinder banks reversed, that is, the number 1, 3, and 5 cylinders are on the right (passenger's) side and the number 2, 4, and 6 cylinders are on the left (driver's) side. The firing order is 1-2-3-4-5-6.

Ford V-type engines have the number 1 cylinder at the right front, figure 2-27. The numbering continues down the right side, then goes to the left side from front to rear. A Ford V-6 firing order is 1-4-2-5-3-6. Ford V-8 firing orders are 1-3-7-2-6-5-4-8 or 1-5-4-2-6-3-7-8, depending on engine size.

The ignition system must deliver ignition voltage to the correct cylinder at the correct time. To maintain the firing order, the spark plug cables must be attached to the distributor cap in the proper order, figure 2-28.

ENGINE-IGNITION SYNCHRONIZATION

During the engine operating cycle, the intake and exhaust valves open and close at specific times. The ignition system delivers a spark when the piston is near the top of the compression stroke and both valves are closed. These actions must all be coordinated or engine damage can occur.

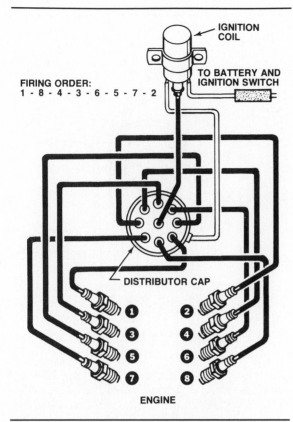

Figure 2-28. The spark plug cables must be connected to the distributor in the proper firing order.

Distributor Drive

The distributor must supply one spark to each cylinder during each cylinder's operating cycle. The distributor cam has as many lobes as the engine has cylinders. In a solid-state system, the trigger wheel has as many teeth as the engine has cylinders. One revolution of the distributor shaft will deliver one spark to each cylinder. Since a cylinder needs only one spark for each *two* crankshaft revolutions, the distributor shaft must turn at one-half engine crankshaft speed. Therefore, the distributor is driven by the camshaft, which also turns at one-half crankshaft speed.

CRANKSHAFT POSITION

The bottom of the piston stroke is called bottom dead center (bdc). The top of the piston stroke is called top dead center (tdc), figure 2-29. The ignition spark occurs near top dead center, as the compression stroke is ending. As the piston approaches the top of its stroke, it is said to be **before top dead center** (btdc). A spark that occurs before top dead center is called an advanced spark. As the piston passes top dead

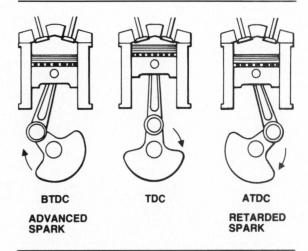

Figure 2-29. Piston position is identified in terms of crankshaft rotation.

center and starts down, it is said to be **after top dead center** (atdc). A spark that occurs after top dead center is called a retarded spark.

Burn Time

Approximately 3 milliseconds (0.003 second) elapse from the instant the air-fuel mixture ignites until its combustion is complete. Remember that this burn time is a function of *time* and not of piston travel or crankshaft degrees. The ignition spark must occur early enough so that the combustion pressure reaches its maximum just after top dead center, when the piston is beginning its downward power stroke. Combustion should be completed by about 10° atdc. If the spark occurs too soon before top dead center, the rising piston will be opposed by combustion pressure. If the spark occurs too late, the force on the piston will be reduced. In both cases, power will be lost. In extreme cases, the engine could be damaged. Ignition must start at the proper instant for maximum power and efficiency.

Before Top Dead Center: The position of a piston as it nears top dead center. Abbreviated: btdc. Usually expressed in degrees, such as 5° btdc.

After Top Dead Center: The position of a piston after it has passed top dead center. Abbreviated: atdc. Usually expressed in degrees, such as 5° atdc.

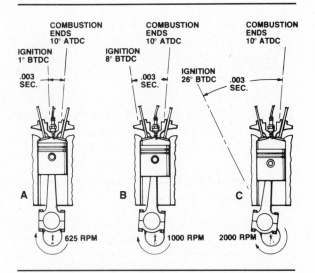

Figure 2-30. As engine speed increases, ignition timing must be advanced.

Engine Speed

As engine speed increases, piston speed increases. If the air-fuel ratio remains relatively constant, the fuel burning time will remain constant. However, at greater engine speed, the piston will travel farther during this burning time. Ignition timing must be changed to ensure that maximum combustion pressure occurs at the proper piston position.

For example, consider an engine, figure 2-30, that requires 0.003 second for the fuel charge to burn and that achieves maximum power if the burning is completed at 10° atdc.

● At an idle speed of 625 rpm, position A, the crankshaft rotates about 11 degrees in 0.003 second. Therefore, timing must be set at 1° btdc to allow ample burning time.
● At 1,000 rpm, position B, the crankshaft rotates 18 degrees in 0.003 second. Ignition should begin at 8° btdc.
● At 2,000 rpm, position C, the crankshaft rotates 36 degrees in 0.003 second. Spark timing must be advanced to 26° btdc.

Change in timing is called spark advance, or ignition advance, and is explained in greater detail in Chapter 19.

INITIAL TIMING

As we have seen, ignition timing must be set correctly for the engine to run at all. This is called the engine's initial, or basic, timing. Initial timing is the correct setting at a specified engine speed. In figure 2-30, initial timing was 1° btdc. Initial timing is normally within a few degrees of top dead center. For many years, most engines were timed at the specified slow

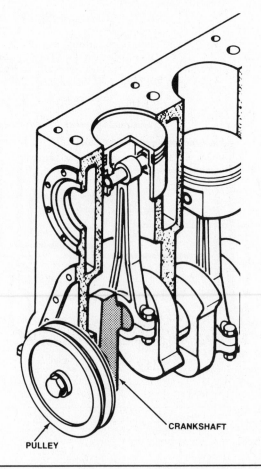

Figure 2-31. Most engines have a pulley bolted to the front end of the crankshaft.

idle speed for the engine. However, since about 1974, a number of engines have required timing at speeds either above or below the slow idle speed. Initial timing can be adjusted to compensate for mechanical wear, slippage, and other factors.

Timing Marks

We have seen that initial timing is related to crankshaft position. To properly time the engine, we must be able to determine crankshaft position. The crankshaft is completely enclosed in the engine block, but most engines have a pulley and vibration damper or harmonic balancer bolted to the front of the crankshaft, figure 2-31. This pulley rotates with the crankshaft and can be considered an extension of the shaft.

Marks on the pulley show crankshaft position. For example, when a mark on the pulley is aligned with a mark on the engine block, the number 1 piston is at top dead center.

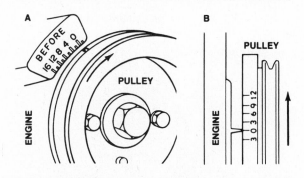

Figure 2-32. Common timing marks.

Timing marks vary widely, even within a manufacturer's product line. There are two common types of timing marks, figure 2-32:

• A mark on the crankshaft pulley and marks representing degrees of crankshaft position on the engine block, position A.

• Marks on the pulley representing degrees of crankshaft position and a pointer on the engine block, position B.

Some engines have timing marks on their flywheel, or a notch on the engine flywheel and a scale on the transmission cover or bellhousing. The flywheel is attached to the end of the crankshaft opposite the pulley. Most front-wheel-drive imported cars have these types of timing marks. Some older domestic engines also have a flywheel timing mark.

Most late-model engines have a special socket used for electromagnetic timing, in addition to traditional timing marks. These can be timed with a timing light or, for greater accuracy, with a special test probe that fits into the socket. The electromagnetic probe works like the pickup coil in an electronic ignition. It produces a signal voltage pulse each time a magnetic particle or notch on the crankshaft passes the probe socket. The signal voltage operates a timing meter that displays the ignition timing.

OTHER ENGINE TYPES

Other engine types besides the 4-stroke engine have been installed in automobiles over the years, but only four have been used with any real success — the 2-stroke, the diesel, the rotary, and the stratified charge engines.

The 2-Stroke Engine

While 4-stroke engines develop *one* power stroke for every *two* crankshaft revolutions, **two-stroke engines** produce a power stroke for *each* revolution. Poppet valves are not used in most automobile and motorcycle 2-stroke engines. The valve work is done instead by the piston, which uncovers intake and exhaust ports in the cylinder as it nears the bottom of its stroke.

Two-Stroke Engine: An engine in which a piston makes only two strokes to complete one operating cycle.

■ **Plastic engines are on the way**

Automotive engines have traditionally been manufactured of cast iron. This alloy of iron, carbon, and silicon is hard, and unlike most metals, it is non-malleable. This means that cast iron cannot be stretched or extended by the high pressures and temperatures inherent in engine operation. It can withstand considerable overheating without permanent damage.

What's wrong with cast iron? Not much, except its weight. Cast iron engines are heavy and in an era when weight is directly related to fuel economy, engineers take every approach they can find to reduce weight.

Enter aluminum. This malleable metal has the advantage of light weight. It's been around for many years now, and engineers are finally beginning to cope with its inherent weaknesses as a substitute for cast iron. Pure aluminum cannot be used, but alloyed with other metals, it has the qualities that make it a contender in today's automotive engine design world that concentrates on smaller 4-cylinder powerplants.

The use of composite materials is in the not-too-distant future. Various fiber-reinforced plastics are capable of withstanding the high temperatures and stress factors involved in an automotive engine. And in the same way that engineers once alloyed one metal with others, they are now combining fiber types to obtain a composite with all the specific qualities they desire. Ford's Special Vehicles Operation (SVO) has been deeply involved in testing nonmetallic composite components in high-performance racing engines for several years.

A company called Polimotor has produced a working prototype of the Ford 2.3-liter 4-cylinder engine manufactured primarily of composite materials. It weighs only 152 pounds (69 kilograms), compared to the 415-pound (188-kilogram) engine from which it was derived. The twin-cam Polimotor composite engine can deliver over 300 horsepower (224 kW) at 9,200 rpm without redlining the engine under 14,000 rpm! Cost is the present drawback, but by the time the technical bugs are all worked out, it should be within reason.

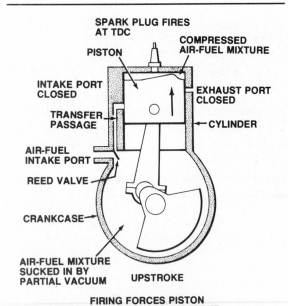

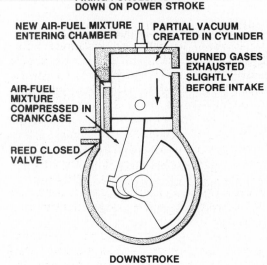

Figure 2-33. Two-stroke cycle engine operation.

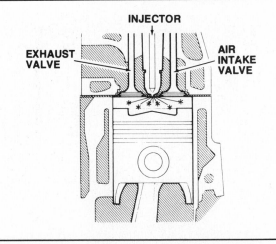

Figure 2-34. Diesel combustion occurs when fuel is injected into the hot, highly compressed air in the cylinder. (Cummins)

Two-stroke gasoline engines used in motorcycles and some small cars draw the air-fuel mixture into the crankcase where it is partially compressed for delivery to the cylinders. It is easy to see that, as a piston moves up, pressure increases in the cylinder above it while pressure decreases in the crankcase below it. By using the crankcase to pull in the air-fuel mixture, the 2-stroke engine combines the intake and compression strokes. Also, by using exhaust ports in the cylinder wall, the power and exhaust strokes are combined. The operating cycle of a 2-stroke engine shown in figure 2-33, works like this:

1. Intake and Compression: As the piston moves up, a low-pressure area is created in the crankcase. A **reed valve** opens, and the air-fuel mixture enters the crankcase. At the same time, the piston compresses a previous air-fuel charge in the cylinder. Ignition occurs in the combustion chamber when the piston is near top dead center.

2. Power and Exhaust: Ignited by the spark, the expanding air-fuel mixture forces the piston downward in the cylinder. As the piston travels down, it uncovers exhaust ports in one side of the cylinder. The exhaust flows through the ports and out of the engine. As the piston continues its downward movement, it compresses the air-fuel charge in the crankcase. The intake reed valve closes to hold the charge in the crankcase. After uncovering the exhaust ports, the piston uncovers similar intake ports nearer the bottom of the stroke. The compressed mixture in the crankcase moves through a transfer passage to the intake ports and flows into the cylinder. The force of the incoming mixture also helps to drive the remaining exhaust gases from the cylinder. A ridge on the top of the piston deflects the intake mixture upward in the cylinder so it will not flow directly out the exhaust ports.

The piston then begins another upward compression stroke, closing off the intake and the exhaust ports, and the cycle begins again. Because the crankcase of a 2-stroke gasoline engine is used for air-fuel intake, it cannot be used as a lubricating oil reservoir. Therefore, engine oil for a 2-stroke gasoline engine must be mixed with fuel.

In theory, the 2-stroke engine should develop twice the power of a 4-stroke engine of the same size, but the 2-stroke design also has its practical limitations. With intake and exhaust

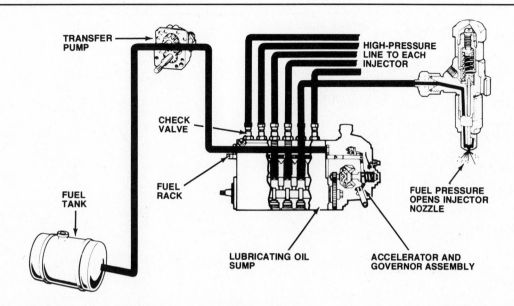

Figure 2-35. Typical automotive diesel fuel injection system.

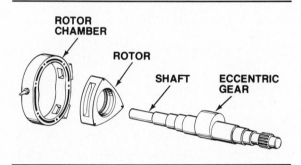

Figure 2-36. The main parts of a Wankel rotary engine are the rotor chamber, the three-sided rotor and the shaft with an eccentric gear.

occurring at almost the same time, it does not breathe or take in air as efficiently. Mixing fresh fuel with unburned fuel preheats the mixture. Because this increases volume, it reduces efficiency.

The Diesel Engine

In 1892, a German engineer named Rudolf Diesel perfected the compression-ignition engine that bears his name. The diesel engine uses heat created by compression to ignite the fuel, so it requires no spark ignition system.

The diesel engine requires compression ratios of 16:1 and higher. Incoming air is compressed until its temperature reaches about 1,000°F (538°C). As the piston reaches the top of its compression stroke, fuel is injected into the cylinder, where it is ignited by the hot air, figure 2-34. As the fuel burns, it expands and produces power.

Diesel engines differ from gasoline-burning engines in other ways. Instead of a carburetor to mix the fuel with air, a diesel uses a precision **injection pump** and individual fuel injectors. The pump delivers fuel to the injectors at a high pressure and at timed intervals. Each injector measures the fuel exactly, spraying it into the combustion chamber at the precise moment required for efficient combustion, figure 2-35. The injection pump and injector system thus perform the fuel delivery job of the carburetor and the ignition timing job of the distributor in a gasoline engine.

The air-fuel mixture of a gasoline engine remains nearly constant — changing only within a narrow range — regardless of engine load or speed. But in a diesel engine, *air* remains constant and the amount of *fuel* injected is varied to control power and speed. The air-fuel mixture of a diesel can vary from as little as 85:1 at idle, to as rich as 20:1 at full load. This higher air-fuel ratio and the increased compression

Reed Valve: A one-way check valve. A reed, or flap, opens to admit a fluid or gas under pressure from one direction, while closing to deny movement from the opposite direction.

Injection Pump: A pump used on diesel engines to deliver fuel under high pressure at precisely timed intervals to the fuel injectors.

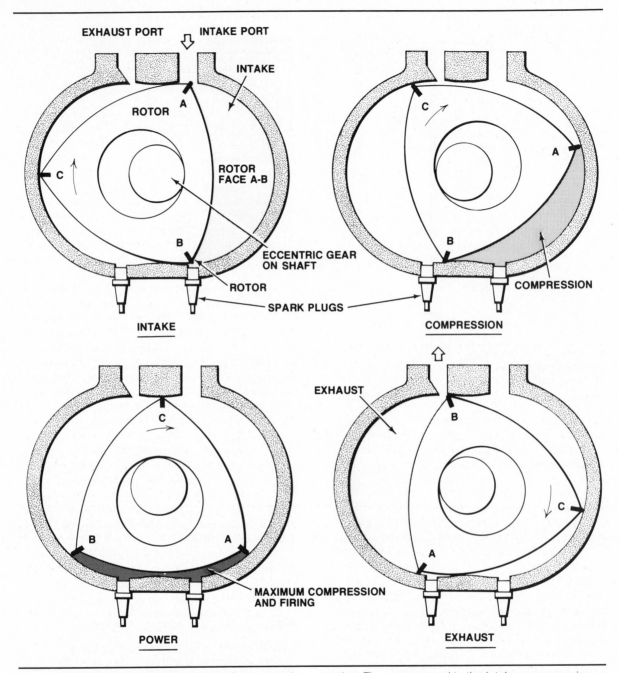

Figure 2-37. This shows the four stages of rotary engine operation. They correspond to the intake, compression, power, and exhaust strokes of a four-stroke reciprocating engine. The sequence is shown for only one rotor face, but each face of the rotor goes through all four stages during each rotor revolution.

pressures make the diesel more efficient in terms of fuel consumption than a gasoline engine.

Like gasoline engines, diesel engines are built in both 2-stroke and 4-stroke versions. The most common 2-stroke diesels are the truck and industrial engines made by the Detroit Diesel Allison Division of General Motors. In these engines, air intake is through ports in the cylinder walls, aided by supercharging. Exhaust is

through valves in the head. Crankcase fuel induction cannot be used in a 2-stroke diesel.

For many years, diesel engines were used primarily in trucks and heavy equipment. Mercedes-Benz, however, has built diesel cars since 1936, and the energy crises of 1973 and 1979 focused attention on the diesel as a substitute for gasoline engines in automobiles.

In the late 1970's, General Motors and Volkswagen developed diesel engines for cars.

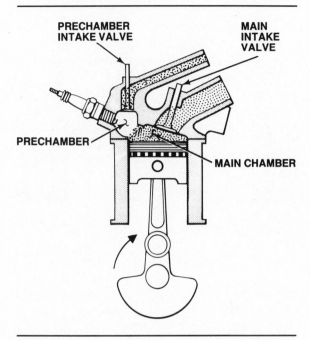

Figure 2-38. Honda CVCC cylinder head, showing main combustion chamber and precombustion chamber (prechamber) with extra intake valve.

They were followed quickly by most major automakers in offering optional diesel engines for their vehicles. By the early 1980's many carmakers were predicting diesel power for more than 30 percent of the domestic auto population. These predictions did not come true, however.

Increasing gasoline supplies and lower prices in the mid-1980's combined with the diesel disadvantages of noise and higher manufacturing costs to reduce the incentives for customers to buy diesel automobiles. Stringent diesel emission regulations added to manufacturing costs and made the engines harder to certify for sale. A few automobile diesel engines, such as the Oldsmobile V-8, were derived from gasoline powerplants and suffered reliability problems. By the late 1980's diesel engine use is mostly limited to truck and industrial applications in the U.S. and Canada, although diesel automobiles continue to sell well in Europe where gasoline costs remain high.

The Rotary (Wankel) Engine

The reciprocating motion of a piston engine is both complicated and inefficient. For these reasons, engine designers have spent decades attempting to devise engines in which the working parts would all rotate on an axis. The major problem with this rotary concept has been the sealing of the combustion chamber. Of the various solutions proposed, only the rotary design of Felix Wankel — as later adapted by NSU, Curtiss-Wright and Toyo Kogyo (Mazda) — has proven practical.

Although the same sequence of events occur in both a rotary and a reciprocating engine, the rotary is quite different in design and operation. A curved triangular rotor moves on an **eccentric**, or off-center, geared portion of a shaft within a long chamber, figure 2-36. As it turns, the rotor's corners follow the housing shape. The rotor thus forms separate chambers whose size and shape change constantly during rotation. The intake, compression, power, and exhaust functions occur within these chambers as shown in figure 2-37. Wankel engines can be built with more than one rotor. Mazda production engines, for example, are 2-rotor engines.

One revolution of the rotor produces three power strokes or pulses, one for each face of the rotor. In fact, each rotor face can be considered the same as one piston. Each pulse lasts for about three-quarters of a rotor revolution. The combination of rotary motion and longer power pulses which overlap results in a smooth-running engine. While the rotary overcomes many of the disadvantages of the piston engine, it has its own disadvantages.

About equivalent in power output to that of a 6-cylinder piston engine, a 2-rotor engine is only one-third to one-half the size and weight. With no pistons, rods, valves, lifters, and other reciprocating parts, the rotary engine has 40 percent fewer parts than a piston engine. But it is also basically a very "dirty" engine. In other words, it gives off a high level of emissions and so it requires additional external devices to clean up the exhaust.

The Stratified Charge Engine

Like the rotary design, the concept of a **stratified charge engine** has been around in many forms for many years. Honda, however, was the first carmaker to produce and use one

Eccentric: Off center. A shaft lobe which has a center different from that of the shaft.

Stratified Charge Engine: An engine that uses 2-stage combustion: first is combustion of a rich air-fuel mixture in a precombustion chamber, then combustion of a lean air-fuel mixture occurs in the main combustion chamber.

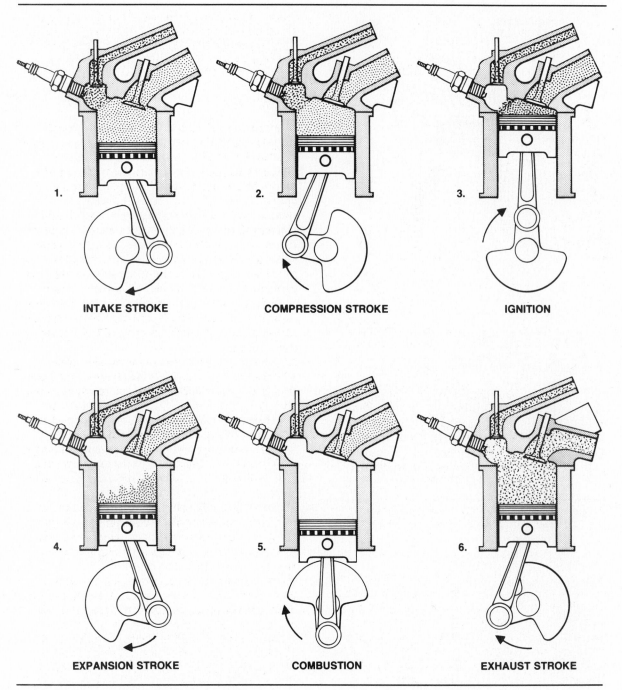

Figure 2-39. Honda CVCC engine operating cycle.

successfully. Honda's Compound Vortex Controlled Combustion (CVCC) design was the first stratified charge gasoline engine used in a mass-produced car.

The CVCC engine has a separate small precombustion chamber located above the main combustion chamber and contains a tiny additional valve, figure 2-38. Except for this feature,

the CVCC is a conventional 4-stroke piston engine. However, it uses a 2-stage combustion process. Figure 2-39 shows the stages in the operating cycle.

The first stage is one of precombustion, in which the air-fuel mixture is ignited in the precombustion chamber. In the second stage, the flame front created moves down into the main

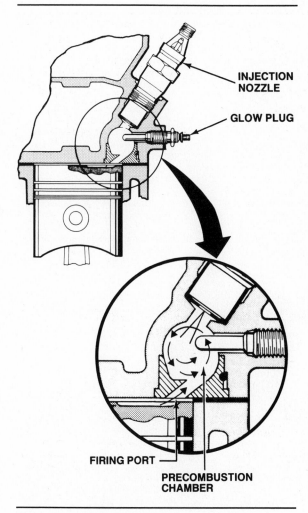

INJECTION NOZZLE

GLOW PLUG

FIRING PORT

PRECOMBUSTION CHAMBER

Figure 2-40. Volkswagen's passenger car diesel engine uses a precombustion swirl chamber.

combustion chamber to ignite a mixture with less fuel in it. The stratified charge engine takes its name from this layering or stratification of the air-fuel mixture just before combustion. At that time, there is a rich mixture (with lots of fuel) near the spark plug, a moderate mixture in the auxiliary combustion chamber, and a lean mixture (with little fuel) in the main chamber. The result is a more complete combustion of the air-fuel mixture, which keeps unburned fuel and emissions to a minimum.

The stratified charge principle is a *method* of controlling the combustion process. It does not represent a *type* of engine construction, such as the reciprocating or rotary engine types. In fact, charge stratification has been applied both to reciprocating diesel engines and to rotary gasoline engines. Most diesel engines used in cars have a precombustion chamber into which the fuel is injected, figure 2-40. This allows the

combustion to occur in two stages: in the precombustion chamber and in the main chamber. This improves cold starting and combustion efficiency and reduces engine noise and vibration.

SUMMARY

Most automobile engines are internal combustion, reciprocating 4-stroke engines. An air-fuel mixture is drawn into sealed combustion chambers by a vacuum created by the downward stroke of a piston. The mixture is ignited by a spark.

Valves at the top of the cylinder open and close to admit the air-fuel mixture and release the exhaust. These valves are driven by a camshaft and synchronized with engine speed. The sequence in which the cylinders fire is the firing order. Several valve designs have been used, including I-head, F-head, and L-head designs. Most engine cylinders have two valves, but some engines have three, four, or five valves per cylinder.

Automobile engines create considerable heat; only about a third of it is transformed into useful energy. The rest is drawn off by the cooling and exhaust systems. Passages in the engine block and cylinder head allow coolant to circulate and carry the heat to the radiator, where it is dissipated. Engine lubrication also is very important. Oil circulates through passages called oil galleries to reduce friction, wear, and rust.

Displacement and compression ratio are two frequently used engine specifications. Displacement indicates engine size, and compression ratio compares total cylinder volume to compression chamber volume.

The ignition interval is the number of degrees between ignition sparks. A 4-cylinder engine commonly has an ignition interval of 180 degrees, although many intervals have been used over the years. Still, each cylinder fires once every 720 degrees of crankshaft rotation. The ignition spark is provided by the distributor or electronic ignition, and is synchronized with the crankshaft rotation.

Timing is essential for the engine to operate. Timing marks on the front of the engine block or the pulley indicate crankshaft position, and can be used to alter the timing from the engine's basic initial timing.

Most automobile engines are 4-stroke gasoline engines, but other types have been used, including the diesel, the rotary (Wankel), the 2-stroke, and the stratified charge.

Review Questions

Choose the single most correct answer.
Compare your answers with the correct answers on page 414.

1. The combustion chamber is usually contained in the:
   a. Engine block
   b. Piston
   c. Cylinder head
   d. Water jacket

2. The bore is the diameter of the:
   a. Connecting rod
   b. Cylinder
   c. Crankshaft
   d. Combustion chamber

3. The four-stroke engine is also
   called the:
   a. Otto cycle engine
   b. Diesel engine
   c. Rotary engine
   d. None of the above

4. The four-stroke cycle operates in
   which order?
   a. Intake, exhaust, power,
      compression
   b. Intake, power, exhaust,
      compression
   c. Compression, power, intake,
      exhaust
   d. Intake, compression, power,
      exhaust

5. Which of these engines is most
   often air cooled?
   a. Rotary
   b. Horizontally opposed
   c. V-type
   d. Incline

6. Valves are opened by:
   a. Camshaft lobes
   b. Connecting rods
   c. The crankshaft
   d. Valve springs

7. Synchronization of camshaft and
   crankshaft rotations is accomplished by:
   a. Gears
   b. A chain and sprockets
   c. A timing belt and sprockets
   d. Any of the above

8. In an eight-cylinder engine, the
   number of power strokes at a
   given instant is:
   a. 8
   b. 2
   c. 4
   d. 1

9. In an I-head engine:
   a. The intake valves are in the
      block; the exhaust valves are in
      the head
   b. All the valves are in the block
   c. All the valves are in the head
   d. None of the above

10. Which of the following is *not*
    used in calculating engine
    displacement?
    a. Stroke
    b. Bore
    c. Number of cylinders
    d. Valve arrangement

11. To change cubic centimeters to
    cubic inches, multiply by:
    a. 0.061
    b. 16.39
    c. 61.02
    d. 1000

12. Compression ratio is:
    a. Piston displacement plus clear-
       ance volume
    b. Total volume times number of
       cylinders
    c. Total volume divided by clear-
       ance volume
    d. Stroke divided by bore

13. A 2-stroke engine will:
    a. Produce a power stroke for
       each crankshaft revolution
    b. Have intake and exhaust ports
    c. Use the crankcase for fuel
       induction
    d. All of the above

14. Diesel engines:
    a. Have no valves
    b. Produce ignition by heat of
       compression
    c. Have low compression
    d. Use special carburetors

15. Because of its fuel injection sys-
    tem, a diesel engine:
    a. Needs no carburetor or
       distributor
    b. Has a constant fuel mixture
    c. Is inefficient
    d. Operates only in a two-stroke
       configuration

16. How many strokes of a piston are
    required to turn the crankshaft
    through 360 degrees?
    a. One
    b. Two
    c. Three
    d. Four the above

17. A "retarded spark" is one that
    occurs:
    a. At top dead center
    b. Before top dead center
    c. After top dead center
    d. At bottom dead center

18. The firing interval of an engine is
    the number of degrees of crank-
    shaft rotation that:
    a. Take place in a 4-stroke engine
    b. Are required to complete one
       full stroke
    c. Occur between ignition sparks
    d. All of the above

19. Which of the following is *not* a nec-
    essary characteristic of a "fast
    burn" combustion chamber?
    a. Centrally located spark plug
    b. High compression
    c. Compact design
    d. 4 valves

3

Engine Air-Fuel Requirements

Automobile engines run on a mixture of gasoline and air. Gasoline has several advantages as a fuel:

1. Vaporization, or evaporation, occurs easily.
2. It burns quickly, but under control, when mixed with air and ignited.
3. It has a high heat value and produces a large amount of heat energy.
4. It is easy to store, handle, and transport.

Gasoline also has certain disadvantages. The chief disadvantage is that combustion produces air pollutants that are given off into the atmosphere through the engine's exhaust.

As a fuel, however, there is no better substitute for gasoline currently available. To understand how the fuel system works in an engine, we must understand the engine's air-fuel requirements. This chapter discusses those requirements and describes how the fuel gets from the fuel tank to the combustion chamber.

AIRFLOW REQUIREMENTS

All gasoline automobile engines share certain air-fuel requirements. For example, a 4-stroke engine can take in only so much air and fuel at any one time. How much it consumes depends upon how much air the engine can take in. This in turn depends upon four major factors:

1. Engine displacement
2. Maximum engine revolutions per minute (rpm)
3. Carburetor airflow capacity
4. Volumetric efficiency.

The first two factors can be used to figure the engine's airflow requirement which the carburetor must provide. This is measured in cubic feet per minute (cfm). To do this, we assume that the engine has 100-percent **volumetric efficiency** (explained later), or what is often called "perfect breathing".

Determining Airflow Requirements

To determine airflow in cubic feet per minute (cfm):

1. Divide the engine displacement by 2.
2. Divide the maximum revolutions per minute by 1,728.
3. Multiply the results of the two previous steps.
4. Multiply again by volumetric efficiency. Assuming it to be 100 percent, use 1 as the multiplier.

The mathematical formula for this procedure reads:

$$\frac{\text{cid}}{2} \times \frac{\text{rpm}}{1{,}728} \times \text{volumetric efficiency} = \text{cfm}$$

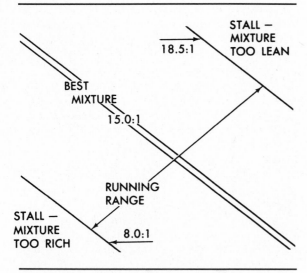

Figure 3-1. Air-fuel ratio limits for a 4-stroke gasoline engine. (Chevrolet)

For example, to determine the maximum airflow capability of a 300-cid engine with a maximum engine speed of 3,600 rpm:

$$\frac{300 \text{ cid}}{2} \times \frac{3,600 \text{ rpm}}{1,728} \times 1 = 312 \text{ cfm}$$

Volumetric Efficiency

Volumetric efficiency is a term used to describe the airflow volume *actually* entering an engine, compared to the engine displacement, which is the maximum volume that it *could* take in. Volumetric efficiency is expressed as a percentage, and it changes with engine speed. For example, an engine might have 75 percent volumetric efficiency at 2,000 rpm. The same engine might be rated at 85 percent at 1,000 rpm and 60 percent at 3,000 rpm.

If the airflow volume is taken in slowly, a cylinder might be filled to capacity. A definite amount of time is required for the airflow to pass through all the curves of the intake manifold and valve port. So, manifold and port design directly relate to the engine's breathing, or volumetric efficiency. Cam timing and exhaust tuning also are important.

If the engine is running fast, the intake valve is not open long enough for a full volume to enter the cylinder. At 1,000 rpm, the intake valve might be open for 1/10 of a second. As engine speed increases, this time is greatly reduced to a point where only a small airflow volume can enter the cylinder. Therefore, volumetric efficiency decreases as engine speed increases. At high speed, it may drop to as low as 50 percent.

To find volumetric efficiency, the airflow volume must be measured at a specified temperature and pressure. This is because the airflow volume will increase as pressure increases and as temperature decreases. Standard pressure for measuring volumetric efficiency is **atmospheric pressure** at sea level, which is 14.7 pounds per square inch (psi), 101 kiloPascals, or 760 millimeters of mercury (mm Hg). Standard temperature is 0°C or 32°F.

Measuring volumetric efficiency is a laboratory exercise, but the idea is valuable for understanding airflow requirements and breathing ability of an engine. Through supercharging — explained in Chapter 16 — and other methods, it is possible to build an engine with more than 100-percent volumetric efficiency. In actual practice, the concept of volumetric efficiency must often be changed to have any practical use.

Although we calculated an engine's cubic feet per minute of airflow at a volumetric efficiency of 100 percent, this figure is seldom if ever reached by a stock engine. For this reason, the 300-cid engine in our example will *not* flow 312 cfm, but quite a bit less. With a stock engine, you can expect a volumetric efficiency of about 75 percent at maximum speed, or 80 percent at the highest torque, or turning force. A high-performance engine will be about 5 percent more efficient, and a racing engine will add another 10 percent to that. Returning to our example, a stock 300-cid engine will actually flow about three-quarters of 312 cfm, or 234 cfm.

AIR-FUEL RATIOS

Because liquid fuel will not burn, it must first be changed into a vapor and mixed with air before it is ignited in the cylinders. For most engines, this is done by the carburetor. For fuel-injected engines, the fuel vaporization and mixing with air is done in the intake manifold or in the combustion chamber before ignition.

In both cases, there is a direct relationship between an engine's airflow and its fuel requirements. This relationship is called the **air-fuel ratio**.

The air-fuel ratio is the proportion by weight of air and gasoline mixed by the carburetor or injection system as required for combustion by the engine. This ratio is important, since there are limits to how rich (with more fuel) or how lean (with less fuel) it can be, and still remain fully combustible for efficient firing. The mixtures with which an engine can operate efficiently range from 8 to 18.5 to 1, figure 3-1. These ratios are usually stated this way: 8 parts

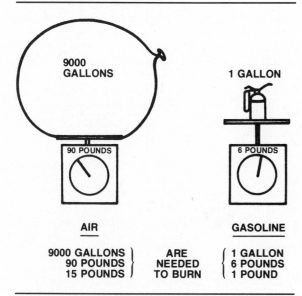

Figure 3-2. The most efficient air-fuel ratio is 15 parts (pounds) of air to 1 part (pound) of gasoline (15:1). (Chevrolet)

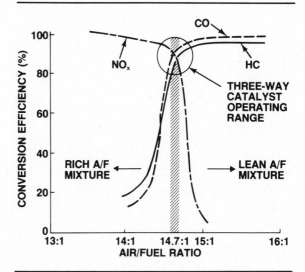

Figure 3-3. All three pollutants are controlled best with an air-fuel ratio of 14.7:1. A leaner mixture increases NO_x; a richer mixture increase HC and CO. (AC-Delco)

of air by weight combined with 1 part of gasoline by weight (8:1) is the richest mixture which an engine can tolerate and still fire regularly; 18.5 parts of air mixed with 1 part of gasoline (18.5:1) is the leanest. Richer or leaner air-fuel ratios will cause the engine to misfire, or not run at all.

To get the best engine efficiency and economy, about 9,000 gallons or liters of air are needed to burn 1 gallon or liter of gasoline, respectively. When expressing this in terms of volume, we find the air-fuel ratio to be 9,000:1. Not only are ratios of this size hard to understand, they are also difficult for engineers to use in their designs and experiments. Therefore, *weight* rather than *volume* is used to calculate air-fuel ratios, since it is easier to work with pounds and kilograms than with gallons and liters.

For example, to convert a volume ratio of 9,000:1 into a more useful weight ratio:
1. 100 gallons of air = 1 pound
2. 9,000 gallons of air = 9,000 ÷ 100 = 90 pounds
3. 1 gallon of gasoline = 6 pounds (approximate).

This means that it requires approximately 15 pounds or kilograms of air to burn one pound or kilogram of gasoline, so our air-fuel ratio is 15:1, figure 3-2.

This relationship between the amounts of air and fuel flow in an engine is sometimes called the fuel-air ratio. Because an engine uses far more air than fuel, the fuel-air ratio is always a number less than one, such as 0.0625. The fuel-air ratio is just a different way of expressing the more familiar air-fuel ratio.

Stoichiometric Air-Fuel Ratio

The ideal mixture or ratio at which all the fuel will blend with all of the oxygen in the air and be *completely burned* is called the **stoichiometric ratio** — a chemically perfect combination. In theory, an air-fuel mixture of about 14.7:1 will produce this ratio, but the exact ratio at which perfect mixture and combustion occurs depends upon the molecular structure of gasoline, which varies somewhat. The stoichiometric ratio is somewhat of a compromise between maximum power and maximum economy, but both are good at 14.7 to 1.

Volumetric Efficiency: The comparison of the *actual* volume of air-fuel mixture drawn into an engine to the *theoretical maximum* volume that could be drawn in. Written as a percentage.

Atmospheric Pressure: The pressure caused by the weight of the earth's atmosphere. At sea level, this pressure is 14.7 psi (101 kPa) at 32°F (0°C).

Air-Fuel Ratio: The ratio of air to gasoline in the air-fuel mixture which enters an engine.

Stoichiometric Ratio: An ideal air-fuel mixture for combustion in which all oxygen and all fuel will be completely burned.

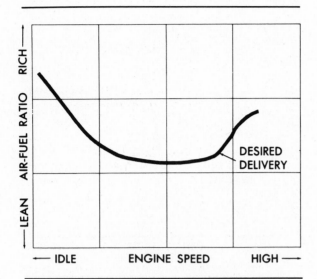

Figure 3-4. The desired air-fuel ratio changes as engine operating conditions change. (Chevrolet)

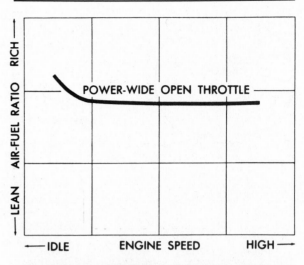

Figure 3-5. The air-fuel ratio needed for maximum power is relatively constant, except at low speed, where it must be slightly richer. (Chevrolet)

Emission control is also optimum at this ratio, if a 3-way oxidation-reduction catalytic converter is used. As the mixture richens, HC and CO conversion efficiency falls off. With leaner mixtures, NO_x conversion efficiency also falls off. As figure 3-3 shows, the conversion efficiency range is very narrow — between 14.65 and 14.75 to 1. A fuel system without feedback control cannot maintain this narrow range.

Engine Air-Fuel Requirements

An automobile engine will work with the air-fuel mixture ranging from 8 to 18.5 to 1. But the ideal ratio would be one that provides both the most power *and* the most economy, while producing the least emissions. But such a ratio does not exist because the fuel requirements of an engine vary widely depending upon temperature, load, and speed conditions.

Research has proved that the best fuel economy is obtained with a 15 to 16:1 ratio, while maximum power output is achieved with a 12.5 to 13.5:1 ratio. A rich mixture is required for idle, heavy load, and high speed conditions; a leaner mixture is required for normal cruising and light load conditions. As you can see, no single air-fuel ratio provides the best fuel economy *and* the maximum power output at the same time, figure 3-4.

Just as outside conditions such as speed, load, temperature, and atmospheric pressure change the engine's fuel requirements, other forces at work inside the engine cause additional variations. Here are three examples:
1. Exhaust gases remain inside the cylinders

and dilute the incoming air-fuel mixture, especially during idle.
2. The mixture is imperfect because complete vaporization of the fuel may not occur.
3. Mixture distribution from a carburetor through the intake manifold to each cylinder is not exactly equal; some cylinders get a richer or leaner mixture than others.

If an engine is to run well under such a wide variety of outside and inside conditions, the carburetor or the injection system must be able to vary the air-fuel ratio quickly, and to give the best mixture possible for the engine's requirements at a given moment.

The best air-fuel ratio for one engine may not be the best ratio for another, even when the two engines are of the same size and design. Engines are mass-produced but will have slight variations in manifolding, combustion chambers, valve timing, and ignition timing. To accurately determine the best mixture, the engine should be run on a **dynamometer** to measure speed, load, and power requirements for all types of driving conditions.

Power Versus Economy

If the goal is to get the most power from an engine, all of the oxygen in the mixture must be burned, because the power output of any engine is limited by the amount of air it can pull in. To be sure that the oxygen combines properly with the available fuel, extra fuel must be provided. This increases the air-fuel ratio (makes it richer in fuel), resulting in some fuel which remains unburned.

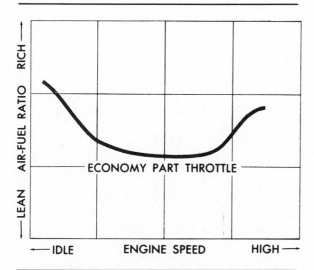

Figure 3-6. The air-fuel ratio for best economy is lean in the middle of the speed range but requires enrichment at high and low speeds. (Chevrolet)

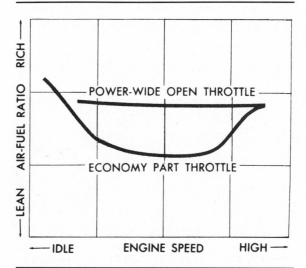

Figure 3-7. The engine must receive lean air-fuel ratios for best economy or rich ratios for maximum power at any given speed. (Chevrolet)

This is also true at idle because of exhaust gases remaining in the cylinders. These tend to dilute the incoming mixture since some of the fuel combines with the exhaust. To make certain that the mixture is properly combustible during idle, more fuel must be delivered to make up for the fuel that combines with the exhaust gases. This makes it more difficult to equally distribute the mixture to the cylinders and creates waste material in the form of carbon monoxide, which is emitted into the atmosphere as a pollutant. To get the best fuel economy and the lowest emissions, the gasoline must be burned as completely as possible in the combustion chamber. This means that the greatest amount of energy (economy) will be produced with the least amount of leftover waste material (emissions). If enough oxygen is to be available to combine with the gasoline, then more air must be provided. This results in a leaner air-fuel mixture (less gasoline) than the ideal ratio.

The air-fuel ratio required to provide maximum power will change very little, except at low speeds, figure 3-5. Reducing speed reduces the airflow into the engine. The result is a poorer mixing of the air and fuel, and less efficiency in its distribution to the cylinders. Thus, at low speeds, a slight enrichment of the mixture is required to make up for this.

The same is true for maximum fuel economy — the leaner air-fuel ratio used will remain virtually the same throughout most of the operating range, figure 3-6. But enrichment will be required during idle and low speeds, as well as during higher speeds and under load — two conditions which require more power.

Enrichment also can occur when it is not required or wanted, as in the case of high altitude driving. As altitude increases, atmospheric pressure drops and the air becomes thinner than it is at sea level. The same amount of air actually weighs less and contains less oxygen at higher altitudes. This means that an engine will take in fewer pounds or kilograms of air and

Dynamometer: A device used to measure mechanical power, such as the power of an engine.

■ Poor Driving Means Poor Mileage

Carmakers have succeeded in producing cars that use less fuel and are more efficient. Still, the driving patterns of individual drivers can make a big difference in how much fuel the car will use.

Here are some interesting points to remember:

● Stopping and restarting a car engine consumes less fuel than running it at idle for one minute.
● Driving at 55 mph (89 kph) instead of 65 mph (105 kph) reduces fuel consumption by about 12 percent. Of course, driving at 40 mph (64 kph) is even *more* economical.
● A minor tune-up will increase mileage by about 10 percent.
● The failure of one spark plug in an 8-cylinder engine can cut 12 percent from the car's mpg. It also can increase HC emissions by as much as 300 percent.
● Turning off the air conditioner can improve gas mileage by up to 10 percent.

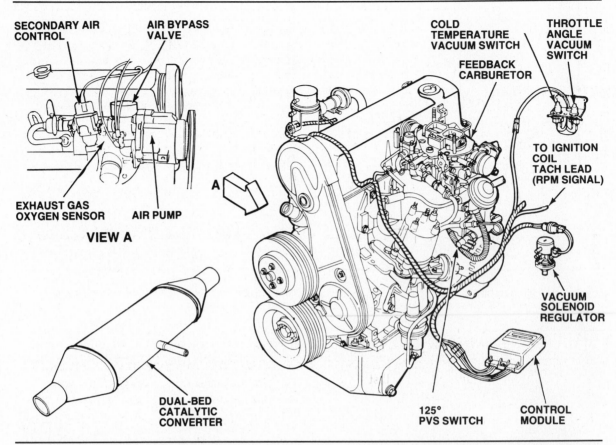

SECONDARY AIR CONTROL

AIR BYPASS VALVE

COLD TEMPERATURE VACUUM SWITCH

THROTTLE ANGLE VACUUM SWITCH

FEEDBACK CARBURETOR

TO IGNITION COIL TACH LEAD (RPM SIGNAL)

A

EXHAUST GAS OXYGEN SENSOR

AIR PUMP

VIEW A

VACUUM SOLENOID REGULATOR

DUAL-BED CATALYTIC CONVERTER

125° PVS SWITCH

CONTROL MODULE

Figure 3-8. Ford's Feedback Carburetor Electronic Control System was one of the first electronically controlled fuel management systems. (Ford)

less oxygen. The result is a richer air-fuel ratio, which must be corrected for efficient high-altitude engine operation. Altitude-compensating carburetors and fuel injection air sensors solve this problem.

For these reasons, the carburetor or injection system must deliver fuel so that the best mileage is provided during normal cruising, with maximum power available whenever the engine is under load, acceleration, or high-speed, figure 3-7.

INTRODUCTION TO ELECTRONIC ENGINE CONTROLS

Electronic engine controls appeared on the automotive scene with some 1977 models. The early control systems regulated only a single function, either ignition timing or fuel metering. However, they were rapidly expanded to incorporate control over both systems, as well as numerous other engine functions.

The basic parts of the first electronically controlled fuel management systems were a feedback carburetor, an electronic control module or microprocessor, an exhaust gas oxygen (EGO) sensor mounted in the exhaust manifold, and a catalytic converter, figure 3-8.

Two types of fuel control actuators are used with carburetors:

• A solenoid or stepper motor mounted on or in the carburetor to directly control the fuel-metering rods or air bleeds, or both

• A remote mounted, solenoid-actuated vacuum valve to regulate carburetor vacuum diaphragms that control the fuel-metering rods and air bleeds.

The fuel management microprocessor constantly monitors the oxygen content of the exhaust gas through signals received from the EGO sensor. The microprocessor sends a pulsed voltage signal to the control device, varying the ratio of on-time to off-time according to the signals received from the EGO sensor. As the percentage of on-time is increased or decreased, the mixture is leaned or richened.

With fuel injection systems, the microprocessor exercises ratio control by switching one or more fuel injectors on and off. The

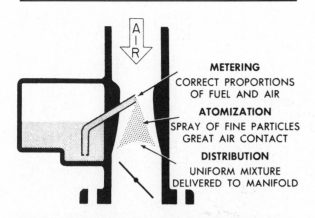

Figure 3-9. The carburetor does the basic job of fuel metering, atomization, and distribution. (Chevrolet)

switching rate is determined by engine speed. The microprocessor varies the length of time the injectors remain open to establish the air-fuel ratio. As the microprocessor receives data from its sensor inputs, it lengthens or shortens the pulse width (on-time) according to engine operating and load conditions. We will study more about fuel injection systems and how they work in Part 4.

FUEL DISTRIBUTION

Before gasoline can do its job as a fuel, it must be metered, atomized, and distributed to each cylinder in the form of a burnable mixture. To do this, a metering device — a carburetor, figure 3-9 — mixes the gasoline with air in the correct ratio and distributes the mixture as required by engine load, speed, throttle plate position, and operating temperature. On most late-model engines, a fuel injection system does the jobs of air-fuel mixing done by a carburetor on older engines. We will begin the study of fuel distribution, however, by concentrating on a basic carbureted engine.

Proper fuel distribution depends on six factors:

1. Correct fuel **volatility**
2. Proper fuel **atomization**
3. Complete fuel **vaporization**
4. Intake manifold passage design
5. Intake throttle plate angle
6. Carburetor or throttle body location on the intake manifold.

Fuel Atomization and Vaporization

Several things are involved in changing gasoline from a liquid into a combustible vapor. In a carbureted engine, the liquid fuel first enters the carburetor where it is sprayed into the incoming air and atomized (reduced to a mist), figure 3-10. The resulting air-fuel mixture then moves into the intake manifold where the mist is changed into a vapor.

Vaporization occurs only when the fuel is hot enough to boil. The boiling point is related to pressure; the higher the pressure, the higher the boiling point; the lower the pressure, the lower the boiling point. Because intake manifold pressure is quite a bit less than atmospheric pressure, the boiling point of gasoline drops when it enters the manifold.

Volatility: The ease with which a liquid changes from a liquid to a gas or vapor.

Atomization: Breaking down into small particles or a fine mist.

Vaporization: Changing a liquid, such as gasoline, into a gas (vapor).

■ Birth Of The Internal Combustion Engine

Like many of man's inventions, the internal combustion engine did not spring full-blown from the mind of one man. Its development was gradual and built on the work of many different men between 1690 and 1876. Historians generally credit the atmospheric engine concept developed by Denis Papin (inventor of the pressure cooker) as the first step toward today's modern gasoline burner. An external combustion engine, the atmospheric engine principles put forth by Papin led directly to the basic steam engine of James Watt and Richard Trevithick.

Meanwhile, the science of heat exchange thermodynamics was being developed on theories postulated by Sadi Carnot, a French physicist, in 1824. During the 1859's, volatile fuels were being refined from petroleum and by 1860, a French engineer with the unusual name of J. J. E. Lenoir modified a steam engine to use illuminating gas as a fuel. The theory of the 4-stroke engine was published by Alphonse Beau De Rochas in 1862, yet De Rochas never built an engine!

Things really started moving with the production of an engine by the firm of Otto and Langen in 1867. Their design used a rack-and-gear device to transmit the power of a free-moving piston to a shaft and flywheel. A free-wheeling clutch in the gear allowed it to rotate freely in one direction and transmit power in the other. The same men built the first modern internal combustion engine, the Otto Silent Engine, in 1876. A 4-stroke design, it was manufactured in the U.S. after 1878 and is said to have inspired the early research of Henry Ford and other automobile pioneers.

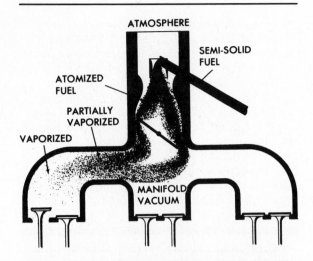

Figure 3-10. Changing liquid fuel to a combustible material is a two-stage process. First, it is atomized to a mist and mixed with air. The air-fuel mixture then must be vaporized. (Chevrolet)

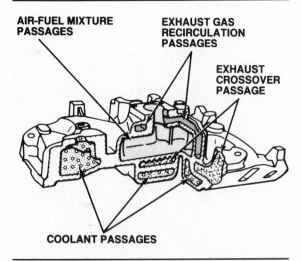

Figure 3-12. A fuel-vaporizing hot spot is created by engine coolant flowing through the intake manifold shown here, or through a spacer between the manifold and carburetor. (Chrysler)

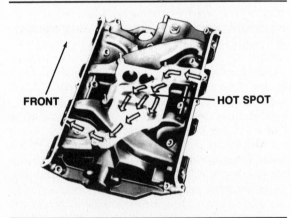

Figure 3-11. Exhaust gases are routed from ports in the cylinder heads through separate passages in the intake manifold to form the manifold hot spot. (Ford)

Heat from the intake manifold floor combines with heat absorbed from air particles surrounding the fuel particles to begin vaporization. It is helped by raising the temperature of the intake manifold, since the higher the temperature, the more complete the vaporization will be. This heated area in the intake manifold is called a "hot spot".

Poor vaporization can be caused by several things:
- A mixture velocity that is too low
- A cold manifold or low manifold vacuum
- Cold incoming air
- Insufficient fuel volatility
- Poor manifold design
- Low carburetor flow capacity.

When poor vaporization occurs, too much liquid reaches the cylinders. Some of this additional fuel is given off as unburned hydrocarbons, and some will wash oil from the cylinder walls, causing engine wear. The rest will be carried past the piston rings as blowby gases.

Intake Manifold Design

The design of an intake manifold has a direct bearing on mixture distribution and volumetric efficiency over the speed range of an engine. The location, size, and surface area of the hot spot on the manifold floor affects vaporization. The hot spot is normally just beneath the carburetor, figure 3-11. Although the hot spot is usually heated by exhaust, engine coolant is sometimes circulated through passages between the carburetor base and the manifold, figure 3-12.

Older engines usually had cast iron manifolds, but the intake manifolds on modern engines are often made of aluminum because of its superior heat conductivity and light weight. Aluminum's heat conductivity helps transfer heat to the air-fuel mixture faster and more uniformly.

Both velocity and heating are also affected by the size of the manifold passages through which the mixture must travel. If the passages are large, the mixture will travel slowly, allowing fuel particles to cling to the manifold wall and thus avoid vaporization. Small passages create a higher velocity, but restrict the travel and distribution of the mixture. The angles at which internal manifold passages turn also can

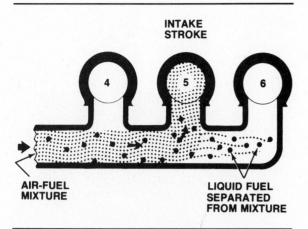

Figure 3-13. An intake manifold with large passages and sharp angles will cause liquid fuel to separate out of the air-fuel mixture. (Chevrolet)

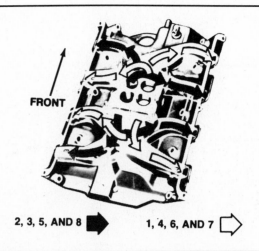

Figure 3-14. Typical V-8 intake manifold fuel passages. (Ford)

be critical, figure 3-13. When they are too sharp, fuel tends to separate out of the mixture.

The air-fuel mixture should be distributed as evenly as possible among the cylinders. Figure 3-14 shows a typical V-8 intake manifold designed for good distribution to all cylinders. If one or more cylinders receives an overly lean mixture, an increase in the overall mixture will be necessary for that cylinder to fire properly. This will cause the other cylinders to receive a mixture that is too rich. Overly lean combustion produces oxides of nitrogen (NO_x), while overly rich combustion produces unburned hydrocarbons (HC) and carbon monoxide (CO). Neither condition is desirable, since they raise emissions and lower fuel economy.

Ford 200- and 250-cid (3.3- and 4.1-liter) car engines had the intake manifold cast into the head to reduce cost and simplify overall engine manufacturing. The last of the inline Chevrolet 6-cylinder engines in the late 1970's had intake manifolds that were integral with the heads. This design was an attempt to improve mixture temperature and distribution, which are hard to control uniformly for an inline 6-cylinder engine.

The trend toward using engines with four valves per cylinder often involves a "split-level" manifold with variable induction and differences in intake valve timing. The manifold usually has an induction **plenum** that stabilizes the incoming air charge and actually allows it to rise slightly in pressure. This provides a uniform air charge that can be distributed equally to each cylinder, unaffected by momentary variations in throttle position and intake turbulence. The plenum feeds air into collector boxes, which, in turn, feed separate intake runners for each intake valve.

The engine computer controls separate throttle valves for intake runners to one of the two intake valves for each cylinder. At low- and mid-range speeds, one set of long runners feeds the intake air to one valve for each cylinder. At high speed, the computer opens the high-speed set of intake throttles to admit air through shorter runners. The principle is that long intake runners increase airflow speed at low engine speed, and short runners accelerate the intake charge at high engine speed.

Many late-model engines with fuel injection have 2-piece, dual-tuned intake manifolds. The design of these manifolds varies considerably according to the type of fuel injection used, figures 3-15 and 3-16, but all share similar features. Individual "tuned" runners connect the plenum chamber to each intake port. These runners are specifically designed for the intake ports to which they connect and provide increased airflow at high speed for maximum power. In figure 3-16, a large-diameter passage located inside the plenum chamber behind the airflow entry point provides secondary tuning. This increases airflow at low speeds for better torque.

Carburetor Size and Placement

Carburetor airflow must be matched to the airflow requirements of the engine. A carburetor that will provide more air and fuel than

Plenum: A chamber that stabilizes the air-fuel mixture and allows it to rise to a pressure slightly above atmospheric pressure.

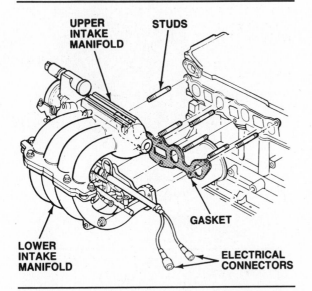

Figure 3-15. One style of intake manifold with tuned runners from the plenum to the intake ports, this Ford EFI manifold is a 2-piece assembly. (Ford)

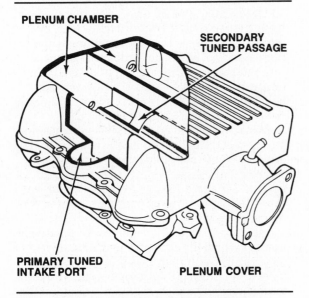

Figure 3-16. This Chrysler intake manifold is manufactured in two pieces that are permanently bonded together. It also has tuned runners. (Chrysler)

the engine requires will produce a rich mixture. This can reduce *both* fuel economy *and* power. A carburetor that provides less fuel and air than the engine needs will cause the engine to work harder to provide the power for any speed and load condition. Again, this means that the engine will not be providing the best combination of economy and power.

The location of the carburetor on the intake manifold is important. Incorrect placement in relation to the manifold passages can interfere

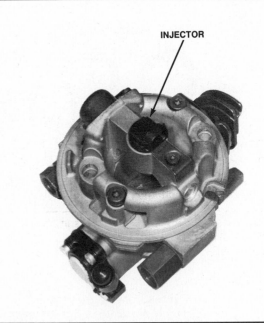

Figure 3-17. A single-bore TBI unit, the Rochester Model 700 uses one Multec fuel injector.

with proper fuel distribution. If the carburetor is located closer to one cylinder than to the others, improper vaporization and distribution to the cylinders may result.

Fuel Injector Location

A carburetor mixes air with fuel and delivers the mixture to the intake manifold. In a fuel injection system, the injectors deliver the fuel to the manifold where it is mixed with the air. Although the role of the injector and carburetor differ somewhat, proper positioning of fuel injectors is as important as the location of the carburetor. We will learn more about fuel injection systems and how they operate in Part 4.

Throttle body injectors

A throttle body injection (TBI) system uses a throttle body, figure 3-17, or fuel charging assembly, figure 3-18, similar in design to a carburetor. In fact, the design of Ford's 2-barrel, 2-injector fuel charging assembly, figure 3-18, bears a strong resemblance to the Motorcraft 2150 carburetor. The TBI unit is positioned on the intake manifold and contains one or two solenoid-operated injectors, figure 3-19, which deliver fuel intermittently to the intake air charge.

Since TBI units deliver fuel for all the cylinders from one or two injectors, their operating cycles and injection duration are calculated to coincide with airflow requirements and engine

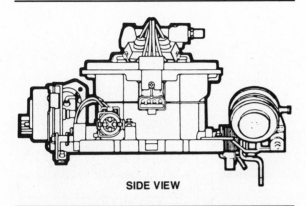

Figure 3-18. Ford's fuel charging assembly looks similar to the Motorcraft 2150 carburetor. (Ford)

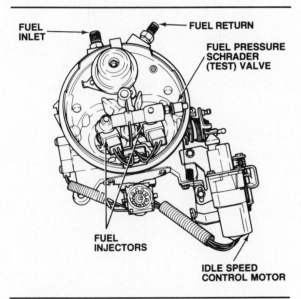

Figure 3-19. A two-bore TBI unit contains one fuel injector in each bore. (Ford)

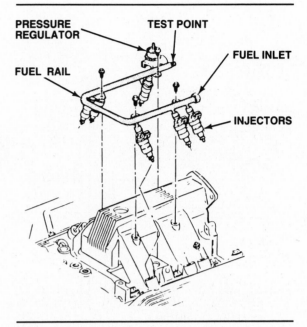

Figure 3-20. Injectors installed at individual ports are connected by a fuel rail. (AC-Delco)

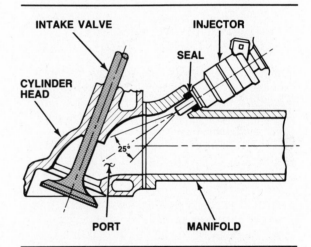

Figure 3-21. A port fuel injector is installed in the intake manifold at each cylinder port. (AC-Delco)

operation. With a 2-injector TBI unit, the injectors can be calibrated differently according to the cylinders they feed. In this way, injector flow can be coordinated with manifold design to improve distribution.

Port injectors

A port, or multipoint, fuel injection system uses a series of individual solenoid-operated injectors connected to a fuel rail, figure 3-20, which supplies fuel under pressure. One injector is positioned in the intake manifold at each cylinder, figure 3-21, and sealed with an O-ring to prevent air leakage that would cause a lean cylinder. Depending upon system design, the injectors may be energized as one or more groups, or individually. When they are energized as groups, half the fuel required by each cylinder is injected every crankshaft revolution and the fuel waits at the intake port for the incoming charge of air. When energized individually, the fuel is delivered with the incoming air charge in time for combustion. Chapter 15 explains different injection sequences in more detail.

FUEL COMPOSITION

Gasoline is a clear, colorless liquid — a complex blend of various basic hydrocarbons (hydrogen and carbon). As a fuel, it has good vaporization

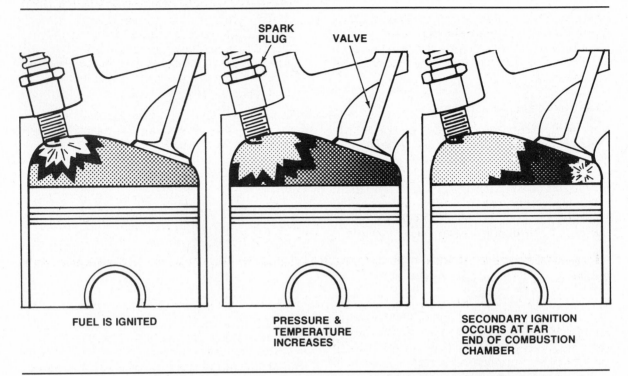

SPARK PLUG   VALVE

FUEL IS IGNITED   PRESSURE & TEMPERATURE INCREASES   SECONDARY IGNITION OCCURS AT FAR END OF COMBUSTION CHAMBER

Figure 3-22. Detonation is a secondary ignition of the air-fuel mixture caused by high cylinder temperatures. It is commonly called "pinging" or "knocking".

qualities and is capable of producing tremendous power when combined with oxygen and ignited. Yet, it is impossible to accurately predict how a certain blend of gasoline will perform in a particular engine, since no two engines are identical. Remember, mass-produced engines are subject to individual variations in production, which can affect fuel efficiency.

In laboratory tests, oil refiners calculate and measure the characteristics most important to produce gasolines suitable for specific jobs. Fuels are blended to meet particular temperature and altitude conditions. The gasoline you use during the summer is not the same blend available in the winter, nor is the gasoline sold in Denver the same as that sold in Death Valley. In addition to temperature and altitude, refiners must consider several other things during the blending process: volatility, chemical impurities, octane rating, and additives.

Volatility

Volatility is a measure of gasoline's ability to change from a liquid to a vapor and is related to temperature and altitude. The more volatile it is, the more efficiently the gasoline will vaporize. As we've seen, efficient vaporization is needed for even fuel distribution to all of the engine's cylinders and for complete combustion.

Volatility is controlled by blending different hydrocarbons that have different boiling points. In this way, it is possible to produce a fuel with a high boiling point for use in warm weather, and one with a lower boiling point for cold-weather driving. Such blending involves some guesswork about weather conditions, so severe and unexpected temperature changes can cause a number of temperature-related problems ranging from hard starting to vapor lock.

Chemical Impurities

Gasoline is refined from crude oil and contains a number of impurities which can harm engines and fuel systems. For example, if the sulfur content is too high, some of it may reach the engine crankcase, where it will combine with water to form sulfuric acid. This substance will corrode engine parts, although proper crankcase ventilation helps to avoid damage. Another impurity, gum, tends to form sticky deposits that will eventually clog carburetor and injector passages and cause piston rings and valves to stick.

To a large extent, the amount of chemical impurities present in gasoline depends upon the type of crude oil used, the refining process, and the oil refiner's desire to keep his production costs low. The more expensive process of **catalytic cracking** usually produces a gasoline with a lower sulfur content than the **thermal cracking** method, which is less expensive and therefore more common.

Octane Rating

When engine compression pressure reaches a certain level, a great deal of heat is generated as the air-fuel mixture is compressed. Unless gasoline is formulated to hold up under such high pressures and temperatures, there will often be a secondary explosion called **detonation**, figure 3-22. This is popularly called "knocking" or "pinging". Detonation causes a loss of power

Catalytic Cracking: An oil refining process which uses a catalyst to break down (crack) the larger components of the crude oil. The gasoline produced usually has a lower sulfur content than gasoline produced by thermal cracking.

Thermal Cracking: A common oil refining process which uses heat to break down (crack) the larger components of the crude oil. The gasoline which is produced usually has a higher sulfur content than gasoline produced by catalytic cracking.

Detonation: Also called knocking or spark knock. An unwanted explosion of an air-fuel mixture caused by high heat and compression.

■ The Future Of Alternative Fuels

As a result of the two energy crises during the 1970's, intense effort was devoted to developing a replacement fuel for gasoline. Ford Motor Company led the way in translating this effort into vehicles that would use such fuels. One result was the propane-fueled 2.3-liter inline 4-cylinder engine that briefly graced 1982 Ford Granada/Mercury Cougar and 1983 Ford LTD/Mercury Marquis cars.

Propane, or liquified natural gas (LNG), is a liquid form of the same clean-burning natural gas used in the home. Liquified by chilling to −258°F (−161°C), it is stored in thermos-type containers. Extensive testing of propane in automobiles showed that exhaust pollutants were practically eliminated. Since many drivers of motor homes were accustomed to using propane gas, it was felt that cars could operate on the same fuel.

The Ford system included a unique air cleaner, a propane carburetor, a fuel lock, and a converter/regulator assembly — all in the engine compartment. Twin propane tanks mounted beneath the trunk floor provided the necessary fuel. The only external sign of the vehicle's power source was a small "PROPANE" emblem on each front fender.

Propane is a flammable substance like gasoline, but is a vapor at normal temperatures and barometric pressures. The Ford system had two relief valves to vent excessive pressure resulting from high ambient temperatures. If a leak developed or the system vented through the relief valves, the propane immediately vaporized and expanded to about 270 times its liquid volume. Since it is heavier than air, propane settles in low spots and gradually dissipates. This created the possibility of a dangerous fire hazard.

For this reason, there were many prohibitions for propane-fueled vehicles:
● Do not vent fuel unnecessarily
● Do not drain the fuel tanks
● Do not use a drying oven when refinishing the paint
● Do not weld near the fuel system tanks or components
● Do not service the vehicle near electrical equipment such as motors or switches that may discharge sparks
● Do not store or service the vehicle over a confined area, such as a lube pit, where vapors might accumulate.

Mechanics were not thrilled with such prohibitions, since the numerous safety precautions involved with propane vehicles interfered with their normal shop operation. Furthermore, these vehicles did not prove popular with the driving public either, and the majority of these Ford cars ended up as fleet sales to companies that were interested in fuel conservation.

The lack of consumer response to Ford's effort was attributed to the difficulty in refueling the propane tanks and a general unavailability of the fuel in many areas. The cars required a greater-than-normal amount of care on the part of drivers and died a quick death in the marketplace.

Gasoline prices started to stabilize at about the same time as the propane-fueled Fords were made available to the public. Most drivers were not willing to cope with the particular problems presented by this alternative fuel source. Thus, the propane Fords passed into the pages of automotive history much as the Chrysler turbine-powered vehicles had a decade earlier.

and overheating of valves, pistons, and spark plugs. The overheating in turn causes more detonation and may eventually damage the engine.

To prevent detonation, gasoline must have a certain **antiknock value**. This characteristic derives from the type of crude oil and the refining processes used to extract the gasoline. It is measured by an **octane rating**. Gasoline with a high octane rating resists detonation during combustion, while one with a low octane value does not.

Additives

Certain chemicals not normally present in gasoline are added during refining to improve its performance:

• Anti-icers are specially treated alcohols which act as antifreeze in the gasoline to prevent moisture in the air from causing carburetor or throttle body icing at low temperatures.
• **Antioxidant inhibitors** are used to prevent the formation of gum.
• Phosphorus compounds prevent spark plug misfiring and preignition.
• Metal deactivators prevent gasoline from reacting chemically with metal storage containers in which it is stored and transported.
• Cleaners and detergents are added to prevent the formation or accumulation of compounds that could clog the many small passages or orifices in a carburetor or fuel injector. Major gasoline refiners use different proprietary chemicals for this task. The use of aftermarket cleaners and detergents is especially desirable when the quality of fuel used is otherwise below standard.
• **Tetraethyl lead** (TEL) is used to prevent detonation and provide lubrication for valve seats and many other moving parts. TEL is a highly toxic substance that is emitted as a particulate in the engine's exhaust. Studies have shown that lead particulate emissions can be a health problem if a large quantity collects in a small geographic area. Lead also destroys the capability of a catalytic converter to work properly. For these reasons, the Environmental Protection Agency instituted a phaseout of lead content in gasoline. Beginning July 1, 1983, the use of lead was restricted to 1.1 grams per gallon. On July 1, 1985, the standard dropped to 0.5 gram, and then to 0.1 gram on January 1, 1986. The EPA further proposed that no more leaded gasoline be produced after January 1, 1988.
• Octane boosters and lead substitutes sold in the automotive aftermarket may provide some octane increase. These additives require some care in selecting an appropriate one, since many are only alcohol solutions, a common but not so desirable octane booster. All are both expensive and impractical for everyday use. Also because of EPA and health regulations, no aftermarket fuel additive contains actual tetraethyl lead.

Alcohol Additives and Fuel Quality

Gasoline blended with alcohol is widely available, although it is not legally required to be labeled as such in many states. A mixture of 10 percent ethanol (ethyl alcohol) and 90 percent unleaded gasoline is called "gasohol". **Gasohol** is a generic term, however, and there are no set standards for the type and amount of alcohol it contains. Several companies now sell premium fuels that use ethanol as the octane booster.

Alcohol improperly blended with gasoline can cause numerous and serious problems with an automotive fuel system, including:

• Corrosion formation on the inside of fuel tanks, steel fuel lines, fuel pumps, carburetors, and fuel injectors.
• Deterioration of the plastic liner used in some fuel tanks, resulting in eventual plugging of the in-tank filter.
• Deterioration and premature failure of fuel line hoses and synthetic rubber or plastic materials such as O-ring seals, diaphragms, inlet needle tips, accelerator pump cups, and gaskets.
• Hard starting, poor fuel economy, lean surge, vapor lock, and other driveability problems.

Antiknock Value: The characteristic of gasoline that helps prevent detonation or "knocking".

Octane Rating: The measurement of the antiknock value of a gasoline.

Antioxidant Inhibitors: A gasoline additive used to prevent oxidation and the formation of gum.

Tetraethyl Lead: A gasoline additive used to help prevent detonation.

Gasohol: A blend of ethanol and unleaded gasoline, usually at a one to nine ratio.

Ethanol: Ethyl alcohol distilled from grain or sugar cane.

Methanol: Methyl alcohol distilled from wood or made from natural gas.

Fuels with an alcohol content tend to absorb moisture from the air. Once the moisture content of the fuel reaches approximately one percent, it combines with the alcohol and separates from the fuel. This water-alcohol mixture then settles at the bottom of the fuel tank where the fuel pickup carries it into the fuel line to the carburetor or fuel injectors, creating a lean surge condition.

All alcohols are solvents. While **ethanol** is relatively mild, **methanol** (methyl alcohol) is highly corrosive. It attacks fuel system components unless properly mixed with corrosion inhibitors and appropriate suspension agents or cosolvents to prevent separation of the water-alcohol combination from the gasoline.

Gasohol has a cleaning effect on service station storage tanks, as well as the vehicle's fuel tank. As a result of this cleaning action, a combination of rust, a jelly-like sludge, and metallic particles passes into the automotive fuel system. These substances cause reduced fuel flow through the filter and will eventually plug the carburetor or injector passageways.

Fuel economy and driveability are other areas of concern with alcohol-gasoline blends. Alcohols contain fewer BTU's of energy per gallon or liter than gasoline, which can result in reduced fuel mileage. In addition to their lower energy content, alcohols are less volatile than gasoline; they require higher temperatures before they will ignite and burn. Since the stoichiometric ratio for alcohol is 6.5 rather the 14.7 of gasoline, an alcohol-gasoline blend creates a lean mixture. In turn, this creates or worsens lean surge in some driving conditions and increases the probability of vapor lock.

The problem of alcohol improperly blended with gasoline has become so common around the United States that automotive tool manufacturers offer alcohol detection kits so that you can determine the quality of fuel being used.

The detection procedure is performed with water as a reacting agent. However, if cosolvents have been used as a suspension agent in alcohol blending, the test will not show the presence of alcohol unless ethylene glycol (automotive antifreeze) is used instead of water as a reacting agent. It is suggested that a gasoline sample be tested twice using the detection kit: first with water and then with ethylene glycol.

The procedure cannot differentiate between types of alcohol (ethanol or methanol), nor is it considered to be absolutely accurate from a scientific standpoint. These tests are accurate enough, however, to determine whether or not there is enough alcohol in the fuel to cause the user to take precautions. Basic instructions on performing a test for alcohol content in gasoline are given in Chapter 3 of your *Shop Manual*.

SUMMARY

Engine size is measured by displacement, or the total volume of all the cylinders. Displacement is a determining factor in the engine's airflow requirement. Stock engines operate at a level below volumetric efficiency, or the maximum volume of air they could take in. Engines will operate efficiently on an air-fuel ratio between 8 and 18.5 to 1, but the exact ratio at which perfect mixture and complete combustion occur is closer to 14.7, or the stoichiometric ratio. The stoichiometric ratio is also the best compromise between power and economy, as well as the point at which the least emissions are produced. Electronic fuel management systems have been designed to provide the engine with the proper air-fuel ratio according to engine speed and load demands.

Several factors influence proper fuel distribution. Gasoline is a volatile liquid, but it must be atomized and vaporized before combustion can occur. Intake manifold design and carburetor or throttle body placement are critical to fuel distribution.

Gasoline must be properly refined to remove chemical impurities and blended with additives to prevent preignition, detonation, carburetor icing, gum and varnish formation, and misfiring. Tetraethyl lead was once the major octane booster, but its use was severely limited and eventually prohibited by the EPA.

Alcohol and gasoline blends are called gasohol. Gasohol has a number of disadvantages, particularly when it is misblended, and can permanently damage a fuel system. Alcohol detection kits allow you to determine the alcohol content, if any, of gasoline.

Review Questions

Choose the single most correct answer.
Compare your answers with the correct answers on page 414.

1. A disadvantage of gasoline is that it:
   a. Vaporizes easily
   b. Burns quickly
   c. Produces pollutants upon combustion
   d. Has a high heat value

2. Which is *not* a factor in determining airflow requirement?
   a. Engine displacement
   b. Maximum rpm
   c. Carburetor size
   d. Volumetric efficiency

3. Volumetric efficiency:
   a. Is the ratio of air entering the engine to engine displacement
   b. Decreases as engine speed increases
   c. Is expressed as a percentage
   d. All of the above

4. At maximum speed, the volumetric efficiency of a stock engine is approximately:
   a. 75%
   b. 50%
   c. 10%
   d. None of the above

5. The richest air-fuel ratio that an internal combustion engine can tolerate is about:
   a. 4:1
   b. 2.5:1
   c. 8:1
   d. 18.5:1

6. To burn one pound of gasoline with maximum efficiency requires about:
   a. 8 pounds of air
   b. 15 pounds of air
   c. 18.5 pounds of air
   d. 17.9 pounds of air

7. A rich air-fuel mixture is needed for:
   a. Idle
   b. Heavy load
   c. Acceleration
   d. All of the above

8. An internal engine condition affecting fuel requirements is:
   a. Engine load
   b. Mixture distribution
   c. Atmospheric pressure
   d. Engine speed

9. Obtaining maximum power results in:
   a. No change in air-fuel ratios
   b. Leather mixtures
   c. Unburned oxygen
   d. Excess unburned fuel

10. Maximum fuel economy requires:
    a. Less air
    b. Leaner air-fuel mixtures
    c. Richer air-fuel mixtures
    d. Higher temperatures

11. For maximum power, the air-fuel ratio:
    a. Becomes leaner at low speeds
    b. Becomes richer at low speeds
    c. Becomes richer at high speeds
    d. None of the above

12. For maximum fuel economy, the air-fuel ratio:
    a. Is lean for middle speeds
    b. Becomes richer at high speeds
    c. Becomes richer at low speeds
    d. All of the above

13. Different fuel blending techniques are used by refiners to:
    a. Obtain the desired volatility and increase the octane rating
    b. Replace impurities with additives
    c. Both a and b
    d. Neither a nor b

14. Mechanic A says that an engine fuel management system is electronically controlled.
    Mechanic B says that an engine fuel management system uses an EGO sensor.
    Who is right?
    a. A only
    b. B only
    c. Both A and B
    d. Neither A nor B

15. Mechanic A says that a TBI injection system uses one injector at each cylinder port.
    Mechanic B says that the TBI unit is installed on the intake manifold where a carburetor would be.
    Who is right?
    a. A only
    b. B only
    c. Both A and B
    d. Neither A nor B

16. Which of the following is *not* a quality of tetraethyl lead?
    a. Prevents detonation
    b. Lubricates valve seats
    c. Can destroy a catalytic converter
    d. Is a non-toxic substance

17. Gasohol is generally regarded as a blend of:
    a. 10 percent ethanol and 90 percent gasoline
    b. 90 percent ethanol and 10 percent gasoline
    c. 50 percent ethanol and 50 percent gasoline
    d. None of the above

PART TWO

The Fuel System

4

Fuel Tanks, Lines, and Evaporative Emission Controls

Part I introduced you to the basic parts, principles, and problems involved in the automobile fuel system and emission controls. Part II covers the major components in the fuel system in more detail. This will help you to develop a working knowledge of the relationships between the fuel system and emission controls. In this chapter, you will learn how the fuel system works. You also will take a detailed look at how each automaker has tackled the problems of evaporative emission controls.

TANKS AND FILLERS

The automobile fuel tank, figure 4-1, is made of two corrosion-resistant steel halves, which are ribbed for additional strength and welded together. Exposed sections of the tank may be made of heavier steel for protection from road damage and corrosion.

Some cars, sports utility vehicles, and light trucks may have an auxiliary fuel tank. A few of these auxiliary tanks have been made of polyethylene plastic. Greater use of composites in fuel tank construction seems likely in the future.

Tank design and capacity are a compromise between available space, filler location, fuel expansion room, and fuel movement. Some late-model tanks deliberately limit tank capacity by extending the filler tube neck into the tank low enough to prevent complete filling, figure 4-1. A vertical **baffle** in this same tank limits fuel sloshing as the car moves.

Regardless of its size and shape, a fuel tank must have the following:

● An inlet or filler tube through which fuel can enter the tank
● A filler cap
● An outlet to the fuel line leading to the fuel pump
● A vent system.

Tank Location and Mounting

Most domestic sedans and coupes generally use a horizontally suspended fuel tank. It is usually mounted below the rear of the floor pan, figure 4-2, between the frame rails, and just ahead of or behind the rear axle. Many station wagons use a vertically positioned tank located on one side of the car between the outer and inner rear fender panels. To prevent squeaks, some cars have felt insulator strips cemented on the top or sides of the tank wherever it contacts the underbody.

Location of the fuel inlet depends on the tank design and filler tube placement. It is usually located behind a filler cap or a hinged door

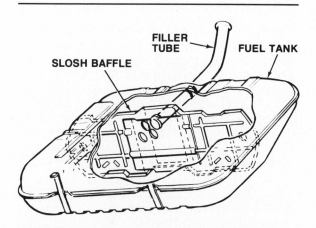

Figure 4-1. The filter tube is located in this tank so that the tank cannot be filled completely. The air space at the top of the tank allows room for fuel expansion. (Oldsmobile)

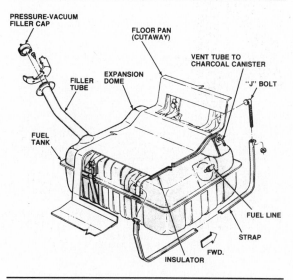

Figure 4-2. Typical fuel tank installation. (Chrysler)

in the center of the rear panel or in the outer side of either rear fender panel. Some older cars have their fuel inlet in other positions. The Type 1 Volkswagen's placement of the fuel inlet under the front hood or in the front body panel is an example. On vehicles with a catalytic converter, a decal reading "Unleaded Fuel Only" is located beside the filler cap.

Fuel tanks generally are held in place by a pair of metal retaining straps. The strap ends are bolted to underbody brackets or support panels. The free ends are drawn underneath the tank to hold it in place, and then bolted to other support brackets or to a frame member on the opposite side of the tank. The retaining straps used to hold station wagon tanks often are fastened between the inner wheel well and quarter panel.

Filler Tubes

Two types of filler pipes are used: a rigid, 1-piece tube soldered to the tank, figure 4-2,

Baffle: A plate or obstruction that restricts the flow of air or liquids. The baffle in a fuel tank keeps the fuel from sloshing as the car moves.

■ Fuel System Development

The first automobiles used a gravity-feed system to provide fuel to the engine. The fuel tank was mounted in the cowl, higher than the engine, and gravity drew fuel out of the tank and carried it to the engine. But front-mounted fuel tanks were limited in capacity and were dangerous.

Moving the fuel tank to the rear of the car solved the problems of safety and storage capacity, but required the use of a vacuum tank. This was a small fuel tank, still positioned above the engine in the cowl, but connected to the rear tank as well. Suction created by engine vacuum provided fuel for the vacuum tank from the larger rear-mounted tank.

If the car was not driven for a long time, the gasoline in the vacuum tank would eventually evaporate. In this case, it was necessary to prime the engine in order to start it and create vacuum which would move fuel

through the system. With the appearance of the mechanical fuel pump after World War I, the vacuum tank was retired.

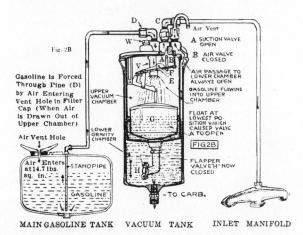

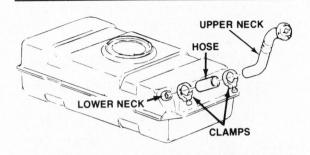

Figure 4-3. Three-piece filler tube assembly. (Oldsmobile)

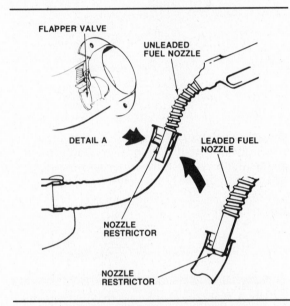

Figure 4-4. Cars that require unleaded fuel have restrictors in the filler tubes to allow only the entry of the smaller unleaded fuel pump nozzles. Restrictors may be spring-loaded flapper valves or simple ring-shaped pieces inside the tube.

and a 3-piece unit, figure 4-3. The 3-piece unit has a lower neck soldered to the tank and an upper neck fastened to the inside of the body sheet metal panel. The two metal necks are connected by a length of hose clamped at both ends.

Cars that require unleaded fuel (cars with catalytic converters and a few without) have a special filler tube, figure 4-4. This filler has a restriction in it so that the only nozzles that can be inserted are the smaller-diameter nozzles of pumps dispensing unleaded fuel.

Tank Venting Requirements

Fuel tanks must be vented, or a **vacuum lock** will prevent fuel delivery. Before 1970, tanks were directly vented to the atmosphere with either a vent line or through the filler cap. But

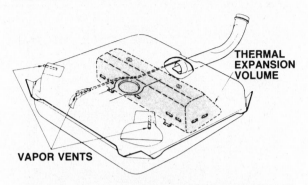

Figure 4-5. This fuel tank has an internal expansion tank to allow for changes in fuel volume due to temperature changes.

those systems added to air pollution by passing fuel vapors into the air. To reduce evaporative hydrocarbon (HC) emissions, controls have been installed on all 1971 and later model cars. In California, 1970 cars also have controls installed).

Because fuel tanks are no longer vented directly to the atmosphere, the tank must allow for fuel expansion, contraction, and overflow that result from changes in temperature. One way to allow for this is to use a separate expansion tank within the main fuel tank, figure 4-5. Another way is to provide a dome in the top of the tank, figure 4-2. As we mentioned earlier, some tanks are limited in capacity by the angle of the fuel filler tube, figure 4-1. The design used on many GM vehicles usually includes a vertical slosh baffle, and reserves about 12 percent of the tank's total capacity for fuel expansion.

Rollover Leakage Protection

All 1976 and later cars have one or more devices to prevent fuel leaks in case of vehicle rollover. Automakers have met this requirement using two different principles:

- A check valve
- A float valve.

Variations of the basic one-way **check valve** may be installed in any number of places between the fuel tank and the carburetor or throttle body injection unit. The valve can be installed in the fuel return line, vapor vent line, fuel tank filler cap, or carburetor fuel inlet filter. Figures 4-6 and 4-7 show typical locations. This type of rollover leakage protection is used primarily by Chrysler, GM, and AMC.

Ford vehicles use a spring-operated **float valve** in the vapor separator, figure 4-8. It closes whenever the car is at a 90-degree angle

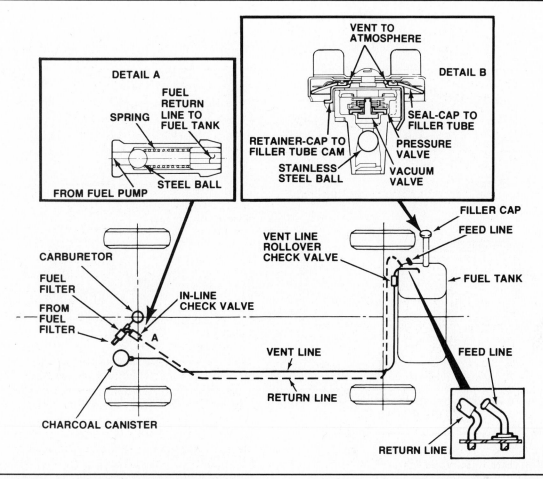

Figure 4-6. One or more rollover leakage protection devices are used in all 1976 and later fuel systems. (AMC)

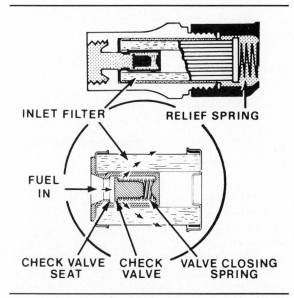

Figure 4-7. The rollover protection check valve is built into the fuel filter used on 1976 and later GM cars. (Chevrolet)

or more. Ford also redesigned its mechanical fuel pump on 1976 and later models to reduce fuel spills during an accident.

Vehicles with electric fuel pumps also use these same rollover protection devices, but they have additional features to ensure that the fuel pump shuts off when an accident occurs. Some pumps depend upon an oil pressure or a distributor (rpm) signal to continue operating; these pumps will turn off whenever the engine dies.

Vacuum Lock: A stoppage of fuel flow caused by insufficient air intake to the fuel tank.

Check Valve: A valve that permits flow in only one direction.

Float Valve: A valve that is controlled by a hollow ball floating in a liquid, such as in the fuel bowl of a carburetor.

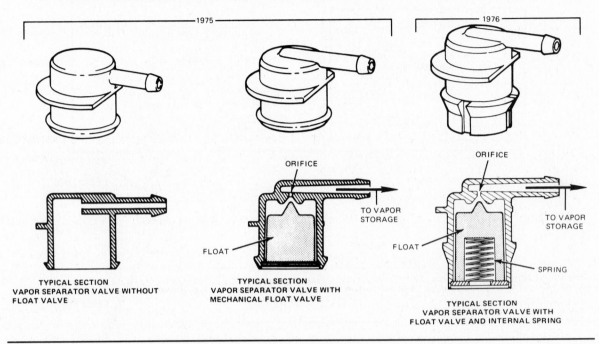

THESE VALVES PUSH INTO POSITION IN A GROMMET-TYPE SEAL, LIKE A PCV VALVE.

SHARP OBJECTS MUST NOT BE USED BETWEEN THE TANK SEAL AND PUSH-IN SEPARATOR DURING REMOVAL OR INSTALLATION.

Figure 4-8. Orifice-type vapor separators used on 1974 and later Ford vehicles contain a mechanical check valve that also provides rollover leak protection. (Ford)

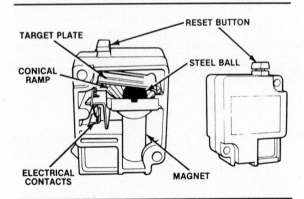

Figure 4-9. Ford uses an inertia switch to turn off the electric fuel pump in an accident.

Late-model Ford vehicles with electronic engine controls and fuel injection have another form of rollover leakage protection. An inertia switch, figure 4-9, is installed in the rear of the vehicle between the electric fuel pump and its power supply. If the car is involved in any sudden impact, the inertia switch contacts will open, shutting off power to the fuel pump. The switch must be reset manually by pushing the button on its top before power can be restored to the pump.

FUEL LINES

The parts of the fuel system are connected by fuel and vapor lines made of steel or nylon tubing and rubber hoses. These are used to supply fuel to the carburetor, throttle body, or fuel rail, to return excess fuel to the tank, and to carry fuel vapors.

The fuel delivery pressure for most carbureted engines is about 5 to 8 psi (34 to 55 kPa). However, the delivery pressure in a low-pressure fuel injection system is between 10 to

■ Try This Tube Tool

Next time you have a piece of steel or copper tubing break off inside an engine block or other casting, try this simple home-made tool to push it out. Use a wrench socket or bushing, a washer, and a self-tapping screw. Place the socket or bushing, with an inside diameter larger than the broken tubing, over the hole in the block. Then place a large flatwasher on the self-tapping screw and insert it through the bushing or socket into the broken tubing. As you tighten the screw, it will pull the tubing out of the hole.

15 psi (69 to 103 kPa); and high-pressure systems often operate with 50 psi (345 kPa) or more. In addition, injection systems retain residual pressure in the lines when the engine is off. All of these factors require special fuel lines. All fuel lines, however, must remain as cool as possible. If any part of the line is located near too much heat, the gasoline passing through it will vaporize more rapidly than the fuel pump can create suction, and **vapor lock** will occur. When it does, the fuel pump will pump only vapor, which passes into the carburetor and out through the bowl vent without the engine receiving any gasoline. Depending on their function, fuel and vapor lines may be either rigid or flexible.

Rigid Lines

All fuel lines that are fastened to the body, frame, or engine are made of seamless steel tubing. Steel springs may be wound around the tubing at certain points to protect against damage.

When rigid fuel line replacement is necessary, only steel tubing should be used. *Copper and aluminum tubing must never be substituted for steel tubing.* These materials will not withstand

Vapor Lock: A condition in which bubbles are formed in a car's fuel system when the fuel gets hot enough to boil. Flow is stopped or restricted as a result.

■ Safety Cells

Most automotive fuel tanks that you service are simple steel tanks that hold liquid fuel. Formed to fit the chassis design and containing some baffles to reduce fuel sloshing, they are pretty straightforward devices. A simple steel tank, however, has some serious disadvantages for vehicles used in hazardous operations. A common fuel tank can be punctured by an impact or leak fuel if overturned. Even with extensive baffling, off-road driving can cause fuel to slosh enough to upset vehicle balance or starve the fuel pickup line. All of these drawbacks to simple fuel tanks led to the development of fuel cells.

A fuel cell is a tank with a rigid shell of steel, aluminum, or some composite material. Inside the shell, a flexible rubber bladder forms a safety liner. The blad-

der is filled with low-density foam material that absorbs the liquid fuel. The fuel stays as a liquid within the foam and can be withdrawn easily by the fuel pump and pickup lines. The foam eliminates sloshing by distributing the fuel evenly throughout the tank regardless of the amount of fuel or vehicle motion. The combination of the rubber bladder and the foam prevents — or reduces — leakage in case of impact or tank rupture. Fuel cells also have check valves in the filler and vent lines to prevent leakage in case of rollover.

Fuel cells are mandatory safety equipment in most racecars and are used in many police cars, fire and rescue trucks, ambulances, and off-road equipment. Special cells with self-sealing, antiballistic ("bulletproof") liners are specified by the U.S. Secret Service as standard equipment in presidential limousines.

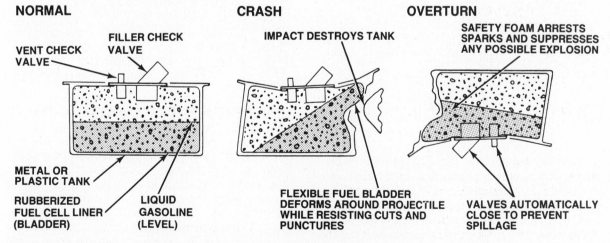

NORMAL
VENT CHECK VALVE
FILLER CHECK VALVE
METAL OR PLASTIC TANK
RUBBERIZED FUEL CELL LINER (BLADDER)
LIQUID GASOLINE (LEVEL)

CRASH
IMPACT DESTROYS TANK
FLEXIBLE FUEL BLADDER DEFORMS AROUND PROJECTILE WHILE RESISTING CUTS AND PUNCTURES

OVERTURN
SAFETY FOAM ARRESTS SPARKS AND SUPPRESSES ANY POSSIBLE EXPLOSION
VALVES AUTOMATICALLY CLOSE TO PREVENT SPILLAGE

Fuel cell construction prevents leakage in case of a crash or a rollover.

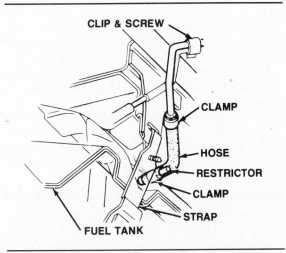

Figure 4-10. Many fuel vent lines have restrictors to control the rate of vapor flow. (Chevrolet)

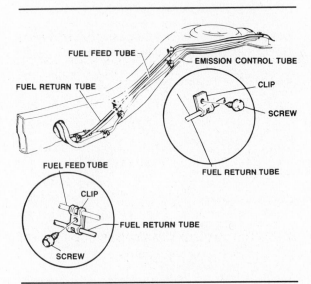

Figure 4-11. Fuel lines are routed along the car frame and secured with clips. Hoses are fastened to steel lines with hose clamps. (Buick)

normal vehicle vibration and they could combine with gasoline to cause a chemical reaction.

In some cars, rigid fuel lines are secured along the car's frame from the tank to a point close to the fuel pump. The gap between frame and pump is then bridged by a short length of flexible hose, which absorbs engine vibrations. Other cars run a rigid line directly from tank to pump. To absorb vibrations, the line crosses 30 to 36 inches (76 to 91 centimeters) of open space between the pump and its first point of attachment to the frame.

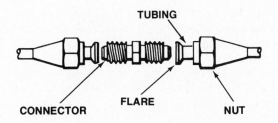

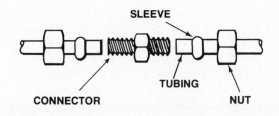

Figure 4-12. Fuel line fittings are either the flare type (top) or the compression type (bottom).

Flexible Lines

In most carbureted fuel systems, synthetic rubber hose sections are used where flexibility is needed. Connections between steel fuel lines and other system components often are made with short hose sections. The inside diameter of fuel delivery hose is generally larger ($5/16$ to $3/8$ inch or 8 to 10 millimeters) than that of fuel return hose ($1/4$ inch or 6 millimeters).

Fuel system hoses must be made of special fuel-resistant material. Ordinary rubber hose such as that used for vacuum lines deteriorates when exposed to gasoline. Only hoses made for fuel systems should be used for replacement. Similarly, vapor vent lines must be made of materials that will resist attack by fuel vapors. Replacement vent hoses are usually marked with the designation EVAP to indicate their intended use.

A metal or plastic restrictor often is used in vent lines to control the vapor flow rate. These may be installed either in the end of the vent pipe, or in the vapor vent hose itself, figure 4-10. When used in the hose instead of the vent pipe, the restrictor must be removed from the old hose and installed in the new one whenever the hose is replaced.

Fuel Line Mounting

Fuel supply lines from the tank to the carburetor are routed to follow the frame along the underbody of the vehicle, figure 4-11. Vapor and return lines may be routed with the fuel

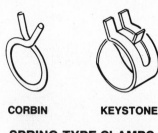

CORBIN **KEYSTONE**

SPRING-TYPE CLAMPS

WORM DRIVE **ROLLED-EDGE**

SCREW-TYPE CLAMPS

Figure 4-13. Various clamps are used on fuel system hoses.

supply lines, but usually are on the frame rail opposite the supply line. All rigid lines are fastened to the frame rail or underbody with screws and clamps or clips.

Carbureted Fittings and Clamps

Brass fittings used in fuel lines are either the flared type or the compression type, figure 4-12, although the flared fitting is more common. The inverted, or SAE 45-degree, flares slip snugly over the connectors to prevent leakage when the nuts are tightened. When replacement tubing is installed, a double flare should be used to ensure a good seal and to prevent the flare from cracking. Compression fittings use a separate sleeve, a tapered sleeve, or a half-sleeve nut to make a good connection.

Various types of clamps are used to secure fuel hoses on carbureted fuel systems, figure 4-13. Spring-type clamps are commonly used for original equipment installation, but only the screw-type (aircraft) clamp should be reused when hoses are changed. Keystone, Corbin, and other spring-type clamps will not hold securely when reused and should be replaced with new ones if they are removed. Screw-type clamps are made in two styles: worm-drive clamps, and those in which the screw and nut stand off from the clamp body, figure 4-13.

Fuel Injection Lines and Clamps

Hoses used for fuel injection systems are made of materials with high resistance to oxidation and deterioration. They also are reinforced to withstand higher pressures than carburetor system fuel hoses. Replacement hoses for injection systems should always be equivalent to original equipment manufacturer (OEM) hoses.

Spring-type clamps must *never* be used on fuel injected engines because they cannot manage the fuel pressures involved. Screw-type clamps are essential on injected engines. Worm-drive clamps are satisfactory for use on carbureted engines, but should not be used on fuel injection systems. The screw teeth can cut and weaken the hose if overtightened. Additionally, hose clamps used on fuel injected engines should have rolled edges to further prevent hose damage.

Fuel Injection Fittings and Nylon Lines

Because of their higher operating pressures, fuel injection systems often use special kinds of fittings to ensure leakproof connections. Some high-pressure fittings on GM cars with port injection systems use O-ring seals instead of the traditional flare connections. Whenever you disconnect such a fitting, inspect the O-ring for damage and replace it if necessary. *Always* tighten O-ring fittings to the specified torque value to prevent damage. Other automakers also use O-ring seals on fuel line connections.

■ **Tools For Making Fuel Lines**

With few exceptions, replacement fuel lines cannot be bought preformed. Tubing is stocked in large rolls and must be shaped and formed however the mechanic wants it. Ordinary hand tools cannot be used to properly make a replacement fuel line. When tubing is cut with a hacksaw, it frequently distorts, and the cut edge will be jagged instead of clean and smooth.

Four special tools are required: a tube cutter, a tube reamer, a tube bender, and a tube flaring device. The tube cutter uses sharpened metal discs to make a smooth, distortion-free cut. After cutting, a tapered reamer is necessary to remove any burrs which could prevent a good seal. The tube bender shapes the tubing without kinking or bending it. Flaring tools are available to make either single or double flares. It is essential to use them to properly shape the connecting ends of any new fuel line.

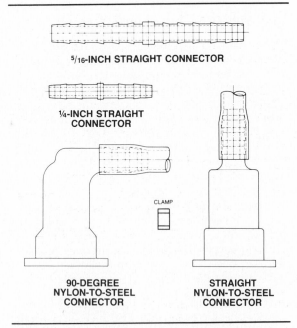

Figure 4-14. Typical Ford connectors for nylon fuel lines. (Ford)

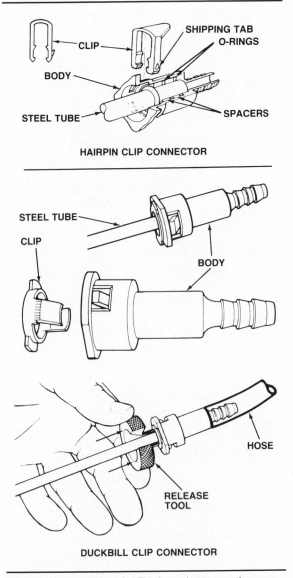

Figure 4-15. Late-model Ford products use these push-connect fuel line fittings. (Ford)

In all cases, the O-rings are made of special materials that can withstand contact with gasoline and alcohol-blend fuels. Some manufacturers specify that the O-rings should be replaced every time the fuel system connection is opened. Whenever you replace one of these O-rings, you *must* use a new part specifically designed for fuel system service. The O-rings used in air conditioning systems are *not* satisfactory.

Ford uses nylon fuel tubing with several unique push-connect fittings. Special barbed connectors are required to join sections of nylon tubing together, figure 4-14. The nylon tubing ends can be softened in hot water before sliding them onto the connectors. Do not soak the tubing in boiling water for a long time. Tubing should not be heated over 212°F (100°C), or it will not return to its original shape and will not grip the connector tightly.

Connectors that join nylon to steel tubing also have barbed ends for the nylon lines, but they use O-rings to seal the steel line, figure 4-15. Ford uses two kinds of retainer clips on these connectors. Fittings for 5/16- and 3/8-inch lines have hairpin clips. Fittings for 1/4-inch lines have duckbill clips.

To remove a duckbill clip, use the special tool shown in figure 4-15, or pliers with thin jaws, to release the clip. Pull the connector apart gently. To remove a hairpin clip, push the shipping tab down to clear the connector body and then spread the clip by hand. Pull the triangular tab to separate the clip from the connector

and gently pull the connector apart. When you reassemble the connector, note that the prongs of the clip are tapered. The tapered sides must face the steel tubing or the connection will be forced apart by fuel pressure. Ford recommends that push connector clips be replaced with new ones whenever a connector is taken apart.

Ford also uses spring-lock connectors to join male and female ends of steel tubing, figure 4-16. The coupling is held together by a garter spring inside a circular cage. The flared end of the female fitting slips behind the spring to lock the coupling together. To open these connectors, a special tool is required that fits around

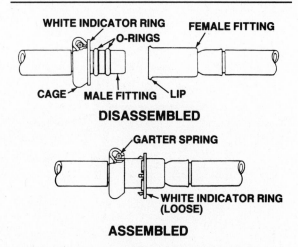

DISASSEMBLED

ASSEMBLED

Figure 4-16. Some Ford metal fuel line connections are made with spring-lock connectors. (Ford)

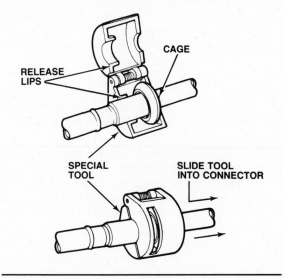

Figure 4-17. Ford spring-lock connectors require this special tool for disassembly. (Ford)

the connector and slides inside the cage to release the spring, figure 4-17. On some vehicles, an indicator ring is left on the fuel line at assembly. To aid reassembly, slide the ring into the cage after you separate the coupling. Reassemble the coupling by hand with a slight twisting motion. The indicator ring, if used, will pop free when the connector is properly seated.

EVAPORATIVE EMISSION CONTROL SYSTEMS

California's stringent emission laws brought the first **evaporative emission controls (EEC)**

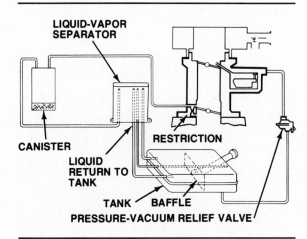

Figure 4-18. General Motors EEC system with liquid-vapor separator and constant-purge canister. (Pontiac)

on 1970 cars sold in that state. Use of EEC systems was extended to all 1971 vehicles, regardless of where they were sold.

The purpose of the EEC system is to trap gasoline vapors that would otherwise escape into the atmosphere. The vapors are routed into the intake airflow and burned.

Common Components

The fuel tank filler caps used on cars with EEC systems differ from those used on non-EEC cars. Some early GM EEC systems use a non-vented cap with a pressure-vacuum relief valve in the line between the fuel tank and the carburetor, figure 4-18, but most EEC caps have pressure-vacuum relief built into them, figure 4-19. Whenever pressure or vacuum exceeds the calibration of the valve, it opens. Once the pressure or vacuum has been relieved, the valve closes. If a sealed cap is used on an EEC system that requires a pressure-vacuum relief design, a vacuum lock may develop in the fuel system, or the fuel tank may be damaged by fuel expansion or contraction.

Fuel tanks are protected in various ways against fuel expansion and overflow caused by heat. An overfill limiter, or temperature expansion tank, was used on many 1970-73 EEC systems to prevent total filling of the tank. This

Evaporative Emission Control (EEC): A way of controlling HC emissions by collecting fuel vapors from the fuel tank and carburetor fuel bowl vents and directing them through an engine's intake system.

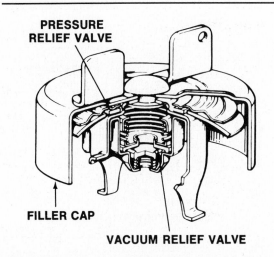

Figure 4-19. Fuel tank caps for EEC systems have vacuum and pressure relief valves.

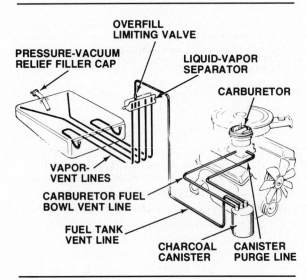

Figure 4-20. This EEC system has a liquid-vapor separator mounted separately from the tank.

limiter is attached to the inside of the fuel tank and contains small holes which open it to the fuel area. When the fuel tank appears to be completely full, that is, it will hold no more and the fuel gauge reads full, the expansion tank remains virtually empty. This provides enough space for fuel expansion and vapor collection if the vehicle is parked in the hot sun after filling the tank.

The dome design of the upper fuel tank section used in some later-model cars, or the overfill limiting valve contained within the vapor-liquid separator, eliminates the need for the overfill limiter tank used in earlier systems.

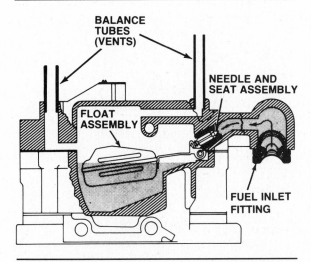

Figure 4-21. Internal carburetor vents. (Chrysler)

Some Ford cars use a **combination valve** which does three things:
1. It isolates the fuel tank from engine pressures and allows vapor to escape from the vapor separator tank to the vapor storage canister.
2. It vents excess fuel tank pressure to the atmosphere in case of a block in the vapor delivery line.
3. It allows fresh air to be drawn into the fuel tank to replace the gasoline as it is used.

All EEC systems use some form of **liquid-vapor separator valve** to prevent liquid fuel from reaching the engine crankcase or the vapor storage canister. Some liquid-vapor separators are built into the tank and use a single vapor vent line from the tank to the vapor canister. When the separator is not built in, figure 4-20, it usually is mounted on the outside of the tank or on the frame near it. In this case, vent lines run from the tank to the separator and are arranged to vent the tank, regardless of whether the car is level or not. Liquid fuel entering the separator will return to the tank through the shortest line.

Carburetor Venting

Carburetors must be vented to keep atmospheric pressure in the fuel bowl. This provides the pressure differential necessary for precise fuel metering. Two types of vents are used: internal and external.

Internal vents
Carburetors are vented internally through the vent tubes that connect the fuel bowl to the airhorn, figure 4-21. The main purpose of these

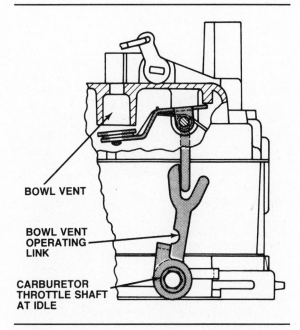

Figure 4-22. External carburetor vent operated by a link from the carburetor throttle shaft. (Chrysler)

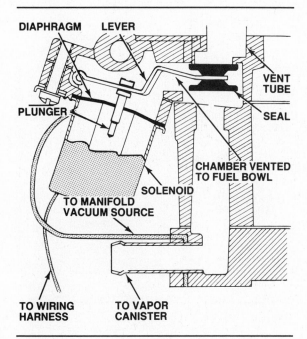

Figure 4-23. A solenoid may be used to switch bowl venting between an internal vent and the canister. (Chrysler)

tubes is to keep atmospheric pressure pushing down on the fuel bowl. This causes the fuel to flow from the bowl, through the circuits and jets, to the lower-pressure area created by the carburetor venturi. The vent tubes also help to compensate for an air pressure drop caused by a dirty air cleaner filter. This helps prevent overly rich air-fuel mixture.

You will learn more about this in Chapter 8. The balance tubes also allow vapors from the fuel bowl to collect in the air cleaner when the engine is off. This helps controls evaporative emissions.

External vents

Many carburetors have external vents for the fuel bowl. On older cars without EEC systems, these vents opened directly to the atmosphere. They released vapors from the fuel bowl to prevent the buildup of vapor pressure, which could cause **percolation**.

The carburetor bowl vent on cars with EEC systems is connected to the vapor storage canister by a rubber hose. External vents often are opened by carburetor linkage, figure 4-22, so that they are closed when the throttle is open, and open when the engine is shut off or idling.

Solenoid-operated vent valves may be used on late-model carburetors. The solenoid shown in figure 4-23 is used on a Carter Thermo-Quad to switch the vent passages between the external vent and the internal vent tube.

Vapor Storage

The EEC system traps gasoline vapors from the fuel tank and the carburetor. These trapped vapors are fed into the engine intake system when it is running, or stored until the engine is started. On almost all late-model vehicles, vapors are stored in a charcoal-granule-filled canister. On a few early EEC systems, vapors were stored in the engine crankcase.

Engine crankcase storage

On many 1970-71 vehicles, the crankcase is used as a vapor storage area. When the engine is started, the stored vapors are drawn from the

Combination Valve: A valve on the fuel tanks of some Ford cars that allows fuel vapors to escape to the vapor storage canister, relieves fuel tank pressure, and lets fresh air into the tank as fuel is withdrawn. Similar to a liquid-vapor separator valve.

Liquid-Vapor Separator Valve: A valve in some EEC fuel systems that separates liquid fuel from fuel vapors.

Percolation: The bubbling and expansion of a liquid. Similar to boiling.

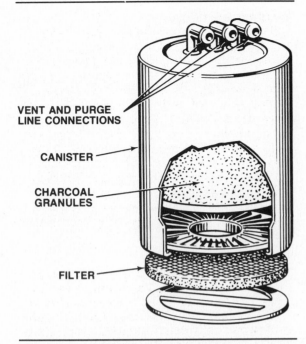

Figure 4-24. A typical vapor storage canister contains 300 or 625 grams of activated charcoal to trap and store fuel vapors.

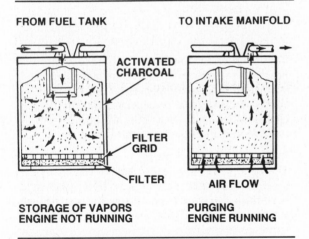

FROM FUEL TANK   **TO INTAKE MANIFOLD**

ACTIVATED CHARCOAL

FILTER GRID

FILTER

AIR FLOW

STORAGE OF VAPORS ENGINE NOT RUNNING   **PURGING ENGINE RUNNING**

Figure 4-25. Typical vapor canister operation. (Chrysler)

crankcase through the **positive crankcase ventilation (PCV)** system and into the engine, where they are burned.

Vapor canister storage

The vapor storage canister appeared on all 1972 domestic cars and has been used on most cars since. The canister is located under the hood, figure 4-20, and is filled with activated charcoal granules which will hold up to one-third their own weight in fuel vapors. The canister is connected to the fuel tank by a vent line. Carbu-

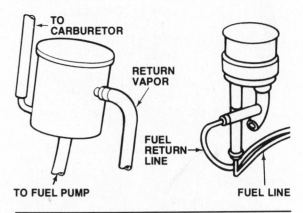

Figure 4-26. Fuel return vapor separators used in some Ford EEC systems. (Ford)

retors with external bowl vents also are vented to the canister. Some Ford and Chrysler vehicles with large or dual fuel tanks may have dual canisters; GM engines may have an auxiliary canister connected to the primary canister's purge air inlet to store vapor overflow.

Activated charcoal is used as a vapor trap because of its great surface area. Each gram of activated charcoal has a surface area of 1,100 square meter, or more than a quarter acre. Typical canisters hold either 300 or 625 grams of charcoal, figure 4-24, with a surface area equivalent to 80 or 165 football fields. Fuel vapor molecules are attached to the carbon surface by **adsorption**. This attaching force is not strong, so the molecules can be removed quite simply by flowing fresh air through the charcoal.

Two methods are used to provide fresh air to the canister for purging. In one design, the bottom of the canister is open to the atmosphere and air enters through a filter, figure 4-25. In this design, purge air is furnished whenever the engine is running. In another design, canisters are closed to the atmosphere and obtain their air from the air injection system. Closed-canister airflow is controlled by a solenoid to purge the vapors during specified engine operating conditions.

To reduce the amount of fuel vapors reaching the carburetor, a small vapor separator, figure 4-26, is installed in the supply line between the fuel pump and the carburetor on many vehicles (particularly Ford products). A vapor return line connects this separator with the fuel tank. Vapors collected in the separator are routed back to the tank to recondense, or they may travel through the regular vent line to the canister. Continuously venting these vapors back to the fuel tank instead of allowing free travel to the carburetor prevents engine surging from fuel overenrichment.

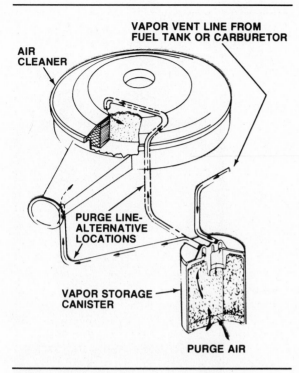

Figure 4-27. Purging the vapor storage canister can be done either through the air cleaner or the carburetor. (Ford)

Vapor Purging

During engine operation, the stored vapors are drawn from the canister to the engine through a hose connected to either the carburetor base or the air cleaner, figure 4-27. This "purging" process sends vapors to be mixed with the air-fuel mixture, which means that the mixture will be richer. To compensate for the mixture enrichment, carburetors used with an EEC system must be calibrated to take vapor purging into account. If the purge rate is not properly controlled to maintain the correct air-fuel ratio under varying engine operating conditions, engine hesitation and surging will result. This purging process can be done in several ways. The flow rate and purge method are determined by two factors:

1. They must reactivate the charcoal.
2. They must have little effect on the air-fuel ratio and driveability.

Constant purge

In this system, the purge rate remains fixed, regardless of engine air consumption. By "tee-ing" into the PCV line at the carburetor, intake manifold vacuum is used to draw vapor from the canister. Even though manifold vacuum fluctuates, an **orifice** in the purge line provides a constant flow rate.

Variable purge

The amount of purge air drawn through the canister is proportional to the amount of fresh air drawn into the engine. In other words, the more air the engine takes in, the more purge air is drawn through the canister. A simple variable purge system is shown in figure 4-27, which illustrated the two ways used to draw the purge air through the canister. This is done either by using a **pressure drop** across the air filter, or by using the velocity of the air moving through the air cleaner snorkel. In both cases, airflow through the air cleaner varies the air flowing through the canister. The simple variable purge often is combined with a constant purge, figure 4-28.

Two-stage purge

If the air cleaner purge flow is not enough, a vacuum-operated **purge valve** may be used, figure 4-29, in addition to the constant airflow to the manifold. **Ported vacuum** from the carburetor controls the purge valve line which opens a second passage from the canister to the intake manifold to provide additional purging.

During idle and low engine speeds, spring tension inside the purge valve holds it closed. As the throttle valve moves beyond the carburetor vacuum port, vacuum is applied to the purge valve diaphragm, causing the valve to lift off its seat.

Positive Crankcase Ventilation (PCV): A way of controlling engine emissions by directing crankcase vapors (blowby) back through an engine's intake system.

Adsorption: A chemical action by which liquids or vapors are gathered on the surface of a material. In a vapor storage canister, fuel vapors are attached (adsorbed) to the surface of charcoal granules.

Orifice: A small opening in a tube, pipe, or valve.

Pressure Drop: A reduction of pressure between two points.

Purge Valve: A vacuum-operated valve used to draw fuel vapors from a vapor storage canister.

Ported Vacuum: Vacuum immediately above the throttle valve in a carburetor.

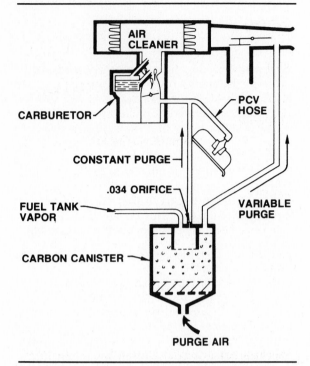

Figure 4-28. In this EEC system, a variable-purge hose runs from the canister to the air cleaner, and a constant-purge hose runs to the intake manifold. (Ford)

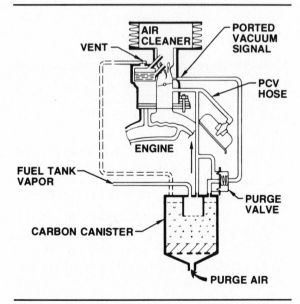

Figure 4-29. The 2-stage purge arrangement in this EEC system uses a vacuum-operated valve to open a second purge line from the canister to the manifold. (Ford)

A carburetor purge port also may be used with the constant purge system. This port is located above the high side of the carburetor

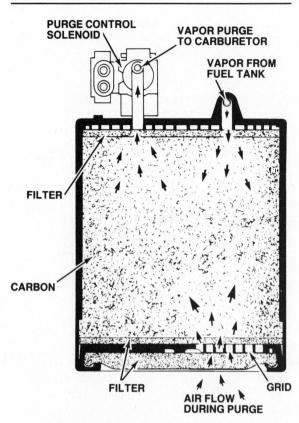

ELECTRIC PURGE VALVE CANISTER

Figure 4-30. In a computer-controlled purging system, the microprocessor controls purge vacuum with a solenoid, (AC-Delco).

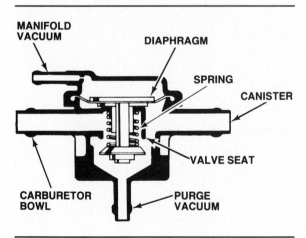

Figure 4-31. A Type 1 GM canister purge valve. (AC-Delco)

throttle plate so that there is no purge flow at idle, but the flow increases as the throttle opens.

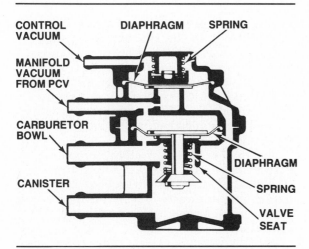

Figure 4-32. A Type 2 GM canister purge valve. (AC-Delco)

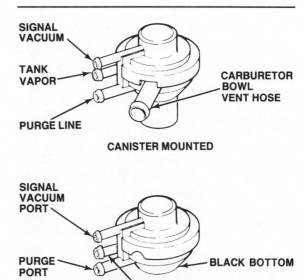

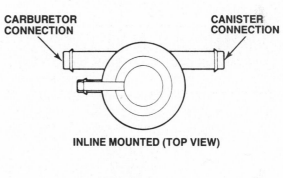

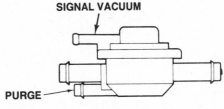

Figure 4-33. Ford canister purge valves are all similar, regardless of their mounting or positioning. (Ford)

Computer-controlled purge

Canister purging on engines with electronic fuel management control systems may be controlled by the engine control computer. Control of this function is particularly important because the additional fuel vapors sent through the purge line can upset the air-fuel ratio provided by a feedback carburetor or fuel injection system. Since air-fuel ratio adjustments are made many times per second, it is critical that vapor purging be controlled just as precisely.

This is done by a microprocessor-controlled vacuum solenoid mounted on top of the canister, figure 4-30, and one or more purge valves. Under normal conditions, most engine control systems only permit purging during closed-loop operation at cruising speeds. During other engine operation conditions, such as open-loop mode, idle, deceleration, or wide-open throttle, the computer prevents canister purging.

General Motors uses various designs, but regardless of their configuration, all use one of two purge valves:

1. The Type 1 valve, figure 4-31, uses spring tension to hold the valve open and permit fuel bowl venting with the engine off. When the engine is running, manifold vacuum closes the valve and permits canister purge.

■ **Is There Life In Your Tank?**

Occasionally you may find a gel-like substance forming in car gasoline tanks, clogging the fuel lines and filters. This is caused by bacteria growth in the fuel tank. This bacteria will grow if there is water present, such as from condensation. To get rid of it, remove the gas tank, clean it with hot water or steam, air dry it, and rinse it with alcohol.

2. The Type 2 valve, figure 4-32, has two vacuum diaphragms. It does the same job as a Type 1 valve and also closes off fuel bowl venting when manifold vacuum provided by the PCV system activates the lower diaphragm with the engine running. When engine speed is increased above idle, control vacuum activates the upper diaphragm to permit canister purging through the PCV system.

Ford EEC systems may use either the EGR solenoid or a separate purge solenoid to control purge valve vacuum. Ford purge valves may be mounted on the canister, remotely mounted, or installed in a vacuum line, figure 4-33. Regardless of their location, all Ford purge valves operate essentially the same as the GM valves just described.

All automakers use variations of the basic system described in this section. The system configuration, components, and locations vary according to the specific fuel system and engine, but all function according to the general principles discussed in this chapter.

SUMMARY

Automotive fuel tanks can be mounted either vertically or horizontally, depending on how much room there is under the vehicle. Filler tubes, besides allowing the tanks to be filled, also are used as fill limiters, leaving room for fuel to expand. Cars requiring unleaded gasoline will not accept a leaded gas nozzle in the tube. Tanks must be vented, but evaporative emission control (EEC) requirements state that the fuel vapors must not be vented to the atmosphere.

Automakers have devised numerous ways to ensure that all vapors remain within the fuel system. They have provided rollover leak protection, pressure-vacuum relief valves, liquid-vapor separators and positive crankcase ventilation.

The EEC systems all use vapor storage canisters. Vapors stored in the canisters are purged into the engine. Each manufacturer has devised slightly different ways to do this, depending on the vehicle and engine requirements. Late-model vehicles with engine control systems have placed the purging function under the control of the computer to ensure that vapor flow does not interfere with air-fuel ratio control or driveability. The most important thing is that all of these EEC systems prevent vapors containing unburned hydrocarbons from reaching the atmosphere, where they pollute the air.

Review Questions

Choose the single most correct answer.
Compare your answers with the correct answers on page 414.

1. Air fuel tanks must:
   a. Be vertically mounted
   b. Be horizontally mounted
   c. Have a vent system
   d. Contain a vertical baffle

2. Filler necks with restricted openings:
   a. Provide better venting
   b. Prevent entry of leaded fuel dispensers
   c. Reduce emissions
   d. Act as rollover check valves

3. Fuel lines fastened to the frame, body, or engine are made of:
   a. Steel
   b. Aluminum
   c. Copper
   d. Rubber

4. Ordinary rubber hose can be used for:
   a. Fuel supply lines
   b. Vapor vent lines
   c. Vacuum lines
   d. Fuel return lines

5. Fuel line fittings:
   a. Are made of copper
   b. Crack easily
   c. Are used to secure rigid lines to the car frame
   d. Are either the flared or compression type

6. Clamps that can be reused when hoses are changed are:
   a. Keystone
   b. Corbin
   c. Screw type
   d. Flat spring

7. EEC systems were used on all domestic cars beginning in:
   a. 1971
   b. 1972
   c. 1973
   d. 1975

8. Liquid vapor separators:
   a. Act as pressure-vacuum relief devices
   b. Are overfill limiters
   c. Prevent vapor lock
   d. Prevent liquid fuel from reaching the crankcase or the vapor storage canister

9. Carburetors are vented:
   a. To allow fuel to return to the fuel tank
   b. To prevent overfill
   c. To maintain atmospheric pressure in the float bowl
   d. All of the above

10. Crankcase vapor storage was not used after:
    a. 1969
    b. 1970
    c. 1971
    d. 1972

11. Activated charcoal is used as a vapor trapping agent because:
    a. It has the area of a football field
    b. It has a huge surface area — 1100 square meters per gram
    c. There are 625 grams of charcoal in a canister
    d. Charcoal is a light material

12. Variable purge:
    a. Takes place only when the engine is off
    b. Is proportional to the air drawn in by the engine
    c. Is controlled by intake manifold vacuum
    d. Requires a graduated orifice

13. Canister purging on engines with electronic fuel management control systems is especially important because:
    a. The additional fuel vapors could upset the air-fuel ratio
    b. The additional fuel vapors could overload the vapor purge canister
    c. The additional fuel vapors could cause a false wide-open throttle signal to be sent to the computer
    d. None of the above

14. Late-model Ford vehicles with electronic engine controls and fuel injection use _____ for rollover protection.
    a. Spring-operated float valves
    b. One-way check valves
    c. Inertia switches
    d. Combination valves

15. Mechanic A says fuel delivery pressure for most carbureted engines is about 10 to 15 psi. Mechanic B says high-pressure fuel injection systems operate at 50 psi or more. Who is right?
    a. A only
    b. B only
    c. Both A and B
    d. Neither A nor B

16. In a 2-stage purge, what usually controls the purge valve line to provide additional purging?
    a. A signal from the computer
    b. Ported vacuum from the carburetor
    c. Pressure drop across the air filter
    d. Velocity of air through the air cleaner snorkel

17. All EEC systems eliminate vapors containing:
    a. NO_x
    b. CO
    c. SO_2
    d. Unburned hydrocarbons

5

Fuel Pumps and Filters

In Chapter 4, we talked about the fuel lines and the fuel system. However, we still have to get the fuel from the tank, through the carburetor or fuel injection system, to the engine. And once it gets there, it must be clean, or the engine will not run properly. In this chapter, we will discuss the fuel pumps that transport the fuel from the tank to the engine, and the filters that remove contamination to ensure the fuel is clean.

PUMP OPERATION OVERVIEW

The fuel pump and the fuel lines, figure 5-1, deliver gasoline from the tank to the carburetor or injection system. The fuel pump moves the fuel with a mechanical action that creates a low-pressure, or suction, area at the pump inlet. This causes the higher atmospheric pressure in the fuel tank to force fuel to the pump. The pump spring also exerts a force on the fuel within the pump and delivers it under pressure to the carburetor or injection system. All pumps, except electric turbine pumps, develop this mechanical action through a reciprocating, "push-pull" motion. The following paragraphs describe various kinds of fuel pumps in detail.

PUMP TYPES

While all pumps deliver fuel through mechanical action, they generally are divided into two groups:
1. Mechanical (driven by the car engine)
2. Electrical (driven by an electric motor or vibrating **armature**).

MECHANICAL FUEL PUMPS

The most common type of fuel pump used by domestic and foreign automakers on carbureted engines is the single-action, diaphragm-type mechanical pump, figure 5-2. The rocker arm is driven by an eccentric lobe on the camshaft. (On some overhead-cam 4-cylinder engines, the eccentric lobe may be on an accessory shaft.) The pump makes one stroke with each revolution of the camshaft. The eccentric lobe (often called simply, "the eccentric") may be part of the camshaft or a pressed steel lobe that is bolted to the front of the camshaft, along with the drive gear, figure 5-3.

In some applications, the rocker arm is driven directly by the eccentric, figure 5-2. Other engines have a pushrod between the eccentric and the pump rocker arm, figure 5-4. The most common example of this arrangement is the small-block Chevrolet V-8 and some 4-cylinder Ford engines.

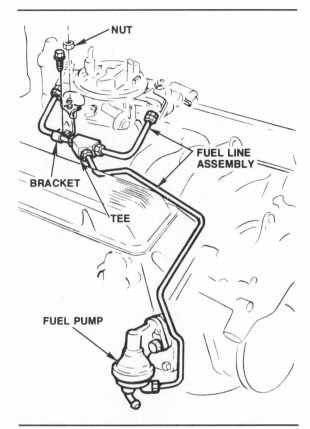

Figure 5-1. Typical fuel pump and line installation on a V-8 engine. (Chevrolet)

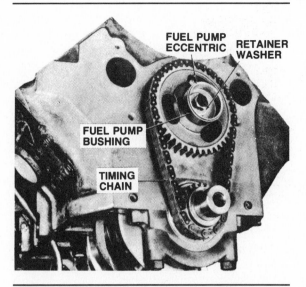

Figure 5-3. This fuel pump eccentric is bolted to the front of the camshaft. (Pontiac)

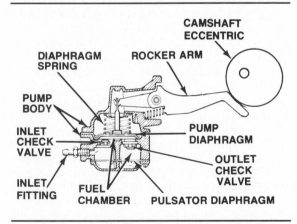

Figure 5-2. Typical diaphragm-type mechanical fuel pump.

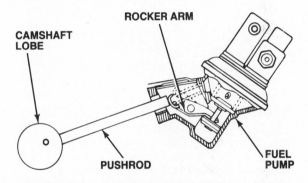

Figure 5-4. Some Chevrolet and Ford engines use a pushrod between the camshaft and pump rocker arm. (Ford)

Mechanical Pump Operation

The fuel intake stroke begins when the rotating camshaft eccentric pushes down on one end of the pump rocker arm. This raises the other end, which pulls the diaphragm up, figure 5-2, and tightens the diaphragm spring. Pulling the diaphragm up creates a vacuum, or low-pressure area, in the fuel chamber. Since there is a constant high pressure in the fuel lines, the inlet check valve in the pump is forced open and fuel enters the fuel chamber.

As the camshaft eccentric continues to turn, it allows the outside end of the rocker arm to "rock" back up. Along with the push given by the diaphragm spring, this allows the diaphragm to relax back down. This is the start of the fuel output stroke. As the diaphragm relaxes, it causes a pressure buildup in the fuel chamber.

Armature: The movable part in a relay. The revolving part in a generator or motor.

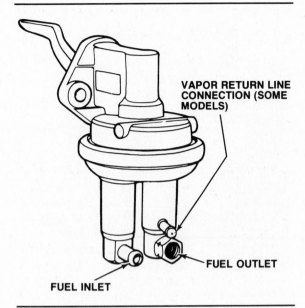

Figure 5-5. This fuel pump has a vapor return line to the fuel tank. (Ford)

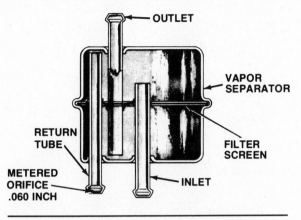

Figure 5-6. A vapor separator is installed between the pump and carburetor to relieve pressure in the fuel line. (Chevrolet)

This pressure closes the inlet check valve and opens the outlet check valve. The fuel flows out of the fuel chamber and into the fuel line on the way to the carburetor. The outlet check valve keeps a constant pressure in the outlet line and prevents fuel from flowing back into the pump.

We measure fuel pump output by the *pressure* and the *volume* of the fuel it delivers. Delivery pressure is controlled by the diaphragm spring. Delivery rate, or volume, is controlled by the carburetor.

The fuel pump delivery rate is proportional to the fuel required by the carburetor. When the carburetor inlet needle valve is open, fuel will flow from the pump, through the lines, and into the carburetor. When the carburetor fuel bowl is full, the needle valve closes and no fuel flows through the lines.

With the needle valve closed and the pressure in the fuel line increasing, the fuel pump diaphragm stays up, even though the rocker arm continues to move up and down in a "free-wheeling" motion. No fuel is pumped until the fuel level in the carburetor bowl drops enough for the inlet needle valve to open again.

During different operating conditions, the fuel level in the carburetor bowl varies. Therefore, the position of the inlet needle valve varies between being fully open and fully closed. The opening of the needle valve and the rate of fuel flowing into the carburetor is always controlled by, and proportional to, the rate of fuel flow out of the carburetor.

Some fuel pumps have a slotted link that operates the diaphragm. The slot fits over the rocker arm and allows a partial stroke of the pump. As the diaphragm responds to outlet line pressure, the slotted link permits partial movement of the diaphragm and a reduced fuel volume.

When an engine with a mechanical fuel pump is shut off, pressure in the fuel line to the carburetor is maintained by the pump diaphragm spring. If engine compartment heat expands the gasoline in the fuel line, the fuel will push the carburetor inlet needle valve open and pass through. The result is too much fuel in the carburetor, and the engine will not re-start easily. This is known as **flooding** the carburetor. Also, since fuel expands when it's hot, it may turn from a liquid into a vapor. This causes vapor lock in the pump and lines. Four methods (described in the following paragraphs) are used to maintain fuel pressure and to prevent flooding and vapor lock.

Some older pumps used an air chamber on the outlet side of the pump to separate and re-circulate vaporized and heated fuel tol the fuel tank through a vapor return line, figure 5-5.

Pumps without the air chamber may use a vapor separator, figure 5-6, in the fuel line between the pump and carburetor. Fuel from the pump fills the vapor separator. The outlet tube in the separator picks up fuel from the bottom of the unit and passes it into the fuel line to the carburetor. Vapor which has gathered rises to the top of the separator, where it is forced through a tube with a metering orifice and into a return line to the fuel tank.

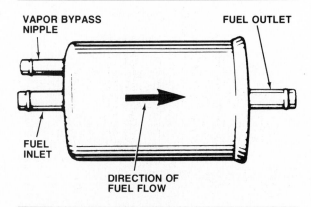

Figure 5-7. A vapor bypass filter combines the fuel filter and vapor relief functions in one unit.

The vapor bypass filter, figure 5-7, has been used mostly on cars with factory-installed air conditioning. It combines the fuel filter and vapor separator into a single unit. Like a vapor separator, a bypass filter uses a restricted nipple and fuel tank return line to relieve vapor pressure buildup. Both the vapor separator and the bypass filter are directional and must be properly installed, or they will not pass fuel to the carburetor.

Many pump designs have a bleed-down system in which tiny holes are drilled through each check valve. This permits pressure buildup in the fuel outlet line to bleed back to the fuel inlet line.

Mechanical Pump Applications

The mechanical fuel pumps used by all domestic cars are manufactured by Carter, AC, or Airtex, all long-time industry suppliers. As a general rule, AMC and Ford use Carter and AC pumps, Chrysler uses Carter and Airtex, and General Motors uses AC pumps.

The diaphragm-type fuel pump is a simple device. Most of them operate in the same way, and the main differences are usually in their exterior appearances. The outside design depends on which engine the pump will be used on and how much room the engine compartment has.

Pumps are so similar in some cases that a production run of the same engine block may use pumps from two different manufacturers. However, replacement pumps must be identical in every respect. Installing a pump that *looks* like the one removed can result in a broken camshaft or accessory shaft as soon as the engine starts.

Mechanical fuel pumps are quite dependable. If they break down, it usually is because of one of the following problems:
- A leaking diaphragm
- Wornout inlet or outlet check valve
- A worn or broken pushrod
- Worn linkage, which reduces the pump stroke.

Flooding: A condition caused by heat expanding the fuel in a fuel line. The fuel pushes the carburetor inlet needle valve open and fills up the fuel bowl even when more fuel is not needed. Also, the presence of too much fuel in the intake manifold.

■ The Dual-Action, Vacuum-Boost Pump

Older cars with vacuum-operated windshield wipers often used double-action diaphragm fuel pumps. These pumps have two diaphragms driven by a single rocker arm. The second diaphragm is a vacuum pump that adds to intake manifold vacuum to run the wipers.

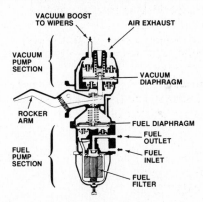

The double-action pump placed a great deal of stress and pressure on the camshaft eccentric. In many cases, it would wear the eccentric enough to hamper pump operation. While the cure for the problem was a new camshaft, many owners replaced the double-action pump with an electric one to save the trouble and cost of installing a new camshaft.

This type of windshield wiper generally disappeared in the 1960's, along with the double-action fuel pump. However, American Motors continued to use this type of pump on some cars through 1971.

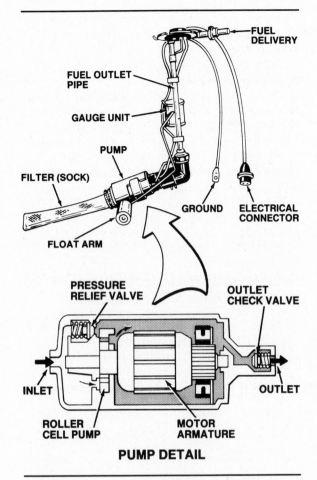

PUMP DETAIL

Figure 5-8. An impeller-type electric fuel pump.

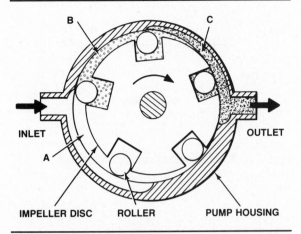

Figure 5-9. The pumping action of an impeller, or rotary vane, pump.

rather than the **pulsating** motion of all other electrical and mechanical pumps.

Figure 5-9 shows the pumping action of a roller vane pump. The pump consists of a cen-

Occasionally, the camshaft or accessory shaft eccentric may wear enough to reduce the pump stroke, or a bolt-on eccentric may come loose from the camshaft. In these cases, the camshaft or accessory shaft or the eccentric must be replaced. It is possible to install an electric fuel pump to bypass a defective mechanical pump.

ELECTRIC FUEL PUMPS

There are four basic kinds of electric fuel pumps: **impeller** (turbine), plunger, diaphragm, and bellows. The impeller or turbine pump, figure 5-8, is driven by a small electric motor. The other three kinds are driven by an electromagnet and vibrating armature and are no longer used as original equipment on standard automobiles.

The impeller-type pump is sometimes called a turbine, roller cell, roller vane, or rotary vane pump. It draws fuel into the pump, then pushes it out through the fuel line to the carburetor or injection system. Since this type of pump uses no valves, the fuel is moved in a steady flow

■ **No One Misses The Good Old Fuel Pump**

Today's fuel pump may seem to be a simple device, but pump manufacturers have worked hard to make it so. Pump designs, capacities, pressures, and performance requirements make the modern pump a rather sophisticated device. This is especially true when you consider that a fuel pump is expected to transport large amounts of gasoline for thousands of miles or kilometers without failure.

Back in the thirties, fuel pump breakdown and replacement was a common occurence every few thousand miles or several thousand kilometers. The fuel pump of the '36 Ford V-87 operated from a pushrod. As the pushrod wore, the pump stroke lessened. Most mechanics and a lot of owners kept the fuel pump operating with a wad of chewing gum or tinfoil stuffed into the push rod cup to compensate for wear. Rather crude, but it worked.

The vacuum booster fuel pump was the first big change in pump design. Since no one could keep their windshield wipers running at a constant speed, pump designers provided additional vacuum with a dual pump design. But super highways, higher horsepower, and emission controls brought new approaches to pump design. Windshield wipers went electric, and the vacuum booster fuel pump disappeared. Intake electric pumps are gradually supplementing and even replacing the traditional mechanical pump design. Automakers now build modern fuel pumps to supply at least 30,000 trouble-free miles or 48 000 crisis-clear kilometers.

tral impeller disc, several rollers that ride in notches in the impeller, and a pump housing that is offset from the impeller centerline. The impeller is mounted on the end of the motor armature and spins whenever the motor is running. The rollers are free to travel in and out within the notches in the impeller to maintain sealing contact with the pump housing. Unpressurized fuel enters the pump and fills the spaces between the rollers, figure 5-9A. As the impeller rotates, a portion of the fuel is trapped between the impeller, the housing, and two rollers, figure 5-9B. Further rotation toward the offset side of the housing compresses the fuel and forces it out of the pump under pressure, figure 5-9C.

Electric Pump Location

The electric fuel pump principally is a pusher unit: it pushes the fuel through the supply line. Because it does not rely on the engine camshaft for power, an electric pump can be mounted in the fuel line anywhere on the vehicle — including inside the fuel tank.

Pusher pumps are most efficient when they are mounted as near as possible to the fuel tank and at or below its level. This allows the pump to use gravity to transfer fuel from the tank to the pump. It also eliminates the problem of vapor lock under all but the most severe conditions. With the pump mounted at the tank, the entire fuel supply line to the carburetor is pressurized. Regardless of how hot the fuel line gets, it is unlikely that vapor bubbles will form to interfere with fuel flow. Having the pump close to or inside the tank also allows the pump to remain cooler because it is away from engine

Impeller: A rotor or rotor blade used to force a gas or liquid in a certain direction under pressure.

Pulsating: To expand and contract rhythmically.

■ Buzzzzz, Click, Smack

Rotary electric fuel pumps are virtually standard equipment on late-model fuel-injected engines. The rotary (roller vane or roller cell) pump delivers the steady volume and pressure needed for reliable injection operation.

The other general type of electric fuel pump — the vibrating-armature pump — has always been a popular aftermarket item for high-performance engines. Many also were used on older imported cars. Vibrating-armature pumps make a distinctive buzzing and clicking sound as they pressurize the fuel lines when the ignition key is turned on.

The vibrating-armature pump operates a plunger, a bellows, or a diaphragm in pulses. The action is similar to that of a mechanical pump. An armature within an electromagnetic coil operates the bellows, plunger, or diaphragm. Current is applied to the coil to move the armature downward and develop pump suction. This opens an inlet check valve, similarly to the action in a mechanical pump.

When the armature nears the bottom of its stroke, it opens electrical contacts that deenergize the coil. A spring then forces the plunger, diaphragm, or bellows upward to force fuel through the outlet check valve.

Fuel output is regulated as it is in a mechanical pump. As pressure rises in the outlet line, it overcomes spring force and keeps the plunger, diaphragm, or bellows from moving on an output stroke.

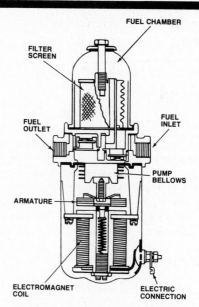

This electric fuel pump uses a bellows driven by a vibrating armature to deliver fuel.

The electrical contacts on some older pumps of this kind would pit and stick together from the continuous arcing as they operated. Many owners of old British sports cars wondered why the small ball peen hammer was included in their car's tool kit. They wondered, that is, until the first time they had to beat on a Lucas electric fuel pump to free the stuck contacts so they could drive on home.

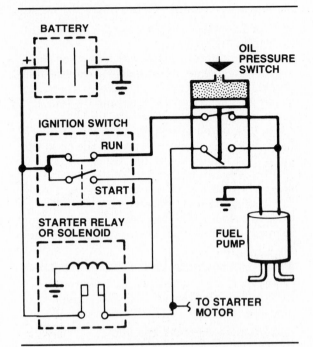

Figure 5-10. During cranking, the fuel pump receives current through the starter relay (solenoid) and the normally closed contacts of the oil pressure switch.

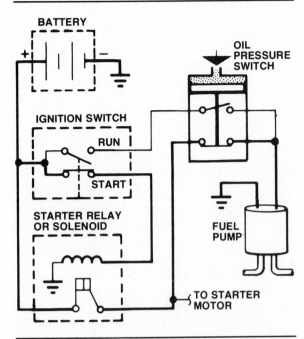

Figure 5-11. When the engine starts, the oil pressure switch opens one set of contacts and closes another. Current then flows through the ignition switch, through the oil pressure switch, and to the fuel pump — as long as the oil pressure remains above a minimum level.

heat. It is therefore less likely to overheat during hot weather.

In-Tank Pumps

Carbureted GM compacts of the Monza, Astre, Skyhawk, and Vega ilk use an impeller-type electric fuel pump located in the fuel tank, as do all fuel-injected engines. This is part of the fuel pickup unit and is similar to the in-tank pump used on early Buick Rivieras and some 1972-74 Fords, figure 5-8.

The Cadillac Seville and all 1975 and later full-size Cadillacs with fuel injection, as well as most fuel-injected Ford vehicles, use two electric fuel pumps: a low-pressure in-tank impeller pump (like that on GM compacts) and a high-pressure, chassis-mounted impeller pump. The in-tank pump provides fuel to the high-pressure pump to prevent vapor lock on the suction side of the fuel system. The in-tank pump is sometimes called a booster pump. The high-pressure pump provides the injection system operating pressure.

Electric Pump Control Circuits

Late-model electric fuel pumps generally receive current through two parallel circuits and may be controlled by the engine fuel management microprocessor.

Oil pressure switch

Most original equipment electric fuel pumps of the 1970's were controlled by a pressure switch in the engine oil system. This switch opens the electric circuit to the pump motor when the engine is off, and controls the operation of the pump when the engine is started and while it is running.

The pressure switch has two sets of contact points. One set is normally closed and allows current to flow from the battery through the starter solenoid or relay to the fuel pump. The other set is normally open. When closed, it allows current to flow from the battery, through the ignition switch, to the fuel pump.

Turning the ignition key to Start energizes the pump by providing current through the normally closed contact points, figure 5-10. Once the engine is running, the pump receives current through the normally open contacts (which have been closed by engine oil pressure), figure 5-11.

Engine oil pressure opens the normally closed contacts and closes the normally open contacts to keep the pump energized. When the ignition switch is turned off, the pump circuit is deenergized. If oil pressure drops below the specified level for any reason (usually 2 psi

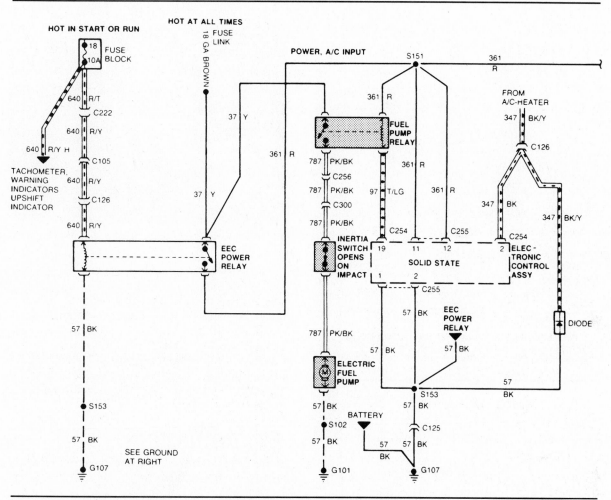

Figure 5-12. Late-model electric fuel pumps used with engine fuel management systems are controlled through a pump relay. This Ford system also uses an inertia switch.

or 14 kPa), contact is broken at the pressure switch and the fuel pump stops immediately.

Computer and relay control

Late-model electric fuel pumps used with engine fuel management systems have a relay, figure 5-12, and use the oil pressure switch circuit as a backup in case the relay malfunctions.

On Chrysler cars, the logic module must receive a distributor (rpm) signal during cranking before it can energize a relay inside the power module to activate the fuel pump, ignition coil, and injectors. If the distributor signal to the logic module is interrupted, the module signals the power module to activate the automatic shutdown relay (ASD) and turn off the pump, coil, and injectors.

Ford and GM systems energize the pump with the ignition switch to initially pressurize the fuel lines, but then deactivate the pump if a distributor (rpm) signal is not received within

one to two seconds. The pump is reactivated as soon as engine cranking is detected. The oil pressure sending unit serves as a backup to the fuel pump relay. In case of pump relay failure, the oil pressure switch will operate the fuel pump once oil pressure reaches about 4 psi (28 kPa).

Inertia safety switch

Fords with fuel injection have an inertia switch between the fuel pump relay and fuel pump, figure 5-12. When the ignition switch is turned to the On position, the electronic engine control (EEC) power relay is energized, providing current to the fuel pump relay and a timing circuit in the EEC module.

If the ignition key is not turned to the Start position within about one second, the timing circuit opens the ground circuit to deenergize the fuel pump relay and shut down the pump. This circuit is designed to pre-pressurize the

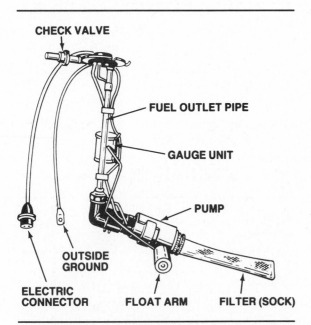

Figure 5-13. Almost all fuel systems have a filter or strainer attached to the pickup tube inside the tank. This one is part of an electric pump and gauge sender assembly.

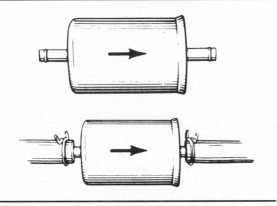

Figure 5-14. Inline fuel filters must be installed so that gasoline flows in the direction indicated by the arrow.

system. Once the key is turned to the Start position, power to the pump is sent through the relay and inertia switch.

The inertia switch opens under a specified impact, such as a collision. When the switch opens, power to the pump is shut off. The switch must be reset manually by depressing the button on its top before current flow to the pump can be restored.

FUEL FILTERS

Despite the care generally taken in refining, storing, and delivering gasoline, some impurities get into the automotive fuel system.

Figure 5-15. Fuel injection systems use large capacity inline fuel filters.

Fuel filters remove dirt, rust, water, and other contamination from the gasoline before it can reach the carburetor or injection system.

The useful life of all filters is limited, although Ford specifies that its filters used with injection systems should last the life of the vehicle. If fuel filters are not cleaned or replaced according to the manufacturer's recommendations, they will become clogged and restrict fuel flow.

Several different types of fuel filters are used, and some systems may contain two or more. Filters can be located in several places within the fuel system.

Fuel Tank Filters and Strainers

A sleeve-type filter of woven Saran is usually fitted to the end of the fuel pickup tube inside the fuel tank, figure 5-13. This filter "sock" prevents sediment, which has settled at the bottom of the tank, from entering the fuel line. It also protects against water contamination by plugging itself up. If enough water somehow enters the fuel tank, it accumulates on the outside of the filter and forms a jelly-like mass. If this happens, the filter must be replaced. Otherwise, no maintenance is required for this filter.

Inline Filters

The inline filter, figure 5-14, is located in the line between the fuel pump and carburetor. This protects the carburetor from contamination, but does not protect the fuel pump. The inline filter usually is a throwaway plastic or metal container with a pleated paper element sealed inside.

Some fuel injection systems use inline filter canisters. These are larger units than are generally used with carbureted engines, figure 5-15.

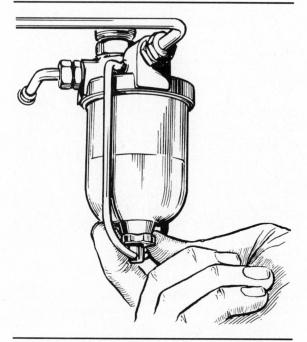

Figure 5-16. Older cars may have sediment bowls that contain a paper, fiber, ceramic, or metal filter element.

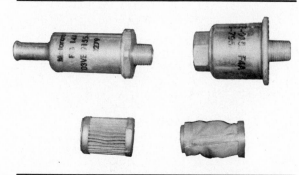

Figure 5-17. A variety of inlet filters used with Motorcraft and Holley carburetors.

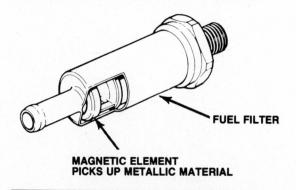

FUEL FILTER

MAGNETIC ELEMENT PICKS UP METALLIC MATERIAL

Figure 5-18. This Ford inlet filter contains a magnetic element to remove metallic contamination from the fuel. (Ford)

Figure 5-19. Rochester carburetors use a pleated paper inlet filter with a check valve.

They may be mounted on a bracket on the fender panel, a shock tower, or another convenient place in the engine compartment; or they may be installed on a frame rail under the vehicle near the electric fuel pump.

Inline filters must be installed so that gasoline flows through them in the direction shown by the arrow, figure 5-14. If an inline filter is installed backwards, it will restrict fuel delivery to the carburetor or injectors.

An inline filter may have a built-in vapor bypass system. These filters have a third nipple, figure 5-7, that connects a fuel return line back to the fuel tank.

Some older domestic and imported cars have a sediment bowl between the fuel pump and carburetor. The bowl contains a pleated paper, ceramic, fiber, or metal filter element. The filter element works much like an inline filter. The bowl cover is held in place by a wire bail and clamp screw. It can be removed for filter cleaning or replacement, figure 5-16. Ceramic and metal elements can be cleaned and reused, if necessary. Paper and fiber filter elements must be replaced when they are dirty.

Carburetor Inlet Filters

Ford and General Motors equip most of their carbureted engines with inlet filters, figure 5-17. The Ford filter is a 1-piece throw-away metal unit containing a filter screen and magnetic washer, figure 5-18, to trap dirt and metal particles. The filter generally screws into the carburetor fuel inlet at one end, and clamps to the inlet hose at the other end. One Ford version used in the mid-1970's, however, was designed specifically as an inline filter.

Some Motorcraft, Holley, and Rochester carburetors use a throw-away pleated paper element. Figure 5-17 (bottom), shows Motorcraft

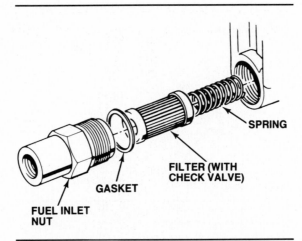

Figure 5-20. The check valve end of the filter must face the fuel line. (Buick)

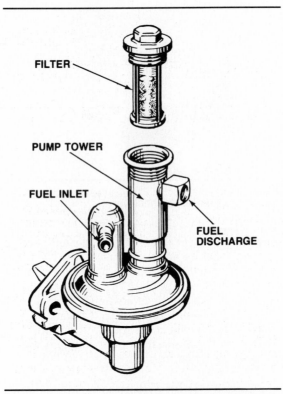

Figure 5-21. Older 6-cylinder Chrysler engines have a filter in the fuel pump outlet tower.

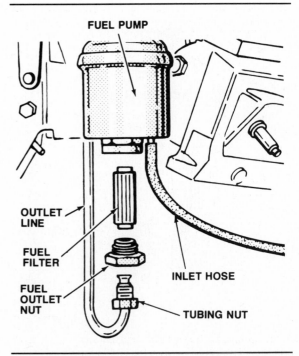

Figure 5-22. Some Cadillacs have a filter in the outlet side of the fuel pump.

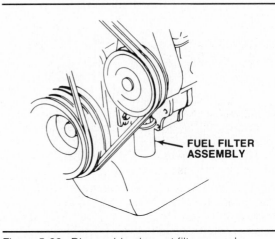

Figure 5-23. Disposable element filters may be mounted on the engine or near the fuel tank.

and Holley elements; figure 5-19 shows Rochester elements. Older cars had a bronze filter element, but reuse is not recommended. Filters on 1976 and later Rochester carburetors must contain the rollover check valve described in Chapter 4. Figure 5-20 shows the proper installation of the Rochester filter with check valve.

Pump Outlet Filters

Some cars have fuel filters in the outlet side of the fuel pump. Those pumps used on Chrysler 6-cylinder engines during the early 1970's, figure 5-21, contain a throwaway filter element installed in the fuel outlet tower. Cadillacs through 1974 use a fuel pump outlet filter located on the bottom of the pump, figure 5-22.

Disposable Element Filters

Screw-on, throw-away element filters, figure 5-23, look much like a replaceable oil filter. Ford

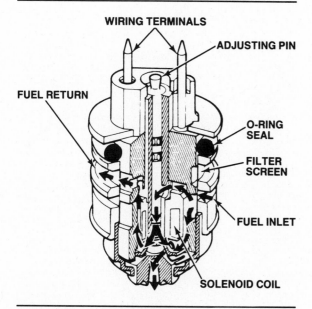

Figure 5-24. Injectors used in throttle body units have one or more external filter screens. (Chrysler)

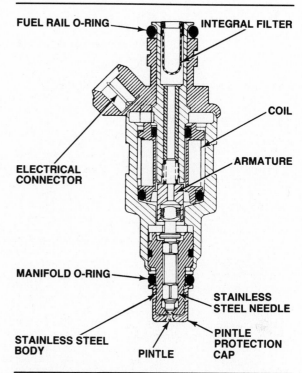

Figure 5-25. Port fuel injectors generally use an integral filter screen. (Ford)

has used this type on the fuel pump of some V-8 engines. The first Cadillac Seville models used a disposable filter mounted to the frame near the left rear wheel. Other fuel-injected Cadillacs have the filter mounted to a bracket on the lower left front of the engine.

Fuel Injection and Filters

Proper filtering of gasoline is essential to fuel injection operation, because particles smaller than one **micron** (0.000025 inch) can interfere with the close tolerances in injectors. Thus, fuel injection systems use various filters:
- The fuel tank filter removes particles larger than 50 microns (0.00125 inch) in size.
- A large-capacity inline filter, figure 5-15, removes particles greater than 10 to 20 microns (0.00025 to 0.00050 inch) in size.

In addition to these filters, some throttle body injection (TBI) units have a filter screen installed in the fuel inlet.

All injectors, throttle body or port, are fitted with one or more filter screens or strainers to remove any particles (generally 10 microns) that might have passed through the other filters. These screens or strainers surround the fuel inlet and thus are external on throttle body injectors, figure 5-24, and internal on port injectors, figure 5-25.

Micron: A unit of length equal to one millionth of a meter, one one-thousandth of a millimeter.

SUMMARY

Fuel pumps move the fuel from the tank to the carburetor or injection system. All pumps do this through a mechanical action that creates a low-pressure area into which the fuel will flow. With check valves and high pressure, the fuel is then forced out of the pump and into the carburetor or injector throttle body or fuel rail.

There are two types of fuel pumps: mechanical and electrical. Mechanical pumps use the engine camshaft or auxiliary shaft eccentric for power. Although there are four types of electrical pumps (plunger, diaphragm, bellows, and impeller), only the impeller type has been used in recent fuel system designs. Electric pumps push the fuel rather than pull it, so they are frequently installed in the fuel tank. Some fuel injection systems use a low-pressure in-tank pump which feeds fuel to a high-pressure external pump that develops system pressure.

Many types of filters are used in the fuel system. They remove contamination from the fuel before it reaches the carburetor or injectors. This is particularly important with fuel injection systems, because of the close tolerances within the injectors. Fuel filters must be replaced or cleaned as directed by the manufacturer. Filters are used as inline filters, at the fuel tank, at the carburetor or throttle body, and on fuel injectors.

Review Questions

Choose the single most correct answer.
Compare your answers with the correct answers on page 414.

1. The most common type of fuel pump is:
   a. The double-action diaphragm type
   b. The single rocker arm type
   c. The single-action diaphragm type
   d. The pushrod-rocker arm type

2. The intake stroke in the fuel pump:
   a. Exerts pressure in the fuel tank line
   b. Creates a vacuum in the fuel chamber
   c. Opens the outlet valve
   d. All of the above

3. The output stroke of the fuel pump:
   a. Increases pressure on the diaphragm spring
   b. Opens the inlet valve
   c. Increases pressure in the pump chamber and opens the outlet valve
   d. Draws fuel into the fuel tank line

4. Fuel pump pressure is controlled by:
   a. The carburetor inlet needle valve
   b. The strength of the diaphragm spring
   c. The carburetor float
   d. None of the above

5. Vapor lock in the pump and lines can be minimized if:
   a. An external air chamber with a vapor return line is used
   b. A vapor separator is installed in the line between the carburetor and pump
   c. A bleed-down system is used
   d. Any of the above

6. When the fuel filter shown below is installed, the arrow must point toward:
   a. The carburetor or fuel injectors
   b. The fuel tank
   c. The vapor canister
   d. The fuel pump

7. Which is *not* a manufacturer of domestic fuel pumps?
   a. Carter
   b. AC
   c. Airtex
   d. GMAC

8. Mechanical fuel pumps:
   a. Are not reliable
   b. Are easily repaired
   c. Can no longer be disassembled
   d. Are all exactly alike

9. Electric fuel pumps are most efficient when located near:
   a. The engine
   b. The fuel tank
   c. The carburetor or injectors
   d. The camshaft

10. Safety control of an electric fuel pump can be provided by:
    a. An oil pressure switch
    b. The starter relay
    c. The ignition switch
    d. Fuel line pressure

11. The fuel pump in an electronic engine fuel management system will operate for one or two seconds and then shut down unless the computer:
    a. Receives an rpm signal
    b. Energizes the ignition switch
    c. Closes the inertia switch
    d. Switches to the back-up mode

12. Inline filters generally:
    a. Are used on both carbureted and fuel injected engines
    b. Are located in the fuel line between the pump and carburetor
    c. Protect the carburetor or injectors but not the pump
    d. All of the above

13. An electric fuel pump has an advantage over a mechanical pump because:
    a. It is lighter
    b. It requires no maintenance
    c. It overcomes vapor lock by rapid fuel delivery
    d. It does not cause wear to the crankshaft eccentric

14. If the fuel pump relay fails in a fuel injection system, the computer continues to operate the fuel pump according to a signal from:
    a. The oil pressure switch
    b. The inertia switch
    c. A timing circuit
    d. The EGO sensor

6

Air Cleaners
and Filters

Gasoline must be mixed with air to form a combustible mixture. Like gasoline, air contains dirt and other materials which cannot be allowed to reach the engine. Just as fuel filters are used to clean impurities from gasoline, an air cleaner and filter, figure 6-1, are used to remove contaminants from the air.

The air cleaner and filter have three main jobs:

1. They clean the air before it is mixed with fuel.
2. They silence intake noise.
3. They act as a flame arrester in case of a backfire.

In 1957, engineers at the Lincoln-Mercury Division of Ford Motor Company were searching for ways to improve driveability under difficult weather conditions. They discovered that driveability improves when the air cleaner is used to provide warm air to the carburetor at low temperatures. It also allows more efficient carburetor adjustments, which in turn:

- Reduce exhaust emissions without reducing engine performance
- Permit leaner air-fuel ratios
- Give better fuel economy
- Reduce **carburetor icing** in cold weather.

Since then, the air cleaner has become a separate emission control system for intake air temperature control. It also has become a part of other emission controls, such as the PCV system.

In this chapter, you will learn the filtering requirements of the automotive engine and how the various air cleaner and filter designs are used to meet them. You also will learn about the temperature-controlled air cleaner's role in emission control and how carmakers put it to work in the fight against pollution.

ENGINE FILTERING REQUIREMENTS

As explained in Chapter 3, the automotive engine burns about 9,000 gallons (34,065 liters) of air for every gallon of gasoline at an air-fuel ratio of 14.7 to 1. With many of today's engines operating on even leaner ratios, the quantity of air consumed per gallon of fuel is closer to 10,000 gallons (37,850 liters). This equals 200,000 gallons (757,000 liters) of air with every 20 gallons (76 liters) of fuel.

This is enough air to fill a good-sized swimming pool, and just like the pool water, the air is filled with particles of dust and dirt which must be removed. Without proper filtering of the air before intake, these particles can affect

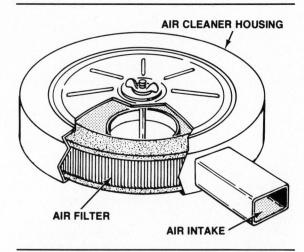

Figure 6-1. A simple air cleaner housing and filter assembly.

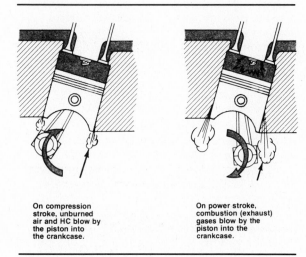

On compression stroke, unburned air and HC blow by the piston into the crankcase.

On power stroke, combustion (exhaust) gases blow by the piston into the crankcase.

Figure 6-2. Crankcase blowby gases.

the operation of the carburetor and upset the air-fuel ratio. Given enough time, they will seriously damage engine parts and shorten engine life.

While abrasive particles cause wear at any place inside the engine where two surfaces move against each other, they first attack piston rings and cylinder walls. Contained in the **blowby** gases, figure 6-2, they pass by the piston rings and into the crankcase. From the crankcase, the particles circulate throughout the engine in the oil. Large amounts of abrasive particles in the oil can damage other moving engine parts.

Although the basic airborne contaminants — dust, dirt and carbon particles — are found whenever a car is driven, they vary in quantity according to the environment. For example, engine air intake of abrasive carbon particles will

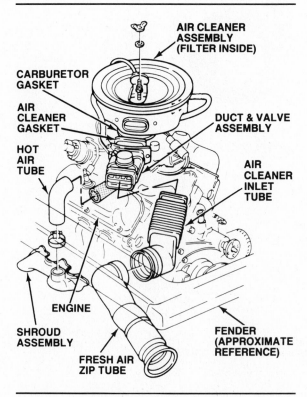

Figure 6-3. Air cleaner housing location on top of the carburetor or throttle body injection (TBI) unit. (Ford)

be far greater in constant bumper-to-bumper traffic. Intake of dust and dirt particles will be greater in agricultural or construction areas.

THE AIR CLEANER AND FILTRATION

Cleaning of the intake air is done by a filter in a 2-piece air cleaner housing made either of stamped steel or composite materials, figure 6-1. The air cleaner housing is located on top of the carburetor or throttle body injection (TBI) unit, figure 6-3, or is positioned to one side of the engine, figure 6-4.

Older carbureted engine air cleaners had a snorkel or air intake tube which drew fresh air into the housing from the engine compartment, figure 6-1. Snorkels are still used, but they are connected by ducting to a fresh air intake which draws air into the housing from outside the vehicle, figure 6-3. Remote air cleaners are connected to the carburetor or the throttle body by similar ducting and also draw fresh air from outside the vehicle, figure 6-4. The air cleaner housing has a removable top section.

Some air filter elements can be cleaned and reused, but most air filter elements are the disposable (throwaway) type, figure 6-5. Certain

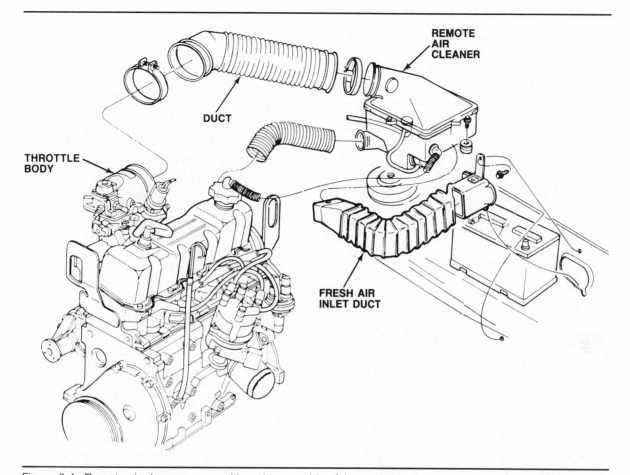

Figure 6-4. Remote air cleaners are positioned to one side of the engine and connected by ducting. (Ford)

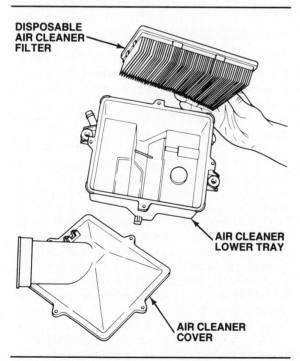

Figure 6-5. Paper air cleaner filter elements are disposable. (Ford)

4-cylinder GM engines, such as those used in the Chevette and Vega-class vehicles, have a throwaway air cleaner housing containing a long-life filter, figure 6-6. After 50,000 miles (80,000 kilometers), the single-piece, welded air cleaner housing is removed from the air intake snorkel and carburetor airhorn, and a new housing is installed.

Filter Replacement

Automakers recommend cleaning or replacing the air filter element at periodic intervals, usually listed in terms of distance driven or months

Carburetor Icing: A condition that is the result of the rapid vaporization of fuel entering a carburetor; the temperature drops enough to freeze the water particles in the airflow.

Blowby: The leakage of combustion gases and unburned fuel past an engine's piston rings.

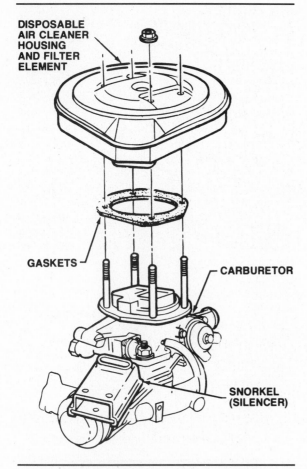

DISPOSABLE AIR CLEANER HOUSING AND FILTER ELEMENT

GASKETS

CARBURETOR

SNORKEL (SILENCER)

Figure 6-6. Disposable air cleaner housing with long-life filter. (Chevrolet)

Figure 6-7. An extreme case: this cam and these valve lifters were destroyed by abrasive particles drawn into an engine operating without an air cleaner.

of service. The distance and time intervals are based on average, or normal, driving. Air filter replacement may be necessary more often when the vehicle is driven under dusty, dirty, or other severe conditions.

Figure 6-8. Typical circular paper air filter element.

It is best to replace a filter element before it becomes too dirty to be effective. A dirty air filter will pass contaminants which cause engine wear, figure 6-7. A dirty air filter element also can change the air-fuel ratio and affect engine performance. The higher the engine speed, the greater the airflow required. Restricted or clogged filters will greatly affect high-speed operation of the engine. If the element becomes so clogged that it does not let through enough air, it can act as a choke to increase fuel consumption. In severe cases, a clogged air filter can even keep the engine from running.

AIR FILTER ELEMENTS

Three general types of air filter elements have been used on cars and light trucks:
1. Paper filters
2. Polyurethane filters
3. Oil bath filters.

Polyurethane filters are available as aftermarket replacements for OEM filters; oil bath filters are heavy, messy, and inefficient, and have not been in general use for over two decades.

The paper air filter element, figure 6-8, is the most common type of filter used on late-model cars and light trucks. It is made of a chemically treated paper stock that contains tiny passages in the fibers. These passages form an indirect path for the airflow to follow. The airflow passes through several fiber surfaces, each of which traps microscopic particles of dust, dirt, and carbon.

Filter paper is pleated and formed into a circle, square, or rectangle (depending upon housing design). Circular filter elements have the top and bottom edges sealed with heat-

Figure 6-9. Typical rectangular paper air filter element.

Figure 6-10. Some paper filters have an outer polyurethane wrapper.

resistant plastic, figure 6-8, to prevent unfiltered air from bypassing the filter. Square and rectangular filter elements, figure 6-9, generally seal only the top edges. A fine wire mesh screen may be used on the inside of the filter ring to reduce the possibility of the element catching fire from an engine backfire. A similar, but coarser, wire mesh screen may be used on the outside of the filter ring for additional strength.

These filter elements generally are made of dry paper, although an oil-dipped paper stock is sometimes used. The light oil coating helps prevent contaminants from working their way through the paper. It also increases the dirt-holding capacity over the same area of dry paper stock. An outer wrapper of polyurethane, figure 6-10, is used sometimes to make the filter work better. Paper element filters are disposable and should be replaced at the recommended intervals. Some technicians attempt to clean a paper element filter by rapping it on a sharp object to dislodge the dirt, or blowing compressed air through the filter. This tends to clog the paper pores and further reduce the airflow capability of the filter.

AIR INTAKE DUCTS AND FRESH AIR INTAKES

The main source of air intake to the carburetor is the air cleaner snorkel or inlet tube. Some air cleaners use a second snorkel to provide additional air intake at full throttle. Two air cleaner snorkels were common on high-performance engines of the 1960's and early 1970's. They

have reappeared on a few high-performance carbureted engine in the 1980's.

The snorkel passes air to the filter and then to the carburetor from the engine compartment. The snorkel also increases the velocity of the air entering the air cleaner housing.

■ Air Cleaner Filter Maintenance

What causes engine wear? Contrary to popular belief, it is not how many miles the engine has been driven, or how old the engine is. If the engine is kept properly lubricated, wear is mainly caused by the dust and dirt that enters it. A teaspoonful of gritty dirt will ruin the piston rings; a cupful will virtually destroy the entire engine. A properly maintained air cleaner filter is the primary line of defense against the gallons of dirty air sucked into the engine with each tiny sip of fuel.

To prevent dirt from entering the engine, a paper filter element should be discarded when dirty. Attempting to clean the element with low-pressure compressed air or rapping it on a hard object to dislodge the dirt and then reusing the filter is not a good idea. These cleaning methods get rid of the dirt which adheres to the outside of the element, but the dirt trapped inside the element cannot be removed and tends to plug up the pores of the paper.

You should avoid rough handling or damage to filters. Washing a paper filter element should be avoided, since even drying a wet filter cannot restore the air passages to normal. Air filter elements should be replaced. In most cases, they are inexpensive compared to the damage that dirt can do if it reaches the engine. At the same time, replace the crankcase ventilation filter, if the air cleaner uses one.

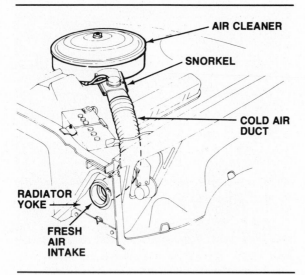

Figure 6-11. Air cleaner with fresh air intake mounted on the radiator yoke.

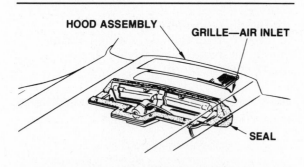

Figure 6-12. Ducted air door located on the rear area of the Corvette hood in the mid-1970's. (Chevrolet)

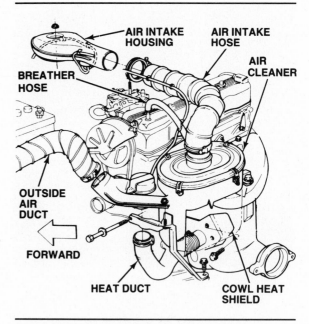

Figure 6-13. Some remote air cleaners are connected to the carburetor or TBI unit by an air intake housing. (Chrysler)

The fresh air intake may be in the cowl or in the rear area of the hood. Ducted hood air doors used on Corvettes in the mid-1970's, figure 6-12, open electrically at full throttle. Pedal linkage closes a switch when the accelerator is pushed to the floorboard. This switch operates a solenoid attached to the air door linkage. The air door provides more intake air at wide-open throttle, just as the second snorkel or intake duct does on some air cleaners.

Remotely Mounted Air Filters and Ducts

Air cleaner and duct design depend on a number of factors such as the size, shape, and location of other engine compartment components, as well as the vehicle body structure. Generally, the air cleaner housing is installed on top of a carburetor or TBI unit, figure 6-3. However, it also can be located away from the engine and connected to the carburetor or TBI unit by an air intake housing, figure 6-13.

Port fuel injection systems generally use a horizontally mounted throttle body. Some systems also have a mass airflow sensor between the throttle body and the air cleaner, figure 6-14. Because placing the air cleaner housing next to the throttle body would cause engine and vehicle design problems, it is more efficient to use this remote air cleaner placement.

Temperatures in the engine compartment will often exceed 200°F (93°C) on a hot day, and hot air can thin out the air-fuel mixture enough to cause detonation and possible engine damage.

Allowing the engine to breath cooler air from outside the engine compartment prevents such problems. Cooler air is provided by a cold air duct or induction (zip) tube. The tube runs from the snorkel to a fresh air intake at the front of the car, figure 6-11. The fresh air intake normally is open at all times, but may have a screen to prevent insects and other foreign matter from being drawn into the air cleaner.

Air cleaner designs used on late-model high-performance engines may have two fresh air inlets, each of which is connected to the air cleaner housing by ducting. This is an updated version of the dual snorkel air cleaner discussed earlier. One inlet provides airflow for general operation; the other opens to provide maximum airflow with the engine at wide-open throttle.

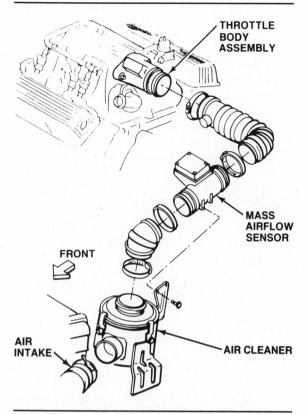

Figure 6-14. Port fuel-injected engines also use a remote air cleaner. If a mass airflow sensor is used, it is inserted in the ducting between the throttle body and the air cleaner. (AC-Delco)

Turbocharged engines present a similar problem. The air cleaner connects to the air inlet elbow at the turbocharger. However, the tremendous heat generated by the turbocharger makes it impractical to place the air cleaner housing too close to the turbocharger. For better protection, a mass airflow sensor is installed between the turbocharger and the air cleaner in some vehicles.

Turbocharger and fuel injection filters and ducts

Remote air cleaners are connected to the turbocharger air inlet elbow or fuel injection throttle body by composite ducting which is usually retained by clamps. The ducting used may be rigid or flexible, but all connections must be airtight.

Filters used in remote air cleaners vary widely in size and shape, but all are similar to the paper element filters described earlier, and should be serviced in a similar manner.

ENGINE AIR TEMPERATURE REQUIREMENTS

Air temperature regulation requirements differ according to engine, carburetor, or fuel injection system used. Sensors are used to ensure proper intake air temperature. These sensors generally are installed in the air cleaner housing and are calibrated to open a vacuum bleed as low as 50°F (10°C) or as high as 120°F (49°C).

■ Foam And Oil Filtration

The disposable paper air filter element is now used as OEM equipment on all vehicles. Two decades ago, however, polyurethane or foam filter elements were used, and the oil bath filter was common on cars, trucks, and some off-road vehicles.

Foam filters consist of a polyurethane wrapper stretched over a metal support screen. Polyurethane contains thousands of pores and interconnecting strands that create a maze-like dirt trap while allowing air to flow through it. Properly maintained, a polyurethane element has a capacity and efficiency equal to that of the dry paper element. It can also be cleaned and reused if necessary. Today, they are popular aftermarket filters, although some automakers include an outer wrapper of polyurethane on their paper filters.

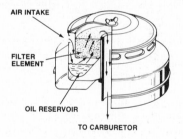

Oil bath air filter.

The oil bath cleaner rests in an oil reservoir in the air cleaner housing. Air entering the housing is deflected downward, where it strikes the oil in the reservoir and deposits heavier particles of dirt. Picking up an oil mist from the reservoir, the air flows back up and across the surface of the filter, where it leaves the mist with finer particles entrapped. The oil then drains back to the reservoir from the filter, carrying the entrapped dirt with it in a self-cleaning action. This type of filtration is most efficient at high airflow rates.

Unfortunately, cleaning polyurethane filters was never popular and servicing the oil bath filter and housing was a dirty, messy chore. For this, more than any other reason, the paper air filter element became king.

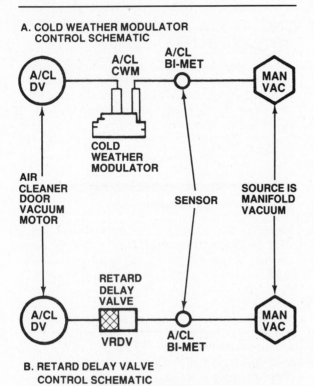

A. COLD WEATHER MODULATOR CONTROL SCHEMATIC

A/CL DV — A/CL CWM — A/CL BI-MET — MAN VAC

COLD WEATHER MODULATOR

AIR CLEANER DOOR VACUUM MOTOR

SENSOR

SOURCE IS MANIFOLD VACUUM

RETARD DELAY VALVE

A/CL DV — VRDV — A/CL BI-MET — MAN VAC

B. RETARD DELAY VALVE CONTROL SCHEMATIC

Figure 6-15. A cold weather modulator (top) or a retard delay valve (bottom) prevents a low vacuum condition from overriding the air cleaner temperature control (Ford)

Ford and some other automakers use a vacuum modulator or a retard delay valve, figure 6-15, to trap vacuum to the air cleaner vacuum motor and hold the air control door in the hot-air position at very low temperatures despite manifold vacuum. Both systems work essentially the same way to prevent low engine vacuum from overriding the temperature control. GM uses a temperature control valve on some engines for the same purpose. Figure 6-16 explains the operation of a vacuum modulator in a typical Ford system.

The retard delay valve, while similar to spark delay valves used in other emission control applications, has an umbrella-type check valve with a **sintered** steel restrictor to delay vacuum release. Most retard delay valve applications are color coded to indicate the release delay time.

THERMOSTATICALLY CONTROLLED AIR CLEANERS

Some form of **thermostatic** control has been used on automobile air cleaners since 1968 to control intake air temperature for improved driveability. These controls became even more important with the need to maintain the precise air-fuel ratios required for exhaust emission control.

The thermostatically controlled air cleaner has the usual sheet metal or composite housing described earlier. Another sheet metal duct, called a heat stove or shroud, is fastened around the exhaust manifold. The heat stove is connected to the air cleaner intake by a flexible hose or metal tube called a hot air tube, or heat duct. Figures 6-3, 6-13 and 6-17 show examples of various designs. Heat radiating from the exhaust manifold is retained by the heat stove and sent to the air cleaner inlet to provide heated air to the carburetor or the throttle body. However, fuel injection systems using a mass airflow sensor do not use temperature control.

An air control valve or damper permits the intake of:

1. Heated air from the heat stove
2. Cooler air from the snorkel or cold-air duct
3. A combination of both.

While the air control valve generally is located in the air cleaner snorkel, it may be in the air intake housing or ducting of remote air cleaners. The air control valve maintains intake air at a specified temperature, usually 90° to 100°F (32° to 38°C). The air control valve is operated by a vacuum motor or diaphragm, although some older domestic and some imported cars use a thermostatic bulb.

Vacuum Motors

Vacuum motor control of air intake temperature is used on all Chrysler products. Some Ford, AMC, and smaller GM engines had thermostatic-bulb air cleaners through the mid-1970's. By the 1980's, almost all engines used vacuum motors to control air cleaner damper operation.

In an air cleaner with a vacuum motor, a **bimetal temperature sensor** and a vacuum bleed in the air cleaner housing regulate vacuum supply to the vacuum motor. Vacuum is supplied from the intake manifold. When intake air temperature is below approximately 100°, the temperature sensor holds the vacuum bleed closed and full manifold vacuum is applied to the vacuum motor. The motor holds the air control valve in the full hot-air position, figure 6-18A.

As intake air warms up, the sensor begins to open the vacuum bleed. This decreases the vacuum sent to the motor. A spring in the motor starts to move the air control valve from the hot-air to the cold-air position, figure 6-18B.

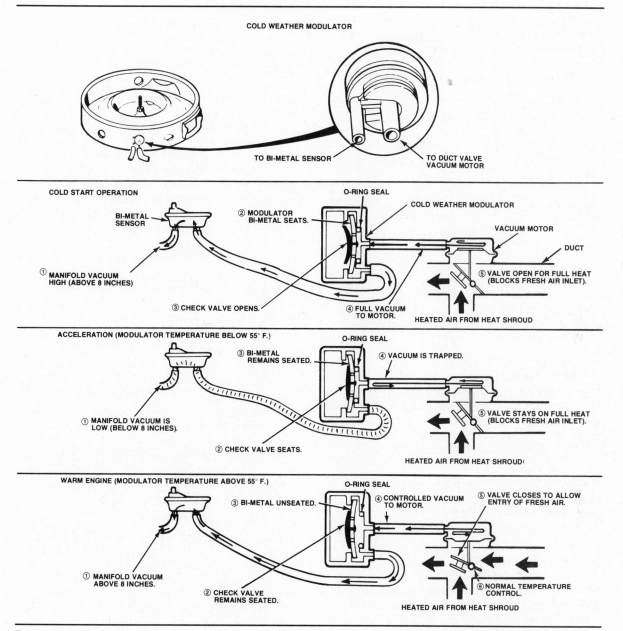

Figure 6-16. Operation of a thermostatically controlled vacuum modulator between the temperature sensor and the vacuum motor. (Ford)

As air temperature continues to rise, the vacuum bleed continues to open, and vacuum to the motor is reduced more. At high air temperatures, vacuum to the air cleaner motor is completely shut off, and the air control valve is in its full cold-air position, figure 6-18C. On some engines, the air control valve also opens to the full cold-air position during heavy acceleration, regardless of air temperature. The opening of the valve provides maximum airflow through the air cleaner to the carburetor or throttle body when it is needed the most.

Sintered: Welded together without using heat, forming a porous material such as the metal disc used in some vacuum delay valves.

Thermostatic: Referring to a device that automatically responds to temperature changes in order to activate a switch.

Bimetal Temperature Sensor: A device made of two strips of metal welded together. When heated, one side will expand more than the other, causing it to bend.

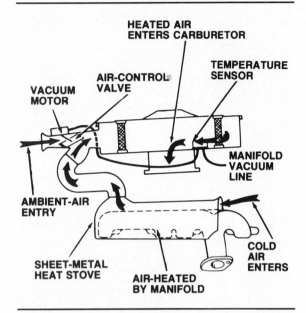

Figure 6-17. A thermostatically controlled air cleaner with a heat stove.

The operating requirements of other engines may be different, however. The vacuum modulator systems described previously trap vacuum in the air motor to hold the damper in the cold-air position. On a cold engine operating close to the 14.7:1 air-fuel ratio, a sudden charge of dense, cold air can cause a lean condition and a hesitation on acceleration.

Thermostatic Bulbs and Coils

Thermostatic bulb operation of the air control was used on many AMC and Ford engines during the 1960's and 1970's. General Motors used a thermostatic coil on small 4-cylinder engines until the mid 1970's. Some import cars also used this method of regulating air intake temperature.

With this type of control, the thermostatic bulb or coil is inside the air cleaner snorkel and connected by linkage to a spring-loaded air control valve. The air control valve is normally held in its closed position by the spring, allowing heated intake air to enter the snorkel, figure 6-19A. As the temperature rises, the thermostatic bulb begins to expand. This expansion exceeds air valve spring tension and the valve gradually opens to its cold-air position, figure 6-19B.

Temperature calibration of the thermostatic bulb or coil differs according to manufacturer and application. On older AMC air cleaners, the air control valve is held in the closed (heat on) position, figure 6-20A, when air entering

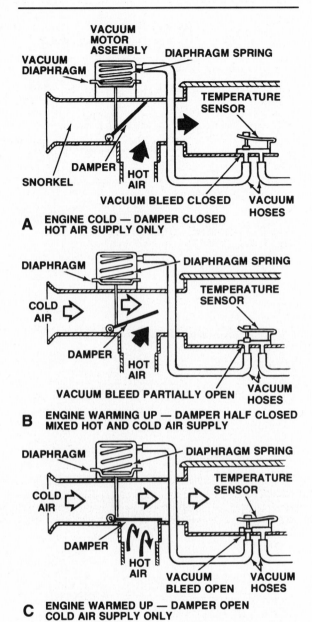

Figure 6-18. A thermostatically controlled air cleaner with vacuum motor control.

the snorkel is less than 105° to 110°F (41° to 43°C). When the air temperature is between 110° and 125°F (43° and 52°C), the air control valve opens partially, allowing a blend of heated manifold air and cooler engine compartment air to enter the air cleaner duct. When intake air temperature reaches 125° to 130°F (52° to 54°C), the air control valve is held in the fully open (heat off) position, figure 6-20B. Air now enters the air cleaner snorkel from the engine compartment or fresh air duct.

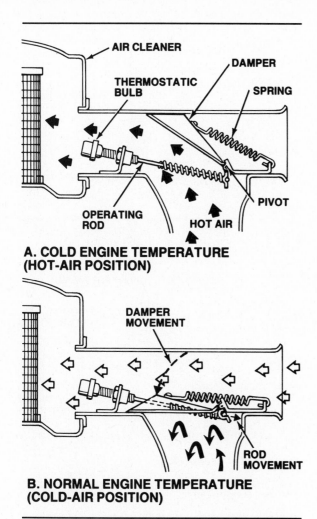

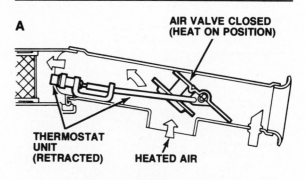

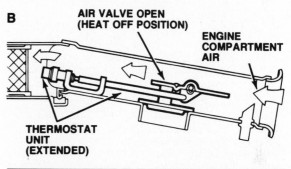

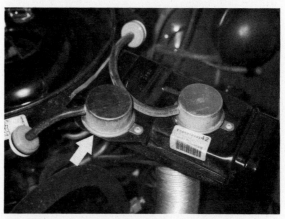

Figure 6-19. A thermostatically controlled air cleaner with thermostatic bulb control. As the bulb expands, it pushes the rod forward to move the damper downward into the hot-air position.

Figure 6-20. Operational cycle of the AMC thermostatically controlled air cleaner. (AMC)

Figure 6-21. A second vacuum motor is used on some late-model AMC air cleaners to operate a trap door or second air control valve in the snorkel.

On 1966-67 Ford engines, the thermostatic bulb holds the air control valve in the closed position (heat on) when air temperature is less then 75°F (24°C). On 1968 and later models, the bulb holds the valve closed until intake air temperature reaches 95° to 100°F (35° to 38°C). At air temperatures of between 85° and 105°F (29° and 41°C) on 1966-67 models, or 100° and 135°F (38° and 57°C) on later models, the bulb opens the valve to allow a blend of hot and cold air through the snorkel. Above these temperatures, the bulb opens the valve to the full cold-air position.

On GM air cleaners with thermostatic coil control, the valve is in the closed (heat on) position at temperatures under 50°F (10°C). Between 50° and 110°F (10° and 43°C), the valve is partly open to allow a blend of warm and cold air. Above 110°F (43°C), the valve is fully open (heat off).

Control Variations

A number of design variations are used by automakers to tailor air cleaner operation for specific engines and operating conditions. Late-model AMC-Jeep 6-cylinder air cleaners have a second vacuum motor to control a second valve in the air cleaner snorkel called a trap door, figure 6-21. This trap door shuts off the air cleaner duct when the engine is off, figure 6-22, to prevent the escape of fuel vapors into the atmosphere. Retard delay valves may be used with either or both vacuum motors.

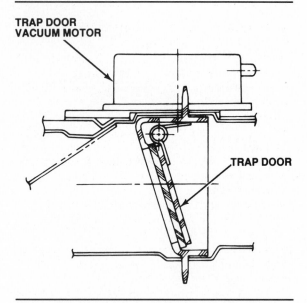

Figure 6-22. By sealing off the air cleaner duct from the atmosphere, the trap door prevents fuel vapors from escaping when the engine is off.

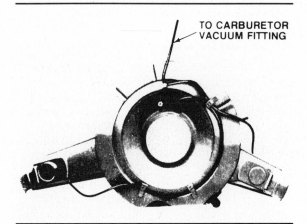

Figure 6-23. Chrysler's dual snorkel air cleaner. (Chrysler)

The dual snorkel air cleaner, figure 6-23, used on some Chrysler V-8 engines, works in the same way as the single snorkel air cleaner. However, only one snorkel receives heated air from the heat stove. The vacuum motor in this snorkel is controlled by a temperature sensor as previously described. The vacuum motor in the second snorkel receives vacuum directly from the intake manifold without passing through a temperature sensor. The valve in the second snorkel is held closed under all operating conditions except heavy acceleration. When manifold vacuum drops under heavy acceleration, the springs in the vacuum motors of both

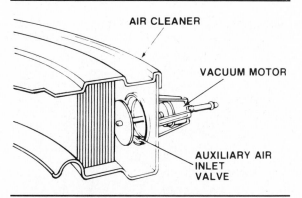

Figure 6-24. Ford's air cleaner with an auxiliary air inlet valve controlled by a vacuum motor.

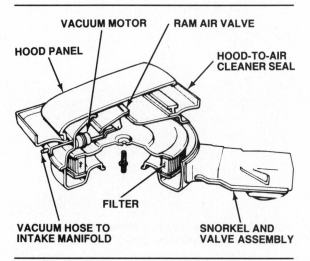

Figure 6-25. Ford's ram air system. (Ford)

snorkels open their valves to provide the maximum flow of cooler air. Ford has used a variation of this system on some late-model high-performance engines.

Some Ford air cleaners have an auxiliary air inlet valve controlled by a vacuum motor, figure 6-24. Located on the rear of the air cleaner housing, it opens to provide more airflow under full-throttle.

Other Ford air cleaners with thermostatic bulb control have an auxiliary vacuum motor under the air cleaner snorkel. The motor is linked to the air control valve by a piston rod. Under full throttle, the vacuum motor takes over from the thermostatic bulb and opens the valve to air from both the heat stove and the cold air intake.

A ram air system, figure 6-25, is used on some Ford engines of the early 1970's. Under full-throttle, a vacuum motor opens an air valve in the functional hood scoop to let in extra outside air.

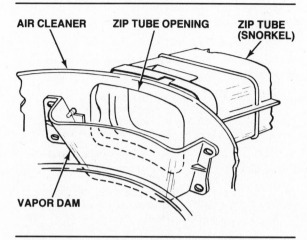

AIR CLEANER   ZIP TUBE OPENING   ZIP TUBE (SNORKEL)

VAPOR DAM

Figure 6-26. Some air cleaners use a vapor dam to trap carburetor fuel vapors when the engine is off. (Ford)

Many Ford engines after 1975 have a thermostatically controlled vacuum modulator between the temperature sensor and the vacuum motor. This keeps the spring in the vacuum motor from opening the air control valve under full throttle with a cold engine. The vacuum modulator, figure 6-16, uses a bimetal thermostatic disc and a check valve to trap vacuum in the vacuum motor. Above 55°F (13°C), the bimetal thermostat in the vacuum modulator opens the vacuum passage through the modulator so that the air cleaner vacuum motor works normally.

The GM temperature control valve (TCV) is located in the air cleaner and works similarly to Ford's vacuum modulator. At temperatures below 80°F (27°C), the valve traps vacuum in the vacuum motor to hold the air cleaner valve closed, even at full throttle. Above 95°F (35°C), the TCV is fully open to permit normal air cleaner operation.

Other Air Cleaner Uses

The air cleaner housing is a convenient place to locate a number of other emission control de-vices. Here are some that may be found on or in the air cleaner housing:

• The PCV system connects to the air cleaner housing to obtain a source of fresh air. The crankcase ventilation filter usually is in the air cleaner, except on some late-model Ford 4-cylinder and V-6 vehicles and Chrysler 6- and 8-cylinder engines, which have the filter in the oil filler cap or hose.
• GM and other automakers attach the manifold absolute pressure (MAP) sensor to the air cleaner housing.
• Chrysler's orifice spark advance control (OSAC) valve is attached to the air cleaner housing on many models.
• Ford has mounted an air injection thermal vacuum switch (TVS) in the air cleaner to control air injection operation.
• A vapor dam, figure 6-26, may be used inside some air cleaners to trap carburetor fuel vapors when the engine is off. Since the vapors are heavier than the air, they remain in the bottom of the air cleaner until the engine is started and they are purged.

SUMMARY

Like fuel, the air used by an engine contains tiny particles of dirt and other contaminants that damage an engine if they are allowed to enter it. Air cleaners and their filters screen out this material, much like fuel filters clean the fuel before it gets to the carburetor or injectors. Air cleaners also are part of the emission control system, since they help reduce emissions and increase performance and fuel economy.

Since 1960, most domestic cars and light trucks have used thermostatically controlled air cleaners, which provide warm air to the carburetor or throttle body at low temperatures. Each carmaker has a slightly different design, but all these devices work essentially the same.

Review Questions

Choose the single most correct answer.
Compare your answers with the correct answers on page 414.

1. Airborne contaminants can:
   a. Change air-fuel ratios
   b. Damage piston rings and cylinder walls
   c. Enter the engine oil
   d. All of the above

2. Which of the following is *not* true about a dirty air filter element?
   a. It will pass contaminants
   b. It does not affect fuel consumption
   c. It will change air-fuel ratio
   d. It may prevent the engine from running

3. Which of the following filters is used almost exclusively as original equipment?
   a. Polyurethane
   b. Dry paper
   c. Oil bath
   d. Oil-dipped paper

4. To prevent a paper air filter element from catching fire:
   a. It is chemically treated
   b. It is soaked in oil
   c. A fine wire mesh screen is used on the inside of the filter ring
   d. A coarse wire mesh screen is used on the outside

5. The primary source of air intake to the carburetor or throttle body is the:
   a. Air cleaner snorkel or intake duct
   b. Venturi
   c. Fuel bowl
   d. Heat stove

6. A heat stove is located on the:
   a. Carburetor
   b. Snorkel
   c. Exhaust manifold
   d. Intake manifold

7. When intake air temperature is low in a thermostatic air cleaner:
   a. The vacuum bleed is closed
   b. The vacuum bleed is open
   c. The air control valve is in the cold air position
   d. Little vacuum is applied to the vacuum motor

8. Mechanic A says that a thermostatically controlled vacuum modulator between the temperature sensor and vacuum motor prevents the air control valve from opening under full throttle with a cold engine.
   Mechanic B says that a vacuum retard valve does the same thing.
   Who is right?
   a. A only
   b. B only
   c. Both A and B
   d. Neither A nor B

9. In the Chrysler dual snorkel air cleaner, the air control valve in the second snorkel is:
   a. Always closed
   b. Always open
   c. Partially open
   d. Open only under full throttle

10. Mechanic A says that heated intake air reduces carburetor icing in cold weather.
    Mechanic B says that a bimetal temperature sensor and vacuum bleed in the air cleaner regulate vacuum to the vacuum motor.
    Who is right?
    a. A only
    b. B only
    c. Both A and B
    d. Neither A nor B

11. Which automaker uses a temperature control valve to hold the air cleaner air control door in the hot-air position at very low temperatures?
    a. Ford Motor Company
    b. General Motors
    c. Chrysler Motors
    d. American Motors

12. An air control valve or damper permits the intake of:
    a. Cool air from the heat shroud
    b. Hot air from the snorkel
    c. Hot air from the manifold
    d. Cool air from the snorkel

7

Intake and Exhaust Manifolds

The intake manifold is a series of enclosed passages that routes the vaporized air-fuel mixture from the carburetor or throttle body to the intake ports in the engine cylinder head. With multipoint fuel injection systems, the intake manifold routes only air. Another series of enclosed passages in the exhaust manifold routes the hot gases resulting from combustion out of the engine to the exhaust system.

In this chapter, you will learn:

- What the intake and exhaust manifolds do
- How their design affects engine performance and emission control
- How add-on devices are used to change manifold operation for better driveability and lower emissions.

INTAKE MANIFOLD PRINCIPLES

The vaporized air-fuel mixture flowing through the carburetor or throttle body must be evenly distributed to each cylinder. The intake manifold does this with a series of carefully designed passages that connect the carburetor or throttle body with the engine's intake valve ports, figure 7-1. To do its job, the intake manifold must provide efficient vaporization and air-fuel delivery.

As you learned in Chapter 3, liquid gasoline is composed of various hydrocarbons which vaporize at different temperatures. If all of the hydrocarbons in gasoline vaporized at the same rate, the job of the intake manifold would be simple. But they do not, and so the intake manifold must be heated to keep the air-fuel mixture properly vaporized. This is done with exhaust manifold heat, engine coolant heat, or both, depending on engine design. Some engines also use electric grid heaters.

V-type engines have the intake manifold in the valley between the cylinder banks, figure 7-2. An exhaust crossover passage inside the manifold carries hot exhaust gases near the base of the carburetor or throttle body, figure 7-3. In a few manifolds, engine coolant is routed through the manifold near the carburetor or throttle body to heat the air-fuel mixture.

Most inline engines have both manifolds on the same side of the engine, with the intake manifold on top of the exhaust manifold, figure 7-4. A chamber between the two manifolds fills with exhaust gases and creates a hot spot to improve fuel vaporization in the intake manifold passages. Some inline engines have the two manifolds mounted on opposite sides of the cylinder head, figure 7-5. This is called a crossflow design. Coolant passages and a heat jacket on the intake manifold supply the heat.

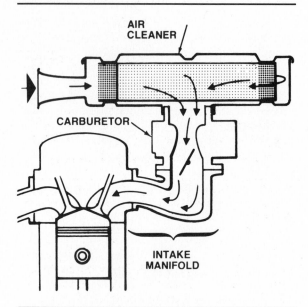

Figure 7-1. Intake manifold passages route the air-fuel mixture from the carburetor (or throttle body) to the intake valve ports.

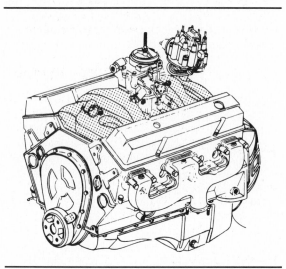

Figure 7-2. V-type engines normally have the intake manifold between the two banks of cylinders (shaded area in the illustration). (Chevrolet)

Even with the help of heat from the exhaust manifold engine coolant, the air-fuel mixture usually does not completely vaporize in the intake manifold. This results in unequal mixture distribution among the cylinders, with some cylinders receiving more fuel and developing more power than others. This problem is greater during engine warmup, when less than normal heat is available to vaporize the fuel.

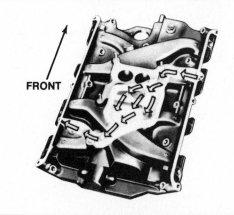

Figure 7-3. Exhaust gases are routed from ports in the cylinder heads through separate passages in the intake manifold to form the manifold hot spot. (Ford)

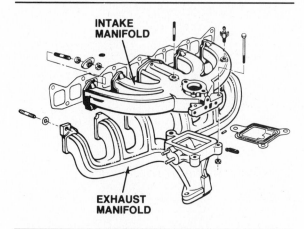

Figure 7-4. Most inline engines have both manifolds on one side of the engine. (Chrysler)

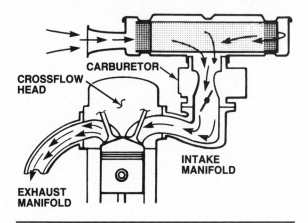

Figure 7-5. A few inline engines have the two manifolds on opposite sides of the engine. (Chrysler)

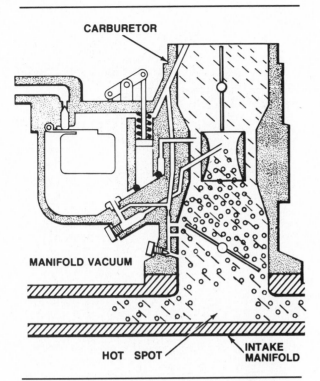

Figure 7-6. The angle of the carburetor throttle plate can affect the flow of the air-fuel mixture above the hot spot.

Causes of Unequal Distribution

The air-fuel mixture reaching the engine cylinders may vary in amount and ratio for several reasons:

● The mixture flow is directed against one side of the manifold by the throttle valve in the carburetor or throttle body, figure 7-6. To some extent, the carburetor choke plate can influence flow in a similar manner.

● Lighter particles of the air-fuel mixture will turn corners in the manifold more easily, while heavier particles tend to continue in one direction, figure 7-7.

● Cylinders closer to the carburetor will receive a richer mixture than those farther away from the carburetor or throttle body. However, this can be minimized by careful carburetor or throttle body placement and good manifold design. Depending on manifold design, using a 2- or 4-barrel carburetor, or 2-bore throttle body, can improve mixture distribution, because each barrel or bore supplies fewer cylinders.

The efficiency of a manifold is determined by the shape, interior surface, and size of its passages. Passages, or manifold **runners**, should

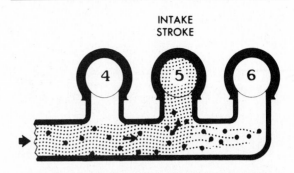

Figure 7-7. An intake manifold with large passages and sharp angles will cause liquid fuel to separate out of the air-fuel mixture. (Chevrolet)

be as short as possible and without sharp corners, bends or turns to interfere with mixture flow. Smooth surfaces speed mixture flow. Rough surfaces aid vaporization by slowing down and breaking up mixture flow. Passages must be large enough in diameter to supply all cylinders with equal amounts of mixture, as passages that are too large or too small will slow down flow. You will learn more about variations in manifold passage design later in this chapter.

BASIC INTAKE MANIFOLD TYPES

Passenger car intake manifolds are made of cast iron or aluminum. The exact design and number of outlets to the engine depends on the engine type, number of cylinders, fuel delivery, and valve port arrangement. Individual intake and exhaust ports may be used, or two cylinders can be supplied by a single port. This is called a **siamesed** port, figure 7-8. Siamesed intake ports are common on inline engines but rare in V-type engines.

Manifold Planes

Production intake manifolds for cars are classified as either single-plane or 2-plane designs. A

Runners: The passages or branches of an intake manifold that connect the manifold's plenum chamber to the engine's inlet ports.

Siamesed: Joined together. A siamesed port on an intake manifold is a single port that supplies the air-fuel mixture to two cylinders.

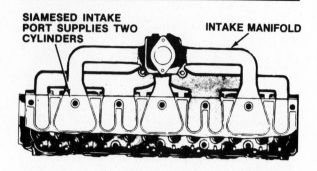

Figure 7-8. By using siamesed ports, this 6-cylinder engine requires only three intake manifold runners. (Pontiac)

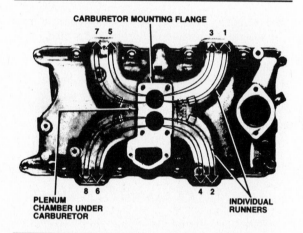

Figure 7-9. This single-plane manifold feeds eight cylinders from a single plenum chamber. (Chrysler)

Figure 7-10. This Chrysler intake manifold has curved runners for better mixture distribution. (Chrysler)

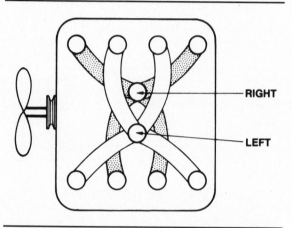

Figure 7-11. The two chambers of a 2-plane manifold are of different heights. (AC-Delco)

single-plane manifold, figure 7-9, uses short branches called runners to connect all of the engine's inlet ports to a single common chamber. This chamber, called the plenum chamber, is simply a storage area for the air-fuel mixture. The mixture accumulates within the chamber until it is drawn out by one of the cylinders. The single-plane design permits more equal cylinder-to-cylinder mixture distribution, but its short runner length causes a drop in mixture velocity, allowing less airflow in low-to-intermediate speed ranges. The single-plane manifold design, however, can produce more horsepower at high rpm than can the 2-plane manifold design.

Inline engines use a single-plane manifold. The carburetor or throttle body usually is placed in the middle of the manifold, leading to the air-fuel flow condition shown in figure 7-7. Cylinders closer to the carburetor or throttle body will receive too much air in the mixture, while those farther away will receive too much fuel. By using siamesed ports, an inline engine

requires fewer manifold runners. Chrysler in-line 6-cylinder engines do not have siamesed intake ports, but the intake manifolds have long, curved runners with no sharp turns, figure 7-10.

V-8 engine intake manifolds may be either a single-plane or a 2-plane design. A 2-plane intake manifold has two separate plenum chambers connected to the intake ports of the engine, figure 7-11. Each chamber feeds two central and two end cylinders. Mixture velocity is greater at low-to-intermediate engine speeds than with the single-plane manifold, but mixture distribution is usually less even than that produced by the single-plane manifold, especially at high speed. When a 4-barrel carburetor is used with a 2-plane manifold, each side of the carburetor (one primary and one secondary barrel) feeds one plenum chamber. When a 2-barrel carburetor is used, each barrel also feeds one chamber.

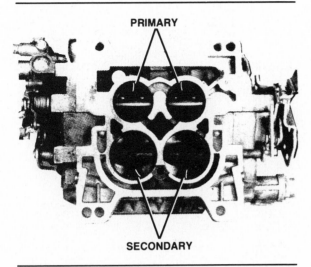

Figure 7-12. A spread-bore carburetor, showing the size difference between the primaries and the secondaries.

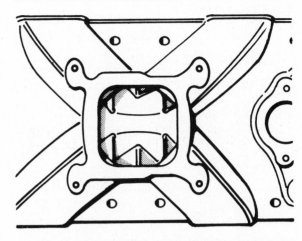

Figure 7-13. The floor of this single-plane manifold is shaped to equalize the mixture flow to all cylinders.

Manifolds for Spread-Bore Carburetors

Until emission control became an important factor in automotive design, all four throats of a 4-barrel carburetor were the same size. Most late-model 4-barrel carburetors are a spread-bore design, with smaller primaries and larger secondaries, figure 7-12. The way the flow goes through spread-bore designs makes it difficult to adapt these carburetors to manifolds not designed for them. It is generally true that 2-plane manifolds work best in the low-to-middle speed ranges, and single-plane designs work better at higher speeds. Engineers have discovered, however, that a single-plane manifold can be designed to incorporate the best features of both types. The design, length, and arrangement of the runners are all important in doing this.

A spread-bore carburetor and a 2-plane manifold distribute a fairly even mixture to all eight cylinders. In effect, each group of four cylinders is fed by similar but independent 2-stage, 2-barrel carburetors. When the same carburetor is fitted to a single-plane manifold, mixture flow tends to seek the shortest path. Each carburetor throat ends up feeding the two closest manifold runners. This means that four cylinders are fed by the larger secondaries, and the other four are fed by the smaller primaries. A lean mixture reaches the front four cylinders, and a rich mixture is fed to the rear four cylinders.

To use a spread-bore carburetor on a single-plane manifold and equalize the air-fuel flow, the plenum chamber floor is grooved and

ridged, figure 7-13. This speeds up the flow to the front cylinders closest to the primaries, and slows down the flow to the rear cylinders closest to the secondaries. This is the most efficient single-plane manifold design to use with a spread-bore carburetor.

Fuel Injection Manifold Systems

Fuel injected engines with throttle body injection (TBI) have manifolding requirements similar to those of a carbureted engine. The TBI unit sits on the intake manifold, and its injector, or injectors, spray fuel into the manifold to be mixed with incoming air. Theoretically, the injector sprays fuel into the area of maximum air velocity to ensure thorough atomization and ideal distribution. Some 2-bore TBI units, however, use individually calibrated injectors to assist in proper distribution. When two TBI

■ **The Good Old Days?**

Well, yes, prices have increased in the automotive industry. Just take a look at this 1926 price list for doing repairs on the hugely popular Ford Model T.

These are the flat-labor rate charges recommended by Ford to all its dealers and mechanics:

Overhaul motor and transmission	$25.00
Overhaul motor only (includes R&R, rebabbitting, reboring)	$20.00
Overhaul transmission only	$14.00
Grind valves, clean carbon	$3.75
Replace head gasket	$1.00
Overhaul carburetor	$1.50
Adjust transmission bands	$0.40

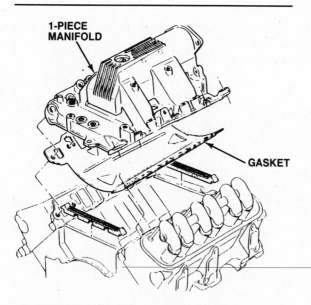

Figure 7-14. The 1-piece manifold used on V-type engines with multi-point fuel injection incorporates the plenum, intake passages, and runners. (Buick)

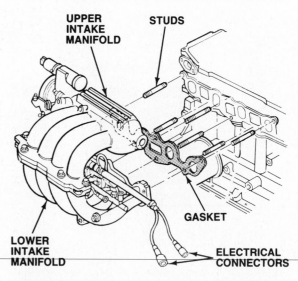

Figure 7-15. A 2-piece manifold used with inline and some V-type engines with multi-point fuel injection has an upper manifold (plenum) and lower manifold with individual runners. (Ford)

units are installed, as in GM's Cross Fire Injection system, the rear injector must be calibrated differently than the front injector for better distribution.

Fuel-injected engines using individual port injectors (also called multi-point or port injection) require special manifold designs. These may be 1- or 2-piece manifolds. The 1-piece design is used with some V-type engines and has the plenum, intake passages, and runners within a single unit, figure 7-14. The 2-piece design is used with inline and some other V-type engines, figure 7-15, and consists of the upper intake manifold or plenum and a lower intake manifold with individual runners.

Since the multi-point or port injection system does not deliver fuel through the manifold, each separate runner can be specifically tailored in size and length (ram-tuned). This helps induction flow to increase cylinder charging and improves fuel distribution among the cylinders. It also eliminates the need for manifold heating or inlet air heating, resulting in a denser air charge. Thus the name "ram tuning".

In a 2-piece manifold design, airflow entering the throttle body passes into the upper intake manifold or intake plenum where it is distributed to the individual runners, figure 7-16. These runners are designed to a specific length according to cylinder requirements. They route the airflow directly to the individual ports where it mixes with the fuel charge just before entering the combustion chamber, figure 7-17.

The difference in performance between a standard manifold and the ram-tuned inlet runners on the Chevrolet 2.8-liter V-6 is shown in figure 7-18. The 600-mm inlet runner, as an example, develops 20 percent more compression, and therefore power, than the standard manifold.

The 1-piece manifold works in basically the same way, except that everything is built into the casting, figure 7-14. The length of the intake passages and the shape and size of the main intake runner are designed to produce a denser air mass at each cylinder just before the intake stroke. In the design shown in figure 7-19, a large diameter passage located inside the plenum chamber behind the airflow entry point produces secondary tuning. This increases airflow at low speeds for better torque.

You learned in Chapter 6 that some port fuel-injected systems do not use an intake air temperature control system. The density of the air charge is increased by eliminating the heated inlet air and manifold heating. Heat is not necessary because there is no requirement to vaporize the fuel in this type of manifolding.

Unlike carbureted or TBI systems, the multi-point or port fuel injection manifolds must provide a place to mount the injectors and fuel rail assembly. This requires precision casting or drilling to properly locate each injector at its required position. Injectors generally are retained in the manifold by clamps and sealed with O-rings. The fuel rail is attached with capscrews.

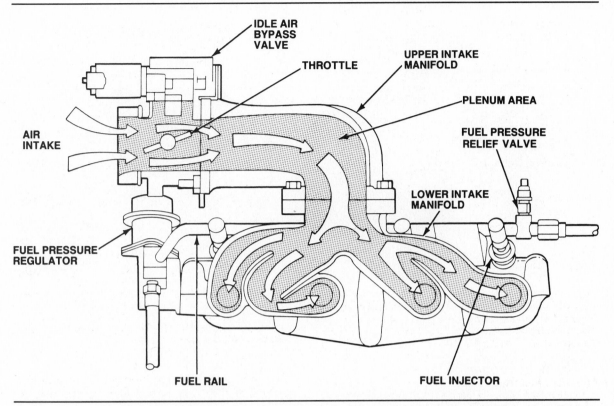

Figure 7-16. Airflow through the upper intake manifold is distributed to individual runners in the lower manifold. (Ford)

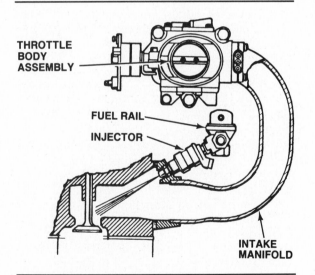

Figure 7-17. Individual runners route the airflow directly to each port where it is mixed with the fuel charge before entering the combustion chamber. (Rochester)

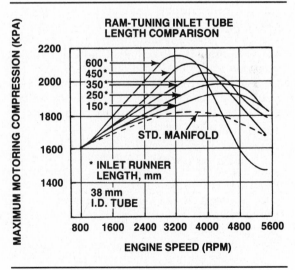

Figure 7-18. This graph shows the effect of ram-tuning the manifold runners. (Rochester)

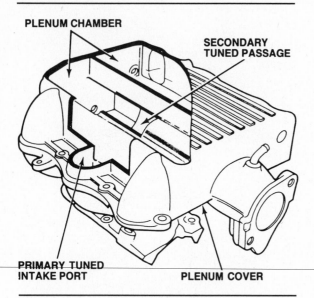

Figure 7-19. Secondary tuning in this Chrysler manifold is provided by a large diameter passage inside the plenum chamber. (Chrysler)

Variable Induction Systems

The tuned length intake runners used in the fuel injection manifolds just described are a low-cost alternative to a multivalve head for increasing an engine's volumetric efficiency. But the increasing number of 4-valve head designs appearing in production (particularly from the Orient) are generally accompanied by equally innovative manifolds designed to fine-tune air induction under the direction of the engine computer (also leading to greater volumetric efficiency). The 4-valve head designs (described in Chapter 2) usually use multiple camshafts and may provide variable valve timing, as required by engine speed and load. To take full advantage of this ability, numerous variable air induction systems have come off the drawing board and into production.

One Nissan design feeds each cylinder through two intake ports to two intake valves. One port in each pair has a butterfly valve which remains closed up to about 3,000 rpm in order to shut off one of the intake paths. The velocity of the air flowing through the single intake path is therefore greater and promotes a high swirl effect. Above 3,000 rpm, the butterfly opens the other intake path and feeds the additional air needed at high engine rpm.

Mazda uses a variable ram-effect induction system (VRIS). This design uses a common plenum chamber with long curved runners (primary ports) for each cylinder. Each runner also has a second shorter branch (secondary

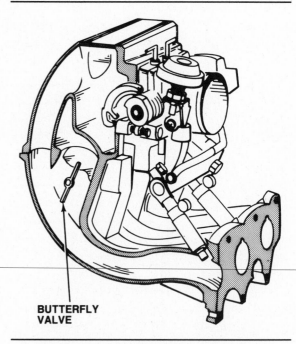

Figure 7-20. Mazda's VRIS system uses an inertia charge effect to improve torque output through interconnected, branched intake paths equipped with butterfly valves.

ports), containing a butterfly valve, figure 7-20. These branches are interconnected by a passage and the valves remain closed below about 5,000 rpm. Above that speed, the computer activates a vacuum actuator through a solenoid to open the secondary port control valve. This shortens the effective length of the passage and interconnects all of the branches. Closing the intake valves creates a positive pressure in the system, which has a domino effect moving from one cylinder to another to pack in more air. Since the interconnecting passage reduces air intake resistance, flow speed is reduced.

Nissan has modified its VG30 V-6 engine used in several vehicles to incorporate variable valve timing and air induction that is controlled by the engine computer, figure 7-21. Dual camshafts are used for each bank of cylinders, one for intake and one for exhaust valves. The intake cam has a helical gear and hydraulic coupling that changes intake valve timing according to a computer-controlled solenoid.

The air induction system is just as ingenious. Air flows into a single plenum chamber, where it passes into a pair of collector boxes. Each collector box contains a throttle chamber which reduces air resistance at high rpm and load conditions. An interconnecting passage between the collector boxes contains a power valve controlled by the computer. At low- and mid-range

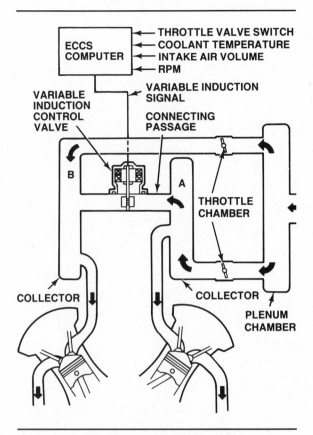

Figure 7-21. Nissan's Induction Control System uses a computer-controlled valve to change the length of the induction path according to engine speed and rpm.

rpm, this valve remains closed to lengthen the intake path. As rpm increases, the valve opens to connect the two collector boxes and increase the volume of airflow.

Toyota also builds several 4-valve engines with dual-level, variable intake manifolds as part of the computer-controlled fuel injection system. The Honda-Acura-Sterling V-6 has a dual intake system with 12 manifold runners for 6 cylinders. A solenoid-controlled vacuum diaphragm opens the secondary runners for more airflow at high speed.

While there are many more such induction designs, the intent behind each is the same — to maintain engine torque across the power band — regardless of the different ways in which the engineers choose to implement it.

The important principles of split, or dual-level, intake manifolds are that the primary (low-speed) runners are longer and narrower than the secondary (high-speed) runners. Long, small cross section runners increase airflow velocity at low engine speed to deliver

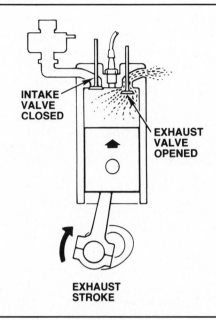

Figure 7-22. Burned gases are pushed out of the combustion chamber by the piston's upward exhaust stroke.

more air, faster at engine speeds below approximately 3,000 rpm. Short, large cross section secondary runners allow airflow to travel a shorter distance at engine speeds above 3,000 rpm. This delivers more air at higher speed than long, narrow runners can. Dual-level manifolds thus avoid the traditional compromises of intake manifold design.

EXHAUST MANIFOLDS

Like the engine block and cylinder head, an exhaust manifold generally is made of cast iron, which offers excellent resistance to changes in heat and temperature. The exhaust manifold bolts to the cylinder head and collects the corrosive exhaust gas from the cylinders, routing it into the exhaust system to be dispersed into the atmosphere. Some late-model exhaust systems, however, are manufactured of stainless steel. This material offers greater resistance to corrosion, faster warmup for quicker converter light-off, and a substantial reduction in weight.

Proper dispersion of exhaust gases through the exhaust manifold design is as important as air-fuel flow through the intake manifold. As the piston moves upward during the exhaust stroke, it forces combustion gas through the open exhaust valve and out of the cylinder, figure 7-22. Whenever gas is pushed through a passageway, turbulence and friction along the sides of the passage cause a resistance called

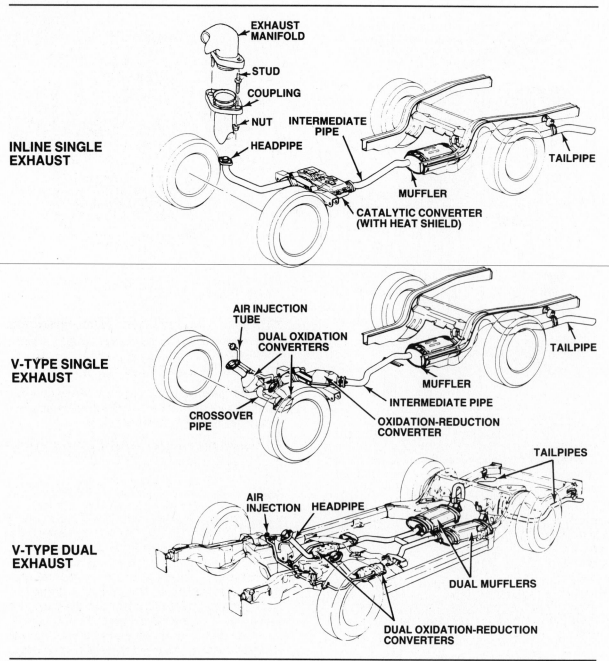

INLINE SINGLE EXHAUST

EXHAUST MANIFOLD
STUD
COUPLING
NUT
HEADPIPE
INTERMEDIATE PIPE
TAILPIPE
MUFFLER
CATALYTIC CONVERTER (WITH HEAT SHIELD)

V-TYPE SINGLE EXHAUST

AIR INJECTION TUBE
DUAL OXIDATION CONVERTERS
TAILPIPE
MUFFLER
INTERMEDIATE PIPE
OXIDATION-REDUCTION CONVERTER
CROSSOVER PIPE

V-TYPE DUAL EXHAUST

AIR INJECTION
HEADPIPE
TAILPIPES
DUAL MUFFLERS
DUAL OXIDATION-REDUCTION CONVERTERS

Figure 7-23. Typical exhaust system configurations. (Ford)

backpressure. A piston encounters backpressure each time it comes up on the exhaust stroke. This resistance causes a power loss which would otherwise go to the engine flywheel. A direct and unrestricted flow of exhaust gas causes less backpressure and prevents this unnecessary power loss.

Backpressure in the exhaust manifold and exhaust system can contaminate the intake manifold's fresh air-fuel mixture. A brief **camshaft overlap** period, combined with back-

pressure in an exhaust manifold, can cause one cylinder's intake stroke to draw exhaust gases from a nearby cylinder. Too much backpressure can also prevent exhaust gases from leaving a cylinder before the next air-fuel charge arrives. In both cases, the air-fuel mixture will be preheated and diluted, resulting in lower engine efficiency.

Dual exhaust systems reduce exhaust backpressure by splitting the exhaust gas flow into two outlet lines. Since the exhaust manifold is

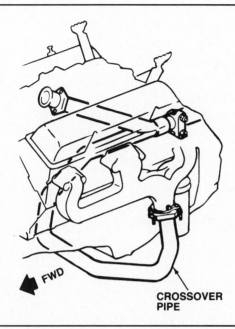

Figure 7-24. A crossover pipe often is used to connect the two exhaust manifolds to a single muffler system. (Chevrolet)

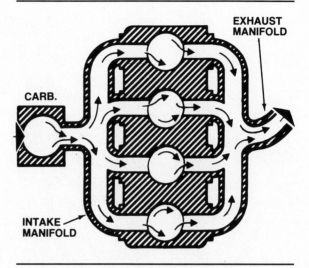

Figure 7-25. Exhaust backpressure is caused in part by the many turns of the exhaust manifold passages.

at the beginning of the flow, manifold design is the most important factor in reducing backpressure. However, good muffler design also can help to minimize restrictions.

There are three basic exhaust system configurations, figure 7-23. Since an inline engine usually has a single exhaust manifold, a single exhaust system generally is used, with the exhaust pipe connecting directly to the exhaust manifold flange. Each cylinder bank on a V-type engine, however, has its own exhaust manifold. When a single exhaust system is used, the two manifolds will connect with a Y-pipe or a crossover pipe, figure 7-24. With dual exhaust systems, each manifold is connected to its own exhaust pipe, muffler, and tailpipe.

Headers and High-Performance Manifolds

Sharp turns and narrow passages in an exhaust manifold will slow the flow of gases from the exhaust ports and increase the amount of backpressure in the system, figure 7-25. Many original equipment manifold designs have duplicated high-performance manifolding by using the sweptback manifold design, figure 7-26.

Backpressure: The resistance, caused by turbulence and friction, that is created as a gas or liquid is forced through a passage.

Camshaft Overlap: The period of crankshaft rotation in degrees during which both the intake and exhaust valves are open.

■ **The Torque Box**

Designing manifolds for production engines is a series of compromises dictated by engine design, emission requirements, driveability, manufacturing costs, and various other factors. In general, the dual plane manifold provides better throttle response, a broader torque curve, and higher maximum torque than a single plane design. The single plane manifold is responsible for high maximum horsepower (more efficiency) at high rpm, but is not good at idle, and can contribute to icing during cold weather.

An attempt to combine the best features of both designs resulted in what engineers call a "torque box". This design uses the long runners from the dual plane manifold with the plenum chamber of the single plane design. The result is good torque characteristics and higher maximum horsepower.

The best known example of a torque box used on production cars is the dual 4-bbl. manifold fitted to early Z/28 Camaros.

The torque box became one of the most common racing manifold designs. Its use on production engines was limited by stricter emission requirements, the use of smaller and less powerful engines, and the emphasis on fuel economy.

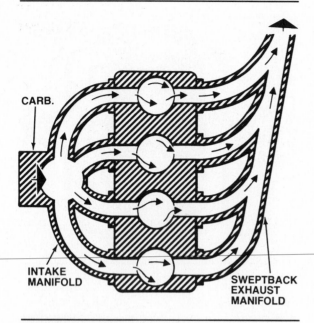

Figure 7-26. Streamlining the exhaust manifold can reduce backpressure.

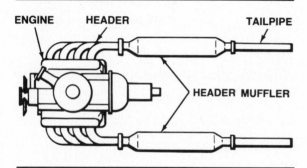

Figure 7-27. Headers used with V-type engines usually have separate muffler systems.

The high performance exhaust manifolds shown in figure 7-27 are called **headers**. Headers are made from heavy steel tubing welded to steel flanges which bolt to the cylinder head. Headers create a slight suction called **scavenging**, which helps to pull the exhaust from each cylinder. Individual header pipes usually differ in length so that low pressure will be present in the headers at different engine speeds.

Although the specialized science of exhaust tuning was once the province of the hot rodder and high-performance vehicles, the problem of exhaust backpressure is an important factor in controlling the emissions of late-model engines. Automotive engineers design backpressure-reducing exhaust systems that can affect the volumetric efficiency of an engine, and yet still leave enough backpressure to regulate the recirculation of exhaust gas in order to reduce

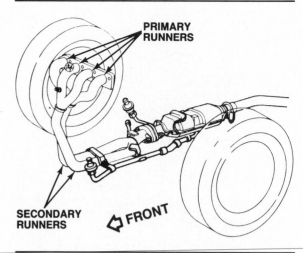

Figure 7-28. Ford's bifurcated intake manifold essentially is a set of stainless steel headers. (Ford)

NO_x and prevent detonation. For this reason, altering the exhaust system on any late-model engine with electronic fuel management can affect EGR operation and can cause driveability problems with which the computer cannot cope.

Small late-model Ford 4-cylinder engines use a **bifurcated** stainless steel exhaust manifold, figure 7-28. The low-restriction, tuned tubular header design consists of four primary runners. These form into two secondary runners and converge into a single outlet that connects to the exhaust system. By routing pressure pulses from adjacent firing cylinders through different pipes, backpressure is reduced by approximately 30 percent.

On some late-model inline 6-cylinder engines, Ford uses two separate cast iron exhaust manifolds, one for the three front cylinders and the other for the three rear cylinders. This design solves two former problems: the single manifold previously used did not seal well against the head and also tended to fail prematurely due to cracking. Like the birfurcated manifold design, the two manifolds connect to a single outlet that leads into the exhaust system.

INTAKE MANIFOLD HEAT CONTROL

When the engine is cold, the incoming air-fuel mixture must be heated for complete vaporization. Hot exhaust gases are the most efficient heating source. A heat control valve, or heat riser, in the exhaust system routes a small part of the exhaust through passages in the intake manifold in order to provide the heat.

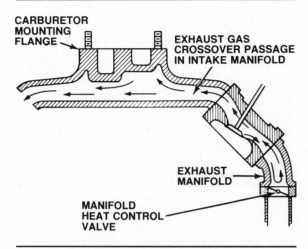

Figure 7-29. A thermostatic heat control valve or heat riser can be installed between the manifold and the exhaust pipe.

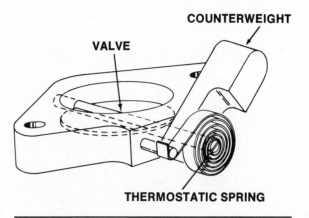

Figure 7-31. The manifold heat control valve can be operated by a thermostatic spring and a counterweight.

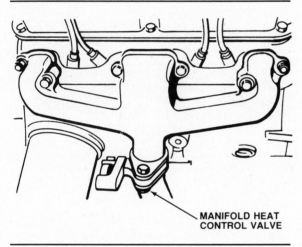

Figure 7-30. The heat control valve also can be within the manifold itself.

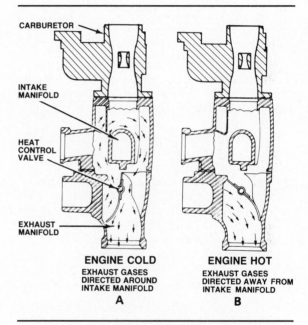

Figure 7-32. In this cross section of an inline engine manifold, the heat control valve forces the exhaust gases to flow either around the intake manifold (A), or directly out the exhaust system (B).

Heat control valves can be operated by:
- A thermostatic spring
- A vacuum diaphragm.

Each of these reacts to engine heat to control the valve operation, as you will see.

Thermostatic Heat Control Valve

Used by automakers for decades, the thermostatic heat control valve is located between the exhaust manifold and the exhaust pipe, figure 7-29, or in the manifold itself, figure 7-30. The valve is held closed by a thermostatic spring when the engine is cold, figure 7-31. This directs the hot exhaust gases around the intake manifold to preheat and help vaporize the air-fuel mixture, figure 7-32A. As the engine warms up, the thermostatic spring unwinds. The

Headers: Exhaust manifolds on high-performance engines that reduce backpressure by using larger passages with gentle curves.

Scavenging: A slight suction caused by a vacuum drop through a well designed header system. Scavenging helps pull exhaust gases out of an engine cylinder.

Bifurcated: Separated into two parts. A bifurcated exhaust manifold has four primary runners that converge into two secondary runners; these converge into a single outlet into the exhaust system.

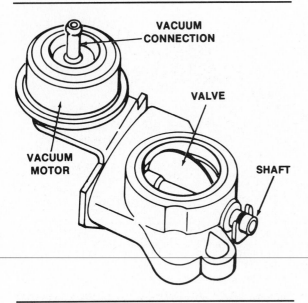

Figure 7-33. A vacuum-operated heat control valve.

counterweight and the pressure of exhaust gas open the valve. Exhaust gases now pass directly out through the exhaust system, figure 7-32B.

The thermostatic heat control valve has a tendency to stick, because the exhaust gases rust and corrode it. If the valve sticks open, it can cause increased fuel consumption, poor performance during warmup, and excessive emissions because the choke remains on. If the valve sticks closed, it can cause poor acceleration, a lack of power, and poor high-speed performance.

Vacuum-Operated Heat Control Valve

GM and Ford introduced vacuum-operated manifold heat control valves, figure 7-33, on some 1975 engines. Chrysler followed in 1977 with vacuum-operated valves on its California V-8 engines. AMC remained with the thermostatic valves on all of its engines until the early 1980's. GM calls its device a vacuum-servo early fuel evaporation (EFE) valve; Ford's is called a vacuum-operated heat control valve (HCV); Chrysler's is a power heat control valve. All work in the same way to provide more precise control of the manifold heat and to reduce emissions while improving driveability.

A rotating flapper valve is contained in a cast iron body, figure 7-34. This valve body is installed between the manifold and the exhaust pipe. The valve shaft extends through the valve body and is linked to a diaphragm in a vacuum motor. Intake manifold vacuum operates the

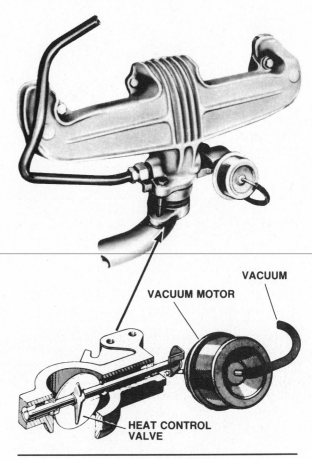

Figure 7-34. A vacuum-operated heat control valve installed in the exhaust manifold. (Chrysler)

vacuum diaphragm. The manifold vacuum source is controlled by a switch that reacts to either coolant temperature or to oil temperature. Ford, Chrysler, and some GM engines have a thermal vacuum switch (TVS) installed in the cooling system; GM's inline 6-cylinder engines of the late 1970's use an oil temperature switch.

When an engine is cold, vacuum is applied to the heat control valve diaphragm to close the valve. This directs exhaust gas through the intake manifold passage. As the engine reaches normal operating temperature, vacuum is shut off from the valve. In systems using a thermostatic vacuum switch, the TVS closes when coolant temperature reaches a specified level, figure 7-35. This blocks vacuum from the valve. Where an oil temperature switch is used, the rising oil temperature opens a normally closed thermostatic switch. This opened switch de-energizes the vacuum solenoid, and a spring in the diaphragm opens the valve. Exhaust now flows through the exhaust system instead of bypassing to the intake manifold.

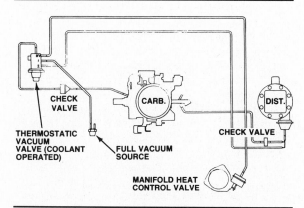

Figure 7-35. A vacuum diagram of the heat control valve installation. This valve responds to engine coolant temperature. (Buick)

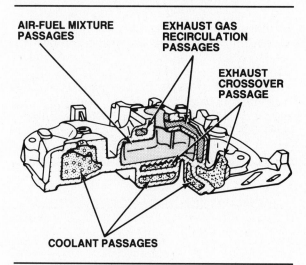

Figure 7-36. Hot engine coolant can also be used to pre-heat the air-fuel mixture. (Chrysler)

Coolant-Heated Intake Manifolds

Engine exhaust is not always used to heat intake manifolds. In some designs, hot engine coolant circulating through passages in the intake manifold preheats the air-fuel mixture, figure 7-36. A thermostat shuts off coolant flow to remove manifold heat when the engine is at normal operating temperature.

Electric Grid Heaters

A rubber insulator containing a ceramic heater grid is installed between the carburetor and the intake manifold of many late-model engines to improve cold engine driveability, figure 7-37. GM also calls this device an early fuel evaporation (EFE) system.

The ceramic heater grid is underneath the primary throttle of the carburetor. When engine

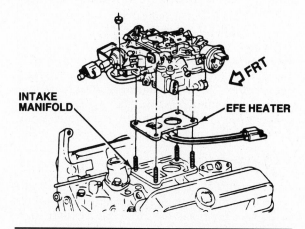

Figure 7-37. An electric heater grid can be installed between the carburetor and the intake manifold to improve cold engine driveability. (Buick)

■ What's Next In Emission Controls?

When the first edition of this book was published in 1978, it made the statement that "emission control devices and systems now in use have been refined about as much as possible". At that time, we considered the thermal reactor to be the next big advance. Obviously, that could not have been further from the truth. A great deal has taken place since that time, revolving mainly around the introduction and increasing sophistication of electronic engine control systems.

One large benefit of adding electronics to the automobile has been the elimination of emission control systems that were once considered permanent fixtures on the automotive scene. Spark timing controls are one example. The entire area of ignition timing, spark advance and retard has been given over to the computer, which does such a precise job that the old control systems have disappeared. A look under the hood of a late-model car shows it to be remarkably lean, compared to the plumbing that "graced" the engine compartment of cars current when this text was first published in 1978.

With fuel injection rapidly making the carburetor a quaint mixing device of the past, automakers have been able to introduce such precision in engine fuel management that heated intake air, air injection, and EGR systems have already disappeared from some engines and may be the next systems to go the way of spark timing controls. As we did nearly a decade ago, we may be going out on a limb, but it appears that the next trend in emission controls is simply fewer of them.

coolant is below a specified temperature, electrical current is sent to the grid by a TVS on noncomputer-controlled engines, or by the engine control computer through a relay on computer-controlled engines. The heater grid temperature is self-regulating at a calibrated value and remains on until the engine coolant reaches a specified temperature, at which time current to the grid is shut off.

SUMMARY

Good manifold design is critical for smooth performance and low emissions required of today's engines. The intake manifold must evenly distribute the air-fuel mixture to each cylinder, and the exhaust manifold must quickly remove the exhaust from the cylinders so that the next incoming air-fuel charge will not be contaminated. Runner length, the smoothness of the inside surfaces, and the arrangement of headers and exhaust pipes are important for the smooth flow of gases.

Multi-point or port fuel injection systems have special manifold requirements. Since this type of injection system does not deliver fuel through the manifold, each separate runner can

be "tuned" to increase efficiency and ensure better distribution. An intake air temperature control system is not used, but multi-point or port fuel injection manifolds must incorporate provisions for mounting the injectors and fuel rail assembly.

Intake manifold designs favor good low-to-middle speed operation, good high-speed operation, or a compromise of the two. A variety of variable air induction systems is beginning to appear on 4-cylinder engines, especially on those from Japan. While they differ in their approach, all are designed primarily to maintain torque across the engine's power band.

The type of fuel delivery system used and its relationship to the manifold has a definite bearing on manifold efficiency. Several methods are used to promote better vaporization in the manifold. Better vaporization can be accomplished through the use of a heat control valve which routes exhaust gases through the intake manifold while a cold engine warms up. Instead of exhaust gases, hot engine coolant may be routed through the manifold. Some late-model engines use an electric heater grid underneath the carburetor for this purpose.

Review Questions

Choose the single most correct answer.
Compare your answers with the correct answers on page 414.

1. The intake manifold is heated by:
   a. Exhaust manifold heat
   b. Engine coolant heat
   c. Both a and b
   d. Neither a nor b

2. Unequal distribution of the air-fuel mixture occurs because of:
   a. Fuel flow from the throttle
   b. Sharp turns in the manifold
   c. Distance of travel to the cylinders
   d. All of the above

3. The most efficient manifold design to use with a spread-bore carburetor is a:
   a. Single-plane
   b. Single-plane with siamesed intake ports
   c. Two-plane
   d. Single-plane with grooved plenum chamber floor

4. Exhaust backpressure:
   a. Causes power loss
   b. Can contaminate air-fuel mixture in the intake manifold
   c. Can be minimized by exhaust manifold streamlining
   d. All of the above

5. Headers:
   a. Are made of aluminum
   b. Cause engines to operate at higher temperatures
   c. Produce a slight suction called scavenging
   d. Increase horsepower by 30%

6. Manifold heat control valves:
   a. Can be located between the carburetor and the intake manifold
   b. Can be located between the exhaust manifold and exhaust pipe
   c. Are open when the engine is cold
   d. None of the above

7. GM's thermostatic vacuum switch reacts to:
   a. Oil temperature
   b. Coolant temperature
   c. Exhaust gases
   d. The vacuum solenoid

8. A ceramic heater grid is positioned under the carburetor:
   a. Primary throttle bore
   b. Secondary throttle bore
   c. Both primary and secondary
   d. Choke plate

9. A rotating flapper valve in the exhaust manifold or exhaust pipe is also called:
   a. A ceramic insulator
   b. A heat riser
   c. An air-fuel preheater
   d. A coolant heater

10. Variable-induction manifolds differ in design, but all have one common goal which is:
    a. To reduce production costs
    b. To reduce engine weight
    c. To maintain torque across the power band
    d. To help designs maintain a low hood profile

11. Heat control valves tend to stick because of:
    a. High temperatures
    b. Rust and corrosion
    c. Poor fuel economy
    d. Exhaust backpressure

PART THREE

Carburetor Systems

8

Basic Carburetion

So far in our study of a car's fuel system, we have discussed fuel tanks, lines, pumps, filters, and manifolds. This is enough to be able to dump raw gasoline into an engine. But putting a spark to this gasoline will not produce the combustion we need to create the power to move the car.

We need something more: we need to be able to change the fuel to a vapor by mixing it with air and then feeding it to the cylinders in precise air-to-fuel ratios. This is the job of the carburetor, figure 8-1.

There is a tremendous variation in carburetor designs, from the simple devices used on older cars to the wildly complex and expensive versions used on racing engines. Regardless of the design, however, all carburetors use the same basic principle: the difference in air pressure.

In this chapter, you will learn how these differences in air pressure apply to a carburetor, how carburetors operate under all types of driving conditions, and how assist devices are used to modify their operation. We will cover the similarities in carburetor designs and discuss how proper carburetor and assist device adjustments can improve driveability and lower the polluting emissions in the engine's exhaust.

Once you understand how the carburetor works and what the similarities are, you can then make carburetor adjustments properly and diagnose carburetor problems more accurately.

PRESSURE DIFFERENTIAL

Since air is a substance, the air outside an engine has a specific weight, and so does the air inside the engine. The weight of air exerts pressure on whatever it touches. The greater this weight, the greater the pressure. When the weight of air outside the engine is greater than the weight of air inside the engine, we say that there is a pressure difference, or differential, between the two.

Atmospheric Pressure

The weight of air is not always the same. It changes with temperature and pressure. For this reason, we must have a reference point when we talk about atmospheric pressure. At sea level under what we call *standard conditions* of 32°F (0°C) with a barometric reading of 760 mm Hg, one cubic foot of air (about 7.5 gallons or 28 liters) weighs about one and a quarter ounces (36 grams). This seems light enough, but remember that the earth's atmosphere is quite thick, figure 8-2. Therefore the column of

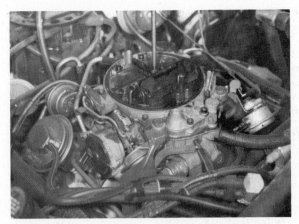

Figure 8-1. The modern carburetor is a complex device, but works on two simple principles: airflow and pressure differential.

air pressing down on an object at sea level is equal to about 14.7 psi (101 kPa).

Effect of temperature

Air expands and becomes lighter as its temperature rises. This reduces the pressure it exerts. As its temperature falls, air contracts. This makes it heavier and increases its pressure. Variations in air temperature account for changing weather conditions. Direct heat from the sun and reflected heat from the earth's surface warm the air. As its temperature increases, air becomes lighter and rises. Cooler air sinks and takes its place, resulting in a constant motion. This motion creates wind and weather patterns.

Effect of altitude

As you climb above sea level, the amount of air pressing down on you becomes smaller. Since a smaller amount of air weighs less, it exerts less

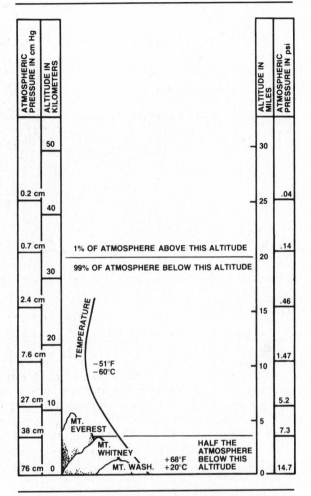

Figure 8-2. The blanket of air surrounding the earth extends for many miles. Atmospheric pressure decreases at higher altitudes.

■ What's A Bar?

Seems like a simple question. A bar is a metal rod, a drinking establishment, something that lawyers pass, or the Bureau of Automotive Repair (in California). A bar also is a unit of pressure measurement.

The term "bar" is short for barometric pressure, which means atmospheric pressure. One bar is one unit of atmospheric pressure. In the older kilogram-meter-second metric system, one standard bar is one kilogram of force applied to one square centimeter. This equals approximately 14.2 pounds per square inch, which is close to the 14.7 pounds per square inch of atmospheric pressure at sea level. Similar to 760 mm Hg of atmospheric pressure, a standard bar is measured at 0°C at sea level. At 14.2 psi, one bar is

close to the other units of atmospheric pressure that you have learned: 14.7 psi, 760 mm Hg, or 101.28 kPa.

You will find fuel pressure and turbocharger pressure specifications given in bar units in many European vehicle service manuals. Using the conversion factor of 14.2 psi = 1 bar, fuel injection pressure of 2.5 bars equals 35.5 psi. Turbocharger boost pressure of 0.7 bar equals 10 psi above atmospheric pressure. (Boost pressure is the pressure *increase* above atmospheric pressure.)

The bar is just another way to measure the common factors of air and fuel pressure in engine operation. It's easy to switch from one measurement unit to another if you remember the conversion factors and that all units generally represent the same quantities.

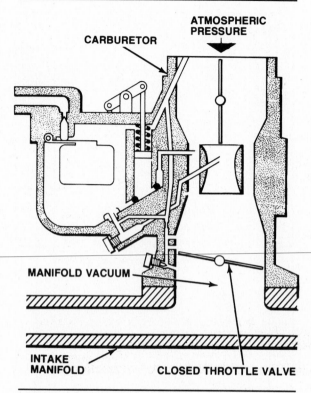

Figure 8-3. Manifold vacuum is the low pressure created in the intake manifold by the downward movement of the engine's pistons.

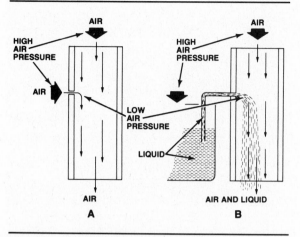

Figure 8-4. Airflow through a tube creates low pressure along the sides of the tube that can draw in more air (A). The low pressure can also draw liquid into the tube (B).

pressure. Air pressure gradually decreases with increased distance above sea level. At 30,000 feet or approximately 9 kilometers above sea level, air pressure is only about 5 psi (34 kPa). A few hundred miles or several hundred kilometers above the earth, the atmosphere ends and is replaced by a vacuum, or a complete lack of pressure.

Manifold Pressure — Vacuum

With each intake stroke of an engine piston in its cylinder, a partial vacuum is produced. As the piston moves down, it creates a larger space in which the air molecules can move. Since the molecules spread out to occupy this increased space, the distance between them increases. The greater the space between the air molecules, the greater the vacuum created.

As the piston moves farther down, it increases the vacuum and lowers the air pressure in the cylinder and intake manifold above it. This causes a **pressure differential** between the air inside and the air outside the engine. To offset this differential, outside air rushes into the engine. As it passes through the carburetor, it is mixed with gasoline to form an air-fuel mixture. This combustible vapor is then drawn by

vacuum through the intake manifold and the open intake valve into the cylinder, figure 8-3. Here it is compressed, burned, and exhausted.

AIRFLOW AND THE VENTURI PRINCIPLE

Opening the carburetor throttle valve causes air to move from the higher pressure area outside the engine, through the carburetor, and into the lower pressure area of the manifold. How much and how fast the air travels is determined by the opening of the throttle valve.

The pressure of air passing rapidly through a carburetor barrel is lower along the sides of the barrel than it is in the center of the airflow. By putting a small hole in the side of the barrel, more air can be drawn into the stream of air rushing through the barrel, figure 8-4A. If a hose is used to connect the hole to a liquid-filled container or bowl, then the liquid will be forced through the hose and into the stream of air rushing by, figure 8-4B.

This is caused by the higher air pressure on the liquid, which forces it to the lower air pressure area inside the barrel. How much liquid passes through the hose depends on the air-flow velocity, or how fast the air is flowing through the inside of the barrel. The higher the velocity (speed) of airflow, the lower the pressure will be at the inlet hole and the more liquid will flow.

If we wish to make the carburetor work better, we must increase the air velocity through the barrel. This can be done by placing a restriction called a **venturi** inside the barrel, figure 8-5. When air flows through the venturi restriction, it speeds up. This speed increase

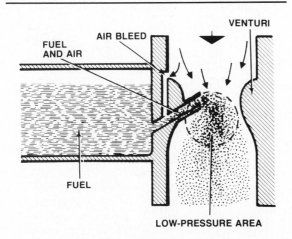

Figure 8-5. Air flowing through the venturi increases in speed. This lowers the pressure within the venturi to draw in more fuel.

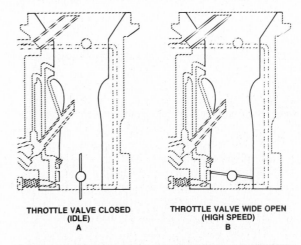

Figure 8-6. The carburetor throttle valve controls engine speed and power by regulating the amount of air and fuel entering the engine.

lowers the pressure inside the carburetor barrel and permits more liquid fuel to be drawn into the airflow.

In addition to mixing liquid fuel with air, the carburetor must also vaporize the liquid as much as possible. To help break up the liquid fuel for better vaporization, a small opening called an air bleed is put in the fuel inlet passage, figure 8-6.

The carburetor also must change the air-fuel mixture automatically. It must deliver a rich mixture for starting, idle and acceleration, and a lean mixture for part-throttle operation. Engine speed and power are regulated by the position of the carburetor throttle valve, which controls the flow of the air-fuel mixture, figure 8-6.

Carburetor Vacuum

There are four measurements of air pressure, or vacuum, that are important when discussing carburetors:

- Atmospheric pressure
- Manifold vacuum
- Venturi vacuum
- Ported vacuum.

Atmospheric pressure is the pressure of the air outside the carburetor. It is always present, and varies within a narrow range, depending upon altitude and atmospheric conditions.

Manifold vacuum is the low pressure beneath the carburetor throttle valve. **Manifold vacuum** is produced by the engine and is always present when the engine is running. Manifold vacuum decreases as the throttle valve is opened.

Venturi vacuum is the low-pressure area created by airflow through the venturi restriction in the carburetor barrel. **Venturi vacuum** increases with the speed of the airflow through the venturi. It is present whenever the throttle valve is open and increases as the throttle is opened.

Ported vacuum is the low-pressure area just above the throttle valve, as you learned in Chapter 4. **Ported vacuum** is present whenever the throttle is opened to expose the port in the lower portion of the carburetor barrel to manifold vacuum. Ported vacuum is absent at idle, high at small throttle openings, and decreases as the throttle is opened farther. Vacuum taken from this point often is used to operate distributor vacuum advance units and other vacuum-operated devices. Small ports, or holes, in the side of the carburetor are connected to hoses, which are connected to the vacuum devices.

Pressure Differential: A difference in pressure between two points.

Venturi: A restriction in an airflow, such as in a carburetor, that speeds the airflow and creates a vacuum.

Manifold Vacuum: Low pressure in an engine's intake manifold, located below the carburetor throttle.

Venturi Vacuum: Low pressure in the venturi of a carburetor, caused by fast airflow through the venturi.

Ported Vacuum: The low-pressure area just above the throttle in a carburetor.

CARBURETION OPERATING PRINCIPLES

All carburetors must perform three vital functions. They must break up the liquid gasoline into a fine mist, change the liquid into a vapor, and distribute the vapor evenly to the cylinders. These three principles of atomization, vaporization, and distribution of fuel are important principles of carburetion.

Gasoline must be atomized, or broken up, into a fine mist if the fuel is to be properly vaporized. Atomization takes place as the fuel travels from the carburetor discharge nozzles into the moving stream of air. You learned about these principles in Chapter 3.

Vaporization starts as the atomized fuel passes the throttle and enters the intake manifold. Complete vaporization cannot occur unless the fuel is hot enough to boil. Vaporization is affected by the following factors:

- Temperature — vaporization increases as the fuel is heated
- Volatility — the greater the volatility of the fuel, the lower the temperature at which it will vaporize and the faster it will vaporize
- Pressure — a decrease in pressure causes fuel to vaporize faster at a lower temperature.

Low volatility, cold intake air, or a cold manifold can cause poor vaporization. As you learned in Chapter 6, thermostatic air cleaners and heated intake manifolds are ways in which the problems of a cold air-fuel mixture can be overcome. Since manifold vacuum creates a low-pressure area, fuel vaporizes more efficiently in the intake manifold. As we discussed in Chapter 7, a poorly designed manifold will result in poor vaporization.

The throttle plate has a direct effect on distribution, since the angle of the throttle sends the mixture against one side of the intake manifold. This tends to feed some cylinders a rich mixture and other cylinders a lean mixture. Cylinders farther away from the carburetor may get less of the mixture than those nearest the carburetor. Engineers must consider the fuel distribution or metering requirements of an engine when they design carburetors and intake manifolds.

Carburetion Air-fuel Ratio Requirements

A carburetor must serve the varying air-fuel ratio needs of an engine for different operating conditions.

During starting, an engine has low intake manifold vacuum and airflow velocity because the engine is turning slowly. Slow cranking speed and a cold engine combine to reduce fuel vaporization. With reduced vaporization, less gasoline reaches the combustion chamber, and the engine needs a richer air-fuel mixture for starting.

At idle, an engine also needs a rich air-fuel mixture. Manifold vacuum is high, but airflow velocity is low. The combined effects of these factors reduce vaporization. Also, some exhaust remains in the cylinders at idle, which dilutes the air-fuel mixture.

As a vehicle accelerates gradually at low speed, engine speed and airflow increase, while vacuum rises in the carburetor. The engine gradually needs a leaner air-fuel mixture for smooth acceleration and economy with low emissions.

At steady cruising speeds with a light engine load, the engine needs a relatively lean air-fuel ratio of 15 or 16 to 1. At cruising speed, the engine operates at a relatively constant speed and load, with steady (relatively high) vacuum and airflow.

For extra power requirements, such as sudden acceleration, hill climbing, or full-throttle operation at any speed, the engine needs a richer air-fuel mixture. Air-fuel ratios of 12.5 to 13.5 allow an engine to develop maximum power for these conditions. Low vacuum accompanies full-power operation, and a carburetor must provide extra fuel with low airflow velocity for acceleration and low-speed, heavy-load conditions. Low vacuum and high airflow accompany high-speed, full-power operation, and a carburetor must respond to these needs as well. Air-fuel ratios for full power do not vary much except at low speed with low airflow velocity. Between the lean ratios of 15 or 16 to 1 and the rich ratios of 12.5 or 13.5 to 1, a modern carburetor must maintain the stoichiometric ratio of 14.7 for the best combination of power, economy, and emission control. These requirements are maintained on late-model vehicles by electronic controls and closed-loop feedback fuel systems.

All of these variable requirements for carburetor operation may seem overly complex at first sight. As you study the basic carburetor systems in the following sections, however, you will see how the systems react to engine conditions and meet all of the variable operating needs. Also, if you understand the engine requirements that a carburetor fulfills with its basic systems, you will understand the corresponding operations of fuel injection systems that are explained in Part 4 of this *Classroom Manual*.

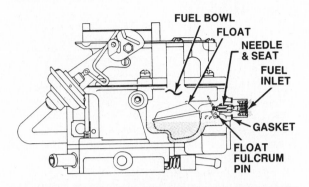

Figure 8-7. Fuel level in the fuel bowl is controlled by the float and needle valve acting against fuel pump pressure.

BASIC CARBURETOR SYSTEMS

To mix fuel and regulate engine speed, the carburetor has a series of fixed and variable passages, jets, ports, and pumps which make up the fuel metering systems of circuits. There are seven basic systems common to all carburetors:

- Float system
- Idle system
- Low-speed system
- High-speed (main metering) system
- Power system
- Accelerator pump system
- Choke system.

Some engineers and technicians think of the idle and low-speed systems as two halves of a single system because the same carburetor passages are used for both. Whether you think of them as one system or two, the important points to understand are the operations that provide a smooth transition from idle to main-metering fuel flow.

Float System

Gasoline from the fuel tank is delivered by the fuel pump to the carburetor fuel bowl, where it is stored for use. Once in the fuel bowl, the gasoline must be kept at a precise, nearly constant level. This level is critical, since it determines the fuel level in all the other passages and circuits within the carburetor. A fuel level that is too high in the bowl will produce an air-fuel mixture that is too rich. A fuel level that is too low will produce an overly lean mixture. For this reason, fuel level is one of the most critical adjustments required by the carburetor.

The main fuel discharge nozzle for the high-speed system is connected directly to the bottom of the fuel bowl. Because liquids seek their own level in any container, the fuel level in the bowl and in the nozzle is the same. If the level

is too high, too much fuel will be drawn into the high-speed system. If the fuel level is too low, too little fuel will be drawn in.

Fuel level is controlled by the float and the inlet needle valve, figure 8-7. As gasoline is drawn from the bowl, the float lowers in the remaining fuel. Fuel pump pressure then opens the needle valve and allows more fuel to enter the bowl. As the fuel level rises, so does the float, until it forces the inlet needle back against its seat. This closes the inlet valve and shuts off both fuel pump flow and pressure to the carburetor bowl. During many operating conditions, fuel flow into and out of the fuel bowl is about equal. The needle stays in a partly open position to maintain the required flow rate.

The float and needle valve regulate fuel flow, as well as fuel level. Since the needle valve is like a door between the carburetor fuel bowl and the fuel pump, it maintains an air space above the fuel in the bowl. This reduces pressure on the fuel to atmospheric pressure. Atmospheric pressure is maintained in the fuel

■ Wick And Surface Carburetors

The float-type carburetor has been standard on automobile engines for more than 80 years. Anyone could be excused for thinking that it is the *only* kind of carburetor ever used for automotive applications. A variety of techniques were used, however, to transmit an explosive air-fuel mixture for early internal combustion engines.

One of the first carburetors was a wick-type device in which a fabric was suspended in the fuel tank with one end submerged in the fuel. Air passing through the saturated fabric vaporized the fuel and carried it into the engine. A control valve on the engine intake regulated the amount of air as a means of leaning or enriching the air-fuel mixture.

The surface carburetor also depended on vapors. Air could either be passed through a fuel tank under pressure so that it bubbled to create vapors, or the fuel could be agitated for the same effect. A control valve on the intake similar to that used with a wick carburetor regulated the mixture ratio. Benzine, naphtha, and ether were commonly used for fuel with both wick and surface carburetors.

Karl Benz adapted the surface carburetor for automotive use in the late 1880's. His design bubbled air through benzine, using a float and needle valve to maintain the fuel level. Exhaust gas routed though a pipe on the bottom of the carburetor increased vaporization. An extractor prevented liquid fuel from entering the engine.

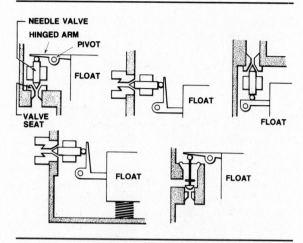

Figure 8-8. Float and needle valve designs vary with different carburetors.

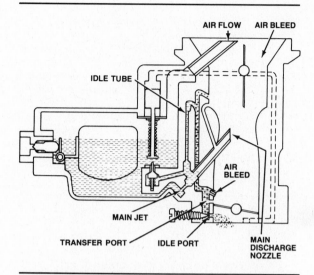

Figure 8-10. Air and fuel for the idle system are mixed inside the carburetor passages and delivered to the idle port below the throttle.

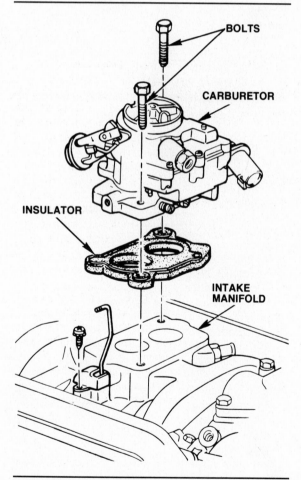

Figure 8-9. This insulator between the carburetor and the intake manifold reduces heat that causes fuel evaporation in the fuel bowl.

bowl by a vent or balance tube venting the bowl to the carburetor airhorn. Atmospheric pressure pushing down on the fuel in the bowl provides the pressure differential needed for precise fuel metering into the venturi vacuum area of the carburetor barrel. If the float and the needle valve do not maintain the correct fuel level in the bowl and too much fuel enters, the carburetor will flood.

Float and needle valve design and location in the fuel bowl vary with different carburetor designs, figure 8-8. Some floats have small springs to prevent them from bobbing up and down when the vehicle travels over rough roads. Many fuel bowls have baffles, which keep the fuel from sloshing on rough roads and sharp turns. The needle valves and their seats in older carburetors were usually made of stainless steel. The steel often attracted metallic particles in the fuel. These particles would collect between the needle and seat, allowing the valve to leak. The needles and seats in most modern carburetors are made of brass, and the needles have tips made of Viton or other plastics that conform to any rough spots on the seat and still provide a good seal when the valve is closed.

When the engine is shut off, engine heat causes the fuel in the bowl to evaporate. This was no problem in pre-emission control days, but with the installation of vapor canister systems, the amount of evaporation from a large fuel bowl can easily overload the canister. Therefore, emission carburetors use a somewhat smaller float bowl. Some carburetors, such as the Carter Thermoquad, use a molded

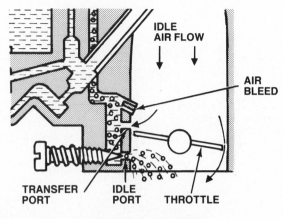

A. IDLE OPERATION

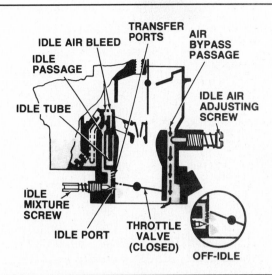

Figure 8-12. Air for the idle circuit in this carburetor passes through a bypass passage and is controlled by an idle air adjusting screw.

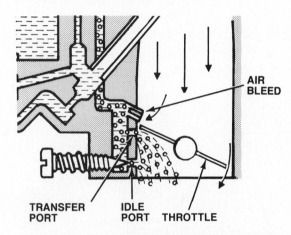

B. LOW SPEED (OFF-IDLE) OPERATION

Figure 8-11. At idle, air flows in through the transfer port to mix with the idle air-fuel mixture. As the throttle opens, flow reverses through the transfer port. Fuel and air now flow out for low-speed operation.

plastic float bowl to reduce heat evaporation because plastic does not conduct heat as well as metal. Others use an insulator, figure 8-9, between the intake manifold and the carburetor to reduce heat.

Idle System

When an engine is idling, the throttle is open only slightly and airflow through the carburetor barrel and venturi is reduced. Since there is little or no venturi effect, no fuel flows from the main discharge nozzle. The idle system, figure 8-10, supplies enough air and fuel to keep the engine running under these conditions.

Intake manifold vacuum is high at idle, so idle ports are located just below the closed throttle. The pressure differential between the

fuel bowl and the vacuum at the idle ports forces fuel through the ports. Gasoline flows from the bowl, through the main jet, to the idle tube. Because the fuel must be well mixed with air for proper distribution, air bleeds in the idle tube let in air for the idle mixture. The air bleeds also prevent fuel **siphoning** at high speeds or when the engine is stopped.

Extra air for the idle air-fuel mixture can be provided in various ways. In many carburetors, the throttle valve does not close completely, but remains slightly open to let in a small amount of air, figure 8-11. A few designs draw air for the idle circuit through a separate air passage in the carburetor body called the idle air bypass, figure 8-12.

Additional small openings called transfer ports, figure 8-11, are located just above the closed throttle valve in the carburetor barrel. At idle, the transfer ports suck air from the barrel into the fuel flow in the idle system. A small amount of air and fuel is released just below the throttle valve. When the engine is under slight acceleration, the throttle valve opens a little and exposes the transfer port to manifold vacuum. This draws the fuel out into the barrel to mix with the air. We will discuss this more completely under the low-speed system.

Siphoning: The flowing of a liquid as a result of a pressure differential, without the aid of a mechanical pump.

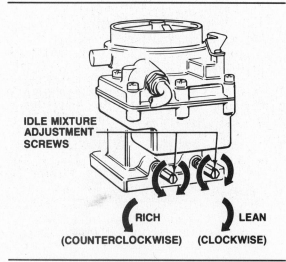

Figure 8-13. Idle mixture screws control the amount of gasoline in the air-fuel mixture.

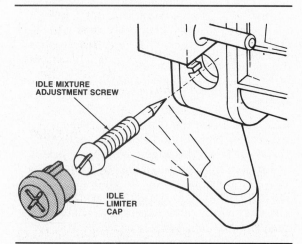

Figure 8-14. Idle limiter caps restrict the amount of adjustment allowed for the idle mixture.

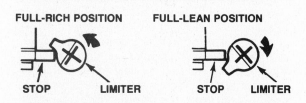

Figure 8-15. Limiter caps allow an adjustment of approximately one turn.

Adjustable needle valves called idle mixture screws, figure 8-13, control the amount of gasoline used in the idle air-fuel mixture. One adjustment screw generally is used for each primary barrel. The screw tips stick out into the idle system passages and are turned inward (clockwise) to create a lean mixture, or outward (counterclockwise) to richen the mixture.

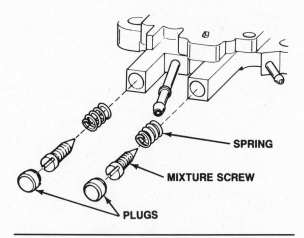

Figure 8-16. The idle mixture screws on late-model carburetors are adjusted and then sealed with plugs or caps. (Carter)

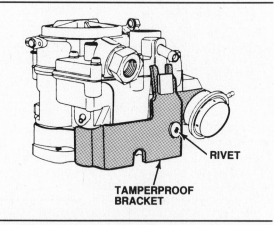

Figure 8-17. Tamper-resistant brackets are used on some carburetors to prevent unauthorized adjustment. (AC-Delco)

The idle mixture screws on carburetors used before emission controls could be adjusted from fully closed to fully open. Most carburetors built from the late 1960's to the late 1970's use plastic limiter caps, figure 8-14, on the idle mixture screws. These caps restrict the amount of adjustment to about one turn in or out, figure 8-15. This prevents excessively rich idle mixtures that contain large amounts of HC and CO. Limiter caps must be replaced whenever a carburetor is overhauled.

Regulations set by the Environmental Protection Agency (EPA) in 1979 required manufacturers to make carburetors tamperproof. To do this, the idle air-fuel ratio of most carburetors made since 1979 is calibrated at the factory, and the mixture screws are covered with caps or plugs, figure 8-16. Other carburetors use a tamper-resistant bracket riveted in place, figure 8-17. The concealment caps, plugs, or brackets

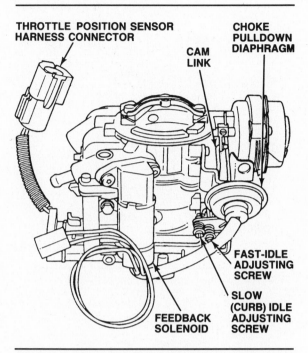

Figure 8-18. The slow (curb) idle screw regulates idle airflow and engine speed by changing the throttle position. (Ford)

Figure 8-19. The idle air bypass speed adjustment screw controls the airflow through a bypass passage to regulate idle speed.

prevent any changes in the air-fuel ratio that would affect idle emission control.

Engine idle speed is adjusted by changing the amount of air going to the idle system. This idle speed adjustment is usually done with a screw that changes the position of the throttle valve in the carburetor, figure 8-18. Carburetors which use idle air bypass passages are adjusted by a large screw that varies the opening in the air passage to change the airflow, figure 8-19. Many late-model feedback carburetors used with electronic engine control systems have an electric motor that controls idle speed and airflow as directed by the system microprocessor.

Low-Speed System

Once the throttle valve begins to open for low-speed operation, the engine needs more fuel than the idle port alone can provide. The airflow passing through the venturi is still not strong enough to develop fuel flow through the main discharge nozzle. To provide more fuel, the transfer port comes into operation as the low-speed system, figure 8-11.

The transfer port is located above the throttle at idle, and the air pressure there is about equal to atmospheric pressure. Air from the barrel flows *into* the transfer port to mix with the fuel going to the idle port. As the throttle opens, the transfer port is exposed to intake vacuum and the flow reverses. Extra fuel flows *out* of

the transfer port to meet the engine's needs during the change from idle to low-speed operation. Fuel continues to flow from the idle port, but at a reduced rate. This permits an almost constant air-fuel mixture during this transition period. Some people consider the idle and the low-speed systems to be a single system because they use the same passages in the carburetor.

High-Speed (Main Metering) System

When the throttle valve opens wider, airflow increases through the carburetor. At the same time, the partial vacuum (low-pressure) area of the intake manifold moves up in the carburetor

■ Funny-Looking Gaskets Aren't Funny

It's easy to overlook gaskets. They are small, not too expensive, and they don't seem very important.

Don't you believe it. Gaskets that don't fit right will not work, and a bad gasket can rob your engine of a lot of power. Here is a tip from the professionals: be careful of how you store gaskets. Store them flat, never standing on edge. Don't hang them on a nail; that would most certainly pull them into funny shapes. Don't store gaskets near heat, since that also could cause them to warp into some interesting shapes. All non-metal gaskets have some water content, so they can't be allowed to dry up. If you do have a dry one, soak it in warm water for just a few minutes. Too much water will, again, cause it to warp.

Best thing to do is store gaskets flat in a protective wrapper. Leave them in a drawer or cabinet where they will not be disturbed, and won't have other parts set on top of them.

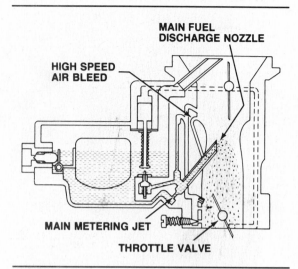

Figure 8-20. Fuel for the high-speed or main metering system flows through the main jet and out the fuel discharge nozzle in the venturi.

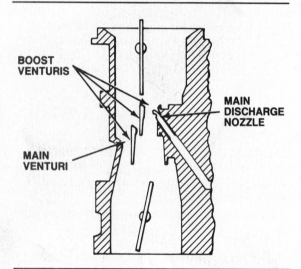

Figure 8-21. Most carburetors have multiple (boost) venturis for better air and fuel mixing.

barrel. This airflow and pressure change strengthens the venturi. This is the high-speed or main metering system, figure 8-20.

For better mixing of the fuel and air, most carburetors have multiple, or boost, venturis placed one inside another, figure 8-21. The main discharge nozzle is located in the smallest venturi to increase the partial vacuum effect on the nozzle. Fuel flows from the bowl through the main jet and main passage into the discharge nozzle. A high-speed air bleed, figure 8-20, mixes air into the fuel before it is discharged from the nozzle.

The primary or upper venturi produces vacuum, which causes the main discharge noz-

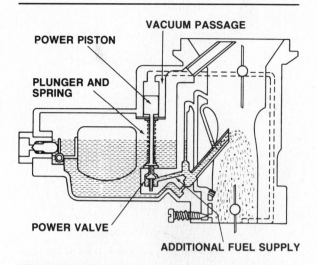

Figure 8-22. This power valve is operated by a vacuum-controlled piston and plunger. When vacuum decreases, the spring moves the plunger to open the power valve.

zle to spray fuel. The secondary venturi creates an air stream which holds the fuel away from the barrel walls where it would slow down and condense. The result is air turbulence, which causes better mixing and finer atomization of the fuel.

As the throttle continues to open wider, fuel flow from the low-speed system tapers off, while flow from the high-speed system increases. The engine's fuel needs are now supplied entirely by the main discharge nozzle during high-speed, light-load cruising.

Power System

The main high-speed system delivers the leanest air-fuel mixture of all the carburetor systems. When engine load increases during high-speed operation, the air-fuel mixture is too lean to deliver the power required by the engine. The extra fuel needed is provided instead by another system called the power system, or power valve. It supplements main metering fuel delivery. The power system or valve can be operated by vacuum or by mechanical linkage. The exact type differs according to carburetor design, but all provide a richer air-fuel mixture.

One type of power valve, figure 8-22, is located in the bottom of the fuel bowl with an opening to the main discharge tube. A spring holds a small poppet valve closed, while a vacuum piston holds a plunger above the valve. Since manifold vacuum decreases as the engine load increases, a large spring moves the plunger downward. This opens the valve and lets more fuel pass to the main discharge nozzle.

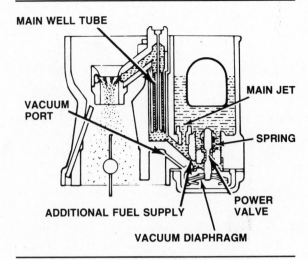

Figure 8-23. The vacuum diaphragm holds the power valve closed. When vacuum decreases, a spring opens the valve to allow more fuel into the main passage.

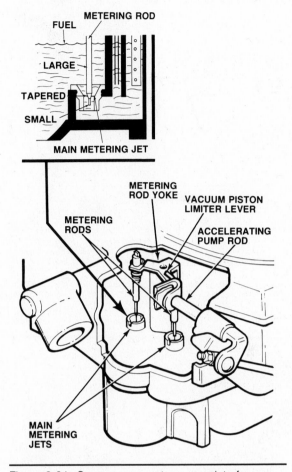

Figure 8-24. Some power systems consist of metering rods placed in the main jets. Mechanical or vacuum linkage moves the rods upward to allow more fuel to flow through the jets when required.

Another type of vacuum-operated power valve uses a diaphragm, figure 8-23. Manifold vacuum against the diaphragm holds the valve closed. As vacuum decreases under an increased load, a spring opens the valve. This sends more fuel through the power system and main discharge nozzle.

Metering rods also can be used as a power system, figure 8-24. These may be controlled by vacuum pistons and springs, or by mechanical linkage connected to the throttle. The ends of the rods installed in the main jet opening are tapered or stepped to increase the extra fuel gradually. The rods restrict the area of the main jets and reduce the amount of fuel that flows through them during light-load operation of the main metering system. Extra fuel for full-throttle power is provided by moving the rods out of the jets to increase the flow through the jets.

Vacuum-controlled metering rods, also called stepup rods, are held in the jets by manifold vacuum applied to pistons attached to the rods. When vacuum drops under heavy load, springs working against the pistons move the rods out of the jets. Mechanically operated metering rods are controlled directly by mechanical linkage connected to the throttle linkage.

Feedback carburetors used with electronic engine control systems generally do not have a separate power system. A mixture control (MC) solenoid operates the metering rods or air bleeds. When a richer mixture is required for additional power, the engine control microprocessor drives the solenoid to its full-rich position.

Accelerator Pump System

This system provides additional fuel for certain engine operating conditions. If the throttle is opened suddenly from a closed, or nearly closed, position, airflow increases faster than fuel flow from the main discharge nozzle. This "dumping" of air into the intake manifold reduces manifold vacuum suddenly and causes a lean air-fuel mixture. This excessively lean mixture results in a brief hesitation or stumble. This is sometimes called a **flat spot**. To keep the mixture rich enough, extra fuel must be provided by the accelerator pump.

Flat Spot: The brief hesitation or stumble of an engine caused by a momentary overly lean air-fuel mixture due to the sudden opening of the throttle.

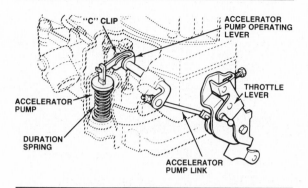

Figure 8-25. Typical plunger-type accelerator pump. (Ford)

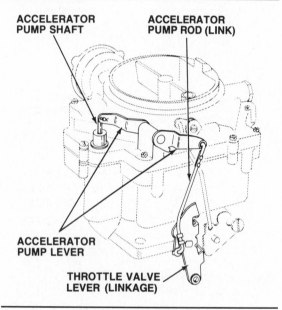

Figure 8-26. The accelerator pump linkage (lever and rod) is connected to the throttle linkage.

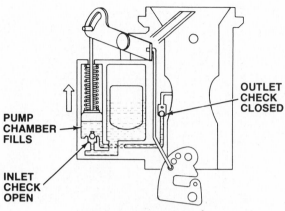

A. PUMP INTAKE STROKE

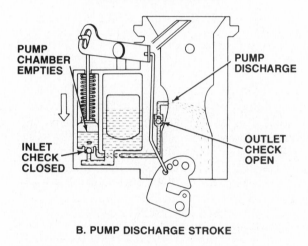

B. PUMP DISCHARGE STROKE

Figure 8-27. Accelerator pump operation.

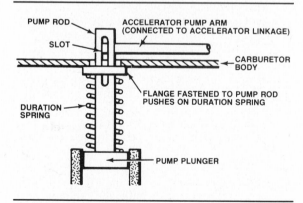

Figure 8-28. The duration spring provides uniform pump delivery regardless of the speed at which the throttle linkage moves.

The accelerator pump, figure 8-25, is a plunger or diaphragm in a separate chamber in the carburetor body. It is operated by a linkage connected to the carburetor throttle linkage, figure 8-26. When the throttle closes, the pump draws fuel into the chamber. An inlet check valve opens to allow fuel into the chamber, figure 8-27A, and an outlet check valve closes so that air will not be drawn through the pump nozzle. When the throttle is opened quickly, the pump moves down or inward to deliver fuel to the nozzle in the barrel, figure 8-27B. The pump outlet check opens, and the inlet check closes. The inlet check ball is usually (but not always) larger than the outlet check ball.

The pump outlet check may be a steel ball or a plunger. The inlet check may be a steel ball, a rubber diaphragm, or part of the pump plunger.

Not all pumps have inlet checks. Some rely on an inlet slot in the pump well, or chamber, that is closed by the plunger on the downward stroke.

Most pump plungers or diaphragms are operated by a duration spring, figure 8-28. The

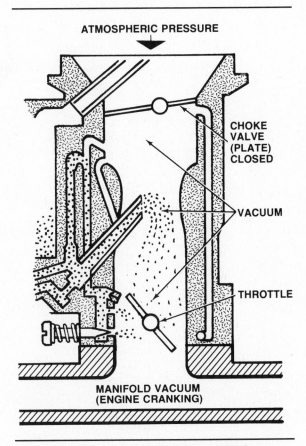

Figure 8-29. Vacuum is present throughout the carburetor barrel below the closed choke. This draws fuel from the idle, low-speed, and high-speed circuits for starting the engine.

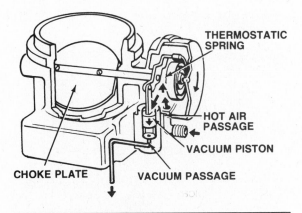

Figure 8-30. An older version of an integral or cap-type choke. The thermostatic bimetal spring is in a housing on the carburetor airhorn. A vacuum piston opens the choke when the engine starts.

Choke System

The choke provides a very rich mixture for starting a cold engine. This extra-rich mixture is needed because:

- Engine cranking speed is slow
- Airflow speed is slow
- Cold manifold walls cause gasoline to condense from the air-fuel mixture, and less vaporized fuel reaches the combustion chambers.

To make the mixture richer, a choke plate or a butterfly valve is positioned above the venturi in the carburetor barrel. This choke plate can be tilted at various angles to restrict the passage of air, figure 8-29. Cranking the engine with the choke plate closed creates a partial vacuum throughout the carburetor barrel below the plate. This airflow reduction and partial vacuum area work together to allow more fuel to be drawn into the mixture.

The choke plate can be controlled manually by a cable running to the driver's compartment, or automatically by a thermostatic coil spring. Chokes on most domestic carburetors since the early 1950's have been operated by a bimetal thermostatic coil spring. The choke plate shaft is connected to the spring by linkage. The bimetal spring normally is located in one of two places:

1. In a round housing on the carburetor airhorn, figure 8-30. This is called an integral choke.

throttle linkage holds the pump in the returned position. When the throttle opens, the linkage releases the pump and the spring moves the plunger for a steady and uniform fuel delivery. The accelerator pump operates during the first half of the throttle travel from the closed to the wide-open position.

During high-speed operation, the vacuum at the pump nozzle in the carburetor barrel may be strong enough to unseat the outlet check and siphon fuel from the pump. This is called pump pullover or siphoning. In a few carburetors, this extra fuel is included in the high-speed system adjustment. In most carburetors, air bleeds are placed in the pump discharge passages to prevent the siphoning. In still other carburetors, an extra weight is added to the outlet check to resist siphoning. The pump plungers in some carburetors have anti-siphon check valves.

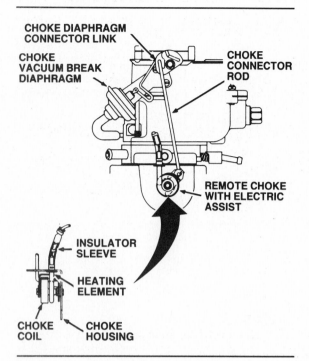

Figure 8-31. In a remote, or well-type choke, the thermostatic bimetal spring is in a heated well on the intake manifold. (Carter)

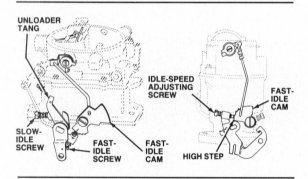

Figure 8-32. The fast-idle cam opens the throttle wider for faster engine speed when the choke is operating. It may work either on the slow-idle speed adjusting screw or on a separate fast-idle screw.

2. Off the carburetor in a well on the intake manifold, figure 8-31. This is called a remote, or well-type, choke.

Regardless of type and location, the thermostatic coil spring forces the choke closed when the engine is cold. Running the engine heats up the spring, which then opens the choke. With an integral choke, hot air from a source near the exhaust manifold, or hot coolant from the cooling system, may be routed to the choke housing to heat the spring. Remote chokes normally are heated by the exhaust routed through a crossover passage in the intake manifold.

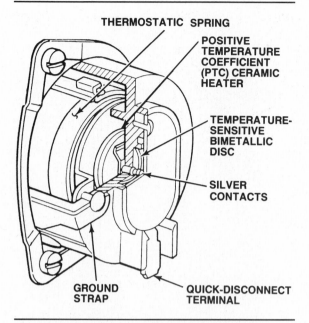

Figure 8-33. An electric choke cap contains a ceramic heater and bimetallic disc to heat the thermostatic spring and release the choke as fast as possible. (Ford)

Most late-model integral and remote chokes have electric heating elements to heat the spring faster and speed the choke opening.

When a cold engine is cranked, the choke must be completely closed. As soon as the engine starts, the choke must open slightly to provide enough airflow. This is done in two ways. First, the choke plate shaft is offset in the carburetor so that airflow will tend to open the plate. Second, manifold vacuum is applied to a vacuum piston or a vacuum-break diaphragm that pulls the choke open a few degrees.

Integral chokes once had a vacuum piston built into the choke housing, figure 8-30, but this design had a tendency to stick. The emission-control requirement for precise choke operation resulted in nearly universal use of the vacuum-break diaphragm on emission carburetors. Most chokes have the vacuum-break diaphragm mounted on the side of the carburetor, figure 8-31.

A cold engine must idle faster than a warm engine, or the lack of air and fuel flow will cause it to stall. A fast-idle cam and screw, figure 8-32, provide enough air and fuel to prevent engine stalling. The cam is linked to the choke plate, and the screw is located on the throttle valve shaft. Depressing the accelerator pedal to start a cold engine allows the choke to close. This moves the fast-idle cam to allow the screw to rest against a high step of the cam.

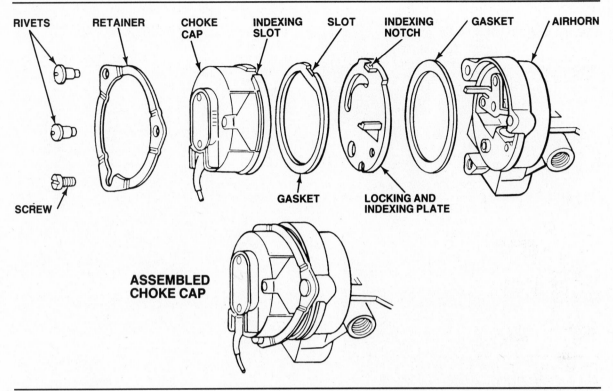

Figure 8-34. Tamperproof carburetor design requires a sealed choke housing to prevent unauthorized adjustment. (Ford)

The cam may contact the normal slow-idle adjusting screw, or a separate fast-idle screw.

In both cases, the throttle is held open slightly more than for a normal slow idle, and idle speed increases between 400 and 800 rpm. As the engine gradually warms up, choke spring tension decreases and a weight pulls the fast-idle cam back, returning engine speed to idle rpm. On engines with computer control systems, a motor or a solenoid controlled by the system microprocessor often regulates fast-idle speed.

A mechanical link or choke unloader, figure 8-32, opens the choke about halfway when the throttle is fully open. If the engine is accidentally flooded during starting, this provides the extra airflow necessary to clear out the fuel.

Because the choke system provides a very rich mixture, it increases HC and CO emissions. To meet emission control standards, late-model engines must get off the choke as soon as possible. This is done in various ways, but the most common method used is the electric choke cap, figure 8-33. This contains a small heating element connected to the alternator, and a temperature-sensing switch. The switch lets the heating element warm the thermostatic coil spring and open the choke as quickly as

possible. Depending upon their design, some carburetors may have 2- or 3-stage choke heaters to deliver specific amounts of heat according to ambient temperature.

The spring tension on older chokes can be adjusted to control the amount of choke closing and the rate at which it opens. Late-model choke housings, however, are sealed with breakaway screws, rivets, or brackets to prevent unauthorized choke adjustment that might change emissions, figure 8-34.

CARBURETOR TYPES

The operation of the basic carburetor systems has been explained in terms of a carburetor which uses a single barrel and throttle valve. But carburetors also are made with two or more barrels. Various carburetor types are used to match fuel flow to engine requirements. Domestic engines all use downdraft carburetors; older imports often used sidedraft carburetor designs. Carburetors are usually classified by the number of barrels or venturis used. The differences are detailed below.

One-Barrel

The 1-barrel carburetor has a single outlet through which all systems feed to the intake

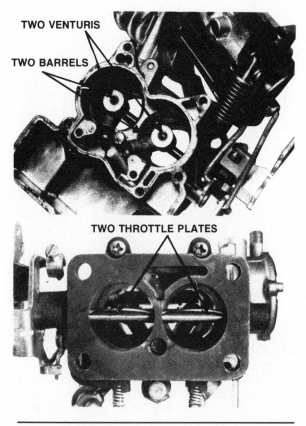

Figure 8-35. A two-barrel carburetor uses one airhorn but contains two venturis and throttle plates.

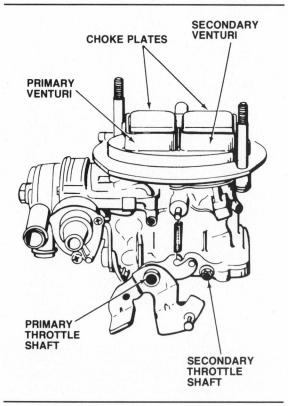

Figure 8-36. The 2-stage, 2-barrel carburetor has a primary and a secondary throttle, operating independently.

manifold. This type of carburetor may also be known as a single-venturi design. These carburetors flow 150 to 300 cubic feet per minute (cfm) and generally are used on 4-cylinder and smaller 6-cylinder engines.

Single-Stage Two-Barrel

This carburetor design contains two barrels and two throttles which operate together, figure 8-35. Since the various fuel discharge passages in each barrel operate at the same time, it can be considered as two 1-barrel carburetors sharing the same body. The two throttle plates are mounted on the same shaft and operate together. The two barrels share a common float, choke, power system, and accelerator pump. Single-stage 2-barrel carburetors are used on many 6- and 8-cylinder engines and generally have an airflow capacity of 200 to 550 cfm.

Two-Stage Two-Barrel

This carburetor design, figure 8-36, resulted from emission control requirements. It differs from the single-stage, 2-barrel design in that its

two throttles operate independently. The primary barrel generally is smaller than the secondary, and handles engine requirements at low-to-moderate speeds and loads. The larger secondary opens when needed to handle heavier load requirements.

The primary stage usually includes the idle, accelerator pump, low-speed, main metering, and power systems. The secondary stage usually has a transfer, main metering, and power system. Both stages draw fuel from the same fuel bowl. Some designs use a common choke for both barrels. In others, only the primary stage is choked.

The 2-stage 2-barrel carburetor has an airflow capacity of 150 to 300 cfm and is used primarily on 4-cylinder and smaller 6-cylinder engines.

Four-Barrel

Used primarily on V-8 engines, the 4-barrel, or quad, carburetor contains two primary and two secondary barrels in a single body, figure 8-37. The two primaries operate like a 1-stage 2-barrel at low-to-moderate engine speeds and loads. The secondary barrels open at about half to three-quarters throttle to provide the increased fuel and airflow required for high-

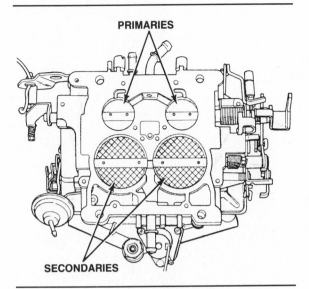

Figure 8-37. Most 4-barrel carburetors have primary and secondary systems that open progressively. (Chrysler)

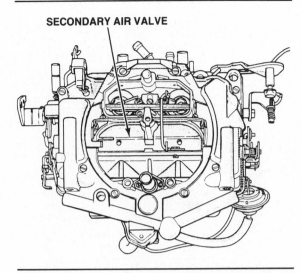

Figure 8-39. Air valves are used to control secondary airflow in many Rochester and Carter carburetors. (Chrysler)

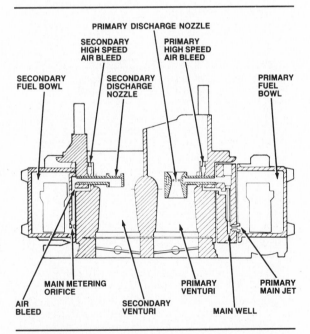

Figure 8-38. Venturi action controls airflow and fuel discharge through the secondary barrels of some 4-barrel carburetors. (Ford)

speed operation. The primary barrels contain the idle, low-speed, and high-speed systems, as well as an accelerator pump and a power system. The secondary barrels have their own high-speed and power systems, and may use their own accelerator system. Some 4-barrel carburetors use separate fuel bowls and fuel supplies for the primary and secondary barrels;

others work with a single fuel bowl and fuel supply for all four barrels.

Two methods are used to provide airflow through the secondary barrels: venturi action, figure 8-38, or air velocity valves, figure 8-39. Air velocity valves look like large choke plates located in the secondary barrels. They are opened by the low pressure created in the secondary barrels when the throttles are opened. Older Rochester and Carter 4-barrel carburetors may have auxiliary air velocity valves inside the secondary barrels. Airflow through the barrels opens the velocity valves; counterweights hold them closed when the throttles are closed. Late-model Rochester Quadrajet carburetors use vacuum diaphragms to modulate air valve movement. Venturis and air valves also may be combined in one carburetor to modulate the airflow through the barrels.

■ **Check For A Cold Carburetor**

Cold outside air can cause carburetors to freeze up, especially when low temperature combines with high humidity. This carburetor icing can, in turn, cause the choke valve to stick or bind in the carburetor. If you are servicing a car whose owner has complained about poor engine performance during cold weather, always be sure to check for carburetor icing. If it's your car, you can throw a blanket over the engine and carburetor when it is left to sit for a long time in cold weather.

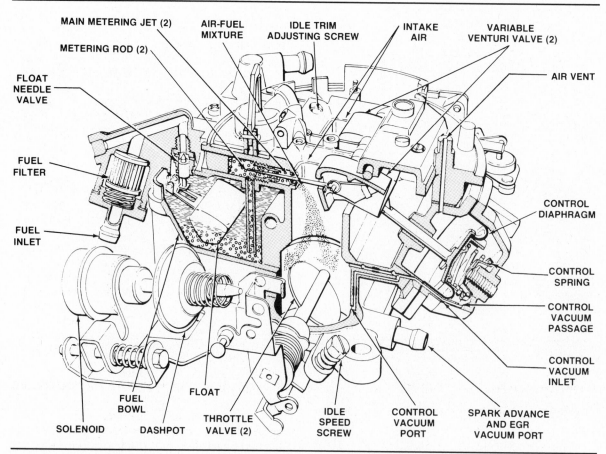

Figure 8-40. The Motorcraft 2700 VV variable-venturi carburetor uses vacuum to control the movement of the venturi valves. (Ford)

The primary barrels supply all eight cylinders during low-to-moderate speeds and loads. The secondary barrels provide additional fuel and airflow for high speeds and heavy loads. The 4-barrel carburetor flows from 400 to more than 900 cfm.

VARIABLE-VENTURI (CONSTANT DEPRESSION) CARBURETORS

As we have seen, carburetors meter fuel by using a venturi to create a partial vacuum in the barrel. Airflow through the venturi increases velocity, which decreases pressure. This pressure drop causes fuel to flow through the discharge nozzle into the barrel. Since both the carburetor barrel and venturi are fixed in size, the volume and velocity of air passing through will be correct for some operating conditions but not correct for others. To produce better performance under *all* operating conditions, auxiliary circuits such as the choke, power, and idle systems must be added to the main metering system.

A carburetor with a venturi whose size changes according to the demands of engine speed and load does not need these extra systems. At the same time, air-fuel mixtures can be controlled more closely for better fuel economy and emission control. Changing the size of the venturi relative to engine speed and load results in an even pressure drop across the venturi under all operating conditions. This gives a variable-venturi carburetor its other name: a "constant depression" carburetor.

Variable-venturi carburetors, such as those manufactured by SU, Solex, Hitachi, and Stromberg, were used on some imported cars for many years. In 1977, Ford Motor Company introduced the Motorcraft 2700 VV, the first variable-venturi carburetor used on domestic cars in 45 years. (Ford's first V-8 engines in 1932 had variable venturi, 1-barrel carburetors.) This was followed by an electronically-controlled feedback version of the same design, the 7200 VV.

The 2700 VV, figure 8-40, Ford's variable-venturi, 2-barrel carburetor, has a fuel inlet sys-

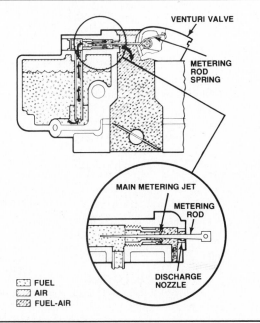

FUEL
AIR
FUEL-AIR

Figure 8-41. The Motorcraft 2700 VV main metering system. (Ford)

tem with a replaceable filter in the inlet housing. Throttle plates and an accelerator pump also are used. However, the variable venturis and the different fuel metering systems make this carburetor unique.

The variable venturis are formed by two rectangular valve plates (actually a single casting) that slide back and forth across the tops of the two barrels. Movement is controlled by a spring-loaded vacuum diaphragm, regulated by a vacuum signal taken below the venturis (but above the throttle plates) in the carburetor barrels. As the throttle opens, the vacuum increases, opening the venturis and allowing more air to enter.

The front edge of each venturi valve has a tapered metering rod. Each rod moves in and out of a fixed main jet on the other side of the barrel, figure 8-41. This arrangement meters fuel in proportion to the airflow through the venturis. Because of the variable venturis, this type of carburetor has fewer fuel metering systems than the fixed venturi design. However,

■ Ford's First Variable Venturi Carburetor

Ford's first variable venturi carburetor was a one-barrel model made by Detroit Lubricator. It was used on the original flathead, Model 18 V-8, in 1932 and early 1933. The variable venturi was formed by two air vanes in the barrel that responded directly to airflow, rather than to vacuum linkage control as in the current Motorcraft 2700VV carburetor.

The air vanes were linked to a movable main jet (metering valve) that slid up and down on a fixed metering rod (pin) in the center of the venturi. The metering rod was adjustable to obtain the proper idle mixture and main system fuel flow.

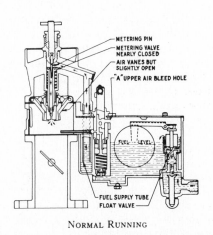

NORMAL RUNNING

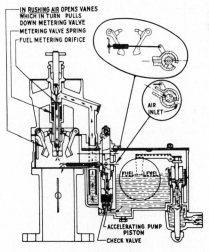

MAXIMUM POWER

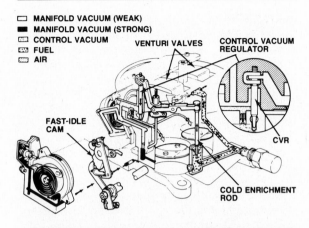

- ▭ MANIFOLD VACUUM (WEAK)
- ▪ MANIFOLD VACUUM (STRONG)
- ▨ CONTROL VACUUM
- ▨ FUEL
- ▨ AIR

VENTURI VALVES

CONTROL VACUUM REGULATOR

FAST-IDLE CAM

CVR

COLD ENRICHMENT ROD

Figure 8-42. The Motorcraft 2700 VV cold enrichment system operation with a cold engine. (Ford)

the main metering system of the 2700 VV carburetor cannot handle all operating conditions without help from secondary systems. For example, at full throttle under heavy load, vacuum may not be strong enough to override the diaphragm spring. In this case, a limiter lever on the throttle shaft pushes the venturi valves fully open. Other auxiliary systems in this carburetor are the accelerator pump, idle trim, cranking enrichment, and cold enrichment systems.

The carburetor's design is innovative, but it requires extremely precise adjustments. A major redesign of some systems took place in 1980, with continuing refinements made through the mid-1980's, when the 2700 VV was discontinued and use of the 7200 VV was restricted to police vehicles.

Cold Enrichment System

The 2700 VV carburetor does not have a traditional choke. Instead, a bimetal thermostatic choke control using an electric-assist heater manages the cold enrichment system, figure 8-42. This system contains a fast-idle cam, a cold enrichment auxiliary fuel passage and metering rod, and a control vacuum regulator rod.

A unique feature of the fast-idle cam on 1977-79 2700 VV models is a vacuum-operated high-cam-speed position (HCSP). When a cold engine is started, a lever slides between the fast-idle cam and the fast-idle lever to provide more throttle opening, figure 8-43A. When the engine starts, vacuum applied to the HCSP diaphragm retracts the lever, figure 8-43B.

When the engine is started below 95°F (35°C), the choke spring pushes the control vacuum regulator rod to block the ported control vacuum and send manifold vacuum to the venturi valve diaphragm. This opens the venturis wider than normal. The fast-idle cam is touched by the HCSP for a wider throttle opening. Redesign of the cranking enrichment system eliminated this system on 1980 and later carburetors.

Cranking Enrichment System

The cranking enrichment system also provides extra fuel only for starting a cold engine. On 1977-79 models, figure 8-44, it uses an electric solenoid energized by the ignition switch to open an auxiliary fuel passage. When the engine starts, the solenoid is deenergized and closes the cranking fuel passage. When this happens, the main metering and cold enrichment systems maintain the fuel flow.

Redesign of the system on 1980 and later models eliminated the solenoid. The cranking enrichment system was changed to provide fuel enrichment both during cranking and cold engine running. This was done with a new linkage which provided an additional stroke for the cranking enrichment rod. The increase in rod travel supplies the additional enrichment fuel that was formerly delivered by the solenoid.

Idle Trim System

On early models, the idle fuel flow is controlled by the main jets, but an additional small amount of fuel is drawn by manifold vacuum through internal passages to discharge ports below the throttles. This is called the idle trim system, figure 8-45, and uses adjustable metering screws. The system was redesigned to eliminate the adjustable metering screws on 1978 carburetors.

CARBURETOR LINKAGE

Vehicle speed is controlled by the accelerator pedal which moves rods, cables, levers, and spring to operate the carburetor throttle valve. The throttle linkage on some vehicles is a combination of solid rods, levers, and links, figure 8-46. Other vehicles use a cable to link the accelerator to the carburetor throttle, figure 8-47.

Other kinds of linkage operate the accelerator pump, the automatic choke, the fast-idle cam, secondary throttle valves, and the automatic transmission downshift points on some cars.

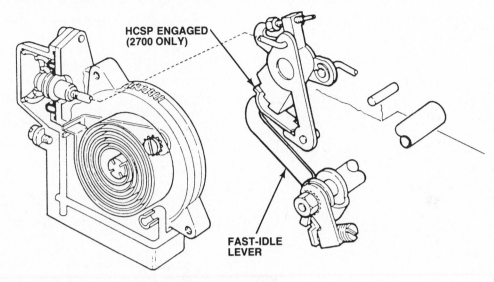

HCSP ENGAGED
(2700 ONLY)

FAST-IDLE
LEVER

A. ENGINE OFF

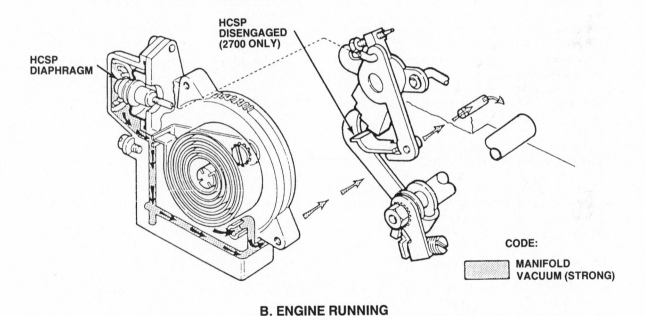

HCSP
DIAPHRAGM

HCSP
DISENGAGED
(2700 ONLY)

CODE:

MANIFOLD
VACUUM (STRONG)

B. ENGINE RUNNING

Figure 8-43. Operation of the HCSP system on 1977-79 Motorcraft 2700 VV carburetors. (Ford)

Accelerator Pump Linkage

The accelerator pump piston or plunger is connected to the throttle by a small rod, or rods, and levers. On some carburetors, the linkage holds the pump in the retracted position against a compressed duration spring. When the throttle opens all the way, the linkage releases the spring and the duration spring moves the pump piston through its stroke.

On some carburetors, the accelerator pump linkage also opens a vent for the fuel bowl when the throttle is closed. On other carburetors, the accelerator pump linkage operates a diaphragm to deliver fuel to the pump system. The accelerator pumps on yet other carburetors are operated directly by mechanical linkage. Pump operation may be balanced by a duration spring and a return spring.

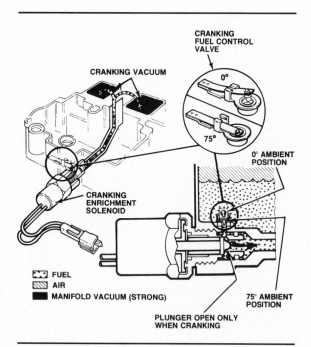

Figure 8-44. The cranking enrichment system used on 1977-79 Motorcraft 2700 VV carburetors. (Ford)

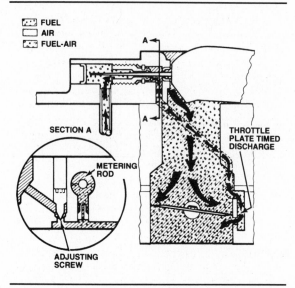

Figure 8-45. The Motorcraft 2700 VV idle trim system. (Ford)

Accelerator pump and linkage designs vary from carburetor to carburetor, but all work on these same principles. On most carburetors, the accelerator pump ends its stroke at the half-throttle position. From this point to the full-throttle position, the high-speed and the power systems can supply enough fuel. The pump linkage on some carburetors can be installed in two or three positions on the carburetor throttle linkage to provide different pump strokes.

Figure 8-46. This throttle linkage is a combination of solid rods and levers.

Figure 8-47. A cable from the accelerator to the carburetor is used for this throttle linkage.

Automatic Choke Linkage

Automatic chokes vary in design, but all require connecting linkage between the thermostatic spring that closes the valve and the vacuum piston or diaphragm that opens it. A remote or well-type choke with a vacuum-break diaphragm is shown in figure 8-48. The rod linkage from the spring to the choke valve closes the choke. The vacuum-break diaphragm opens the choke through its linkage as soon as the engine starts.

With most integral, piston-type chokes, the thermostatic spring acts directly on a lever on the end of the choke shaft inside the choke housing, figure 8-32. The vacuum piston also acts directly on this lever to open the choke when the engine starts. Some carburetors are built with an integral choke housing mounted away from the choke valve. This arrangement requires an external linkage rod from the thermostatic spring to the choke valve, as well as a separate vacuum-break diaphragm.

Fast-Idle Cam Linkage

The fast-idle cam is linked to the choke valve, figure 8-48, to provide a faster than normal idle and prevent stalling when the engine is cold. A fast-idle operating lever is attached to the choke shaft and connected to the fast-idle cam by a link. When the choke closes, the linkage turns the fast-idle cam so that a high step of the cam touches the idle speed adjusting screw or a separate fast-idle screw. As the choke opens, the

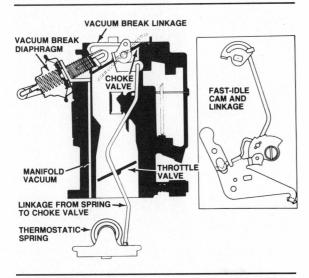

Figure 8-48. Several pieces of linkage are used on this choke system. (Chevrolet)

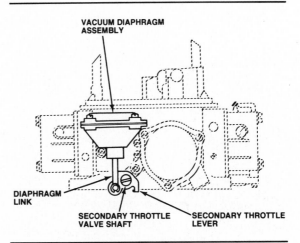

Figure 8-50. Vacuum-operated secondary throttle are controlled by a vacuum diaphragm.

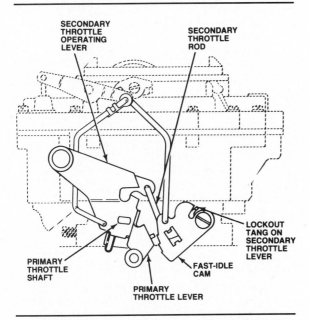

Figure 8-49. Mechanical secondary throttle linkage on a 4-barrel carburetor.

fast-idle cam continues to follow the choke movement and reduces idle speed step by step. When the choke is fully open, gravity keeps the fast-idle cam away from the idle speed screw. However, when the engine is shut off, the idle speed screw blocks the cam and keeps it from turning back to the fast-idle position as long as the throttle is closed. This also keeps the choke thermostatic spring from closing the choke valve as the engine cools. To begin the choking operation with a cold engine, the accelerator

must be depressed to release the fast-idle cam and linkage and allow the choke to close.

Secondary Throttle Linkage

The secondary throttles of 4-barrel and 2-stage, 2-barrel carburetors may be operated by mechanical linkage or by vacuum. Secondary throttle valves operated by mechanical linkage have an operating rod to connect the primary throttle shaft to the secondary throttle shaft, figure 8-49. Secondary throttles begin to open when the primary throttles are about half open. The secondaries continue to open along with the primaries, but at a faster rate. The primary and secondary throttles then reach the fully open position at the same time.

On some carburetors, the primary and secondary throttle shafts are mechanically linked, but the secondary throttles are not visible through the carburetor airhorn. This type of carburetor has secondary air velocity valves, called auxiliary throttle valves, figure 8-39. These valves have offset shafts and counterweights so that they remain closed until air velocity through the carburetor barrels is strong enough to open them. These auxiliary throttle valves operate only when the mechanical secondary throttles are open, but they do not rely on the mechanical movement of the secondary throttles.

Vacuum-operated secondary throttle valves are controlled by a vacuum diaphragm mounted on the side of the carburetor, figure 8-50. At low cruising speeds, the secondary throttle valves are closed and the engine's air-fuel requirements are met by the primary half of the

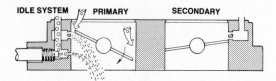

A. THROTTLE PLATES AT IDLE

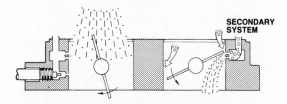

**B. SECONDARY PROGRESSION
AS THROTTLE IS OPENED**

Figure 8-51. Secondary throttle valves are closed during low cruising speeds but open progressively as more fuel and air are required at higher speeds. (Ford)

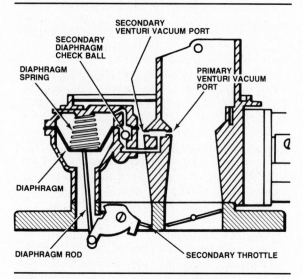

Figure 8-52. The secondary diaphragm responds to vacuum from ports within the carburetor barrels.

carburetor, figure 8-51A. At higher speeds, when more fuel and air are needed, the secondary throttles are opened by linkage connected to the vacuum diaphragm, figure 8-51B.

The vacuum diaphragm responds to increasing vacuum within the primary venturi as engine airflow increases, figure 8-52. Linkage from the diaphragm opens the secondary throttles. Diaphragm action is changed by another vacuum port, or air bleed, in the secondary barrels.

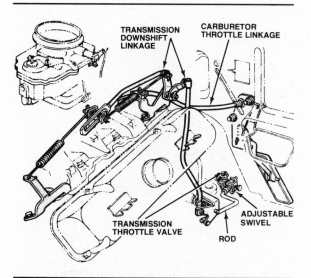

Figure 8-53. Chrysler products with automatic transmissions have mechanical linkage between the carburetor and transmission throttle valve to control shift points. (Chrysler).

The amount and rate at which the secondary throttles open are determined by the vacuum signal at the diaphragm. When the secondaries are closed, the air bleed in the secondary barrels weakens the vacuum signal at the diaphragm a set amount. As the secondaries open, the air bleed becomes a vacuum port as vacuum develops within the secondary barrels. This vacuum signal is then added to the vacuum from the port within the primary venturis to open the secondary throttles completely.

Sudden secondary throttle opening is prevented by a ball check valve in the vacuum chamber passage which allows a gradual vacuum buildup. As engine speed decreases, the weaker vacuum signal allows the diaphragm spring to close the secondary throttles.

All secondary throttle linkage, whether mechanically or vacuum operated, includes a secondary throttle or air valve lockout device to keep the secondary throttle from opening when the choke is closed, figure 8-49.

Transmission Linkage

Some vehicles with an automatic transmission or transaxle may use an adjustable throttle rod, figure 8-53, or a cable linkage, figure 8-54, between the transmission and the carburetor or throttle body. This controls shift points and shift quality. Other automatic transmissions do this with vacuum control.

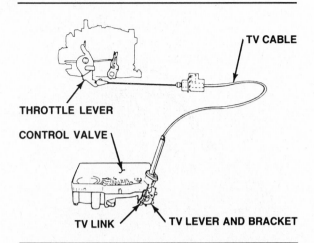

Figure 8-54. Some GM vehicles control automatic transmission shifting by a cable between the carburetor and the transmission throttle valve (TV). (Buick)

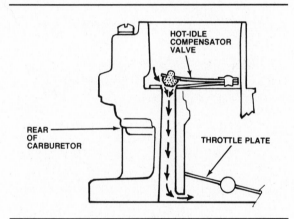

Figure 8-55. The hot-idle compensator is a thermostatic valve that opens at high temperature to admit more air to the idle circuit.

CARBURETOR CIRCUIT VARIATIONS AND ASSIST DEVICES

Variations in the basic carburetor systems we just discussed are used by all automakers. One or more add-on devices also may be used to improve economy, driveability, and emission control. Those most commonly used are covered in the following sections.

Hot-Idle Compensator Valves

High carburetor inlet air temperature causes gasoline to vaporize rapidly, which can cause an overly rich idle mixture. To prevent this, many carburetors use a hot-idle compensator

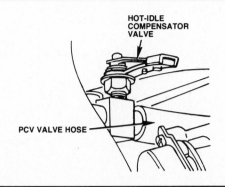

Figure 8-56. This hot-idle compensator is located in the PCV valve hose away from the carburetor.

valve, figure 8-55. The compensator is a thermostatic valve consisting of a bimetal spring, bracket, and small poppet. The compensator valve usually is located either in the carburetor barrel or in a chamber on the rear of the carburetor bowl. A dust cover is placed over the chamber. A third location (used primarily in Autolite 2-barrel carburetors on air-conditioned Ford vehicles of the late 1960's) is an external mounting in the PCV valve hose near the carburetor, figure 8-56.

The hot-idle compensator valve normally is closed by spring tension and engine vacuum. As temperature rises, the bimetal strip bends. This uncovers an auxiliary air passage or air bleed through which air enters the carburetor below the throttle plates. As this extra air mixes with excess fuel to lean out the idle mixture, it prevents stalling and rough idling. Once the carburetor temperature returns to normal, the compensator valve closes to shut off the extra air. If the valve does not close fully, it causes a high idle speed with high CO emissions.

Idle Enrichment Valves

Emissions carburetors run on leaner mixtures. For good cold-engine operation, the idle mixture must be enriched in some cases. Some 1975 and later carburetors used by Chrysler have an idle enrichment system. This system works opposite to a hot-idle compensator valve.

A small vacuum diaphragm mounted near the carburetor top, figure 8-57, controls idle circuit air. When control vacuum is applied, the diaphragm reduces idle system air. This increases fuel and reduces the air in the air-fuel mixture. Diaphragm vacuum is controlled by a temperature switch in the radiator. As the engine warms, this switch stops the vacuum signal, returning the air-fuel mixture to its normal lean level.

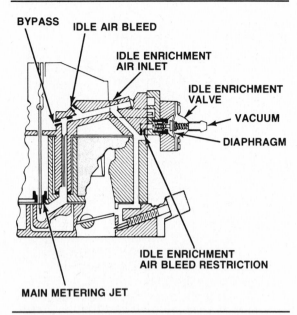

BYPASS IDLE AIR BLEED

IDLE ENRICHMENT
AIR INLET

IDLE ENRICHMENT
VALVE

VACUUM

DIAPHRAGM

IDLE ENRICHMENT
AIR BLEED RESTRICTION

MAIN METERING JET

Figure 8-57. The Chrysler idle enrichment valve.

Fast-Idle Pulloff (Choke Pulloff)

The rich air-fuel mixtures resulting from long periods of choke and fast idle can damage catalytic converters. Some converter-equipped GM cars use a fast-idle pulloff to avoid converter overheating. In one system, manifold vacuum acts on a vacuum-break diaphragm at the rear of the carburetor. This diaphragm will drop the fast idle cam to a lower step 35 seconds after engine coolant temperature reaches 70°F (21°C). Vacuum to the diaphragm is controlled by a vacuum solenoid operated by a coolant temperature switch and delay timer.

Another system uses the front vacuum diaphragm to pull the throttle down one step on the fast-idle cam as soon as engine coolant temperature reaches 150°F (66°C). Vacuum to the diaphragm is controlled by a temperature vacuum switch.

A third method uses an electric solenoid instead of a vacuum diaphragm to open the choke. This pulls the fast-idle screw off the cam whenever the engine is started with the coolant below a specified temperature. A temperature switch on the engine and a firewall-mounted relay provide current for the pulloff solenoid.

Regardless of the system used, fast-idle pulloff has no effect on engine warmup during ordinary operation because normal throttle movement will disengage the fast-idle cam. The fast-idle pulloff system only operates when the engine is warming up while parked.

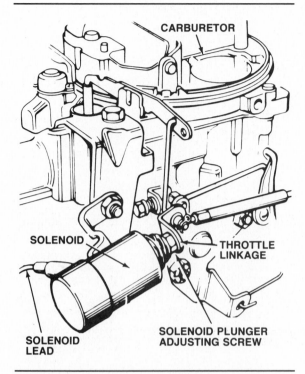

CARBURETOR

SOLENOID

THROTTLE
LINKAGE

SOLENOID
LEAD

SOLENOID PLUNGER
ADJUSTING SCREW

Figure 8-58. The throttle stop solenoid holds the throttle open for normal slow idle and allows the throttle to close farther when the engine is shut off.

Throttle Stop Solenoids

Engine **dieseling**, or after-run, results when combustion chamber temperatures remain hot enough to ignite an idle air-fuel mixture after the ignition is turned off. Dieseling is caused by several aspects of late-model engines:

- Higher operating temperatures
- Faster idle speeds
- Retarded ignition timing at idle
- Lean air-fuel mixtures.

Closing the throttle more than it would close for the engine's normal slow idle speed will prevent dieseling. Shutting off airflow past the throttle valve closes the idle circuit.

A throttle stop solenoid, figure 8-58, provides the new stop position for the throttle during normal slow idle. Turning on the ignition energizes the solenoid, and its plunger moves out to contact the idle speed adjusting screw or a bracket on the throttle shaft. This holds the throttle open slightly for a normal slow idle until the ignition is shut off. The solenoid is then deenergized, its plunger retracts, and the throttle closes to block airflow.

Since a variety of different throttle solenoids or positioners have been used over the years,

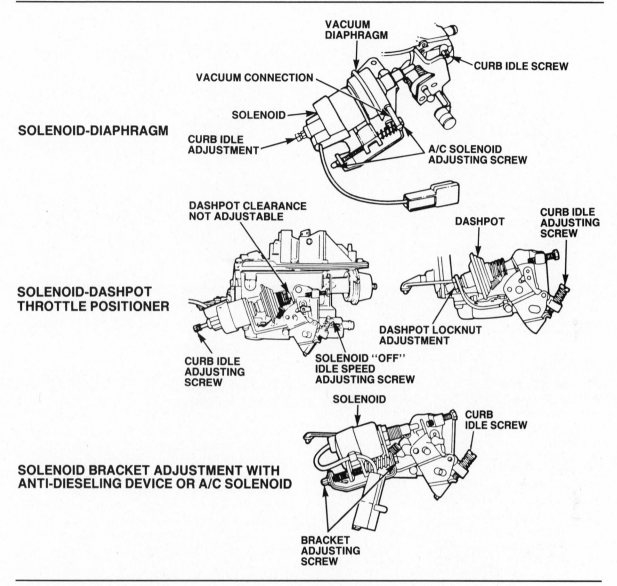

VACUUM
DIAPHRAGM

CURB IDLE SCREW

VACUUM CONNECTION

SOLENOID

SOLENOID-DIAPHRAGM

CURB IDLE
ADJUSTMENT

A/C SOLENOID
ADJUSTING SCREW

DASHPOT CLEARANCE
NOT ADJUSTABLE

DASHPOT

CURB IDLE
ADJUSTING
SCREW

SOLENOID-DASHPOT
THROTTLE POSITIONER

DASHPOT LOCKNUT
ADJUSTMENT

CURB IDLE
ADJUSTING
SCREW

SOLENOID "OFF"
IDLE SPEED
ADJUSTING SCREW

SOLENOID

CURB
IDLE SCREW

SOLENOID BRACKET ADJUSTMENT WITH
ANTI-DIESELING DEVICE OR A/C SOLENOID

BRACKET
ADJUSTING
SCREW

Figure 8-59. Some of the large number of throttle solenoids or positioners used by Ford. (Ford)

figure 8-59, you should check the manufacturer's adjustment procedures before attempting to service any throttle solenoid.

Air Conditioning Throttle Solenoids

Many late-model vehicles with air conditioning may use a solenoid that looks exactly like a throttle solenoid, figure 8-60. In many cases, it is the same solenoid and may even carry the same part number. However, it should not be confused with the throttle stop solenoid just discussed, since it performs an entirely different function.

This solenoid is energized *only* through the air conditioner switch. Its plunger moves forward to contact a bracket on the throttle shaft

only when the air conditioning is turned on. This maintains or slightly raises engine idle speed to prevent the engine from stalling due to the increased load. It also helps to prevent overheating from the air conditioning condenser heat load by speeding up the radiator fan.

Dieseling: A condition in which extreme heat in an engine's combustion chamber continues to ignite fuel after the ignition has been turned off.

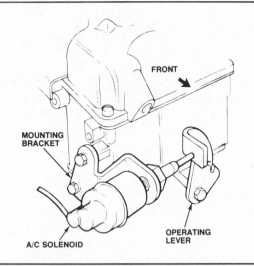

Figure 8-60. An air conditioning throttle solenoid opens the throttle slightly when the air conditioner is on. This maintains a uniform idle speed, even with the increased engine load.

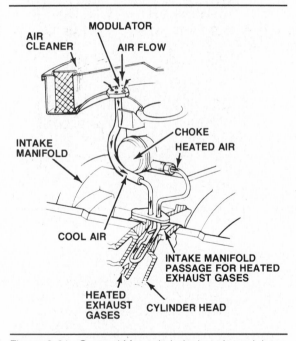

Figure 8-61. General Motors' choke hot air modulator.

Choke Hot Air Modulators

Some GM engines of the mid-1970's used a choke hot air modulator check valve (CHAM-CV) in the air cleaner, figure 8-61. The valve is closed at air cleaner temperatures below 68°F (20°C). Air that is to be heated by the heater coil passes through a tiny hole in the modulator. This restricts hot airflow over the bimetal thermostatic coil and results in a slower choke

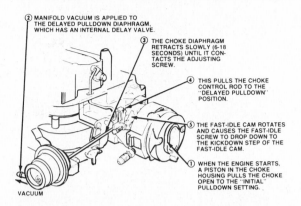

Figure 8-62. The delayed choke pulldown diaphragm provides rich initial choking, rapid choke release, and fast-idle modulation. (Ford)

warmup. When air cleaner temperature rises above 68°F (20°C), the modulator opens to permit more airflow for a faster choke warmup.

Staged Choke Pulldowns

Found only on 1972 Ford engines, this system has a small double-chamber housing connected to the choke linkage at one end and to manifold vacuum at the other. A divider with an orifice separates the two chambers inside the housing. The chamber facing the carburetor contains a spring and silicone fluid; the other chamber holds a diaphragm and bimetal valve. The valve controls the application of manifold vacuum to the diaphragm. The orifice controls the time interval of choke opening and varies between 15 to 54 seconds, depending upon engine model.

Delayed Choke Pulldowns

A delayed choke pulldown operated by a vacuum diaphragm is used on some Motorcraft carburetors beginning in 1975. This opens the choke to a wider setting 6 to 18 seconds after the engine starts. As the pulldown diaphragm operates, the fast-idle screw is pulled from the top to the second step of the fast-idle cam to reduce cold-engine idle speed. Figure 8-62 shows the exact sequence of operation.

Temperature-Controlled Vacuum Breaks

Many Rochester carburetors since 1975 use two vacuum-break diaphragms to provide better mixture control by choking the engine more when it is cold and less when warm. Slightly

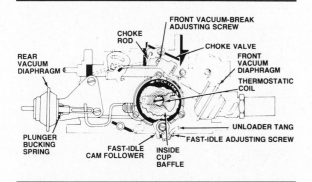

Figure 8-63. Temperature-controlled vacuum break used on some GM V-8 engines. (Pontiac)

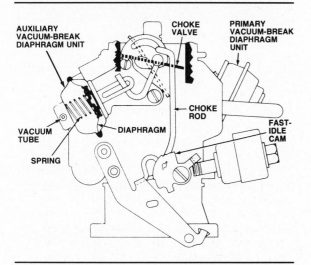

Figure 8-64. Temperature-controlled vacuum break used on some GM 6-cylinder engines. (Chevrolet)

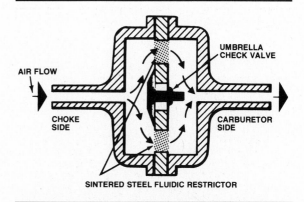

Figure 8-65. Ford's choke delay valve.

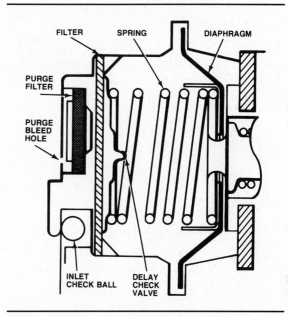

Figure 8-66. General Motors' choke delay valve.

different systems are used on GM V-8 engines, figure 8-63, and 6-cylinder engines, figure 8-64.

On V-8 engines, the primary (front) diaphragm opens the choke to keep the engine from stalling when first started. The secondary (rear) diaphragm opens the choke wider when air cleaner temperature exceeds 62°F (17°C). The rear diaphragm operates on manifold vacuum provided by a temperature vacuum valve on the air cleaner. The vacuum-break diaphragm has an inside restriction to delay diaphragm movement by several seconds.

On 6-cylinder engines, the primary diaphragm (choke coil side) opens the choke to keep the engine from stalling when first started. The auxiliary diaphragm (throttle lever side) opens the choke wider when engine coolant temperature exceeds 80°F (27°C). The auxiliary diaphragm operates on manifold vacuum provided by a two-nozzle temperature vacuum switch on the cylinder head.

Choke Delay Valves

Choke delay valves have various designs, but all do essentially the same thing as the vacuum-break diaphragm. They delay the opening of the choke for a period of time to improve driveability and cold engine warmup.

Ford uses the same valve for choke delay, figure 8-65, that it uses for spark delay in the ignition system. The valves are color coded to indicate delay interval and can be used for either purpose. When used as a choke delay valve, it is installed in a hose between the intake manifold and the choke vacuum piston or diaphragm. The black side must face the vacuum source (manifold) and the colored side faces the choke.

GM uses an internal bleed check valve in the rear vacuum break diaphragm, figure 8-66. This

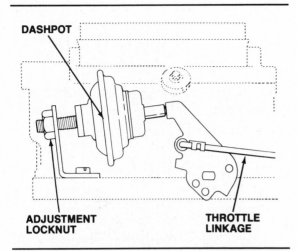

Figure 8-67. The dashpot slows throttle closing.

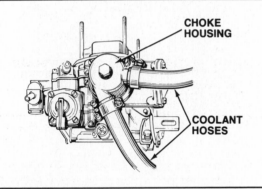

Figure 8-68. This type of hot-water heated choke routes engine coolant through the choke housing.

Figure 8-69. Other chokes are heated by routing engine coolant through a hose clipped to the choke cap.

delays choke opening beyond the amount allowed by the front vacuum break for a few seconds until the engine can run at a leaner mixture. Vacuum acting on the diaphragm draws filtered air through the bleed hole to purge any fuel vapor or contamination.

A slightly different system is used on some 2.5L (151-cid) 4-cylinder GM engines. The engine has a choke vacuum-break and a vacuum-delay valve. The valve delays manifold vacuum against the choke vacuum-break unit for about 40 seconds when starting the engine. If the engine stalls after being started, a relief feature in the valve permits immediate release of vacuum to let the choke close quickly.

Dashpots

These small chambers containing a spring-loaded diaphragm and plunger have been used for over 30 years. Before that hydraulic and magnetic dashpots were used.

As the throttle closes, a link from the throttle contacts the dashpot plunger, figure 8-67. As force is applied to the plunger, air slowly bleeds out of the diaphragm chamber through a small hole.

Dashpots originally were used to prevent an excessively rich mixture on deceleration which can cause stalling. They now function as emission control devices on late-model engines by reducing HC emissions on deceleration.

Choke Heaters

The need to keep exhaust emissions low means that choking the engine can only be done for a brief time. Prompt choke release can be done in several ways, all of which apply heat to the thermostatic coil spring to warm it up quickly.

Hot water choke

One way to apply heat to the thermostatic coil spring is by routing part of the engine coolant through the choke housing, figure 8-68. Once the engine reaches normal operating temperature, coolant heat helps release the choke. Older Ford models used an external coolant by-pass hose held against the choke housing cap by a spring clip, figure 8-69. This supplemented the exhaust manifold air passed through the choke housing.

Hot water chokes reduce over-choking when an engine is restarted. Since water holds heat longer than air, the choke coil will remain warm longer when exposed to heat from hot coolant. This reduces the amount of choking on a restart.

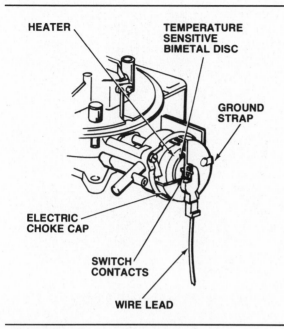

Figure 8-70. The Ford and AMC automatic choke electric heater.

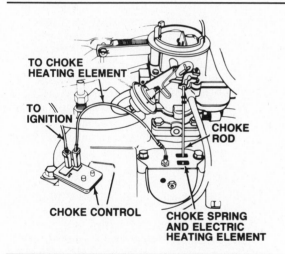

Figure 8-71. Chrysler's electric-assist choke.

Electric-assist choke

Electrically assisted heater elements are used on almost all late-model engines. These speed up the choke opening when underhood temperature is above about 60°F (16°C), by heating the choke bimetal thermostatic coil spring.

Ford and older AMC choke heater elements are located in the integral choke cap on the carburetor choke housing, figure 8-70. Electric current is supplied continuously from the alternator directly to the choke cover temperature-sensing switch. At underhood temperatures above 60° to 65°F (16° to 18°C), the switch closes

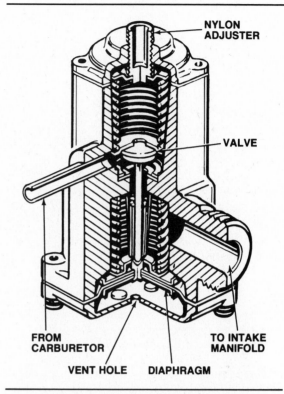

Figure 8-72. Fuel deceleration (decel) valve. (Ford)

to pass current to the choke heater. The circuit is grounded through a strap connected to the carburetor.

The Chrysler electric-assist choke heater is located in the intake manifold choke well and is regulated by a control switch which receives power from the ignition switch, figure 8-71. The control switch energizes the heater at temperatures above 63° to 68°F (17° to 20°C), and deenergizes it when the switch warms to between 110° and 130°F (43° to 54°C). A 2-stage heater control is used on some engines to provide three levels of heat, depending upon ambient temperature. The low heat level is provided by a resistor on the control switch.

Some GM engines from 1975 on use an electric choke heater. This 2-stage heater receives current from the engine oil pressure switch. Below specified air temperatures (usually 50° to 70°F or 10° to 21°C), a bimetal sensor in the cover turns off current to the larger heater stage. Both stages operate at higher temperatures. This choke heater receives current as long as the engine is running.

Fuel Deceleration (Decel) Valves

The decel valve, figure 8-72, was used primarily by Ford on some 4-cylinder and V-6 engines

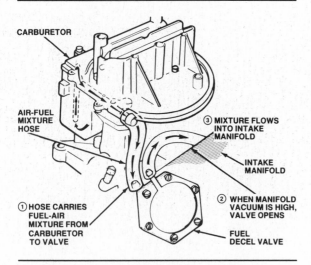

Figure 8-73. Decel valve operation. (Ford)

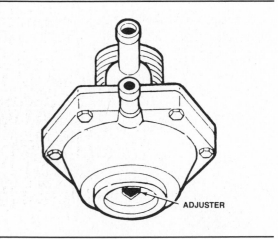

Figure 8-74. Ford's decel valve was redesigned in 1975 but its operation remained unchanged. (Ford)

during the 1970's to momentarily provide extra fuel and air during deceleration. Decel valves also have been used on some GM and imported-car engines. The valve prevents cylinders from misfiring during deceleration and sending an unburned charge of hydrocarbons through the exhaust. The extra air and fuel provided by the valve ensure complete combustion.

When the throttle is closed, increased manifold vacuum opens the valve diaphragm, figure 8-73. This, in turn, opens a passage between the carburetor and the manifold, allowing an additional air-fuel charge to enter the intake system. The extra air-fuel charge slows deceleration speed and reduces HC emissions. When vacuum drops, spring action and an air bleed to the diaphragm reseat the valve.

Some decel valves are adjustable. A nylon screw in the valve top, figure 8-72, can be adjusted to control the valve's opening time and duration. A round valve design, first used in 1974, figure 8-74, was not adjustable but was modified to permit adjustment on 1975 and later engines.

Deceleration Throttle Openers

This is a solenoid which holds the throttle open during deceleration to prevent excess fuel from being pulled through the idle system by high manifold vacuum. It is used on catalytic converter-equipped cars to prevent converter overheating. GM cars with the combination emission control (CEC) system use a throttle positioning solenoid that is energized on deceleration. Some Chrysler Corporation cars with converters have a solenoid to keep the throttle from closing completely at engine speeds above 2,000 rpm.

ALTITUDE-COMPENSATING CARBURETORS

Earlier in this chapter, you learned that atmospheric pressure is greatest at (or below) sea level. Suppose that you drive a car from a low elevation into an area where the elevation is 7,000 feet (2134 meters). As the altitude increases, atmospheric pressure decreases and less air enters the carburetor. This means that the air-fuel mixture passing into the engine becomes richer as altitude increases. The result is poor driveability and high CO emissions at the higher elevation. For the driver who is only passing through high-elevation areas, the poor driveability is mainly a temporary inconvenience. But for driving at that elevation for an extended time, the car will run better if the engine is tuned for high elevation by leaning the air-fuel mixture.

However, if the retuned engine is driven back to lower elevations without once again adjusting the mixture, performance will suffer. As the car descends from the higher elevation to sea level, the air-fuel mixture receives increasing amounts of air, leaning the mixture too much. If the engine is to operate properly at the lower elevation, it will have to be retuned to restore the proper air-fuel mixture. Unfortunately, it is not always possible or even desirable to tune the engine for such driving conditions. As a result, performance suffers, driveability is impaired, and emissions are excessive.

To maintain an appropriate air-fuel mixture while the car is driven in an altitude other than that for which it is adjusted, GM and Chrysler introduced the altitude compensating carburetor on some 1975 models. During the 1977

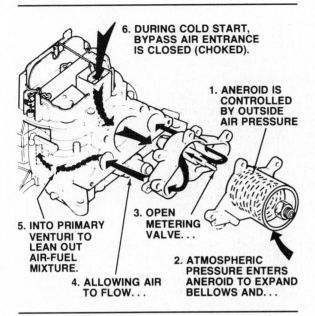

6. DURING COLD START, BYPASS AIR ENTRANCE IS CLOSED (CHOKED).

1. ANEROID IS CONTROLLED BY OUTSIDE AIR PRESSURE

5. INTO PRIMARY VENTURI TO LEAN OUT AIR-FUEL MIXTURE.

3. OPEN METERING VALVE...

4. ALLOWING AIR TO FLOW...

2. ATMOSPHERIC PRESSURE ENTERS ANEROID TO EXPAND BELLOWS AND...

Figure 8-75. Motorcraft 2150 and 4350 automatic altitude compensator.

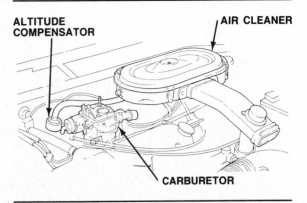

ALTITUDE COMPENSATOR

AIR CLEANER

CARBURETOR

Figure 8-76. Some Ford and Chrysler engines use remote altitude compensators to bleed air from the clean side of the air cleaner into the idle fuel system. (Ford)

model year, the Environmental Protection Agency designated 167 counties in 10 western states (not including California) as high-altitude emission control areas. These counties are entirely above 4,000 feet (1219 meters) in elevation. These high-altitude emission control requirements were suspended in 1978 but were reintroduced later in a modified form.

Because the major problem at high altitude is CO emission due to richer mixtures, the center of the special emission controls is the carburetor. Feedback carburetors used with electronic engine control systems continuously adjust the air-fuel ratio and automatically compensate for changes in atmospheric pressure, as we will see when we study them later in this chapter. But nonfeedback carburetors used in high altitude areas require auxiliary systems or devices to provide more air or less fuel when operating at higher elevation than when operating at sea level. Most of the altitude-compensating systems are automatic, responding to changes in atmospheric pressure. There are, however, systems that require manual adjustment or operation.

Automatic Compensation

The most widely used altitude compensating device is the **aneroid bellows**. An aneroid bellows is an accordion-shaped bellows that responds to changes in atmospheric pressure by expanding and contracting. As pressure decreases at high altitude, the bellows expands.

The air bypass passage on Motorcraft 2150 and 4350 altitude carburetors has its own air intake and choke valve, figure 8-75. The aneroid bellows on these carburetors have adjusting screws and locknuts that appear to be for adjusting the tension on the bellows. However, these screws are for original factory adjustments only, and should not be changed while adjusting or overhauling the carburetor. The bellows and air valve can be removed for carburetor cleaning without upsetting the adjustment.

The altitude compensator used with Motorcraft 740 carburetors is a unit remotely mounted on the bulkhead, figure 8-76. It leans the mixture by drawing air from the clean side of the air cleaner filter and bleeding it into the primary and idle fuel systems at altitudes above

Aneroid Bellows: An accordion-shaped membrane exposed to barometric (atmospheric) pressure. Pressure changes cause the bellows to flex. On a carburetor, this flexing actuates a valve or metering rods to change fuel flow.

■ A Tube Tip

Sometimes, when you are working on a carburetor still in the engine, you'll bust your knuckles trying to reach the idle mixture screws. Most mechanics know that one way to get at those screws is to use a length of rubber tubing. Slip it tightly over the idle screws to make the adjustments. Similar tubing, but with a bit larger inside diameter, can be used to loosen or tighten hard-to-get-at spark plugs.

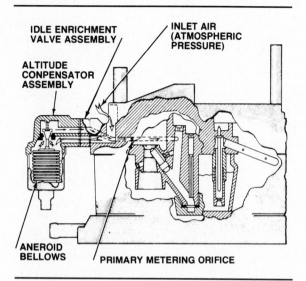

Figure 8-77. The Carter Thermo-Quad automatic altitude compensator. (Chrysler)

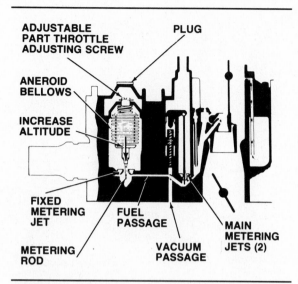

Figure 8-78. The Rochester Quadrajet automatic altitude compensator.

3,000 feet (914 meters). Chrysler uses a similar system with many late-model 4-cylinder engines.

The altitude compensator on the Carter Thermo-Quad, figure 8-77, was introduced on 1975 models. It also is automatic and requires no service.

The aneroid bellows is also part of the Rochester M4MEA Quadrajet used on 1976 and later high-altitude engines. The aneroid bellows controls the position of the metering rods in the primary main jets, figure 8-78. At high altitude, the bellows expands and moves the rods into the jets to reduce fuel flow and keep the mixture from becoming excessively rich. This unit requires adjustment only when it is replaced.

Manual Compensation

American Motors' high altitude 6-cylinder engines from the mid- and late 1970's use a Carter YF 1-barrel carburetor with a manually adjusted auxiliary air bleed. The adjustment plug is located on the side of the airhorn near the fuel inlet.

Some Carter BBD 2-barrel carburetors also have manually adjusted air bleeds. The adjustment screw for this air bleed is inside the airhorn above the venturi clusters.

Holley 5200 2-stage, 2-barrel carburetors used on some 1977 Ford 2.3-liter engines have a driver-controlled fuel valve to compensate for altitude changes. Located in the carburetor base, the fuel valve is controlled by a 2-position lever (SEA LEVEL and ALTITUDE) mounted under the instrument panel. A cable and swivel

linkage opens and closes the fuel valve. Under normal use, the system requires no adjustment.

FEEDBACK-CONTROLLED CARBURETORS

The first engine fuel management systems appeared in 1978 on California engines. Ford's feedback carburetor electronic control system (FCECS), figure 8-79, was used on 2.3-liter Pinto and Bobcat engines. The system contained a modified Holley 5200 carburetor (redesignated the Model 6500) in which the power and fuel enrichment pistons were replaced by a metering valve and diaphragm controlled by the system's electronic control unit (ECU). Other components were vacuum and temperature switches, a vacuum solenoid regulator, exhaust gas oxygen (EGO) sensor, and a dual-bed catalytic converter.

The system was called "feedback" because the EGO sensor installed in the exhaust manifold sent a voltage signal to the ECU indicating amount of oxygen in the exhaust. The ECU responded to this signal by cycling a vacuum solenoid regulator on and off. The solenoid in turn regulated vacuum to the feedback metering valve diaphragm which established the metering valve position, thus controlling the fuel flow into the carburetor's main well tube. This solenoid cycling action allowed the ECU to maintain the air-fuel ratio at approximately 14.7 to 1.

The GM electronic fuel control (EFC) system introduced on 1978 California engines contained similar parts and worked about the same to control the air-fuel ratio.

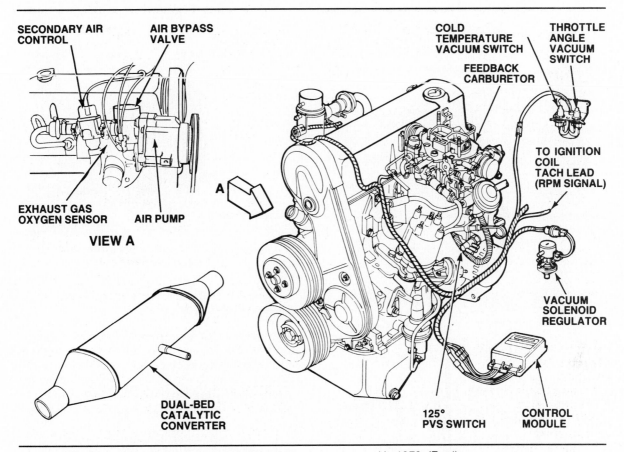

Figure 8-79. Ford's first electronic fuel management system appeared in 1978. (Ford)

Principles of Operation

Electronic control of fuel metering involves the ability of the engine control computer to turn a solenoid on and off more rapidly than any mechanical device can. This solenoid on-off sequence is called a cycle. The part of the time the solenoid is on is called the **duty cycle**. The solenoid can be designed to operate at any number of cycles per second. A complete cycle requires a specific length of time, but the duty cycle is a variable percentage of the complete cycle.

For example, a solenoid may operate 10 times per second. Each operating cycle is thus 1/10 of a second. But of that 1/10 second portion of the cycle, the solenoid may be on 10 percent of the time and off 90 percent, or on 90 percent of the time and off 10 percent, or any combination in between. This variable percentage of the complete cycle during which the solenoid is *on* is the solenoid's duty cycle.

The duty cycle is used by the engine control computer to precisely meter air (air bleed control) or fuel (fuel flow control) into the mixture according to the input it receives from the exhaust gas oxygen (EGO) sensor. The metering

of air or fuel can be done by either a stepper motor or a solenoid. The EGO sensor tells the computer how much the mixture varies from the desired 14.7 ratio. The computer then signals the carburetor to correct its performance, if necessary.

This process is repeated many times per second, allowing the computer to fine-tune carburetor operation and the air-fuel mixture, according to engine speed and load conditions. This makes it possible for the catalytic converter to work effectively to reduce exhaust emissions. Since the carburetor constantly responds to information about its past performance (feedback information from the EGO sensor to the computer), it is called a feedback carburetor.

The computer does not control the air-fuel mixture under all conditions at all times

Duty Cycle: The percentage of total time in one complete on-off cycle during which a solenoid is energized.

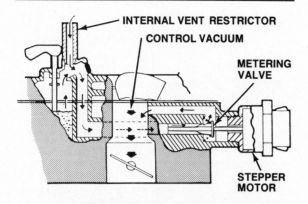

Figure 8-80. Early models of the Motorcraft 7200 VV carburetor use a stepper motor to vary air pressure on the fuel bowl.

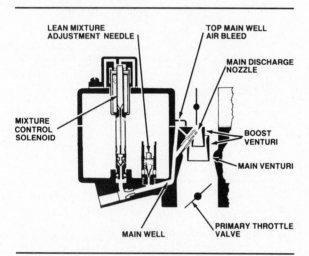

Figure 8-81. The mixture control solenoid in Rochester carburetors controls metering rod position. (Buick)

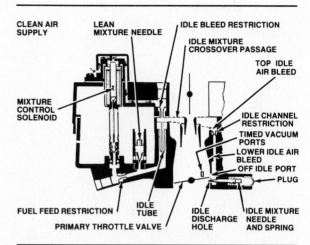

Figure 8-82. The same idle mixture solenoid may also control an idle air bleed passage. (Buick)

Stepper Motors

A **stepper motor** is used with some feedback carburetors. This dc motor moves in a specified number of incremental steps (usually 100 to 120) according to the voltage applied. Each step is tiny (approximately 0.004 inch or 0.1 mm), with motor speed varying from 12 to 100 steps per second.

The stepper motor is connected to tapered metering pins positioned either in air bleed orifices or in the main metering jets. When power is applied to the stepper motor, it moves the metering pins inward (rich position) to an end stop. This gives the engine control computer a stable reference. The stepper motor then backs the pins out to a position calculated to deliver a 14.7 to 1 air-fuel ratio and the computer takes over its operation. This is called initialization.

Stepper motors are used in early Motorcraft 7200VV and some Carter carburetors. In the 7200 VV, the stepper motor moves a valve which lets control vacuum into the fuel bowl. Since this lowers the pressure above the fuel bowl, less fuel is pushed into the main metering system, figure 8-80. The result is a leaner air-fuel mixture. This design is sometimes called the "backsuction" system. The Carter carburetors use a stepper motor to position the metering pins in the air bleeds. This controls the amount of air in the air-fuel ratio.

Mixture Control Solenoids

Mixture control (MC) solenoids are used to control the air-fuel ratio in various ways, but all work on the variable duty cycle principle just discussed. The MC solenoid installed in

through the duty cycle. For example, when the engine is first started, the fuel management system operates in an **open-loop** mode. This means that the computer ignores any signals from the EGO sensor and operates the stepper motor or solenoid with a programmed, fixed duty cycle. Various fixed duty cycles may be programmed into the computer for open-loop conditions, ranging from cold starts to hot wide-open throttle acceleration. When certain conditions are met, such as normal engine coolant temperature, the microprocessor switches system operation into the **closed-loop** mode. At this time, it evaluates the EGO sensor signals and varies the duty cycle of the stepper motor or solenoid to manage the air-fuel ratio. You will learn more about the operation of an engine management system in Part 4.

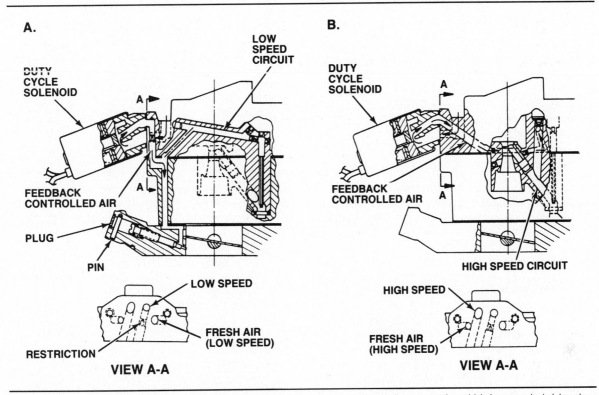

Figure 8-83. The Chrysler Thermo-Quad pulse solenoid opens and closes low-speed and high-speed air bleeds. (Chrysler)

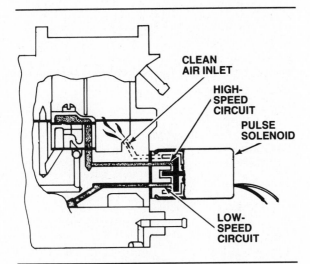

Figure 8-84. The Chrysler BBD pulse solenoid bleeds air to both the low- and high-speed circuits. (Carter)

Rochester carburetors regulates fuel flow directly by controlling a metering rod in the main jet, figure 8-81. The MC solenoid also controls a rod, figure 8-82, that opens and closes an idle air bleed.

Some Carter Thermo-Quad carburetors use the variable duty cycle of an MC solenoid to open and close the low-speed air bleed, figure 8-83A, and the high-speed air bleed, figure 8-83B. The Carter BBD carburetor uses a similar solenoid, figure 8-84, which bleeds air to both the low-speed and the high-speed circuits when energized, or shuts off air to both circuits when deenergized. Since solenoids used in this application are either on or off, they are called pulse solenoids.

Open-Loop: An operational mode in which the engine control microprocessor adjusts the system to function according to predetermined instructions and does not respond to feedback signals from the EGO sensor.

Closed-Loop: An operational mode in which the engine control microprocessor reads and responds to feedback signals from the EGO sensor, adjusting system operation accordingly.

Stepper Motor: A direct current motor that moves in incremental steps from deenergized to fully energized.

Vacuum Control Solenoids

Other carburetors may use an external solenoid that controls vacuum to an internal diaphragm in the carburetor. The internal diaphragm may regulate fuel flow by changing the position of the metering rods in the carburetor jets. It can also open and close air bleeds used in the idle and main metering circuits.

SUMMARY

The carburetor is the all-important device that converts air and gasoline into an air-fuel mixture that can be burned in the cylinders. Carburetors must operate under all types of conditions and in all temperatures. Although there are dozens of carburetor designs available, all operate on the same basic principle of pressure differential. Pressure differential is the difference in air pressure between the relatively high pressure of the atmosphere outside the engine, and the low pressure of the carburetor and intake manifold. A partial vacuum is created by the downward stroke of the piston, which draws air in through the manifold and the carburetor.

For a carburetor to operate efficiently, the gasoline must be properly atomized, vaporized, and distributed to the cylinders through the intake manifold.

In spite of different designs, most carburetors have the same seven basic systems of passages, ports, jets, and pumps. Venturis, or restrictions, help speed the airflow and draw more fuel into the airflow. For each barrel, or throat, on a carburetor, there is also a throttle valve.

Most carburetors made since 1979 are of a tamperproof design in which the idle mixture and choke adjustment are set at the factory and sealed according to U.S. government regulations. This was done to prevent unauthorized adjustments that would affect emissions.

Carburetors are used with various assist devices to improve driveability, reduce emissions, help in cold-weather starting, and avoid overheating. Other assist devices provide better fuel mixture during the full range of driving conditions, including high-altitude driving.

Many late-model carburetors are part of an electronic fuel management system. An exhaust gas oxygen (EGO) sensor in the exhaust manifold measures the oxygen content in the exhaust gas and signals the engine control computer. To maintain the desired 14.7 to 1 air-fuel ratio, the computer signals a mixture control (MC) solenoid or stepper motor in the carburetor to richen or lean the mixture as required. Since they respond to this feedback information from the EGO sensor through the computer, they are called feedback carburetors.

Review Questions
Choose the single most correct answer.
Compare your answers with the correct answers on page 414.

1. Which of the following is *not* true of air pressure?
   a. It is measured in millimeters of mercury (mm Hg)
   b. It is measured in pounds per square inch (psi)
   c. It is always constant
   d. It results from the weight of air pressing on a surface

2. Air pressure:
   a. Increases with height
   b. Decreases with height
   c. Remains the same regardless of height
   d. Increases with warming temperature

3. A pressure differential is created between outside air and engine air by:
   a. An increase in cylinder volume
   b. A decrease in cylinder volume
   c. Air rushing through the carburetor
   d. The intake manifold

4. The restriction in an airflow tube or barrel is called:
   a. A throttle
   b. An air bleed
   c. A vaporizer
   d. A venturi

5. The air pressure along the sides of a barrel in which there is an airflow:
   a. Is higher than at the center of the flow
   b. Is higher than atmospheric or outside pressure
   c. Is lower than at the center of the flow
   d. Increases with airflow velocity

6. Ported vacuum:
   a. Is the low-pressure area in the carburetor just above the throttle valve
   b. Is the low pressure beneath the throttle valve
   c. Is the air pressure in the venturi
   d. Is equal to atmospheric pressure

7. At idle:
   a. Venturi vacuum is high
   b. Manifold vacuum is high
   c. Ported vacuum is high
   d. All of the above

8. Which is *not* a fuel-metering system?
   a. Float system
   b. Idle system
   c. Power system
   d. EEC system

9. The carburetor float controls fuel level:
   a. By closing the needle valve when the level is low
   b. By opening the needle valve when the level is high
   c. By closing the needle valve when the level is high
   d. All of the above

10. At engine idle:
    a. Venturi vacuum is high
    b. Fuel enters the carburetor barrel above the throttle
    c. Air is provided by an air bleed in the idle tube
    d. The throttle is one-third open

11. Which is *not* used to provide extra air at idle?
    a. The throttle valve
    b. The choke
    c. Transfer ports
    d. Idle air bypass

12. At low off-idle speeds, extra fuel is provided by:
    a. The transfer port
    b. The main nozzle
    c. The idle air bleed
    d. The idle air adjust screw

13. In a high-speed system, better fuel and air mixtures are obtained with:
    a. Transfer ports
    b. A single venturi
    c. Multiple venturis
    d. None of the above

14. Power circuits are operated by:
    a. Vacuum diaphragms
    b. Vacuum pistons
    c. Mechanical metering rods
    d. All of the above

15. Which is *not* part of the accelerator pump circuit?
    a. The metering rod
    b. The inlet check
    c. The outlet check
    d. The duration spring

16. Which carburetor circuit makes the fuel mixture richer when starting an engine?
    a. Power circuit
    b. Choke circuit
    c. High-speed circuit
    d. Accelerator pump circuit

17. The illustration shows:
    a. A power valve diaphragm
    b. An integral choke
    c. A remote choke
    d. None of the above

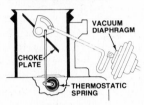

18. The choke unloader:
    a. Releases the fast-idle cam after a specified time
    b. Directs warm air to the thermostatic spring
    c. Opens a vacuum bleed in the vacuum break diaphragm
    d. Opens the choke valve when the throttle is open fully

19. The fast-idle cam:
    a. Opens the throttle
    b. Closes the throttle
    c. Opens the choke valve
    d. Closes the choke valve

20. Independently operated throttle are found in:
    a. One-barrel carburetors
    b. Single-stage two-barrel carburetors
    c. Two-stage two-barrel carburetors
    d. All of the above

21. Which is *not* true of four-barrel carburetors?
    a. The primaries act like single-stage two-barrel at low speeds
    b. Airflow through the secondary barrels is through venturi action *or* air velocity valves
    c. The primaries feed four cylinders and the secondaries feed the other four
    d. The secondary barrels open between half and three-quarter throttle

22. Variable-venturi carburetors:
    a. Maintain constant throttle
    b. Maintain a uniform pressure drop across the venturi
    c. Have been used on domestic cars for 45 years
    d. Are formed by two parallel valve plates

23. Main fuel metering in a variable-venturi carburetor is done by:
    a. Metering rods
    b. A dashpot
    c. EGR port vacuum
    d. The idle trim system

24. Which is *not* part of the cold enrichment system of a 2700VV carburetor?
    a. A fast-idle cam
    b. A solenoid
    c. A control vacuum regulator rod
    d. An auxiliary fuel passage and metering rod

25. The accelerator pump is controlled by:
    a. Vacuum break linkage
    b. Thermostatic spring linkage
    c. Throttle linkage
    d. All of the above

26. Which of the following allows feedback carburetors used with electronic fuel management systems to compensate for changes in altitude?
    a. A remote compensator
    b. An aneroid bellows
    c. The microprocessor
    d. An EGO sensor

27. The early fuel management systems contained a microprocessor, a catalytic converter, vacuum and temperature switches, a solenoid vacuum regulator, an EGO sensor and:
    a. A feedback carburetor
    b. A non-feedback carburetor
    c. An altitude-compensating carburetor
    d. An emissions carburetor

28. Feedback carburetors control fuel flow with some type of:
    a. Hot-idle compensator
    b. Mixture control solenoid
    c. Decel valve
    d. Aneroid bellows

9

Autolite-Motorcraft Carburetors

Autolite-Motorcraft carburetors are used on Ford Motor Company vehicles and American Motors vehicles through the late 1970's and early 1980's. When using the *Shop Manual* to overhaul specific carburetors, you may find the exploded drawings and carburetor descriptions in this chapter useful. The following carburetors are included:

- Models 1100 and 1101 One-Barrel
- Model 1250 One-Barrel
- Models 2100 and 2100-D Two-Barrel (AMC Model 6200)
- Models 2150 and 2150-A Two-Barrel
- Model 4100 Four-Barrel
- Models 4300, 4300-A, and 4300-D Four-Barrel
- Model 4350 Four-Barrel
- Model 2700 VV Two-Barrel
- Model 7200 VV Two-Barrel
- Models 740 and 5740 Two-Barrel.

AUTOLITE 1100 AND 1101 ONE-BARREL

The Autolite 1100 and Autolite-Motorcraft 1101, figure 9-1, have five fuel metering circuits: idle, main, accelerator pump, power enrichment, and choke. The upper body houses the idle, main, and power enrichment circuits. The float chamber vent and the fuel inlet system are also in the upper body. The lower body contains the fuel bowl and accelerator pump circuit. A hydraulic dashpot also may be included in the lower body. Cold starting on some models is helped by a manifold-heated, piston-type choke attached to the lower body. Other models have a manual choke. Carburetors used with the improved-combustion (IMCO) exhaust emission system use an external dashpot mounted on the airhorn by a bracket. All other models use an integral dashpot.

AUTOLITE 1250 ONE-BARREL

The single-barrel Model 1250, figure 9-2, is used only on the English Ford 1600-cc engine in the early Pinto. It has five fuel metering circuits: idle, main, accelerator pump, choke, and power enrichment. The upper body contains the fuel inlet and bowl vent, the thermostatic choke and the main metering system. The throttle body contains the fuel bowl and the idle and accelerator pump circuits. Cold starting is aided by a water-heated thermostatic choke attached to the upper body.

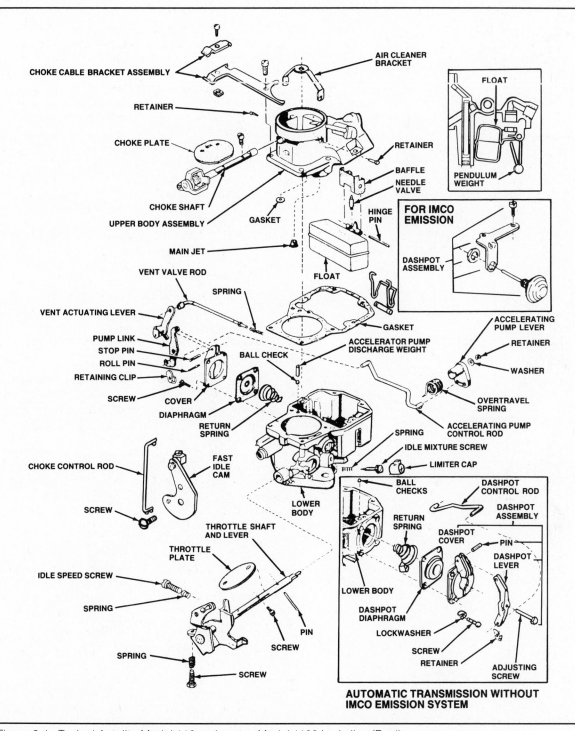

Figure 9-1. Typical Autolite Model 110 carburetor. Model 1100 is similar. (Ford)

AUTOLITE 2100 AND 2100-D TWO-BARREL

Early versions were designated Model 56200 when used by American Motors. The Autolite 2100 or 6200, figure 9-3, has five fuel metering circuits: idle, main, accelerator pump, choke, and power enrichment. The airhorn is the main body cover and contains the choke valve and the internal and external fuel vents. All fuel metering circuits are in the main body, to which the thermostatic choke and dashpot are attached.

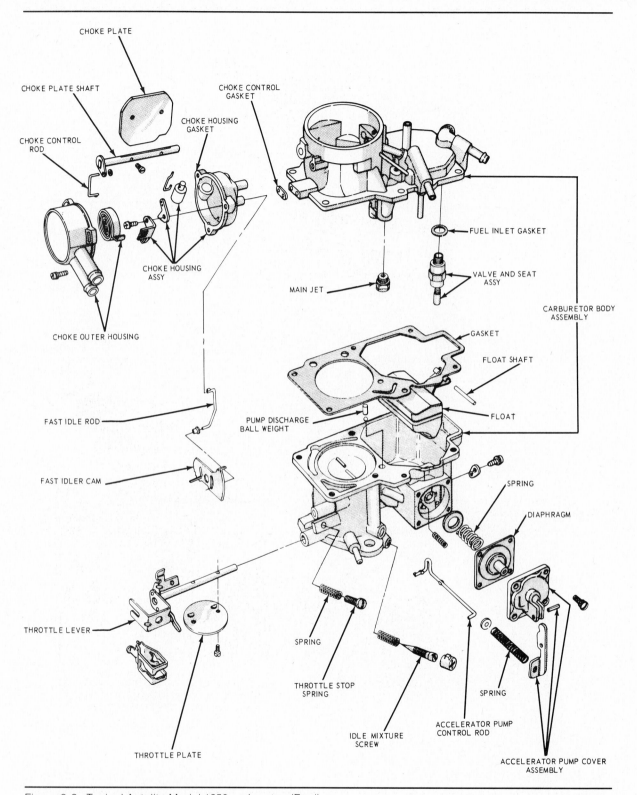

Figure 9-2. Typical Autolite Model 1250 carburetor. (Ford)

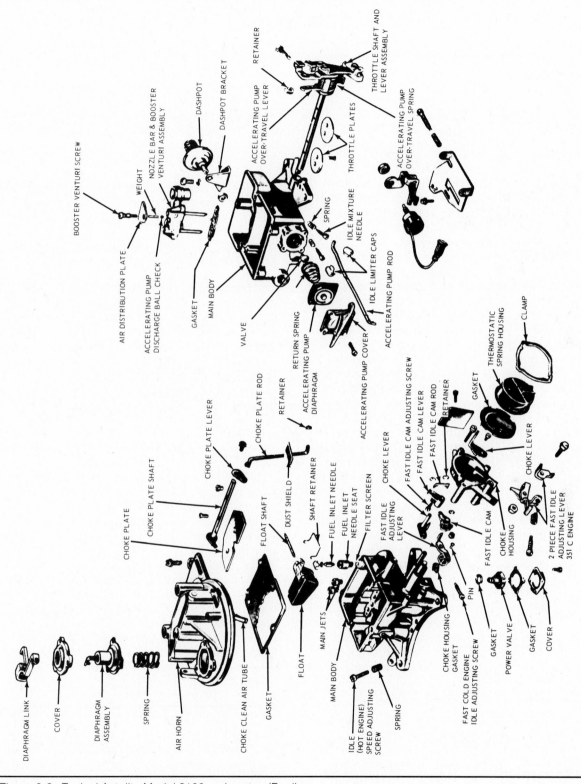

Figure 9-3. Typical Autolite Model 2100 carburetor. (Ford)

Model 2100-D is vented to a charcoal canister and has a choke diaphragm assembly in the airhorn. Some Model 2100 carburetors use a staged choke pulldown system, explained in Chapter 8, while later versions have an electric-assist choke.

MOTORCRAFT 2150 AND 2150-A TWO-BARREL

Introduced in 1975, this more sophisticated version of the Model 2100, figure 9-4, uses a pull-over enrichment system to provide extra fuel when airflow through the airhorn is high. Fuel is drawn through orifices in the fuel bowl and passes into the airhorn through air bleeds. All versions use the electric-assist choke.

The airhorn of carburetors used on Ford's 2.8-liter V6 engine during the late 1970's contained a fuel deceleration metering system. Each booster venturi contains high-speed bleed orifices. These work with a mechanical bleed control system, which is controlled by a cam on the throttle shaft. This system uses reverse-tapered metering rods in the high-speed bleed orifices to provide more precise high-speed operation and better low-speed response.

An altitude compensator is used on some 1976 and later Model 2150 carburetors installed on high-altitude applications. Intake air enters a bypass intake and is metered into the airflow above the throttle valves to provide the leaner air-fuel mixture needed at higher altitudes. The airflow is controlled by a valve operated by an aneroid bellows at the rear of the carburetor.

Model 2150-A is a feedback version used on 1983 and later models. A pulse solenoid attached to the side of the main body controls the carburetor duty cycle by metering air into both the idle and main vacuum passages on direction from the EEC computer. High-altitude versions of this 1983 carburetor have an integral gradient (aneroid) altitude compensator mounted in place of the feedback solenoid. The altitude compensation function was incorporated in the EEC computer on 1984 and later models.

AUTOLITE 4100 FOUR-BARREL

Model 4100, figure 9-5, uses six fuel metering circuits: idle, primary and secondary main, accelerator pump, power enrichment, and choke. The airhorn assembly is the main body cover and contains the choke plate, the fuel inlet, a hot-idle compensator, primary and secondary fuel bowl vents, secondary throttle vacuum tubes, and an automatic choke clean air pickup tube. The choke heat chamber in the right exhaust manifold is connected to the clean air pickup tube by a rubber hose and steel tube.

The main body contains the twin fuel bowls and all fuel metering circuits. Each of two primary (front) barrels has a main and a booster venturi, main fuel discharge, idle fuel discharge, accelerator pump discharge, and a primary throttle valve. The two secondary (rear) barrels contain a main fuel discharge and vacuum-operated throttle valves. The primary and secondary barrels are of equal size. A manifold-heated, piston-type thermostatic choke attached to the main body assists in cold starts.

AUTOLITE 4300, 4300-A, AND 4300-D FOUR-BARREL

The Model 4300, figure 9-6, is virtually a complete revision of the original 4100. It uses a 3-piece body rather than the 2-piece body of the 4100. Other major differences include a single fuel bowl with a pontoon float, internal balance and mechanical atmospheric vents, secondary throttle valves mechanically linked to the primaries, and secondary barrels with larger throats than the primaries. The manifold-heated, piston-type thermostatic choke housing is built into the throttle body. Many Model 4300 carburetors have an electric-assist heater built into the thermostatic choke cap.

Model 4300-A carburetors have auxiliary air valve plates above the secondary main venturis. An integral hydraulic dashpot is used to dampen any sudden movements of the air valve plates, preventing flutter and erratic engine operation.

The secondary air valve plates on 4300-D carburetors, figure 9-7, are located above the secondary bore. On some 4300-D carburetors, the choke is heated by both manifold air and engine coolant flowing through a hose held against the choke cap by a spring clip. Other 4300-D carburetors have electric-assist chokes. A staged choke release is operated by a vacuum diaphragm.

MOTORCRAFT 4350 FOUR-BARREL

The Model 4350, figure 9-8, was introduced in 1975 and differs from the 4300 chiefly in its use of vacuum piston rods in the primary fuel metering system. These provide more precise control over enrichment. A vacuum-operated delayed choke pulldown diaphragm controls the top-step pulloff of the fast-idle cam. Some versions also use a mechanical vent valve for the fuel bowl.

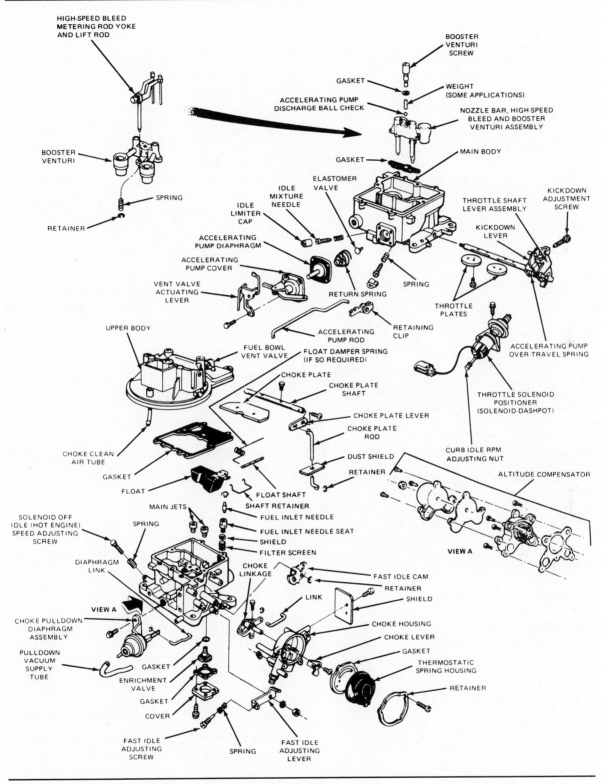

Figure 9-4. Typical Motorcraft Model 2150 carburetor. (Ford)

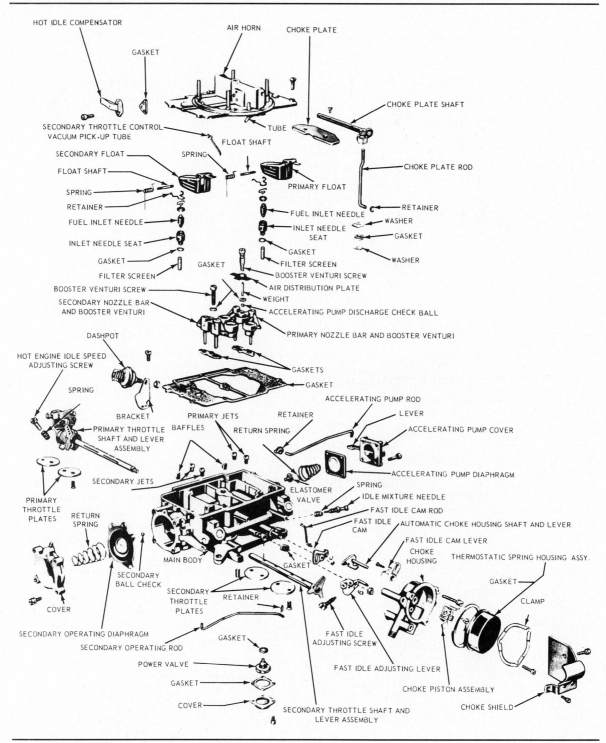

Figure 9-5. Typical Autolite Model 4100 carburetor. (Ford)

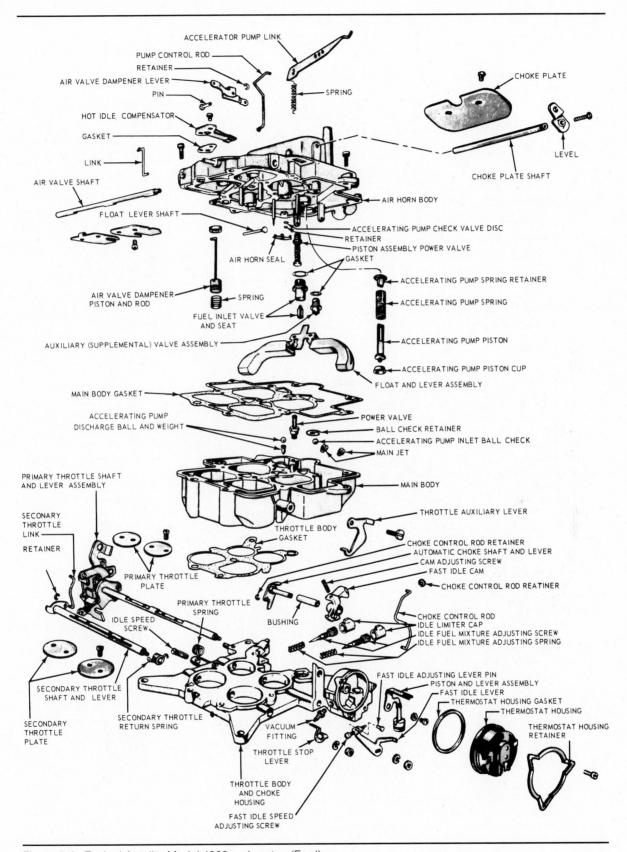

Figure 9-6. Typical Autolite Model 4300 carburetor. (Ford)

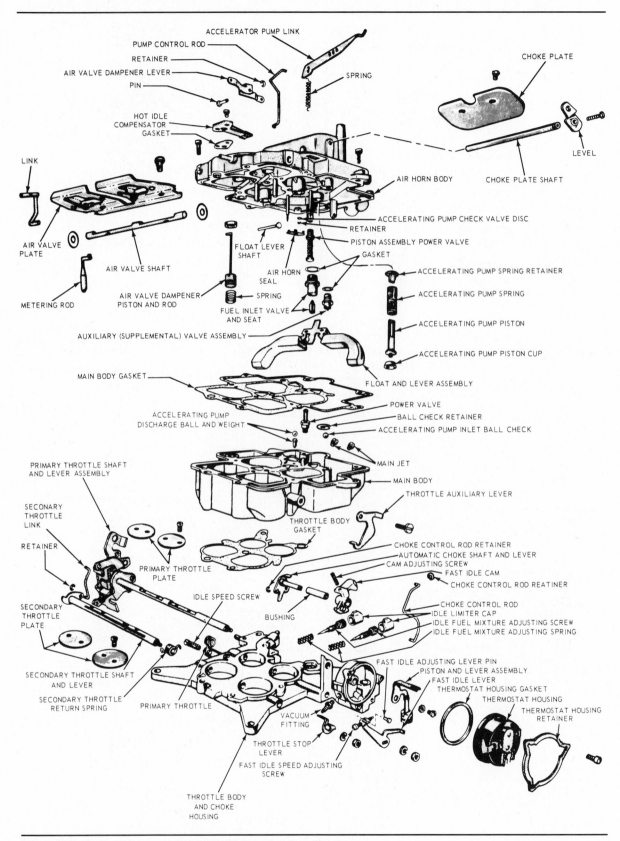

Figure 9-7. Typical Motorcraft Model 4300-D carburetor. (Ford)

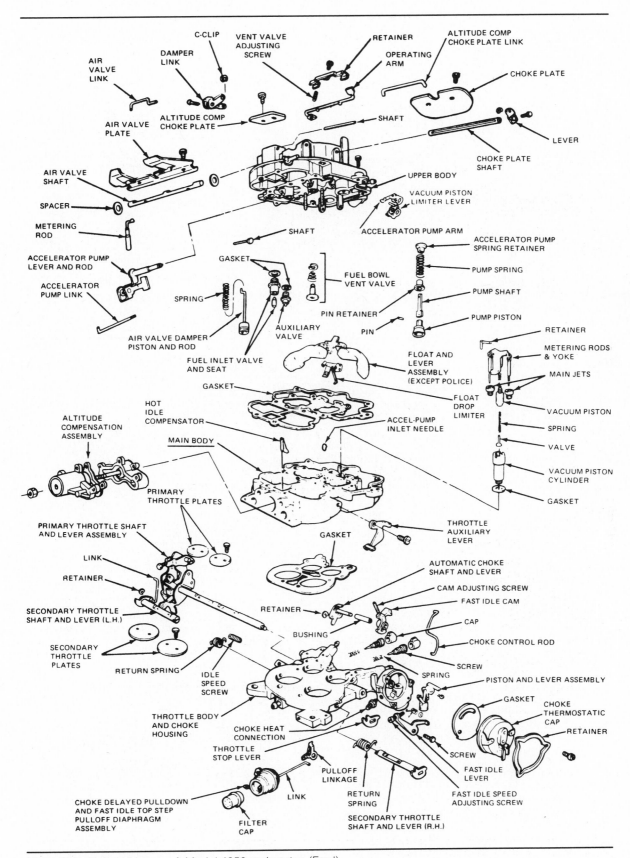

Figure 9-8. Typical Motorcraft Model 4350 carburetor. (Ford)

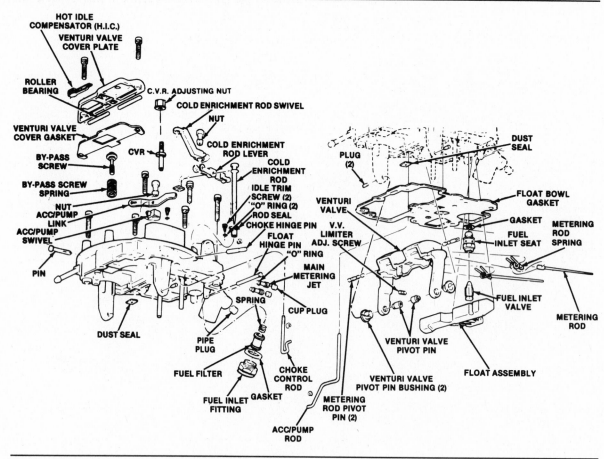

Figure 9-9. Typical Motorcraft Model 2700 VV and 7200 VV carburetor, main body. (Ford)

An altitude compensator was added to 1976 and later Model 4350 carburetors. Like the Model 2150, intake air entering the bypass intake is metered into the air flow above the throttle valves to provide the leaner air-fuel mixture required at higher altitudes. The air flow is controlled by a valve operated by an aneroid bellows in the rear of the carburetor housing.

MOTORCRAFT 2700 VV TWO-BARREL

This 2-barrel carburetor, figures 9-9, 9-10, and 9-11, varies its venturi area according to the demands of engine speed and load. All other Autolite-Motorcraft carburetors have fixed venturis. The Model 2700 VV uses a movable venturi valve which moves in and out of the airflow through the carburetor throats.

Engine vacuum and throttle position control the venturi valve, which is connected to two main metering rods to control fuel flow. Air speed through the carburetor remains just

about constant, instead of varying as in fixed-venturi designs. Additional circuits such as the idle, power enrichment, and choke systems are not required.

Other features include a system which trims the fuel flow at idle, an accelerator pump system similar to that used on the Model 4350 carburetor, cranking and cold-enrichment systems, a control vacuum regulator and an external vent.

MOTORCRAFT 7200 VV TWO-BARREL

The Model 7200 VV introduced in 1979 is the only variable-venturi carburetor remaining in the Ford lineup, now found primarily on police vehicles. Essentially the same design as the 2700 VV, it differs primarily in the addition of a feedback control motor, a metering valve, and a revised internal vent system.

The feedback control motor positions the backsuction metering valve according to signals from the EEC computer. To lean out an excessively rich mixture, the control motor opens the

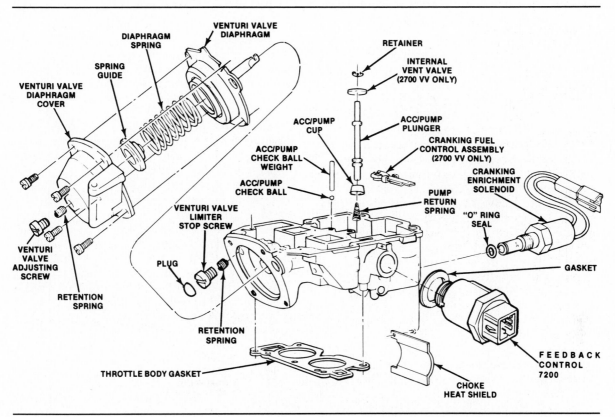

Figure 9-10. Typical Motorcraft Model 2700 VV and 7200 VV carburetor, upper body. (Ford)

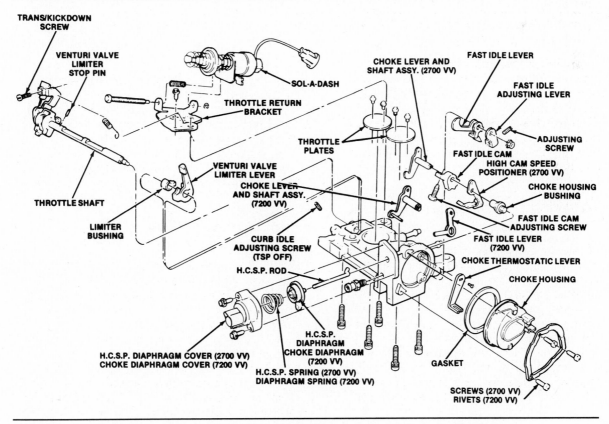

Figure 9-11. Typical Motorcraft Model 2700 VV and 7200 VV carburetor, throttle body. (Ford)

metering valve to expose more control vacuum to the fuel bowl. Conversely, it enriches the mixture by closing the metering valve to reduce the amount of control vacuum affecting the fuel bowl.

The accelerator pump stem on Model 7200 VV carburetors does not have an internal vent valve as do the 2700 VV models. The internal vent valve on Model 2700 VV carburetors seals off the fuel bowl when the engine is off, sending the bowl vapors through the external vent to the canister. Model 7200 VV carburetors, however, must have an unrestricted opening to the fuel bowl if the control vacuum signal is to operate properly. For this reason, the 7200 VV internal vent restrictor functions as a calibrated air bleed, raising the bowl vent opening over the external vent opening. Fuel bowl vapors thus pass through the lower external vent opening into the canister when the engine is off. In both carburetor designs, bowl vapors are directed into the venturi airstream during engine operation.

The external fuel bowl vent on Model 7200 VV carburetors does not use the external vent valve disk found on Model 2700 VV carburetors. For proper feedback operation, the external vent must remain closed to atmospheric pressure or the carburetor operates at its rich limit.

MOTORCRAFT 740 AND 5740 TWO-BARREL

The Motorcraft-Weber 740, figure 9-12, is a 2-stage, 2-barrel design with equal-size primary and secondary barrels. It was initially used on Ford's Fiesta in 1977 and then on early Escort engines. The secondary throttle is mechanically operated by linkage from the primary throttle. Five fuel metering circuits are used: idle, main, accelerator pump, power enrichment, and choke systems. A fuel shutoff solenoid is used to prevent dieseling. The bowl vent also is operated by a solenoid.

The Model 5740, figure 9-13, is an updated version of the original 740. It uses a remotely mounted fuel bowl vent valve, a remotely mounted altitude compensator, and a manifold-vacuum-operated idle speed control, or throttle positioner, which holds a specified idle speed under varying engine load conditions. Some models also have a wide-open throttle (WOT) cutout switch, which disengages the air conditioning compressor during wide-open throttle operation.

1. COVER (ASSY.)
2. CHOKE SHAFT BUSHINGS
3. CHOKE PLATE
4. CHOKE PLATE SCREWS
5. SECONDARY CHOKE SHAFT
6. SECONDARY CHOKE LINK
7. CHOKE LINKAGE RETAINING CLIPS
8. PRIMARY CHOKE SHAFT
9. PRIMARY CHOKE LINK DIRT SEAL
10. DIRT SEAL RETAINER
11. PRIMARY CHOKE LINK
12. CHOKE BIMETAL SHAFT BUSHING
13. FAST IDLE CAM SPRING
14. CHOKE BIMETAL LEVER
15. CHOKE BIMETAL SHAFT
16. CHOKE ASSIST SPRING
17. ELECTRIC CHOKE RETAINING SCREWS
18. ELECTRIC CHOKE RETAINING RING
19. ELECTRIC CHOKE UNIT
20. CHOKE HOUSING DIRT SHIELD
21. CHOKE HOUSING SCREWS
22. CHOKE PULLDOWN SPRING
23. CHOKE PULLDOWN DIAPHRAGM COVER
24. CHOKE PULLDOWN ADJUSTING SCREW
25. CHOKE PULLDOWN ADJUSTING SCREW SEAL
26. CHOKE PULLDOWN DIAPHRAGM ASSEMBLY
27. CHOKE HOUSING (ASSY.)
28. CHOKE HOUSING VACUUM SEAL (O RING)
29. CHOKE LEVER
30. CHOKE BIMETAL SHAFT LOCK WASHER
31. CHOKE BIMETAL SHAFT NUT
32. COVER GASKET
33. FUEL BOWL FLOAT
34. HIGH SPEED AIR BLEEDS
35. WELL TUBES
36. IDLE JET HOLDER
37. IDLE JET
38. MAIN JET
39. TEFLON SHAFT SEAL
40. TEFLON SHAFT SEAL
41. SECONDARY THROTTLE SHAFT SPACER
42. SECONDARY THROTTLE SHAFT
43. THROTTLE SHAFT BUSHINGS
44. SECONDARY THROTTLE STOP SCREW
45. PRIMARY SHAFT LOCATOR WASHERS
46. THROTTLE LEVER
47. FAST IDLE SPEED ADJUSTING SCREW
48. FAST IDLE SPEED ADJUSTING SCREW LOCK NUT
49. CHOKE PULL DOWN DIAPHRAGM COVER SCREW
50. PRIMARY THROTTLE SHAFT NUT

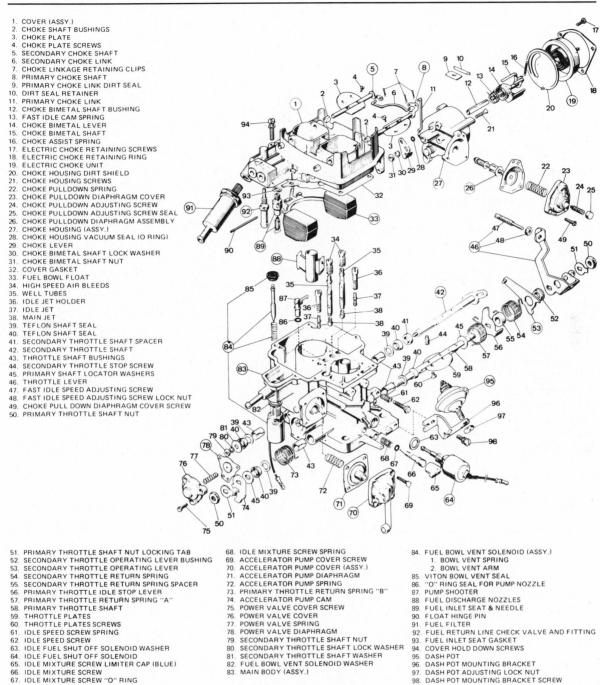

51. PRIMARY THROTTLE SHAFT NUT LOCKING TAB
52. SECONDARY THROTTLE OPERATING LEVER BUSHING
53. SECONDARY THROTTLE OPERATING LEVER
54. SECONDARY THROTTLE RETURN SPRING
55. SECONDARY THROTTLE RETURN SPRING SPACER
56. PRIMARY THROTTLE IDLE STOP LEVER
57. PRIMARY THROTTLE RETURN SPRING "A"
58. PRIMARY THROTTLE SHAFT
59. THROTTLE PLATES
60. THROTTLE PLATES SCREWS
61. IDLE SPEED SCREW SPRING
62. IDLE SPEED SCREW
63. IDLE FUEL SHUT OFF SOLENOID WASHER
64. IDLE FUEL SHUT OFF SOLENOID
65. IDLE MIXTURE SCREW LIMITER CAP (BLUE)
66. IDLE MIXTURE SCREW
67. IDLE MIXTURE SCREW "O" RING

68. IDLE MIXTURE SCREW SPRING
69. ACCELERATOR PUMP COVER SCREW
70. ACCELERATOR PUMP COVER (ASSY.)
71. ACCELERATOR PUMP DIAPHRAGM
72. ACCELERATOR PUMP SPRING
73. PRIMARY THROTTLE RETURN SPRING "B"
74. ACCELERATOR PUMP CAM
75. POWER VALVE COVER SCREW
76. POWER VALVE COVER
77. POWER VALVE SPRING
78. POWER VALVE DIAPHRAGM
79. SECONDARY THROTTLE SHAFT NUT
80. SECONDARY THROTTLE SHAFT LOCK WASHER
81. SECONDARY THROTTLE SHAFT WASHER
82. FUEL BOWL VENT SOLENOID WASHER
83. MAIN BODY (ASSY.)

84. FUEL BOWL VENT SOLENOID (ASSY.)
  1. BOWL VENT SPRING
  2. BOWL VENT ARM
85. VITON BOWL VENT SEAL
86. "O" RING SEAL FOR PUMP NOZZLE
87. PUMP SHOOTER
88. FUEL DISCHARGE NOZZLES
89. FUEL INLET SEAT & NEEDLE
90. FLOAT HINGE PIN
91. FUEL FILTER
92. FUEL RETURN LINE CHECK VALVE AND FITTING
93. FUEL INLET SEAT GASKET
94. COVER HOLD DOWN SCREWS
95. DASH POT
96. DASH POT MOUNTING BRACKET
97. DASH POT ADJUSTING LOCK NUT
98. DASH POT MOUNTING BRACKET SCREW

Figure 9-12. Typical Motorcraft-Weber Model 740 carburetor. (Ford)

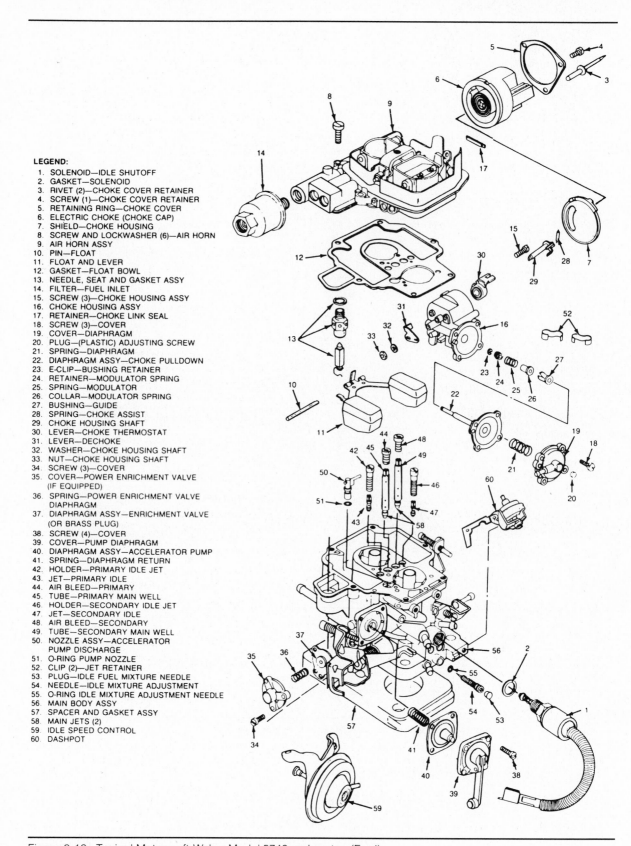

LEGEND:
1. SOLENOID—IDLE SHUTOFF
2. GASKET—SOLENOID
3. RIVET (2)—CHOKE COVER RETAINER
4. SCREW (1)—CHOKE COVER RETAINER
5. RETAINING RING—CHOKE COVER
6. ELECTRIC CHOKE (CHOKE CAP)
7. SHIELD—CHOKE HOUSING
8. SCREW AND LOCKWASHER (6)—AIR HORN
9. AIR HORN ASSY
10. PIN—FLOAT
11. FLOAT AND LEVER
12. GASKET—FLOAT BOWL
13. NEEDLE, SEAT AND GASKET ASSY
14. FILTER—FUEL INLET
15. SCREW (3)—CHOKE HOUSING ASSY
16. CHOKE HOUSING ASSY
17. RETAINER—CHOKE LINK SEAL
18. SCREW (3)—COVER
19. COVER—DIAPHRAGM
20. PLUG—(PLASTIC) ADJUSTING SCREW
21. SPRING—DIAPHRAGM
22. DIAPHRAGM ASSY—CHOKE PULLDOWN
23. E-CLIP—BUSHING RETAINER
24. RETAINER—MODULATOR SPRING
25. SPRING—MODULATOR
26. COLLAR—MODULATOR SPRING
27. BUSHING—GUIDE
28. SPRING—CHOKE ASSIST
29. CHOKE HOUSING SHAFT
30. LEVER—CHOKE THERMOSTAT
31. LEVER—DECHOKE
32. WASHER—CHOKE HOUSING SHAFT
33. NUT—CHOKE HOUSING SHAFT
34. SCREW (3)—COVER
35. COVER—POWER ENRICHMENT VALVE
    (IF EQUIPPED)
36. SPRING—POWER ENRICHMENT VALVE
    DIAPHRAGM
37. DIAPHRAGM ASSY—ENRICHMENT VALVE
    (OR BRASS PLUG)
38. SCREW (4)—COVER
39. COVER—PUMP DIAPHRAGM
40. DIAPHRAGM ASSY—ACCELERATOR PUMP
41. SPRING—DIAPHRAGM RETURN
42. HOLDER—PRIMARY IDLE JET
43. JET—PRIMARY IDLE
44. AIR BLEED—PRIMARY
45. TUBE—PRIMARY MAIN WELL
46. HOLDER—SECONDARY IDLE JET
47. JET—SECONDARY IDLE
48. AIR BLEED—SECONDARY
49. TUBE—SECONDARY MAIN WELL
50. NOZZLE ASSY—ACCELERATOR
    PUMP DISCHARGE
51. O-RING PUMP NOZZLE
52. CLIP (2)—JET RETAINER
53. PLUG—IDLE FUEL MIXTURE NEEDLE
54. NEEDLE—IDLE MIXTURE ADJUSTMENT
55. O-RING IDLE MIXTURE ADJUSTMENT NEEDLE
56. MAIN BODY ASSY
57. SPACER AND GASKET ASSY
58. MAIN JETS (2)
59. IDLE SPEED CONTROL
60. DASHPOT

Figure 9-13. Typical Motorcraft-Weber Model 5740 carburetor. (Ford)

10

Carter Carburetors

Carter carburetors are used on Chrysler Corporation, Ford Motor Company, and older American Motors vehicles. Some models have also been used on a few General Motors cars. This chapter contains descriptions and exploded drawings of the following Carter models:

- Models YF, YFA, and YFA-FB One-Barrel
- Model BBS One-Barrel
- Models BBD and BBD-FB Two-Barrel
- Model WCD Two-Barrel
- Model AFB Four-Barrel
- Model AVS Four-Barrel
- Model Thermo-Quad (TQ) Four-Barrel.

CARTER YF, YFA, AND YFA-FB ONE-BARREL

Many variations of this basic Carter design have been used over the years. All models have three sections: the airhorn, main body, and throttle body. Figures 10-1 and 10-2 are exploded views of the YFA and YFA-FB; the YF is similar to the YFA.

The airhorn is the main body cover. It contains the choke plate and choke control assembly, the fuel inlet fitting, the needle valve and seat, and the float and lever. The main body contains the idle, main, and accelerator pump systems. The throttle body holds the throttle valve and shaft assembly and the idle mixture adjusting screw.

Some versions have an antistall dashpot or solenoid throttle positioner attached to the airhorn. Model YFA uses a choke hot air inlet adapter and is vented to a charcoal canister by a mechanical external bowl vent. Models used on high-altitude applications in the late 1970's have altitude compensators to automatically lean the air-fuel mixture at higher elevations.

The YFA-FB version, figure 10-2, uses a feedback solenoid attached to the airhorn near the bowl vent tube. The solenoid meters air into both the idle and main circuits on direction from the engine control computer. Late models also use a throttle position sensor and an idle-speed-control device. The main body used with some applications may contain a temperature-compensated accelerator pump that sends part of the fuel back to the fuel bowl during pump operation at temperatures above 71°F (22°C).

CARTER BBS ONE-BARREL

The single-barrel Model BBS, figure 10-3, has three sections: the airhorn, the main body, and the throttle body. The airhorn is the main body cover and contains the choke plate and accelerator pump plunger. The main body contains the fuel bowl, fuel inlet, hot-idle compensator

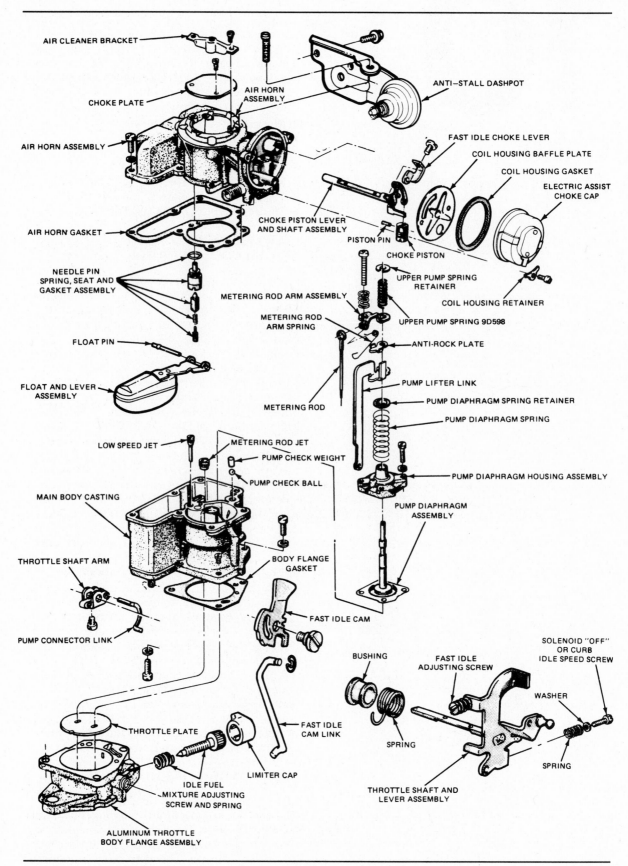

Figure 10-1. Typical Carter Model YFA carburetor. (Ford)

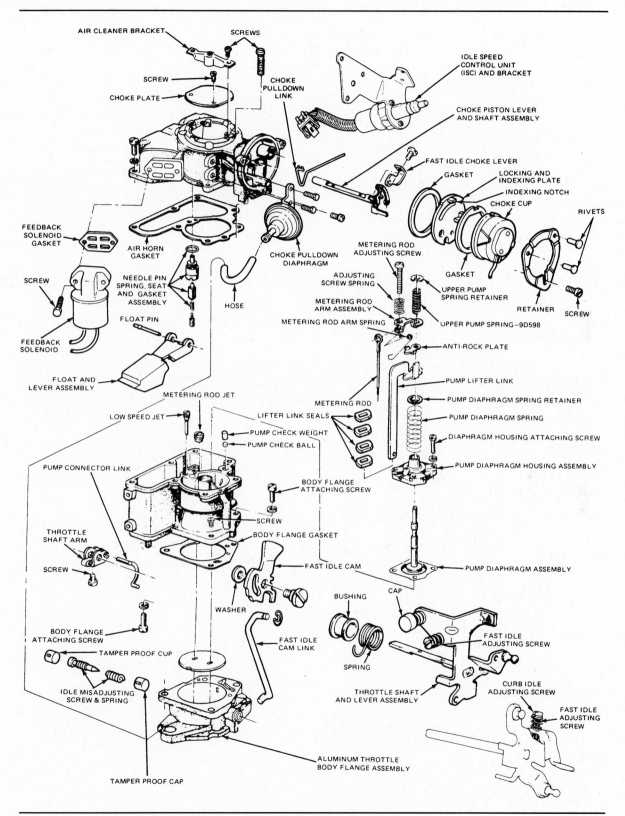

Figure 10-2. Typical Carter Model YFA-FB carburetor with feedback solenoid.

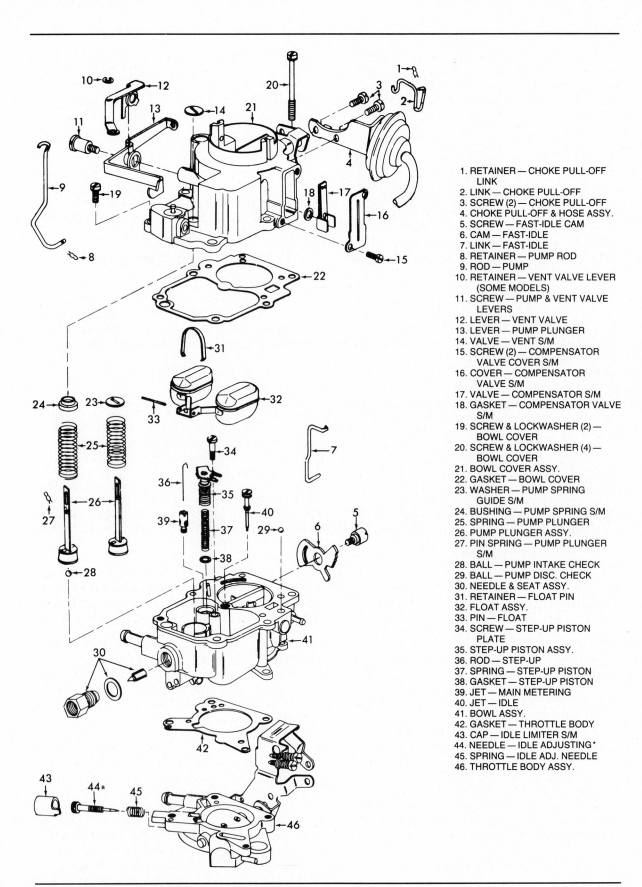

1. RETAINER — CHOKE PULL-OFF LINK
2. LINK — CHOKE PULL-OFF
3. SCREW (2) — CHOKE PULL-OFF
4. CHOKE PULL-OFF & HOSE ASSY.
5. SCREW — FAST-IDLE CAM
6. CAM — FAST-IDLE
7. LINK — FAST-IDLE
8. RETAINER — PUMP ROD
9. ROD — PUMP
10. RETAINER — VENT VALVE LEVER (SOME MODELS)
11. SCREW — PUMP & VENT VALVE LEVERS
12. LEVER — VENT VALVE
13. LEVER — PUMP PLUNGER
14. VALVE — VENT S/M
15. SCREW (2) — COMPENSATOR VALVE COVER S/M
16. COVER — COMPENSATOR VALVE S/M
17. VALVE — COMPENSATOR S/M
18. GASKET — COMPENSATOR VALVE S/M
19. SCREW & LOCKWASHER (2) — BOWL COVER
20. SCREW & LOCKWASHER (4) — BOWL COVER
21. BOWL COVER ASSY.
22. GASKET — BOWL COVER
23. WASHER — PUMP SPRING GUIDE S/M
24. BUSHING — PUMP SPRING S/M
25. SPRING — PUMP PLUNGER
26. PUMP PLUNGER ASSY.
27. PIN SPRING — PUMP PLUNGER S/M
28. BALL — PUMP INTAKE CHECK
29. BALL — PUMP DISC. CHECK
30. NEEDLE & SEAT ASSY.
31. RETAINER — FLOAT PIN
32. FLOAT ASSY.
33. PIN — FLOAT
34. SCREW — STEP-UP PISTON PLATE
35. STEP-UP PISTON ASSY.
36. ROD — STEP-UP
37. SPRING — STEP-UP PISTON
38. GASKET — STEP-UP PISTON
39. JET — MAIN METERING
40. JET — IDLE
41. BOWL ASSY.
42. GASKET — THROTTLE BODY
43. CAP — IDLE LIMITER S/M
44. NEEDLE — IDLE ADJUSTING*
45. SPRING — IDLE ADJ. NEEDLE
46. THROTTLE BODY ASSY.

Figure 10-3. Typical Carter Model BBS carburetor. (Borg-Warner)

and the idle, main, and accelerator metering systems. The throttle body holds the throttle valve and shaft assembly, as well as the idle mixture screw. Fuel supply is controlled by a stepup metering rod and a spring-operated accelerator pump.

CARTER BBD AND BBD-FB TWO-BARREL

The Model BBD is basically a 2-barrel version of the Model BBS. It has the same three main sections: the airhorn, main body, and throttle body. All models manufactured through 1973 are an air-bleed design using downhill nozzles, figure 10-4. Later models are a solid-fuel design with uphill nozzles, figure 10-5, which allows more precise fuel metering to meet emission control requirements.

The airhorn is the main body cover and contains the choke plate and choke control assembly, the bowl vent, and the accelerator pump plunger.

The main body contains twin fuel bowls, the fuel inlet and needle valve assembly, two pontoon floats, and a hot-idle compensator. Also located here are the five fuel metering circuits: idle, main, accelerator, power, and choke.

The throttle body contains the throttle valve and shaft assembly and two idle mixture screws. Many assist devices have been used with BBD carburetors: choke pulloff diaphragm, electric assist choke, wide-open throttle (WOT) dump valve, dashpot, idle speed solenoid, throttle positioner solenoid, transducer/ground switch, and many others.

The feedback version of the BBD (Model BBD-FB) has a variable air bleed controlled either by a stepper motor or a pulse solenoid.

CARTER WCD TWO-BARREL

Popular on Chrysler vehicles of the early 1960's, the Model WCD, figure 10-6, has four sections: the airhorn, the main body cover, the main body, and the throttle body. Four fuel metering circuits are used: idle, main, accelerator, and choke. These are contained in the main body along with twin fuel bowls and floats. The airhorn contains the choke valve and shaft assembly. The fuel inlet, metering rods, and accelerator pump piston are all housed in the main body cover. The throttle body holds the throttle valve and shaft assembly and two idle mixture adjusting screws.

CARTER AFB FOUR-BARREL

Used on GM and Chrysler vehicles during the 1960's and early 1970's, the Model AFB is a 3-piece carburetor of aluminum castings, figure 10-7. Four fuel metering circuits are located in the main body: idle, main, accelerator, and choke. Twin fuel bowls each feed a primary and a secondary nozzle on one side of the carburetor. Step-up rods, pistons, and springs can be serviced without removing the airhorn from the carburetor, or the carburetor from the engine. Both the primary and secondary barrels feature removable venturi clusters with high-speed and low-speed circuit parts. Mechanical secondaries linked to the primaries work with offset air velocity valves.

CARTER AVS FOUR-BARREL

Quite similar in appearance to the AFB, the Model AVS, figure 10-8, shares the same features. However, the AVS has no venturis in the secondary barrels. Airflow is controlled instead by air velocity valves at the tops of the barrel.

CARTER THERMO-QUAD (TQ) FOUR-BARREL

A 3-piece carburetor, the Thermo-Quad, figure 10-9, differs somewhat from other Carter designs. A molded plastic fuel bowl and suspended metering system result in lower fuel temperatures. This in turn means more precise fuel metering and better emission control. Twin metering rods are operated by a combination mechanical-vacuum system according to engine load. A cam on the primary throttle shaft raises the rods as the throttle starts to open; then vacuum control takes over.

The airhorn or upper body contains all fuel metering circuits: the choke valve and shaft assembly; secondary air valves and controls; the fuel inlet system; the low-speed, high-speed, and accelerator pump circuits; a vacuum-controlled step-up piston; fuel discharge nozzles, and all air bleeds. The throttle body contains the hot-idle compensator, the throttle valve and shaft assemblies, and the idle mixture adjusting screws. The main body is simply the molded plastic fuel bowl.

Fuel mixes with air *after* leaving the nozzle, instead of *before*. Mechanical secondaries are linked to the primaries. The vacuum-operated air valve is controlled by a spring on the mounting shaft and a diaphragm with two purposes. It holds back initial air valve operation when the throttle is opened quickly and provides choke pulldown when a cold engine is started.

A feedback version of the Thermo-Quad uses a duty-cycle solenoid to provide limited regulation of the air-fuel ratio on direction from the engine control computer. The solenoid meters airflow, working in conjunction with a conventional fixed main metering jet.

1. PIN SPRING
1A. RETAINER
2. SPACER
3. CHOKE CONNECTOR ROD
4. THROTTLE CONNECTOR ROD
5. PIN SPRING (SMALL)
6. CHOKE DIAPHRAGM
   CONNECTOR LINK
7. HOSE
8. CHOKE SHAFT LEVER
9. CHOKE DIAPHRAGM SCREWS
10. CHOKE DIAPHRAGM
    ASSEMBLY
11. AIRHORN SCREWS (SHORT)
12. AIRHORN SCREWS (LONG)
13. DASHPOT AND BRACKET
    ASSEMBLY
14. AIRHORN
15. AIRHORN GASKET
16. PUMP PLUNGER ASSEMBLY
17. PUMP PLUNGER WASHER
18. PUMP PLUNGER BUSHING
19. PUMP PLUNGER SPRING
20. PIN SPRING (PLUNGER ROD)
20A. PLUNGER SHAFT RETAINER
21. PUMP ARM SCREW
22. PUMP ARM
23. VENTURI CLUSTER SCREW
24. VENTURI COVER
25. VENTURI COVER GASKET
26. VENTURI CLUSTER ASSEMBLY
27. VENTURI CLUSTER GASKET
28. PUMP INTAKE CHECK BALL
    (LARGE)
29. PUMP DISCHARGE CHECK
    BALL (SMALL)
30. STEP-UP PISTON PLATE
    SCREW
31. STEP-UP PISTON PLATE
32. STEP-UP PISTON ROD (2)
33. STEP-UP PISTON SPRING (2)
34. STEP-UP PISTON GASKET
35. NEEDLE & SEAT ASSEMBLY
36. FLOAT LEVER PIN RETAINER
37. FLOAT & LEVER ASSEMBLY
38. FLOAT LEVER PIN
39. MAIN JETS
40. COMPENSATOR VALVE SCREW
41. COMPENSATOR VALVE COVER
42. COMPENSATOR VALVE
43. COMPENSATOR GASKET
44. MAIN BODY CASTING
45. BODY FLANGE SCREW
46. BODY FLANGE GASKET
47. IDLE LIMITER CAP
48. IDLE MIXTURE SCREW
49. THROTTLE SPEED SCREW
50. FLANGE ASSEMBLY
A. RETAINER (2)
B. WASHER
C. PUMP ARM SCREW
D. PUMP ARM
E. COVER PLATE SCREW
F. COVER PLATE
G. VENT VALVE SPRING
H. VENT VALVE
J. COVER PLATE GASKET
K. PUMP PLUNGER
L. PUMP PLUNGER SPRING

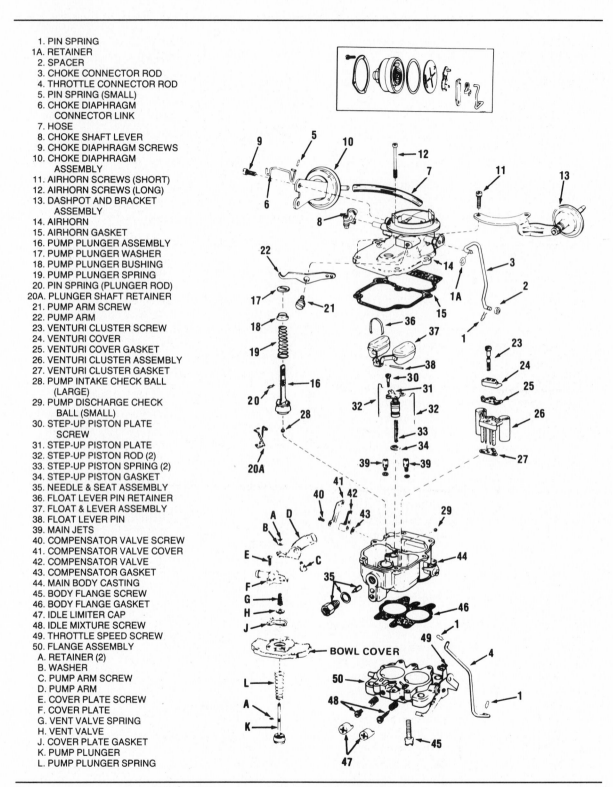

Figure 10-4. Typical Carter Model BBD carburetor (air-bleed design). (Carter)

1. CHOKE SHAFT LEVER SCREW
2. CHOKE SHAFT LEVER
3. CHOKE PULL-OFF ROD
4. CHOKE PULL-OFF BRACKET SCREW
5. CHOKE PULL-OFF AND BRACKET
5A. CHOKE PULL-OFF HOUSING — IF EQUIPPED
5B. CHOKE PULL-OFF HOUSING RIVETS — IF EQUIPPED
6. "E" RETAINER
7. THROTTLE CONNECTOR ROD
8. FAST-IDLE CAM SCREW
9. FAST-IDLE CAM
10. FAST-IDLE ROD
11. DUST COVER SCREW
12. DUST COVER
13. DUST COVER GASKET
14. PUMP AND METERING ROD ARM SCREW (2)
15. PUMP AND METERING ROD ARM WASHER (2)
16. PUMP COUNTER SHAFT
17. METERING ROD ARM
18. PUMP ARM
19. PUMP "S" LINK
20. VACUUM PISTON ASSEMBLY
21. METERING ROD (2)
22. VENT VALVE GROMMET SEAL — IF EQUIPPED
23. CHOKE PULL-OFF HOSE
24. E.G.R. DUMP VALVE HOSE
25. BOWL COVER SCREW
26. CARBURETOR IDENTIFICATION TAG
27. BOWL COVER AND BRACKET SCREW (2)
28. BOWL COVER (2)
29. BOWL COVER GASKET
30. SOLENOID & BRACKET
30A. VACUUM MODULATOR
30B. TRANSDUCER, BRACKET & IDLE GROUND POST — IF EQUIPPED
31. IDLE ENRICHMENT COVER SCREW (3)
32. IDLE ENRICHMENT COVER
33. IDLE ENRICHMENT COVER SPRING
34. IDLE ENRICHMENT COVER GASKET
35. IDLE ENRICHMENT DIAPHRAGM
36. PLUNGER SPRING
37. PLUNGER ASSEMBLY
38. INTAKE CHECK BALL (LARGE)
39. VACUUM PISTON SPRING
40. NEEDLE, SEAT, AND GASKET
41. BAFFLE
42. FLOAT PIN RETAINER
43. FLOAT
44. FLOAT PIN
45. MAIN METERING JETS (2)
46. VENTURI CLUSTER SCREW (2)
47. VENTURI COVER ASSEMBLY
47A. VENTURI COVER ASSEMBLY (ALT.)
48. VENTURI COVER GASKET
49. VENTURI CLUSTER ASSEMBLY
50. VENTURI GASKET
51. DISCHARGE CHECK BALL (SMALL)

52. BODY FLANGE SCREW
53. BODY FLANGE
54. BODY FLANGE GASKET
55. MAIN BODY CASTING
56. E.G.R. DUMP VALVE BRACKET SCREW
57. E.G.R. DUMP VALVE AND BRACKET
58. LIMITER CAP (2)
59. IDLE MIXTURE SCREW (2)
60. IDLE MIXTURE SCREW SPRING (2)
61. IDLE MIXTURE SCREW PLUGS (2)

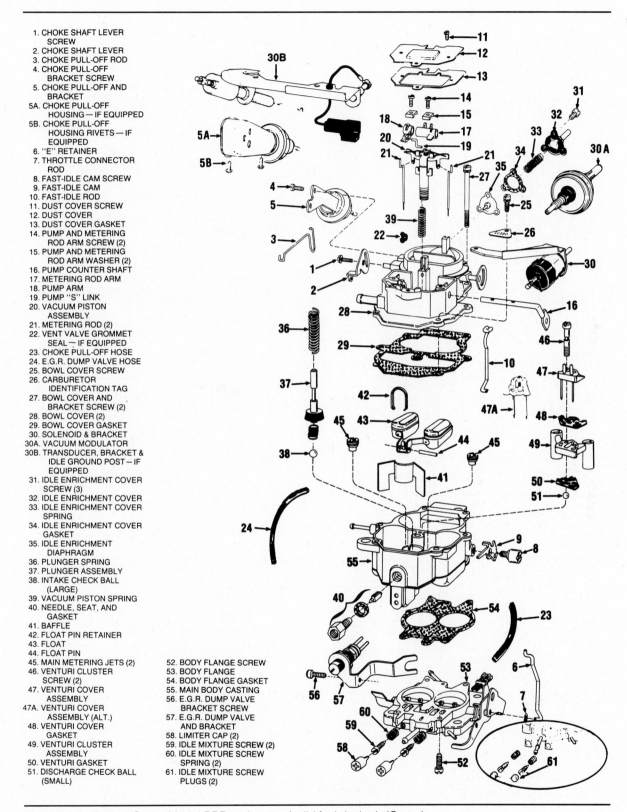

Figure 10-5. Typical Carter Model BBD carburetor (solid fuel design). (Carter)

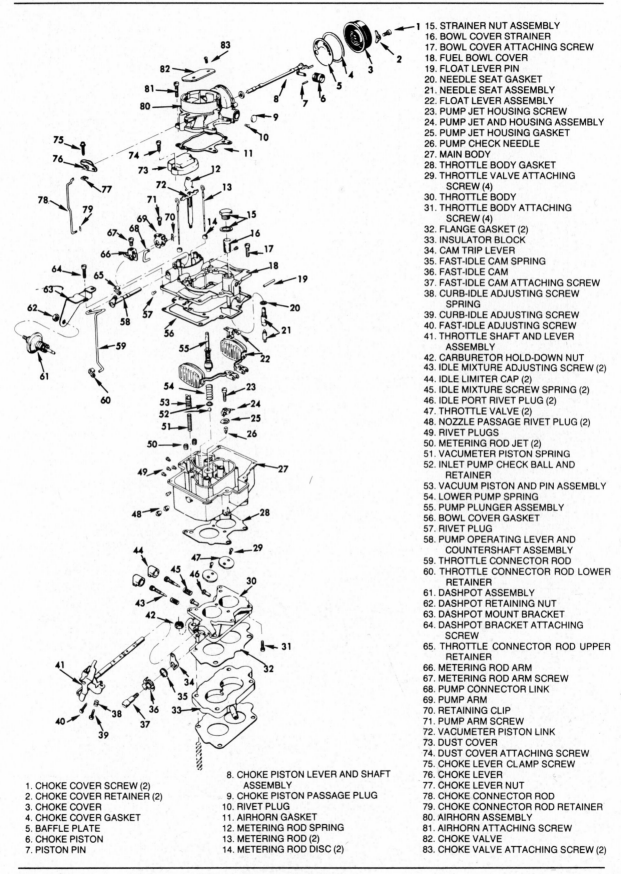

15. STRAINER NUT ASSEMBLY
16. BOWL COVER STRAINER
17. BOWL COVER ATTACHING SCREW
18. FUEL BOWL COVER
19. FLOAT LEVER PIN
20. NEEDLE SEAT GASKET
21. NEEDLE SEAT ASSEMBLY
22. FLOAT LEVER ASSEMBLY
23. PUMP JET HOUSING SCREW
24. PUMP JET AND HOUSING ASSEMBLY
25. PUMP JET HOUSING GASKET
26. PUMP CHECK NEEDLE
27. MAIN BODY
28. THROTTLE BODY GASKET
29. THROTTLE VALVE ATTACHING
    SCREW (4)
30. THROTTLE BODY
31. THROTTLE BODY ATTACHING
    SCREW (4)
32. FLANGE GASKET (2)
33. INSULATOR BLOCK
34. CAM TRIP LEVER
35. FAST-IDLE CAM SPRING
36. FAST-IDLE CAM
37. FAST-IDLE CAM ATTACHING SCREW
38. CURB-IDLE ADJUSTING SCREW
    SPRING
39. CURB-IDLE ADJUSTING SCREW
40. FAST-IDLE ADJUSTING SCREW
41. THROTTLE SHAFT AND LEVER
    ASSEMBLY
42. CARBURETOR HOLD-DOWN NUT
43. IDLE MIXTURE ADJUSTING SCREW (2)
44. IDLE LIMITER CAP (2)
45. IDLE MIXTURE SCREW SPRING (2)
46. IDLE PORT RIVET PLUG (2)
47. THROTTLE VALVE (2)
48. NOZZLE PASSAGE RIVET PLUG (2)
49. RIVET PLUGS
50. METERING ROD JET (2)
51. VACUMETER PISTON SPRING
52. INLET PUMP CHECK BALL AND
    RETAINER
53. VACUUM PISTON AND PIN ASSEMBLY
54. LOWER PUMP SPRING
55. PUMP PLUNGER ASSEMBLY
56. BOWL COVER GASKET
57. RIVET PLUG
58. PUMP OPERATING LEVER AND
    COUNTERSHAFT ASSEMBLY
59. THROTTLE CONNECTOR ROD
60. THROTTLE CONNECTOR ROD LOWER
    RETAINER
61. DASHPOT ASSEMBLY
62. DASHPOT RETAINING NUT
63. DASHPOT MOUNT BRACKET
64. DASHPOT BRACKET ATTACHING
    SCREW
65. THROTTLE CONNECTOR ROD UPPER
    RETAINER
66. METERING ROD ARM
67. METERING ROD ARM SCREW
68. PUMP CONNECTOR LINK
69. PUMP ARM
70. RETAINING CLIP
71. PUMP ARM SCREW
72. VACUMETER PISTON LINK
73. DUST COVER
74. DUST COVER ATTACHING SCREW
75. CHOKE LEVER CLAMP SCREW
76. CHOKE LEVER
77. CHOKE LEVER NUT
78. CHOKE CONNECTOR ROD
79. CHOKE CONNECTOR ROD RETAINER
80. AIRHORN ASSEMBLY
81. AIRHORN ATTACHING SCREW
82. CHOKE VALVE
83. CHOKE VALVE ATTACHING SCREW (2)

1. CHOKE COVER SCREW (2)
2. CHOKE COVER RETAINER (2)
3. CHOKE COVER
4. CHOKE COVER GASKET
5. BAFFLE PLATE
6. CHOKE PISTON
7. PISTON PIN

8. CHOKE PISTON LEVER AND SHAFT
   ASSEMBLY
9. CHOKE PISTON PASSAGE PLUG
10. RIVET PLUG
11. AIRHORN GASKET
12. METERING ROD SPRING
13. METERING ROD (2)
14. METERING ROD DISC (2)

Figure 10-6. Typical Carter Model WCD carburetor. (Borg-Warner)

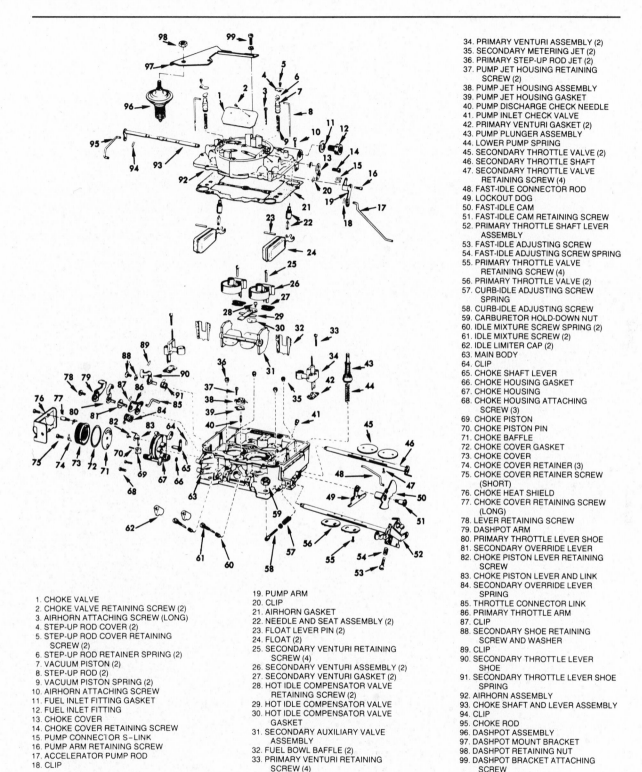

34. PRIMARY VENTURI ASSEMBLY (2)
35. SECONDARY METERING JET (2)
36. PRIMARY STEP-UP ROD JET (2)
37. PUMP JET HOUSING RETAINING SCREW (2)
38. PUMP JET HOUSING ASSEMBLY
39. PUMP JET HOUSING GASKET
40. PUMP DISCHARGE CHECK NEEDLE
41. PUMP INLET CHECK VALVE
42. PRIMARY VENTURI GASKET (2)
43. PUMP PLUNGER ASSEMBLY
44. LOWER PUMP SPRING
45. SECONDARY THROTTLE VALVE (2)
46. SECONDARY THROTTLE SHAFT
47. SECONDARY THROTTLE VALVE RETAINING SCREW (4)
48. FAST-IDLE CONNECTOR ROD
49. LOCKOUT DOG
50. FAST-IDLE CAM
51. FAST-IDLE CAM RETAINING SCREW
52. PRIMARY THROTTLE SHAFT LEVER ASSEMBLY
53. FAST-IDLE ADJUSTING SCREW
54. FAST-IDLE ADJUSTING SCREW SPRING
55. PRIMARY THROTTLE VALVE RETAINING SCREW (4)
56. PRIMARY THROTTLE VALVE (2)
57. CURB-IDLE ADJUSTING SCREW SPRING
58. CURB-IDLE ADJUSTING SCREW
59. CARBURETOR HOLD-DOWN NUT
60. IDLE MIXTURE SCREW SPRING (2)
61. IDLE MIXTURE SCREW (2)
62. IDLE LIMITER CAP (2)
63. MAIN BODY
64. CLIP
65. CHOKE SHAFT LEVER
66. CHOKE HOUSING GASKET
67. CHOKE HOUSING
68. CHOKE HOUSING ATTACHING SCREW (3)
69. CHOKE PISTON
70. CHOKE PISTON PIN
71. CHOKE BAFFLE
72. CHOKE COVER GASKET
73. CHOKE COVER
74. CHOKE COVER RETAINER (3)
75. CHOKE COVER RETAINER SCREW (SHORT)
76. CHOKE HEAT SHIELD
77. CHOKE COVER RETAINING SCREW (LONG)
78. LEVER RETAINING SCREW
79. DASHPOT ARM
80. PRIMARY THROTTLE LEVER SHOE
81. SECONDARY OVERRIDE LEVER
82. CHOKE PISTON LEVER RETAINING SCREW
83. CHOKE PISTON LEVER AND LINK
84. SECONDARY OVERRIDE LEVER SPRING
85. THROTTLE CONNECTOR LINK
86. PRIMARY THROTTLE ARM
87. CLIP
88. SECONDARY SHOE RETAINING SCREW AND WASHER
89. CLIP
90. SECONDARY THROTTLE LEVER SHOE
91. SECONDARY THROTTLE LEVER SHOE SPRING
92. AIRHORN ASSEMBLY
93. CHOKE SHAFT AND LEVER ASSEMBLY
94. CLIP
95. CHOKE ROD
96. DASHPOT ASSEMBLY
97. DASHPOT MOUNT BRACKET
98. DASHPOT RETAINING NUT
99. DASHPOT BRACKET ATTACHING SCREW

1. CHOKE VALVE
2. CHOKE VALVE RETAINING SCREW (2)
3. AIRHORN ATTACHING SCREW (LONG)
4. STEP-UP ROD COVER (2)
5. STEP-UP ROD COVER RETAINING SCREW (2)
6. STEP-UP ROD RETAINER SPRING (2)
7. VACUUM PISTON (2)
8. STEP-UP ROD (2)
9. VACUUM PISTON SPRING (2)
10. AIRHORN ATTACHING SCREW
11. FUEL INLET FITTING GASKET
12. FUEL INLET FITTING
13. CHOKE COVER
14. CHOKE COVER RETAINING SCREW
15. PUMP CONNECTOR S-LINK
16. PUMP ARM RETAINING SCREW
17. ACCELERATOR PUMP ROD
18. CLIP

19. PUMP ARM
20. CLIP
21. AIRHORN GASKET
22. NEEDLE AND SEAT ASSEMBLY (2)
23. FLOAT LEVER PIN (2)
24. FLOAT (2)
25. SECONDARY VENTURI RETAINING SCREW (4)
26. SECONDARY VENTURI ASSEMBLY (2)
27. SECONDARY VENTURI GASKET (2)
28. HOT IDLE COMPENSATOR VALVE RETAINING SCREW (2)
29. HOT IDLE COMPENSATOR VALVE
30. HOT IDLE COMPENSATOR VALVE GASKET
31. SECONDARY AUXILIARY VALVE ASSEMBLY
32. FUEL BOWL BAFFLE (2)
33. PRIMARY VENTURI RETAINING SCREW (4)

Figure 10-7. Typical Carter Model AFB carburetor. (Borg-Warner)

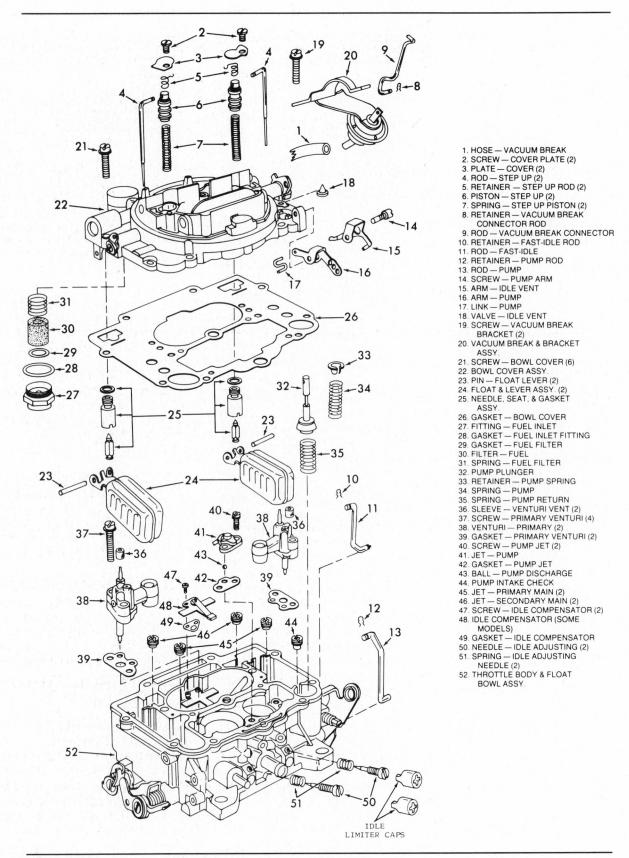

1. HOSE — VACUUM BREAK
2. SCREW — COVER PLATE (2)
3. PLATE — COVER (2)
4. ROD — STEP UP (2)
5. RETAINER — STEP UP ROD (2)
6. PISTON — STEP UP (2)
7. SPRING — STEP UP PISTON (2)
8. RETAINER — VACUUM BREAK
    CONNECTOR ROD
9. ROD — VACUUM BREAK CONNECTOR
10. RETAINER — FAST-IDLE ROD
11. ROD — FAST-IDLE
12. RETAINER — PUMP ROD
13. ROD — PUMP
14. SCREW — PUMP ARM
15. ARM — IDLE VENT
16. ARM — PUMP
17. LINK — PUMP
18. VALVE — IDLE VENT
19. SCREW — VACUUM BREAK
    BRACKET (2)
20. VACUUM BREAK & BRACKET
    ASSY.
21. SCREW — BOWL COVER (6)
22. BOWL COVER ASSY.
23. PIN — FLOAT LEVER (2)
24. FLOAT & LEVER ASSY. (2)
25. NEEDLE, SEAT, & GASKET
    ASSY.
26. GASKET — BOWL COVER
27. FITTING — FUEL INLET
28. GASKET — FUEL INLET FITTING
29. GASKET — FUEL FILTER
30. FILTER — FUEL
31. SPRING — FUEL FILTER
32. PUMP PLUNGER
33. RETAINER — PUMP SPRING
34. SPRING — PUMP
35. SPRING — PUMP RETURN
36. SLEEVE — VENTURI VENT (2)
37. SCREW — PRIMARY VENTURI (4)
38. VENTURI — PRIMARY (2)
39. GASKET — PRIMARY VENTURI (2)
40. SCREW — PUMP JET (2)
41. JET — PUMP
42. GASKET — PUMP JET
43. BALL — PUMP DISCHARGE
44. PUMP INTAKE CHECK
45. JET — PRIMARY MAIN (2)
46. JET — SECONDARY MAIN (2)
47. SCREW — IDLE COMPENSATOR (2)
48. IDLE COMPENSATOR (SOME
    MODELS)
49. GASKET — IDLE COMPENSATOR
50. NEEDLE — IDLE ADJUSTING (2)
51. SPRING — IDLE ADJUSTING
    NEEDLE (2)
52. THROTTLE BODY & FLOAT
    BOWL ASSY.

IDLE
LIMITER CAPS

Figure 10-8. Typical Carter Model AVS carburetor. (Borg-Warner)

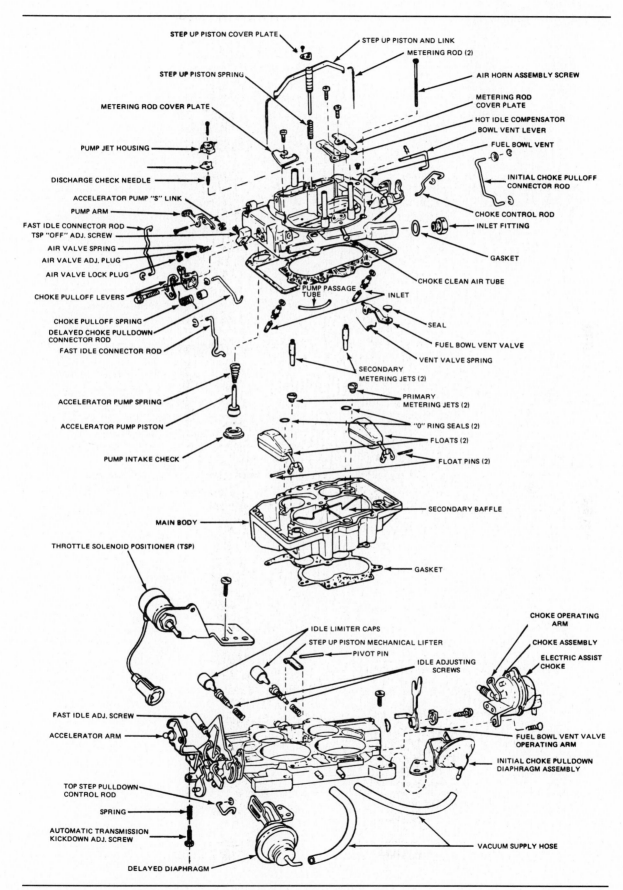

Figure 10-9. Typical Carter Thermo-Quad (TQ) carburetor. (Chrysler)

Chapter

11

Holley Carburetors

Holley carburetors have been used by all domestic automakers. Basic carburetor features are described in this chapter with a typical exploded drawing provided to help you understand the design. The following carburetors are included:

- Model 1920 One-Barrel
- Model 1931 One-Barrel
- Models 1940, 1945, and 6145 One-Barrel
- Model 1946 One-Barrel
- Models 1949, 1949-C, and 6149 One-Barrel
- Model 2209 Two-Barrel
- Models 2210 and 2245 Two-Barrel
- Models 2280 and 6280 Two-Barrel
- Model 2300 Two-Barrel
- Models 5200, 5210-C, 5220, 6500, 6510-C, and 6520 Two-Barrel
- Models 4150, 4160, and 4180 Four-Barrel
- Models 4165 and 4175 Four-Barrel
- Model 4360 Four-Barrel.

HOLLEY 1920 ONE-BARREL

The Model 1920, figure 11-1, is a single casting with detachable fuel bowl, main well, and power valve (economizer) assemblies. Four fuel metering circuits are used: idle, main, power enrichment, and accelerator pump. The vacuum-operated, spring-staged choke valve is contained within the carburetor bore and connected by linkage to a remote choke device in the intake manifold. A power valve in the metering body is operated by manifold vacuum. Some versions use a dashpot mounted on the body assembly to retard throttle return to the idle position.

HOLLEY 1931 ONE-BARREL

The single-barrel Model 1931, figure 11-2, has a 1-piece casting with detachable fuel bowl and an integral choke. Four fuel metering circuits are used: idle, main, power enrichment, and accelerator pump. The choke valve is contained within the carburetor bore and is connected to the thermostatic spring in the choke housing by a shaft and lever assembly. Manifold vacuum is used to operate the power valve in the metering body. Fuel bowl venting is by a fixed internal vent to the airhorn. An external vent opens at idle speeds.

HOLLEY 1940, 1945, AND 6145 ONE-BARREL

This 3-piece design used primarily by Chrysler, figure 11-3, has an airhorn, main body, and throttle body. All versions use four basic fuel

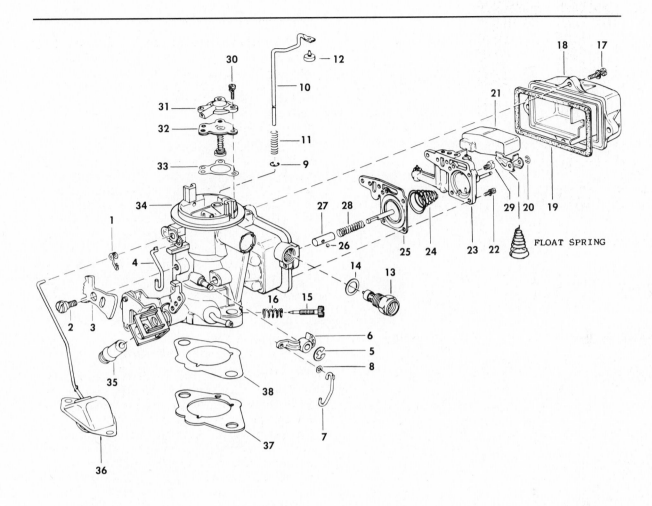

1. CHOKE ROD RETAINER CLIP
2. FAST-IDLE CAM SCREW
3. FAST-IDLE CAM
4. FAST-IDLE ROD
5. PUMP LEVER RETAINER ("E" WASHER)
6. PUMP LEVER
7. PUMP LINK
8. PUMP-LINK WASHER
9. BOWL VENT ROD RETAINER ("E" WASHER)
10. BOWL VENT ROD
11. BOWL VENT ROD SPRING
12. BOWL VENT VALVE
13. NEEDLE AND SEAT ASSEMBLY
14. NEEDLE SEAT GASKET
15. IDLE-MIXTURE ADJUSTING NEEDLE
16. IDLE ADJUSTING NEEDLE SPRING
17. SCREW AND LOCKWASHER ASSEMBLY
18. FUEL BOWL
19. FUEL-BOWL GASKET

20. FLOAT LEVER RETAINER ("E" WASHER)
21. FLOAT ASSEMBLY
22. SCREW AND LOCKWASHER ASSEMBLY
23. MAIN WELL AND ECONOMIZER BODY ASSY.
24. PUMP DIAPHRAGM SPRING
25. PUMP DIAPHRAGM ASSEMBLY
26. PUMP PUSH-ROD SLEEVE BALL
27. PUMP PUSH-ROD SLEEVE
28. PUMP PUSH-ROD SPRING
29. MAIN-METERING JET
30. SCREW AND LOCKWASHER ASSEMBLY
31. ECONOMIZER DIAPHRAGM COVER
32. ECONOMIZER DIAPHRAGM ASSEMBLY
33. ECONOMIZER DIAPHRAGM GASKET
34. BODY ASSEMBLY
35. THROTTLE-ROD INSULATOR BUSHING
36. AUTOMATIC CHOKE ASSY. (WELL TYPE)
37. MANIFOLD FLANGE GASKET, THICK
38. MANIFOLD FLANGE GASKET, THIN

Figure 11-1. Typical Holley Model 1920 carburetor. (Borg-Warner)

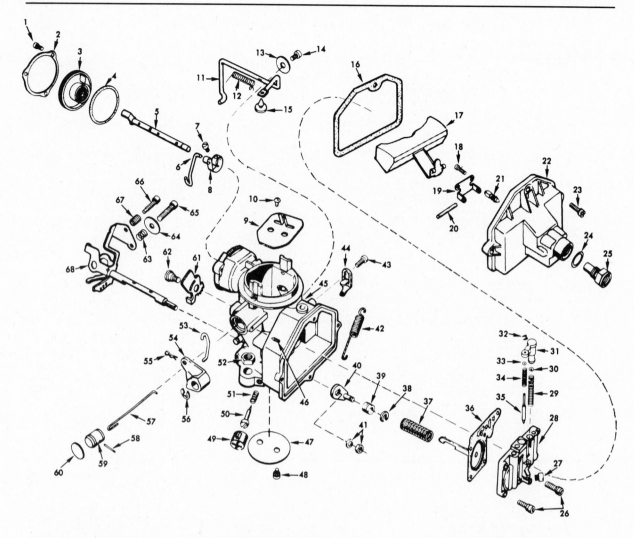

Figure 11-2. Typical Holley Model 1931 carburetor. (Holley)

1. CHOKE COVER SCREW (3)
2. CHOKE COVER RETAINER
3. CHOKE COVER ASSEMBLY
4. CHOKE COVER GASKET
5. CHOKE SHAFT ASSEMBLY
6. FAST-IDLE CONNECTOR ROD
7. CHOKE LEVER SCREW
8. CHOKE LEVER
9. CHOKE VALVE
10. CHOKE VALVE SCREW (2)
11. BOWL VENT ROD
12. BOWL VENT ROD SPRING
13. VENT ROD RETAINER WASHER (2)
14. VENT ROD RETAINER SCREW (2)
15. BOWL VENT VALVE
16. FUEL BOWL GASKET
17. FLOAT ASSEMBLY
18. FLOAT SHAFT BRACKET SCREW (2)
19. FLOAT SHAFT RETAINING BRACKET
20. FLOAT SHAFT
21. FUEL INLET NEEDLE
22. FUEL BOWL
23. FUEL BOWL RETAINING SCREW (4)

24. FUEL INLET GASKET
25. FUEL INLET FITTING
26. METERING BLOCK SCREW (5)
27. MAIN JET
28. METERING BLOCK
29. POWER VALVE PISTON SPRING
30. POWER VALVE SHIM
31. POWER VALVE PISTON
32. POWER VALVE PISTON RETAINER
33. POWER VALVE SPRING WASHER
34. POWER VALVE SPRING
35. POWER VALVE
36. PUMP DIAPHRAGM ASSEMBLY
37. PUMP DIAPHRAGM SPRING
38. RETAINING WASHER
39. LEVER PIN COLLAR
40. RETURN SPRING LEVER
41. RETURN LEVER LOCKWASHER AND NUT
42. THROTTLE RETURN SPRING
43. PERCH BRACKET RETAINING SCREW
44. PERCH BRACKET
45. MAIN BODY
46. CHOKE PISTON STOP SCREW

47. THROTTLE VALVE
48. THROTTLE VALVE SCREW (2)
49. IDLE LIMITER CAP
50. IDLE MIXTURE ADJUSTMENT SCREW
51. IDLE MIXTURE SCREW SPRING
52. CARBURETOR HOLD-DOWN NUT
53. PUMP LINK
54. PUMP OPERATING LEVER
55. PUMP LINK RETAINER
56. PUMP OPERATING LEVER RETAINER
57. CHOKE PISTON LINK
58. CHOKE PISTON PIN
59. CHOKE PISTON
60. CHOKE PISTON PLUG
61. FAST-IDLE CAM
62. FAST-IDLE CAM SCREW
63. CURB-IDLE ADJUSTING SCREW SPRING
64. VENT ROD OPERATING WASHER
65. CURB-IDLE ADJUSTING SCREW
66. FAST-IDLE ADJUSTING SCREW
67. FAST-IDLE ADJUSTING SCREW
68. THROTTLE SHAFT ASSEMBLY

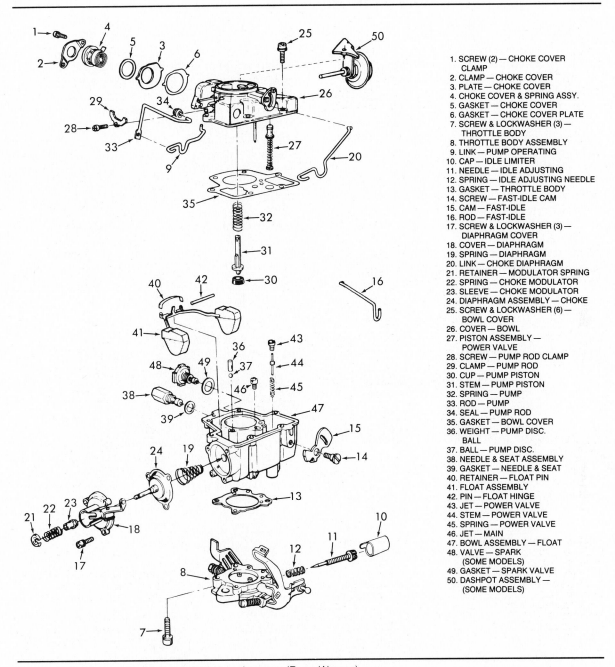

Figure 11-3. Typical Holley Model 1940 carburetor. (Borg-Warner)

1. SCREW (2) — CHOKE COVER CLAMP
2. CLAMP — CHOKE COVER
3. PLATE — CHOKE COVER
4. CHOKE COVER & SPRING ASSY.
5. GASKET — CHOKE COVER
6. GASKET — CHOKE COVER PLATE
7. SCREW & LOCKWASHER (3) — THROTTLE BODY
8. THROTTLE BODY ASSEMBLY
9. LINK — PUMP OPERATING
10. CAP — IDLE LIMITER
11. NEEDLE — IDLE ADJUSTING
12. SPRING — IDLE ADJUSTING NEEDLE
13. GASKET — THROTTLE BODY
14. SCREW — FAST-IDLE CAM
15. CAM — FAST-IDLE
16. ROD — FAST-IDLE
17. SCREW & LOCKWASHER (3) — DIAPHRAGM COVER
18. COVER — DIAPHRAGM
19. SPRING — DIAPHRAGM
20. LINK — CHOKE DIAPHRAGM
21. RETAINER — MODULATOR SPRING
22. SPRING — CHOKE MODULATOR
23. SLEEVE — CHOKE MODULATOR
24. DIAPHRAGM ASSEMBLY — CHOKE
25. SCREW & LOCKWASHER (6) — BOWL COVER
26. COVER — BOWL
27. PISTON ASSEMBLY — POWER VALVE
28. SCREW — PUMP ROD CLAMP
29. CLAMP — PUMP ROD
30. CUP — PUMP PISTON
31. STEM — PUMP PISTON
32. SPRING — PUMP
33. ROD — PUMP
34. SEAL — PUMP ROD
35. GASKET — BOWL COVER
36. WEIGHT — PUMP DISC. BALL
37. BALL — PUMP DISC.
38. NEEDLE & SEAT ASSEMBLY
39. GASKET — NEEDLE & SEAT
40. RETAINER — FLOAT PIN
41. FLOAT ASSEMBLY
42. PIN — FLOAT HINGE
43. JET — POWER VALVE
44. STEM — POWER VALVE
45. SPRING — POWER VALVE
46. JET — MAIN
47. BOWL ASSEMBLY — FLOAT
48. VALVE — SPARK (SOME MODELS)
49. GASKET — SPARK VALVE
50. DASHPOT ASSEMBLY — (SOME MODELS)

metering circuits: idle-transfer, main, accelerator pump, and power enrichment. An insulating gasket between the main body and throttle body prevents heat transfer to the fuel bowl. The fuel bowl completely surrounds the venturi and uses dual floats. The main body contains the fuel inlet and all metering systems. The choke system is built into the Model 1940, while the more sophisticated Model 1945 uses a remote choke. The Model 1945 design is vented to a charcoal canister. Some versions have an idle enrichment valve.

The Model 6145 feedback version uses a mixture control solenoid mounted in the airhorn. This solenoid meters airflow on command from the engine control computer and works in conjunction with the conventional fixed main metering jet.

HOLLEY 1946 ONE-BARREL

The 3-piece Model 1946, figure 11-4, is found primarily on Ford inline 6-cylinder engines of the late 1970's and early 1980's. The carburetor has five basic fuel metering circuits: idle, main, power enrichment, accelerator pump, and choke. The airhorn contains the fuel bowl vent, accelerator pump and enrichment valve pistons, and integral choke housing.

Fuel inlet is through the main body, which also contains twin fuel bowls with dual pontoon floats and a hot-idle compensator. The throttle body contains the throttle valve and shaft assembly and the idle mixture screw.

Model 1945 carburetors may use a throttle solenoid positioner or a 2-stage throttle kicker to prevent dieseling and maintain a suitable idle speed with air conditioning. Altitude compensation is accomplished by carburetor calibration.

HOLLEY 1949, 1949-C, AND 6149 TWO-BARREL

This single-barrel booster style carburetor, figure 11-5, is found on Ford vehicles with the 4-cylinder high-swirl combustion (HSC) engine. The 1949 and 1949-C are used on Canadian models and the 6149 feedback version is used on U.S. Ford products.

The 1949 and 1949-C use eight fuel metering circuits: idle and idle transfer, main metering, vacuum-gradient power-enrichment, auxiliary main metering, wide-open throttle (WOT) power enrichment, accelerator, and choke. The 6149 lacks the vacuum-gradient power-enrichment but has feedback control of the idle and main metering circuits.

All carburetors use a hot-idle compensator and a mechanical bowl vent. Altitude compensation is a function of the engine control computer with the Model 6149; the Model 1949 and 1949-C carburetors use a remotely mounted aneroid altitude compensator.

A remotely mounted duty-cycle solenoid controls vacuum to a fuel control diaphragm and actuator mounted in the airhorn. Movement of the actuator affects the position of a tapered metering valve needle in the main body, regulating the flow of fuel as determined by the computer to maintain the correct air-fuel ratio.

HOLLEY 2209 TWO-BARREL

Figure 11-6 illustrates this 3-piece design consisting of an airhorn, a main body, and a throttle body. The Model 2209 contains four basic fuel metering circuits: idle, main, accelerator pump, and power enrichment. The integral choke is connected by linkage to the choke valve in the airhorn. A thermostatic spring in the choke housing and an external diaphragm assembly control choke operation. Fuel inlet is through the airhorn, which also contains the power valve piston. The throttle body houses the throttle valve and shaft assembly and two idle mixture adjusting screws. A dashpot is used on some models to retard throttle closing at idle.

HOLLEY 2210 AND 2245 TWO-BARREL

The Models 2210 and 2245 are a 3-piece design used by Chrysler, figure 11-7, and consist of an airhorn, a main body, and a throttle body. The design has five basic fuel metering circuits: basic idle, idle enrichment, main, accelerator pump, and power enrichment. Fuel inlet is through the airhorn, with a single fuel bowl and float feeding both barrels. Both versions use a remote choke device linked to the choke valve in the airhorn. The throttle body contains the throttle valve and shaft assembly and two idle mixture adjusting screws. Some versions use a hot-idle compensator.

HOLLEY 2280 AND 6280 TWO-BARREL

The Models 2280 and 6280 also are a 3-piece carburetor design with four basic metering systems: idle and low-speed, main metering, accelerator pump, and power enrichment. They are more similar in appearance to the Carter BBD design than to other Holley 2200-series carburetors. The power enrichment system uses both mechanical and vacuum-operated power valves to deliver mixture enrichment whenever higher power is required. The mechanical power valve is located on the choke side of the main body; the vacuum valve is installed on the throttle lever side. Fuel inlet is through the main body. Both versions use a remote choke linked to the choke valve in the airhorn. The throttle body contains the throttle valve and shaft assembly and two idle mixture adjusting screws under concealment plugs.

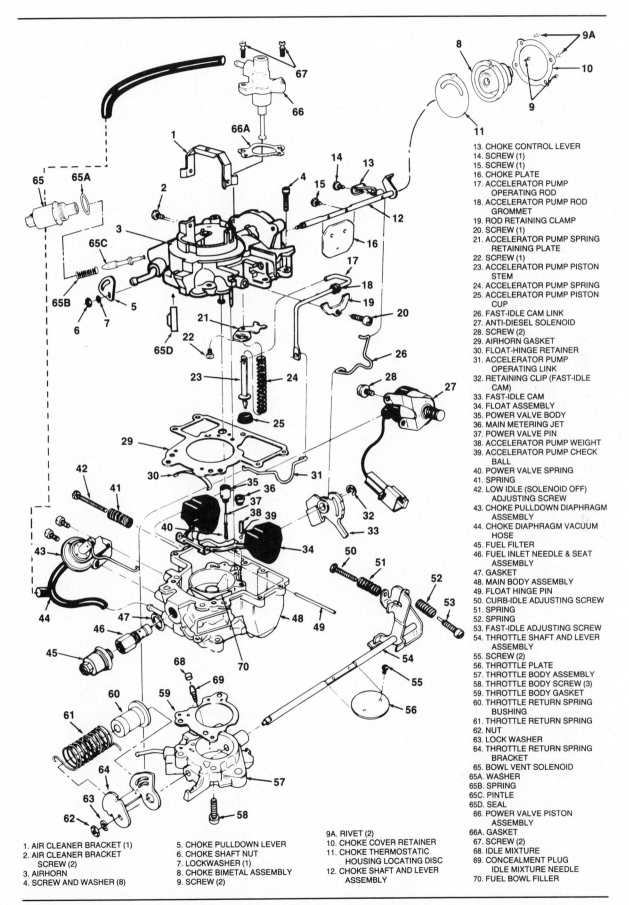

1. AIR CLEANER BRACKET (1)
2. AIR CLEANER BRACKET SCREW (2)
3. AIRHORN
4. SCREW AND WASHER (8)
5. CHOKE PULLDOWN LEVER
6. CHOKE SHAFT NUT
7. LOCKWASHER (1)
8. CHOKE BIMETAL ASSEMBLY
9. SCREW (2)

9A. RIVET (2)
10. CHOKE COVER RETAINER
11. CHOKE THERMOSTATIC HOUSING LOCATING DISC
12. CHOKE SHAFT AND LEVER ASSEMBLY

13. CHOKE CONTROL LEVER
14. SCREW (1)
15. SCREW (1)
16. CHOKE PLATE
17. ACCELERATOR PUMP OPERATING ROD
18. ACCELERATOR PUMP ROD GROMMET
19. ROD RETAINING CLAMP
20. SCREW (1)
21. ACCELERATOR PUMP SPRING RETAINING PLATE
22. SCREW (1)
23. ACCELERATOR PUMP PISTON STEM
24. ACCELERATOR PUMP SPRING
25. ACCELERATOR PUMP PISTON CUP
26. FAST-IDLE CAM LINK
27. ANTI-DIESEL SOLENOID
28. SCREW (2)
29. AIRHORN GASKET
30. FLOAT-HINGE RETAINER
31. ACCELERATOR PUMP OPERATING LINK
32. RETAINING CLIP (FAST-IDLE CAM)
33. FAST-IDLE CAM
34. FLOAT ASSEMBLY
35. POWER VALVE BODY
36. MAIN METERING JET
37. POWER VALVE PIN
38. ACCELERATOR PUMP WEIGHT
39. ACCELERATOR PUMP CHECK BALL
40. POWER VALVE SPRING
41. SPRING
42. LOW IDLE (SOLENOID OFF) ADJUSTING SCREW
43. CHOKE PULLDOWN DIAPHRAGM ASSEMBLY
44. CHOKE DIAPHRAGM VACUUM HOSE
45. FUEL FILTER
46. FUEL INLET NEEDLE & SEAT ASSEMBLY
47. GASKET
48. MAIN BODY ASSEMBLY
49. FLOAT HINGE PIN
50. CURB-IDLE ADJUSTING SCREW
51. SPRING
52. SPRING
53. FAST-IDLE ADJUSTING SCREW
54. THROTTLE SHAFT AND LEVER ASSEMBLY
55. SCREW (2)
56. THROTTLE PLATE
57. THROTTLE BODY ASSEMBLY
58. THROTTLE BODY SCREW (3)
59. THROTTLE BODY GASKET
60. THROTTLE RETURN SPRING BUSHING
61. THROTTLE RETURN SPRING
62. NUT
63. LOCK WASHER
64. THROTTLE RETURN SPRING BRACKET
65. BOWL VENT SOLENOID
65A. WASHER
65B. SPRING
65C. PINTLE
65D. SEAL
66. POWER VALVE PISTON ASSEMBLY
66A. GASKET
67. SCREW (2)
68. IDLE MIXTURE
69. CONCEALMENT PLUG IDLE MIXTURE NEEDLE
70. FUEL BOWL FILLER

Figure 11-4. Typical Holley Model 1946 carburetor. (Ford)

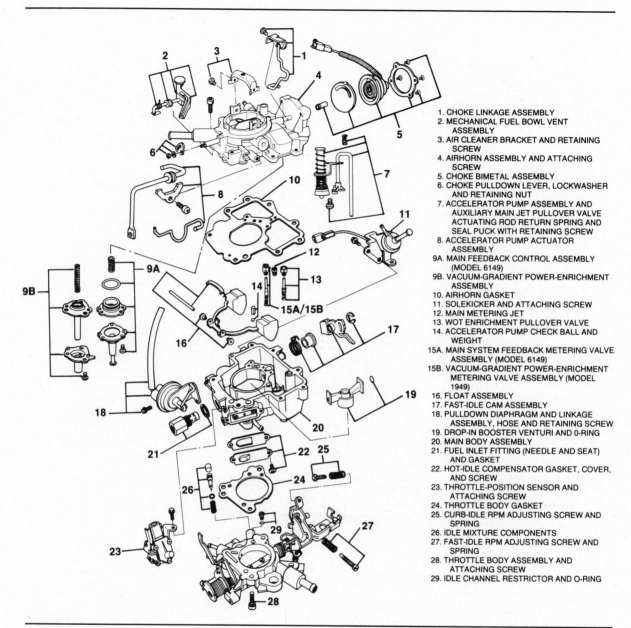

1. CHOKE LINKAGE ASSEMBLY
2. MECHANICAL FUEL BOWL VENT ASSEMBLY
3. AIR CLEANER BRACKET AND RETAINING SCREW
4. AIRHORN ASSEMBLY AND ATTACHING SCREW
5. CHOKE BIMETAL ASSEMBLY
6. CHOKE PULLDOWN LEVER, LOCKWASHER AND RETAINING NUT
7. ACCELERATOR PUMP ASSEMBLY AND AUXILIARY MAIN JET PULLOVER VALVE ACTUATING ROD RETURN SPRING AND SEAL PUCK WITH RETAINING SCREW
8. ACCELERATOR PUMP ACTUATOR ASSEMBLY
9A. MAIN FEEDBACK CONTROL ASSEMBLY (MODEL 6149)
9B. VACUUM-GRADIENT POWER-ENRICHMENT ASSEMBLY
10. AIRHORN GASKET
11. SOLEKICKER AND ATTACHING SCREW
12. MAIN METERING JET
13. WOT ENRICHMENT PULLOVER VALVE
14. ACCELERATOR PUMP CHECK BALL AND WEIGHT
15A. MAIN SYSTEM FEEDBACK METERING VALVE ASSEMBLY (MODEL 6149)
15B. VACUUM-GRADIENT POWER-ENRICHMENT METERING VALVE ASSEMBLY (MODEL 1949)
16. FLOAT ASSEMBLY
17. FAST-IDLE CAM ASSEMBLY
18. PULLDOWN DIAPHRAGM AND LINKAGE ASSEMBLY, HOSE AND RETAINING SCREW
19. DROP-IN BOOSTER VENTURI AND O-RING
20. MAIN BODY ASSEMBLY
21. FUEL INLET FITTING (NEEDLE AND SEAT) AND GASKET
22. HOT-IDLE COMPENSATOR GASKET, COVER, AND SCREW
23. THROTTLE-POSITION SENSOR AND ATTACHING SCREW
24. THROTTLE BODY GASKET
25. CURB-IDLE RPM ADJUSTING SCREW AND SPRING
26. IDLE MIXTURE COMPONENTS
27. FAST-IDLE RPM ADJUSTING SCREW AND SPRING
28. THROTTLE BODY ASSEMBLY AND ATTACHING SCREW
29. IDLE CHANNEL RESTRICTOR AND O-RING

Figure 11-5. Typical Holley Model 1949/6149 carburetor. (Ford)

HOLLEY 2300 TWO-BARREL

This model, figure 11-8, has a horizontally mounted fuel bowl and metering block assembly attached to the main body, resting on the throttle body. Four fuel metering circuits are used: idle, main, accelerator pump, and power enrichment. These are assisted by the fuel inlet and choke systems. When used in 2-mount and 3-mount versions, only the center carburetor contains the choke, idle, accelerator pump, and power enrichment circuits. The outboard carburetors use a throttle control vacuum diaphragm and are mechanically connected to the center carburetor's slotted throttle lever by adjustable

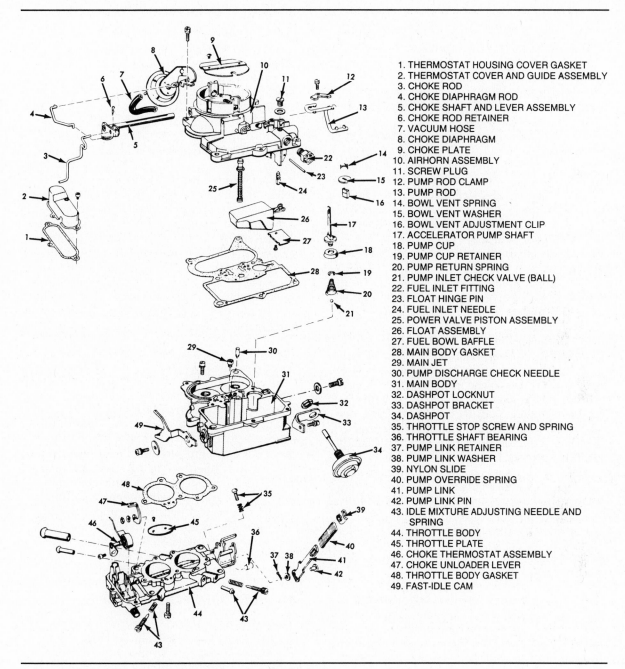

1. THERMOSTAT HOUSING COVER GASKET
2. THERMOSTAT COVER AND GUIDE ASSEMBLY
3. CHOKE ROD
4. CHOKE DIAPHRAGM ROD
5. CHOKE SHAFT AND LEVER ASSEMBLY
6. CHOKE ROD RETAINER
7. VACUUM HOSE
8. CHOKE DIAPHRAGM
9. CHOKE PLATE
10. AIRHORN ASSEMBLY
11. SCREW PLUG
12. PUMP ROD CLAMP
13. PUMP ROD
14. BOWL VENT SPRING
15. BOWL VENT WASHER
16. BOWL VENT ADJUSTMENT CLIP
17. ACCELERATOR PUMP SHAFT
18. PUMP CUP
19. PUMP CUP RETAINER
20. PUMP RETURN SPRING
21. PUMP INLET CHECK VALVE (BALL)
22. FUEL INLET FITTING
23. FLOAT HINGE PIN
24. FUEL INLET NEEDLE
25. POWER VALVE PISTON ASSEMBLY
26. FLOAT ASSEMBLY
27. FUEL BOWL BAFFLE
28. MAIN BODY GASKET
29. MAIN JET
30. PUMP DISCHARGE CHECK NEEDLE
31. MAIN BODY
32. DASHPOT LOCKNUT
33. DASHPOT BRACKET
34. DASHPOT
35. THROTTLE STOP SCREW AND SPRING
36. THROTTLE SHAFT BEARING
37. PUMP LINK RETAINER
38. PUMP LINK WASHER
39. NYLON SLIDE
40. PUMP OVERRIDE SPRING
41. PUMP LINK
42. PUMP LINK PIN
43. IDLE MIXTURE ADJUSTING NEEDLE AND
    SPRING
44. THROTTLE BODY
45. THROTTLE PLATE
46. CHOKE THERMOSTAT ASSEMBLY
47. CHOKE UNLOADER LEVER
48. THROTTLE BODY GASKET
49. FAST-IDLE CAM

Figure 11-6. Typical Holley Model 2209 carburetor. (Holley)

connector rods. Outboard carburetor throttle valves open by vacuum and close mechanically.

HOLLEY 5200, 5210-C, 5220, 6500, 6510-C, AND 6520 TWO-BARREL

These 2-piece carburetors manufactured under license from Weber are 2-stage designs with smaller primary bores than the secondaries. Figure 11-9 (early 5200 and 6510-C) and figure 11-10 (later 5200 and 6500) are typical exploded views. The secondary is mechanically linked to the primary on 5200, 5210-C, 6500, and 6510-C models. The 5220 and 6520 models have a vacuum-controlled secondary stage.

The primary stage uses five basic fuel metering circuits: idle, transfer and lower speed, main, accelerator pump, and power enrichment. The secondary stage uses a transfer circuit, a main metering circuit, and a power-enrichment circuit. The primary and secondary stages are fed by a single fuel bowl vented to a charcoal canister. A hot-idle compensator is found on some versions.

Models 5200 (nonfeedback) and 6500 (feedback) are used by Ford, models 5210-C (nonfeedback) and 6510-C (feedback) are found on GM products, and Chrysler uses the 5220 (nonfeedback) and 6520 (feedback) carburetors. Early 5200 versions use a hot-water-heated integral choke, but later models have an electric assist built into the choke housing cap. The 5210-C and 6510-C have a hot-water-heated integral choke with vacuum-break diaphragms. Chrysler models use an integral choke with electric assist.

All of these carburetors may be found with various throttle stop solenoids and dashpots, according to model year and application. Those used on 1977 Ford vehicles have a mechanical altitude compensator as explained in Chapter 8. Chrysler models may use a remotely mounted altitude compensator, which functions much like the system used with the Model 1949 and 6949 carburetors.

Feedback versions have the mixture control solenoid mounted in the airhorn. This solenoid meters fuel flow on command from the engine control computer and works in conjunction with the conventional fixed main metering jet.

HOLLEY 4150, 4160, AND 4180 FOUR-BARREL

Numerous variations of this Holley design have been used on Ford and GM vehicles over the years. The basic carburetor consists of a main body attached to a throttle body and fitted with primary and secondary fuel bowls and metering blocks, figure 11-11. The fuel inlet system uses an external distribution tube to route fuel from the primary inlet to the secondary inlet. The primary fuel bowl is vented with a vent valve operated by a throttle shaft lever.

The primary stage has a fuel bowl and vent, a metering block, and an accelerator pump circuit. The power system is also contained within the primary metering block. Each primary bore has a primary venturi and a booster venturi, a main fuel discharge nozzle, an idle fuel passage, and a throttle valve. The primary stage furnishes air-fuel mixture throughout the engine operational range.

The secondary stage of the 4150 has a fuel bowl, a metering block, and a secondary throttle-operating diaphragm. Each secondary bore contains a primary and a booster venturi, idle fuel passages, a transfer system, a main secondary fuel discharge nozzle, and a throttle valve. The secondary state provides extra air-fuel mixture when engine load demands it. The primary and secondary throats are the same size.

The various Models 4160 and 4180 carburetors are similar to the 4150, except that a simpler secondary metering plate is used instead of a metering block, figure 11-12. The 4160-C is a service replacement for Motorcraft Models 4300 and 4350. The 4160 and 4180 designs differ from the 4150 primarily in the idle system. The 4160 uses idle mixture screws in the primary metering block; the 4180 has tamperproof idle mixture screws installed on the primary side of the throttle body.

HOLLEY 4165 AND 4175 FOUR-BARREL

The Models 4165 and 4175, figure 11-13, are 4-barrel spread-bore carburetors designed to replace the Rochester Quadrajet 4M and Carter Thermo-Quad carburetors. Both models are the same basic design as the 4150 and work in the same way. While they are very similar in appearance to the Models 4150 and 4160, parts and gaskets are not interchangeable. The 4165 has mechanically operated secondaries and an accelerator pump in both the primary and secondary sides. The 4175 uses a vacuum-operated secondary system and does not need an accelerator pump on the secondary side.

HOLLEY 4360 FOUR-BARREL

This spread-bore model, figure 11-14, was introduced in 1976 as another replacement for the Rochester Quadrajet. Unlike other Holley 4-barrel carburetors, it has a single fuel bowl built into the main body casting. The carburetor consists of an airhorn, a main body, and a throttle body. The main metering circuit for the primary side is similar to the one used in the Models 5200 and 5210-C. Both the primary and the secondary sides have idle systems, but only the primary side is adjustable. A spring-driven, piston-type accelerator pump is used. The secondary throttles are mechanically operated. The main jets have metric threads.

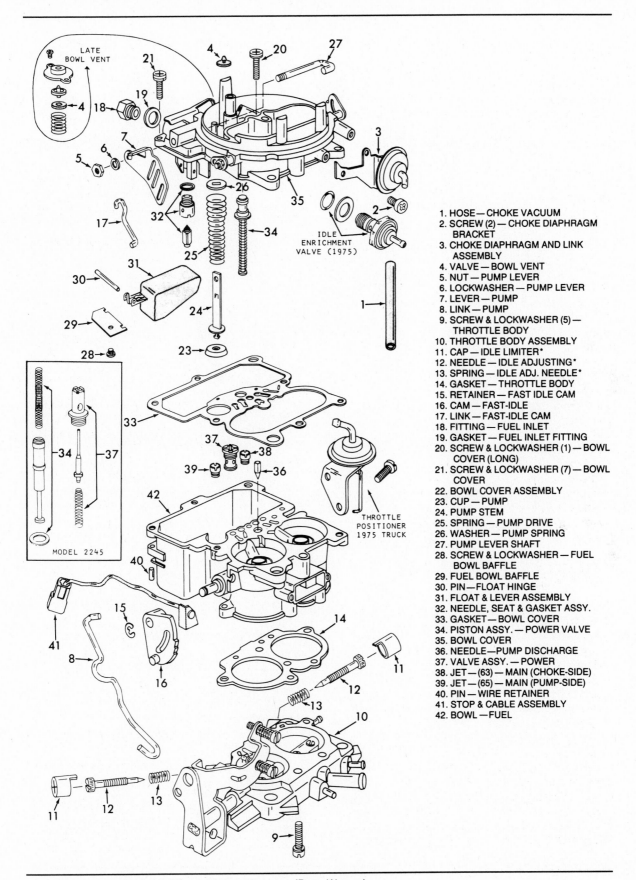

LATE BOWL VENT

IDLE ENRICHMENT VALVE (1975)

THROTTLE POSITIONER 1975 TRUCK

MODEL 2245

1. HOSE — CHOKE VACUUM
2. SCREW (2) — CHOKE DIAPHRAGM BRACKET
3. CHOKE DIAPHRAGM AND LINK ASSEMBLY
4. VALVE — BOWL VENT
5. NUT — PUMP LEVER
6. LOCKWASHER — PUMP LEVER
7. LEVER — PUMP
8. LINK — PUMP
9. SCREW & LOCKWASHER (5) — THROTTLE BODY
10. THROTTLE BODY ASSEMBLY
11. CAP — IDLE LIMITER*
12. NEEDLE — IDLE ADJUSTING*
13. SPRING — IDLE ADJ. NEEDLE*
14. GASKET — THROTTLE BODY
15. RETAINER — FAST IDLE CAM
16. CAM — FAST-IDLE
17. LINK — FAST-IDLE CAM
18. FITTING — FUEL INLET
19. GASKET — FUEL INLET FITTING
20. SCREW & LOCKWASHER (1) — BOWL COVER (LONG)
21. SCREW & LOCKWASHER (7) — BOWL COVER
22. BOWL COVER ASSEMBLY
23. CUP — PUMP
24. PUMP STEM
25. SPRING — PUMP DRIVE
26. WASHER — PUMP SPRING
27. PUMP LEVER SHAFT
28. SCREW & LOCKWASHER — FUEL BOWL BAFFLE
29. FUEL BOWL BAFFLE
30. PIN — FLOAT HINGE
31. FLOAT & LEVER ASSEMBLY
32. NEEDLE, SEAT & GASKET ASSY.
33. GASKET — BOWL COVER
34. PISTON ASSY. — POWER VALVE
35. BOWL COVER
36. NEEDLE — PUMP DISCHARGE
37. VALVE ASSY. — POWER
38. JET — (63) — MAIN (CHOKE-SIDE)
39. JET — (65) — MAIN (PUMP-SIDE)
40. PIN — WIRE RETAINER
41. STOP & CABLE ASSEMBLY
42. BOWL — FUEL

Figure 11-7. Typical Holley Model 2210 carburetor. (Borg-Warner)

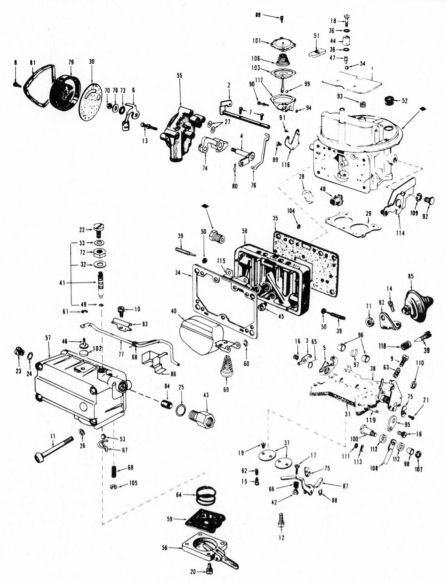

59. PUMP DIAPHRAGM ASSEMBLY
60. FLOAT SPRING RETAINER
61. AIR VENT RETAINER
62. FAST-IDLE CAM LEV. SCR. SPRING
63. THROTTLE STOP SCREW SPRING
64. PUMP DIAPHRAGM RETURN SPRING
65. FAST-IDLE CAM LEV. SPRING
66. PUMP OPER. LEV. ADJ. SPRING
67. PUMP INLET CHECK BALL RETAINER
68. AIR VENT ROD SPRING
69. FLOAT SPRING
70. CHOKE THERMOSTAT SHAFT NUT
71. DASHPOT SCREW NUT
72. FUEL VALVE SEAT ADJ. NUT
73. CHOKE THERM. LEVER SPACER
74. FAST-IDLE CAM ASSEMBLY
75. PUMP CAM
76. CHOKE ROD
77. AIR VENT ROD
78. CHOKE THERM. SHAFT NUT L. W.
79. THERMOSTAT HOUSING ASSEMBLY
80. CHOKE ROD RETAINER
81. THERMOSTAT HOUSING CLAMP
82. DASHPOT BRACKET
83. AIR VENT ROD CLAMP
84. FILTER SCREEN ASSEMBLY
85. DASHPOT ASSEMBLY
86. BAFFLE PLATE
87. PUMP OPERATING LEVER
88. PUMP OPERATING LEV. RETAINER
89. ADAPTER MOUNTING & DIAPHRAGM COVER ASSY. SCREW
90. THROT. DIAPHRAGM HSG. SCR.
91. ADAPTER PASSAGE SCREW
92. CHOKE BRACKET SCREW
93. AIR ADAPTER HOLE PLUG
94. THROT. DIAPHRAGM HSG. GSKT.
95. THROTTLE LEVER
96. THROTTLE SHAFT BEARING
97. THROTTLE SHAFT BRG. (CENTER)
98. THROTTLE CONNECTOR PIN BUSHING
99. DIAPHRAGM CHECK BALL
100. THROTTLE CONNECTOR PIN
101. DIAPHRAGM HOUSING COVER
102. AIR VENT CAP
103. DIAPHRAGM ASSEMBLY
104. DIAPHRAGM LINK RETAINER
105. AIR VENT ROD SPG. RETAINER
106. DIAPHRAGM SPRING
107. THROTTLE LINK CONNECTOR PIN NUT
108. THROTTLE CONNECTOR BAR
109. CHOKE BRKT. SCR. LOCK WASHER
110. THROT. LINK CONNECTOR PIN WASHER
111. THROT. CONNECTOR PIN WASHER
112. THROTTLE CONNECTOR PIN SPACER
113. THROT. CONNECTOR PIN RETAINER
114. CHOKE CONTROL LEVER BRACKET
115. METERING BODY VENT BAFFLE
116. THROT. DIAPHRAGM ADAPTER
117. DIAPHRAGM HOUSING
118. IDLE ADJ. NEEDLE SPRING
119. PUMP OPER. LEVER STUD

1. CHOKE PLATE
2. CHOKE SHAFT ASSEMBLY
3. FAST-IDLE PICK-UP LEVER
4. CHOKE HSG. SHAFT & LEV. ASSY.
5. FAST-IDLE CAM LEVER
6. CHOKE THERM. LEV., LINK & PISTON ASSEMBLY
7. CHOKE PLATE SCREW
8. THERM. HSG. CLAMP SCREW
9. THROTTLE STOP SCREW
10. AIR VENT ROD CLAMP SCR. & L. W.
11. FUEL BOWL TO MAIN BODY SCREW
12. THROT. BODY SCR. & L. W.
13. CHOKE HSG. SCR. & L. W.
14. DASHPOT BRKT. SCR. & L. W.
15. FAST-IDLE CAM LEVER SCREW
16. FAST-IDLE CAM LEV. & THROT. LEV. SCREW & L. W.
17. PUMP OPER. LEV. ADJ. SCREW
18. PUMP DISCHARGE NOZZLE SCREW
19. THROTTLE PLATE SCREW
20. FUEL PUMP COV. ASSY. SCR. & L. W.

21. PUMP CAM LOCK SCR. & L. W.
22. FUEL VALVE SEAT LOCK SCREW
23. FUEL LEVEL CHECK PLUG
24. FUEL LEVEL CHECK PLUG GASKET
25. FUEL INLET FITTING GASKET
26. FUEL BOWL SCREW GASKET
27. CHOKE HOUSING GASKET
28. POWER VALVE BODY GASKET
29. THROTTLE BODY GASKET
30. CHOKE THERM. HOUSING GASKET
31. FLANGE GASKET
32. FUEL VALVE SEAT ADJ. NUT GSKT.
33. FUEL VALVE SEAT LOCK SCR. GSKT.
34. FUEL BOWL GASKET
35. METERING BODY GASKET
36. PUMP DISCHARGE NOZZLE GASKET
37. THROTTLE PLATE
38. THROT. BODY & SHAFT ASSY.
39. IDLE ADJUSTING NEEDLE
40. FLOAT & HINGE ASSY.

41. FUEL INLET VALVE & SEAT ASSY.
42. PUMP OPER. LEV. ADJ. SCR. FITTING
43. FUEL INLET FITTING
44. PUMP DISCHARGE NOZZLE
45. MAIN JET
46. AIR VENT VALVE
47. PUMP DISCHARGE NEEDLE VALVE OR CHECK BALL WEIGHT
48. POWER VALVE ASSEMBLY
49. FUEL VALVE SEAT "O" RING SEAL OR GASKET
50. IDLE NEEDLE SEAL
51. CHOKE ROD SEAL
52. CHOKE COLD AIR TUBE GROMMET
53. PUMP INLET CHECK BALL
54. PUMP DISCHARGE CHECK BALL
55. CHOKE HSG. & PLUGS ASSY.
56. FUEL PUMP COVER ASSY.
57. FUEL BOWL & PLUGS ASSY.
58. MAIN METERING BODY & PLUGS ASSY.

Figure 11-8. Typical Holley Model 2300 carburetor. (Holley)

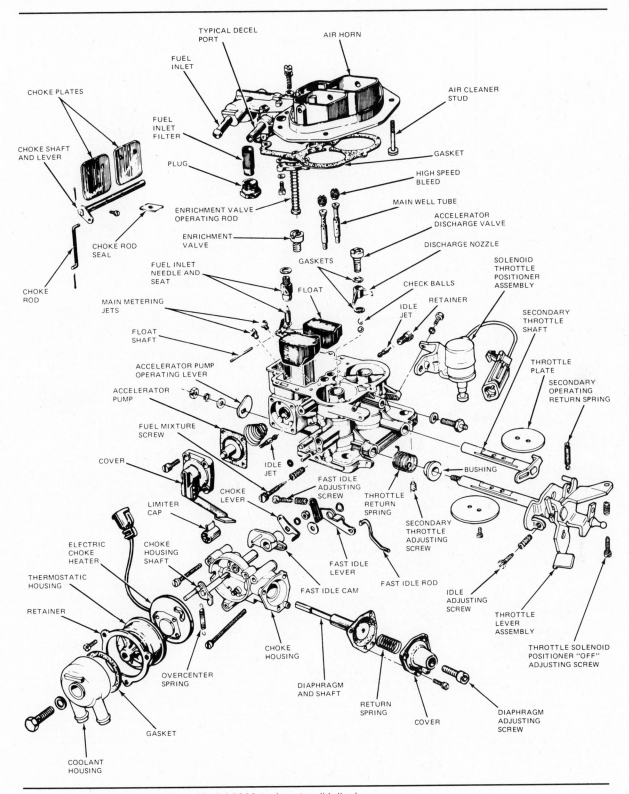

Figure 11-9. Typical early Holley Model 5200 carburetor. (Holley)

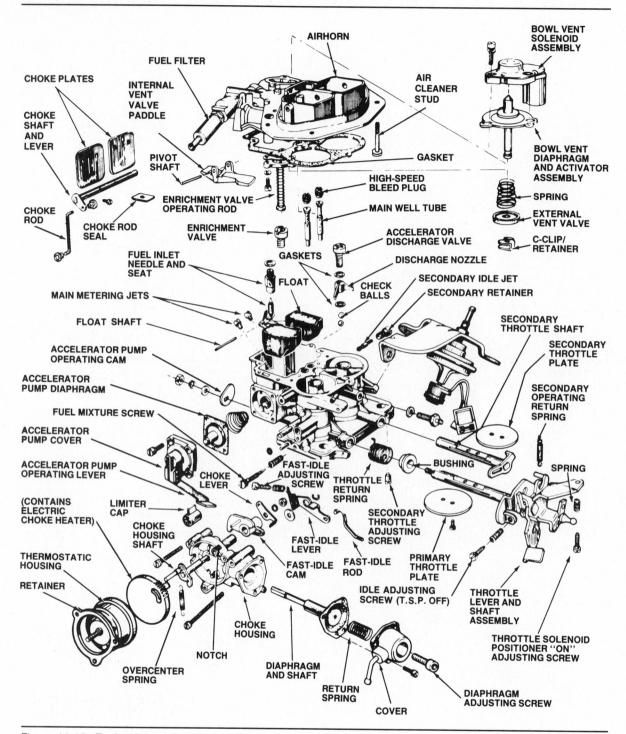

Figure 11-10. Typical late Holley Model 5200/5210-C carburetor. (Ford)

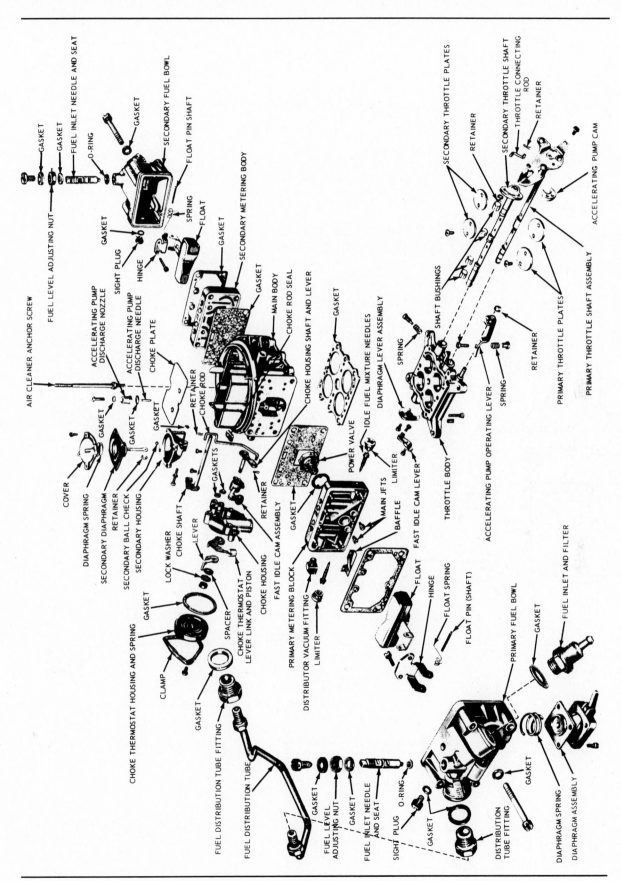

Figure 11-11. Typical Holley Model 4150 carburetor. (Holley)

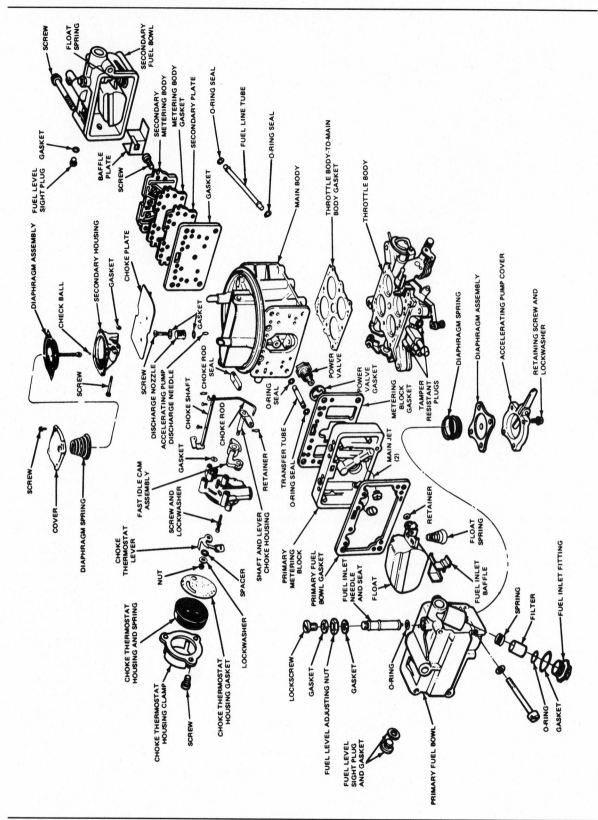

Figure 11-12. Typical Holley Model 4180-C carburetor. (Ford)

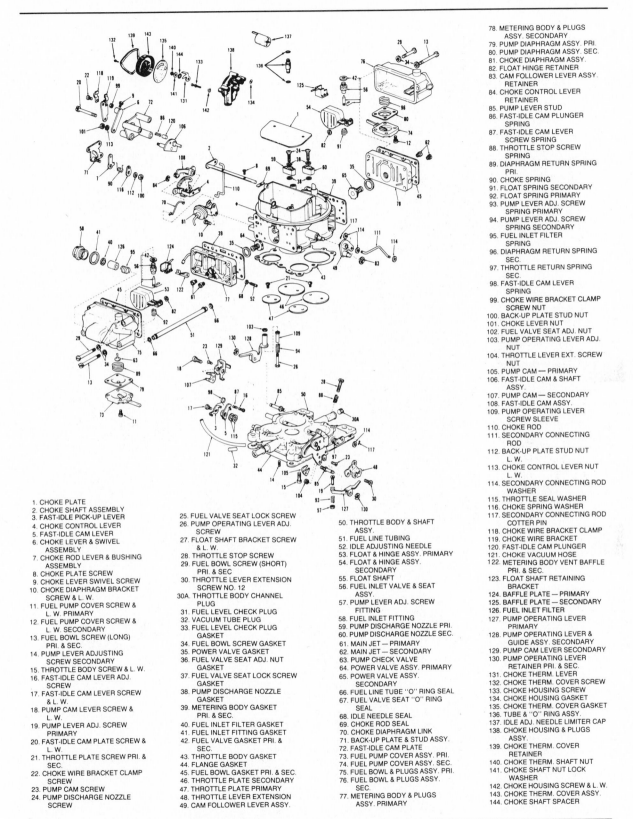

78. METERING BODY & PLUGS ASSY. SECONDARY
79. PUMP DIAPHRAGM ASSY. PRI.
80. PUMP DIAPHRAGM ASSY. SEC.
81. CHOKE DIAPHRAGM ASSY.
82. FLOAT HINGE RETAINER
83. CAM FOLLOWER LEVER ASSY. RETAINER
84. CHOKE CONTROL LEVER RETAINER
85. PUMP LEVER STUD
86. FAST-IDLE CAM PLUNGER SPRING
87. FAST-IDLE CAM LEVER SCREW SPRING
88. THROTTLE STOP SCREW SPRING
89. DIAPHRAGM RETURN SPRING PRI.
90. CHOKE SPRING
91. FLOAT SPRING SECONDARY
92. FLOAT SPRING PRIMARY
93. PUMP LEVER ADJ. SCREW SPRING PRIMARY
94. PUMP LEVER ADJ. SCREW SPRING SECONDARY
95. FUEL INLET FILTER SPRING
96. DIAPHRAGM RETURN SPRING SEC.
97. THROTTLE RETURN SPRING SEC.
98. FAST-IDLE CAM LEVER SPRING
99. CHOKE WIRE BRACKET CLAMP SCREW NUT
100. BACK-UP PLATE STUD NUT
101. CHOKE LEVER NUT
102. FUEL VALVE SEAT ADJ. NUT
103. PUMP OPERATING LEVER ADJ. NUT
104. THROTTLE LEVER EXT. SCREW NUT
105. PUMP CAM — PRIMARY
106. FAST-IDLE CAM & SHAFT ASSY.
107. PUMP CAM — SECONDARY
108. FAST-IDLE CAM ASSY.
109. PUMP OPERATING LEVER SCREW SLEEVE
110. CHOKE ROD
111. SECONDARY CONNECTING ROD
112. BACK-UP PLATE STUD NUT L. W.
113. CHOKE CONTROL LEVER NUT L. W.
114. SECONDARY CONNECTING ROD WASHER
115. THROTTLE SEAL WASHER
116. CHOKE SPRING WASHER
117. SECONDARY CONNECTING ROD COTTER PIN
118. CHOKE WIRE BRACKET CLAMP
119. CHOKE WIRE BRACKET
120. FAST-IDLE CAM PLUNGER
121. CHOKE VACUUM HOSE
122. METERING BODY VENT BAFFLE PRI. & SEC.
123. FLOAT SHAFT RETAINING BRACKET
124. BAFFLE PLATE — PRIMARY
125. BAFFLE PLATE — SECONDARY
126. FUEL INLET FILTER
127. PUMP OPERATING LEVER PRIMARY
128. PUMP OPERATING LEVER & GUIDE ASSY. SECONDARY
129. PUMP CAM LEVER SECONDARY
130. PUMP OPERATING LEVER RETAINER PRI. & SEC.
131. CHOKE THERM. LEVER
132. CHOKE THERM. COVER SCREW
133. CHOKE HOUSING SCREW
134. CHOKE HOUSING GASKET
135. CHOKE THERM. COVER GASKET
136. TUBE & "O" RING ASSY.
137. IDLE ADJ. NEEDLE LIMITER CAP
138. CHOKE HOUSING & PLUGS ASSY.
139. CHOKE THERM. COVER RETAINER
140. CHOKE THERM. SHAFT NUT
141. CHOKE SHAFT NUT LOCK WASHER
142. CHOKE HOUSING SCREW & L. W.
143. CHOKE THERM. COVER ASSY.
144. CHOKE SHAFT SPACER

1. CHOKE PLATE
2. CHOKE SHAFT ASSEMBLY
3. FAST-IDLE PICK-UP LEVER
4. CHOKE CONTROL LEVER
5. FAST-IDLE CAM LEVER
6. CHOKE LEVER & SWIVEL ASSEMBLY
7. CHOKE ROD LEVER & BUSHING ASSEMBLY
8. CHOKE PLATE SCREW
9. CHOKE LEVER SWIVEL SCREW
10. CHOKE DIAPHRAGM BRACKET SCREW & L. W.
11. FUEL PUMP COVER SCREW & L. W. PRIMARY
12. FUEL PUMP COVER SCREW & L. W. SECONDARY
13. FUEL BOWL SCREW (LONG) PRI. & SEC.
14. PUMP LEVER ADJUSTING SCREW SECONDARY
15. THROTTLE BODY SCREW & L. W.
16. FAST-IDLE CAM LEVER ADJ. SCREW
17. FAST-IDLE CAM LEVER SCREW & L. W.
18. PUMP CAM LEVER SCREW & L. W.
19. PUMP LEVER ADJ. SCREW PRIMARY
20. FAST-IDLE CAM PLATE SCREW & L. W.
21. THROTTLE PLATE SCREW PRI. & SEC.
22. CHOKE WIRE BRACKET CLAMP SCREW
23. PUMP CAM SCREW
24. PUMP DISCHARGE NOZZLE SCREW

25. FUEL VALVE SEAT LOCK SCREW
26. PUMP OPERATING LEVER ADJ. SCREW
27. FLOAT SHAFT BRACKET SCREW & L. W.
28. THROTTLE STOP SCREW
29. FUEL BOWL SCREW (SHORT) PRI. & SEC
30. THROTTLE LEVER EXTENSION SCREW NO. 12
30A. THROTTLE BODY CHANNEL PLUG
31. FUEL LEVEL CHECK PLUG
32. VACUUM TUBE PLUG
33. FUEL LEVEL CHECK PLUG GASKET
34. FUEL BOWL SCREW GASKET
35. POWER VALVE GASKET
36. FUEL VALVE SEAT ADJ. NUT GASKET
37. FUEL VALVE SEAT LOCK SCREW GASKET
38. PUMP DISCHARGE NOZZLE GASKET
39. METERING BODY GASKET PRI. & SEC.
40. FUEL INLET FILTER GASKET
41. FUEL INLET FITTING GASKET
42. FUEL VALVE GASKET PRI. & SEC.
43. THROTTLE BODY GASKET
44. FLANGE GASKET
45. FUEL BOWL GASKET PRI. & SEC.
46. THROTTLE PLATE SECONDARY
47. THROTTLE PLATE PRIMARY
48. THROTTLE LEVER EXTENSION
49. CAM FOLLOWER LEVER ASSY.

50. THROTTLE BODY & SHAFT ASSY.
51. FUEL LINE TUBING
52. IDLE ADJUSTING NEEDLE
53. FLOAT & HINGE ASSY. PRIMARY
54. FLOAT & HINGE ASSY. SECONDARY
55. FLOAT SHAFT
56. FUEL INLET VALVE & SEAT ASSY.
57. PUMP LEVER ADJ. SCREW FITTING
58. FUEL INLET FITTING
59. PUMP DISCHARGE NOZZLE PRI.
60. PUMP DISCHARGE NOZZLE SEC.
61. MAIN JET — PRIMARY
62. MAIN JET — SECONDARY
63. PUMP CHECK VALVE
64. POWER VALVE ASSY. PRIMARY
65. POWER VALVE ASSY. SECONDARY
66. FUEL LINE TUBE "O" RING SEAL
67. FUEL VALVE SEAT "O" RING SEAL
68. IDLE NEEDLE SEAL
69. CHOKE ROD SEAL
70. CHOKE DIAPHRAGM LINK
71. BACK-UP PLATE & STUD ASSY.
72. FAST-IDLE CAM PLATE
73. FUEL PUMP COVER ASSY. PRI.
74. FUEL PUMP COVER ASSY. SEC.
75. FUEL BOWL & PLUGS ASSY. PRI.
76. FUEL BOWL & PLUGS ASSY. SEC.
77. METERING BODY & PLUGS ASSY. PRIMARY

Figure 11-13. Typical Holley Model 4165 carburetor. (Holley)

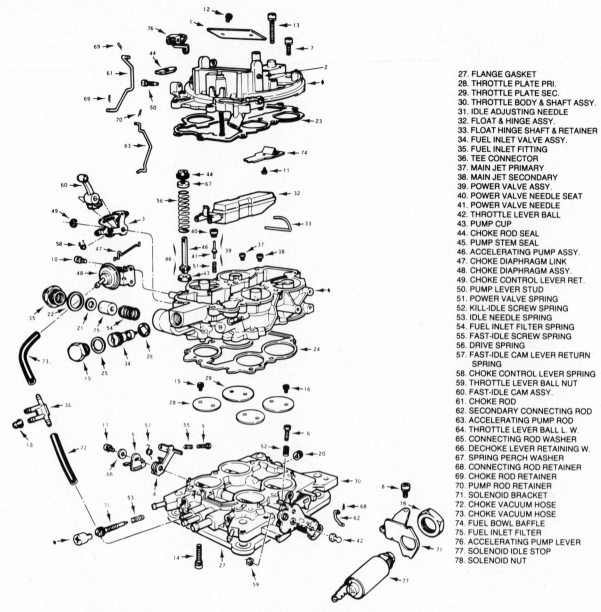

27. FLANGE GASKET
28. THROTTLE PLATE PRI.
29. THROTTLE PLATE SEC.
30. THROTTLE BODY & SHAFT ASSY.
31. IDLE ADJUSTING NEEDLE
32. FLOAT & HINGE ASSY.
33. FLOAT HINGE SHAFT & RETAINER
34. FUEL INLET VALVE ASSY.
35. FUEL INLET FITTING
36. TEE CONNECTOR
37. MAIN JET PRIMARY
38. MAIN JET SECONDARY
39. POWER VALVE ASSY.
40. POWER VALVE NEEDLE SEAT
41. POWER VALVE NEEDLE
42. THROTTLE LEVER BALL
43. PUMP CUP
44. CHOKE ROD SEAL
45. PUMP STEM SEAL
46. ACCELERATING PUMP ASSY.
47. CHOKE DIAPHRAGM LINK
48. CHOKE DIAPHRAGM ASSY.
49. CHOKE CONTROL LEVER RET.
50. PUMP LEVER STUD
51. POWER VALVE SPRING
52. KILL-IDLE SCREW SPRING
53. IDLE NEEDLE SPRING
54. FUEL INLET FILTER SPRING
55. FAST-IDLE SCREW SPRING
56. DRIVE SPRING
57. FAST-IDLE CAM LEVER RETURN
    SPRING
58. CHOKE CONTROL LEVER SPRING
59. THROTTLE LEVER BALL NUT
60. FAST-IDLE CAM ASSY.
61. CHOKE ROD
62. SECONDARY CONNECTING ROD
63. ACCELERATING PUMP ROD
64. THROTTLE LEVER BALL L. W.
65. CONNECTING ROD WASHER
66. DECHOKE LEVER RETAINING W.
67. SPRING PERCH WASHER
68. CONNECTING ROD RETAINER
69. CHOKE ROD RETAINER
70. PUMP ROD RETAINER
71. SOLENOID BRACKET
72. CHOKE VACUUM HOSE
73. CHOKE VACUUM HOSE
74. FUEL BOWL BAFFLE
75. FUEL INLET FILTER
76. ACCELERATING PUMP LEVER
77. SOLENOID IDLE STOP
78. SOLENOID NUT

1. CHOKE PLATE
2. CHOKE SHAFT & LEVER
   ASSEMBLY
3. CHOKE CONTROL LEVER
4. FAST-IDLE CAM LEVER
5. DECHOKE LEVER
6. KILL-IDLE ADJUSTING SCREW
7. AIRHORN TO MAIN BODY
   SCREW SHORT
8. SOLENOID BRACKET SCREW &
   L. W.
9. FAST-IDLE ADJUSTING SCREW
10. CHOKE DIAPHRAGM BRACKET
    SCREW
11. FUEL BOWL BAFFLE SCREW
12. CHOKE PLATE SCREW

13. AIRHORN TO MAIN BODY
    SCREW LONG
14. THROTTLE BODY TO MAIN BODY
    SCREW & L. W.
15. THROTTLE PLATE SCREW PRI.
16. THROTTLE PLATE SCREW SEC.
17. DECHOKE LEVER SCREW & L. W.
18. TEE PLUG
19. FUEL INLET PLUG
20. POWER BRAKE PLUG
21. FUEL INLET FILTER GASKET
22. FUEL INLET FITTING GASKET
23. MAIN BODY GASKET
24. THROTTLE BODY GASKET
25. FUEL INLET PLUG GASKET
26. FUEL VALVE SEAT GASKET

PARTS NOT SHOWN ON
ILLUSTRATION
_____

PCV TUBE PLUG
THROTTLE LEVER BALL L.W.
THROTTLE LEVER BALL NUT
TRANS KICK-DOWN STUD
TRANS KICK-DOWN NUT

Figure 11-14. Typical Holley Model 4360 carburetor. (Holley)

Chapter

12

Rochester Carburetors

Although primarily used by General Motors, Rochester carburetors have been used by American Motors, Chrysler, and Ford Motor Company. The following carburetors are described in this chapter:

- Models M, MV, 1MV, ME, 1ME, and 1MEF (Monojet) One-Barrel
- Models 2G, 2GV, 2GC, and 2GF Two-Barrel
- Models 2MC, M2MC, M2ME, E2MC, and E2ME (Dualjet) Two-Barrel
- Models 2SE and E2SE (Varijet) Two-Barrel
- Model 4GC Four-Barrel
- The Quadrajet Four-Barrel (Models 4M and M4M open-loop variations)
- The Quadrajet Four-Barrel (Model E4M and closed-loop variations).

ROCHESTER M, MV, 1MV, ME, 1ME, AND 1MEF MONOJET ONE-BARREL

Monojet carburetors, figure 12-1, are used on inline 6-cylinder engines. Carburetors designated simply as M use a manual choke. Models MV or 1MV use a remote choke and the ME and 1ME designations have an integral cap-type electric choke. The 1MV and 1ME versions use an electrically operated idle stop solenoid. The 1MEF, figure 12-2, contains a factory-adjusted metering rod screw in the airhorn to control the position of the enrichment part of the metering rod in the jet.

The basic design consists of an airhorn, a main body, and a throttle body. The throttle body is an aluminum casting for improved heat distribution. It is used with an insulating body-to-bowl gasket to prevent excessive heat from reaching the float bowl.

Four conventional fuel metering circuits are used: idle, main, accelerator pump, and power enrichment. A triple venturi is used with a plain tube nozzle. Fuel flow through the main metering system is controlled by mechanical and vacuum devices. A main well air bleed and variable orifice are also used. The fuel bowl is internally vented to the airhorn, with an external idle vent on some versions. A hot-idle compensator may also be used.

ROCHESTER 2G, 2GV, 2GC, AND 2GF TWO-BARREL

These 3-piece carburetors consist of an airhorn, fuel bowl, and throttle body. Figure 12-3 shows the early 2G and 2GC; figure 12-4 shows the late model 2G and 2GF (with governor). The 2G uses a manual choke, the 2GV a remote automatic choke, and the 2GC an integral automatic choke. The 2GF is used on 1985 and later truck

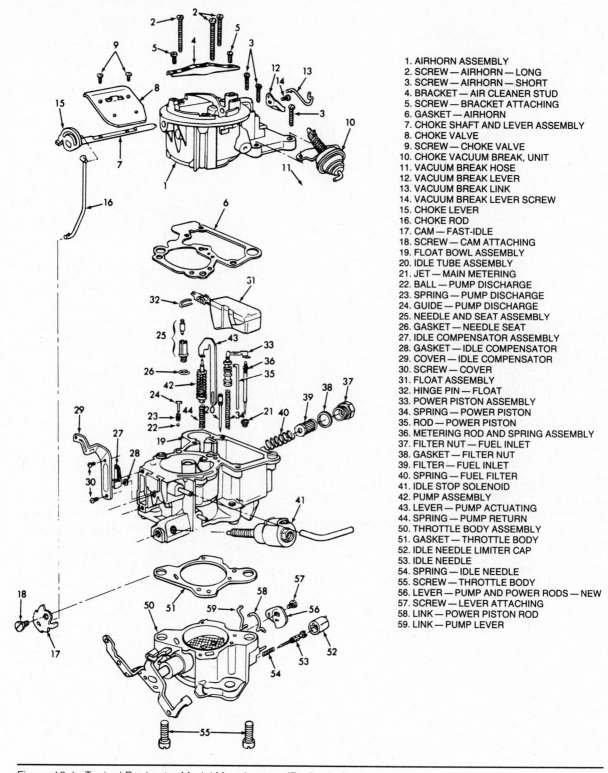

Figure 12-1. Typical Rochester Model M carburetor. (Rochester)

1. AIRHORN ASSEMBLY
2. SCREW — AIRHORN — LONG
3. SCREW — AIRHORN — SHORT
4. BRACKET — AIR CLEANER STUD
5. SCREW — BRACKET ATTACHING
6. GASKET — AIRHORN
7. CHOKE SHAFT AND LEVER ASSEMBLY
8. CHOKE VALVE
9. SCREW — CHOKE VALVE
10. CHOKE VACUUM BREAK, UNIT
11. VACUUM BREAK HOSE
12. VACUUM BREAK LEVER
13. VACUUM BREAK LINK
14. VACUUM BREAK LEVER SCREW
15. CHOKE LEVER
16. CHOKE ROD
17. CAM — FAST-IDLE
18. SCREW — CAM ATTACHING
19. FLOAT BOWL ASSEMBLY
20. IDLE TUBE ASSEMBLY
21. JET — MAIN METERING
22. BALL — PUMP DISCHARGE
23. SPRING — PUMP DISCHARGE
24. GUIDE — PUMP DISCHARGE
25. NEEDLE AND SEAT ASSEMBLY
26. GASKET — NEEDLE SEAT
27. IDLE COMPENSATOR ASSEMBLY
28. GASKET — IDLE COMPENSATOR
29. COVER — IDLE COMPENSATOR
30. SCREW — COVER
31. FLOAT ASSEMBLY
32. HINGE PIN — FLOAT
33. POWER PISTON ASSEMBLY
34. SPRING — POWER PISTON
35. ROD — POWER PISTON
36. METERING ROD AND SPRING ASSEMBLY
37. FILTER NUT — FUEL INLET
38. GASKET — FILTER NUT
39. FILTER — FUEL INLET
40. SPRING — FUEL FILTER
41. IDLE STOP SOLENOID
42. PUMP ASSEMBLY
43. LEVER — PUMP ACTUATING
44. SPRING — PUMP RETURN
50. THROTTLE BODY ASSEMBLY
51. GASKET — THROTTLE BODY
52. IDLE NEEDLE LIMITER CAP
53. IDLE NEEDLE
54. SPRING — IDLE NEEDLE
55. SCREW — THROTTLE BODY
56. LEVER — PUMP AND POWER RODS — NEW
57. SCREW — LEVER ATTACHING
58. LINK — POWER PISTON ROD
59. LINK — PUMP LEVER

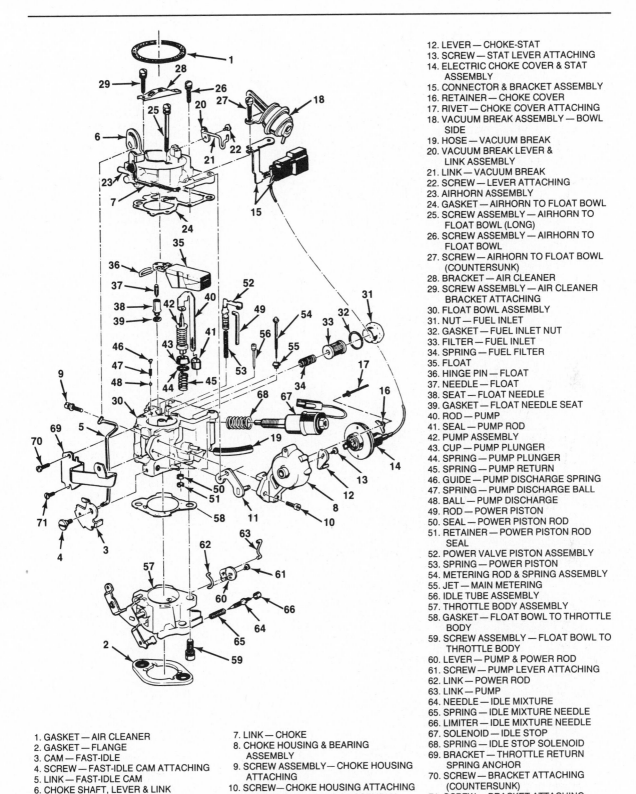

12. LEVER — CHOKE-STAT
13. SCREW — STAT LEVER ATTACHING
14. ELECTRIC CHOKE COVER & STAT ASSEMBLY
15. CONNECTOR & BRACKET ASSEMBLY
16. RETAINER — CHOKE COVER
17. RIVET — CHOKE COVER ATTACHING
18. VACUUM BREAK ASSEMBLY — BOWL SIDE
19. HOSE — VACUUM BREAK
20. VACUUM BREAK LEVER & LINK ASSEMBLY
21. LINK — VACUUM BREAK
22. SCREW — LEVER ATTACHING
23. AIRHORN ASSEMBLY
24. GASKET — AIRHORN TO FLOAT BOWL
25. SCREW ASSEMBLY — AIRHORN TO FLOAT BOWL (LONG)
26. SCREW ASSEMBLY — AIRHORN TO FLOAT BOWL
27. SCREW — AIRHORN TO FLOAT BOWL (COUNTERSUNK)
28. BRACKET — AIR CLEANER
29. SCREW ASSEMBLY — AIR CLEANER BRACKET ATTACHING
30. FLOAT BOWL ASSEMBLY
31. NUT — FUEL INLET
32. GASKET — FUEL INLET NUT
33. FILTER — FUEL INLET
34. SPRING — FUEL FILTER
35. FLOAT
36. HINGE PIN — FLOAT
37. NEEDLE — FLOAT
38. SEAT — FLOAT NEEDLE
39. GASKET — FLOAT NEEDLE SEAT
40. ROD — PUMP
41. SEAL — PUMP ROD
42. PUMP ASSEMBLY
43. CUP — PUMP PLUNGER
44. SPRING — PUMP PLUNGER
45. SPRING — PUMP RETURN
46. GUIDE — PUMP DISCHARGE SPRING
47. SPRING — PUMP DISCHARGE BALL
48. BALL — PUMP DISCHARGE
49. ROD — POWER PISTON
50. SEAL — POWER PISTON ROD
51. RETAINER — POWER PISTON ROD SEAL
52. POWER VALVE PISTON ASSEMBLY
53. SPRING — POWER PISTON
54. METERING ROD & SPRING ASSEMBLY
55. JET — MAIN METERING
56. IDLE TUBE ASSEMBLY
57. THROTTLE BODY ASSEMBLY
58. GASKET — FLOAT BOWL TO THROTTLE BODY
59. SCREW ASSEMBLY — FLOAT BOWL TO THROTTLE BODY
60. LEVER — PUMP & POWER ROD
61. SCREW — PUMP LEVER ATTACHING
62. LINK — POWER ROD
63. LINK — PUMP
64. NEEDLE — IDLE MIXTURE
65. SPRING — IDLE MIXTURE NEEDLE
66. LIMITER — IDLE MIXTURE NEEDLE
67. SOLENOID — IDLE STOP
68. SPRING — IDLE STOP SOLENOID
69. BRACKET — THROTTLE RETURN SPRING ANCHOR
70. SCREW — BRACKET ATTACHING (COUNTERSUNK)
71. SCREW — BRACKET ATTACHING

1. GASKET — AIR CLEANER
2. GASKET — FLANGE
3. CAM — FAST-IDLE
4. SCREW — FAST-IDLE CAM ATTACHING
5. LINK — FAST-IDLE CAM
6. CHOKE SHAFT, LEVER & LINK ASSEMBLY
7. LINK — CHOKE
8. CHOKE HOUSING & BEARING ASSEMBLY
9. SCREW ASSEMBLY— CHOKE HOUSING ATTACHING
10. SCREW—CHOKE HOUSING ATTACHING
11. CHOKE SHAFT & LEVER ASSEMBLY

Figure 12-2. Typical Rochester Model 1MEF carburetor. (Rochester)

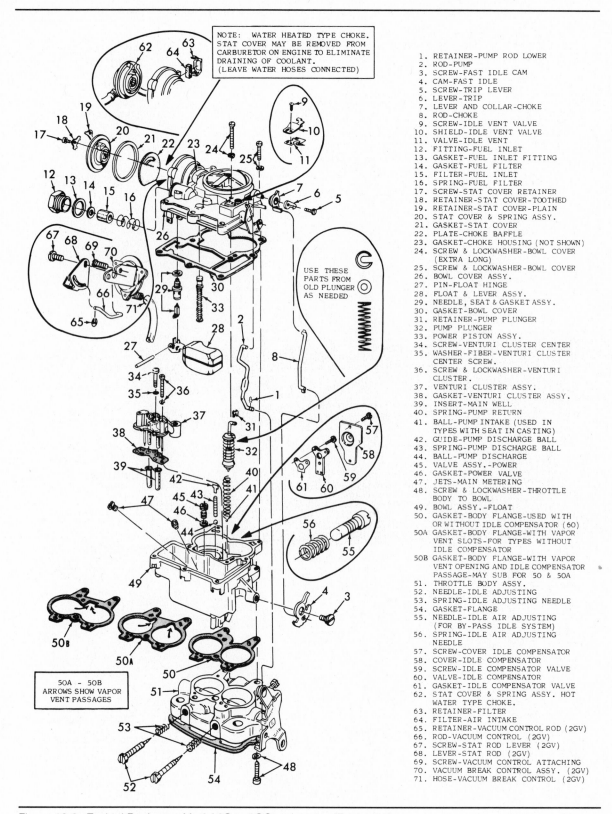

NOTE: WATER HEATED TYPE CHOKE.
STAT COVER MAY BE REMOVED FROM
CARBURETOR ON ENGINE TO ELIMINATE
DRAINING OF COOLANT.
(LEAVE WATER HOSES CONNECTED)

USE THESE
PARTS FROM
OLD PLUNGER
AS NEEDED

50A - 50B
ARROWS SHOW VAPOR
VENT PASSAGES

1. RETAINER-PUMP ROD LOWER
2. ROD-PUMP
3. SCREW-FAST IDLE CAM
4. CAM-FAST IDLE
5. SCREW-TRIP LEVER
6. LEVER-TRIP
7. LEVER AND COLLAR-CHOKE
8. ROD-CHOKE
9. SCREW-IDLE VENT VALVE
10. SHIELD-IDLE VENT VALVE
11. VALVE-IDLE VENT
12. FITTING-FUEL INLET
13. GASKET-FUEL INLET FITTING
14. GASKET-FUEL FILTER
15. FILTER-FUEL INLET
16. SPRING-FUEL FILTER
17. SCREW-STAT COVER RETAINER
18. RETAINER-STAT COVER-TOOTHED
19. RETAINER-STAT COVER-PLAIN
20. STAT COVER & SPRING ASSY.
21. GASKET-STAT COVER
22. PLATE-CHOKE BAFFLE
23. GASKET-CHOKE HOUSING (NOT SHOWN)
24. SCREW & LOCKWASHER-BOWL COVER
    (EXTRA LONG)
25. SCREW & LOCKWASHER-BOWL COVER
26. BOWL COVER ASSY.
27. PIN-FLOAT HINGE
28. FLOAT & LEVER ASSY.
29. NEEDLE, SEAT & GASKET ASSY.
30. GASKET-BOWL COVER
31. RETAINER-PUMP PLUNGER
32. PUMP PLUNGER
33. POWER PISTON ASSY.
34. SCREW-VENTURI CLUSTER CENTER
35. WASHER-FIBER-VENTURI CLUSTER
    CENTER SCREW.
36. SCREW & LOCKWASHER-VENTURI
    CLUSTER.
37. VENTURI CLUSTER ASSY.
38. GASKET-VENTURI CLUSTER ASSY.
39. INSERT-MAIN WELL
40. SPRING-PUMP RETURN
41. BALL-PUMP INTAKE (USED IN
    TYPES WITH SEAT IN CASTING)
42. GUIDE-PUMP DISCHARGE BALL
43. SPRING-PUMP DISCHARGE BALL
44. BALL-PUMP DISCHARGE
45. VALVE ASSY.-POWER
46. GASKET-POWER VALVE
47. JETS-MAIN METERING
48. SCREW & LOCKWASHER-THROTTLE
    BODY TO BOWL
49. BOWL ASSY.-FLOAT
50. GASKET-BODY FLANGE-USED WITH
    OR WITHOUT IDLE COMPENSATOR (60)
50A GASKET-BODY FLANGE-WITH VAPOR
    VENT SLOTS-FOR TYPES WITHOUT
    IDLE COMPENSATOR
50B GASKET-BODY FLANGE-WITH VAPOR
    VENT OPENING AND IDLE COMPENSATOR
    PASSAGE-MAY SUB FOR 50 & 50A
51. THROTTLE BODY ASSY.
52. NEEDLE-IDLE ADJUSTING
53. SPRING-IDLE ADJUSTING NEEDLE
54. GASKET-FLANGE
55. NEEDLE-IDLE AIR ADJUSTING
    (FOR BY-PASS IDLE SYSTEM)
56. SPRING-IDLE AIR ADJUSTING
    NEEDLE
57. SCREW-COVER IDLE COMPENSATOR
58. COVER-IDLE COMPENSATOR
59. SCREW-IDLE COMPENSATOR VALVE
60. VALVE-IDLE COMPENSATOR
61. GASKET-IDLE COMPENSATOR VALVE
62. STAT COVER & SPRING ASSY. HOT
    WATER TYPE CHOKE.
63. RETAINER-FILTER
64. FILTER-AIR INTAKE
65. RETAINER-VACUUM CONTROL ROD (2GV)
66. ROD-VACUUM CONTROL (2GV)
67. SCREW-STAT ROD LEVER (2GV)
68. LEVER-STAT ROD (2GV)
69. SCREW-VACUUM CONTROL ATTACHING
70. VACUUM BREAK CONTROL ASSY. (2GV)
71. HOSE-VACUUM BREAK CONTROL (2GV)

Figure 12-3. Typical Rochester Model 2G or 2GC carburetor. (Rochester)

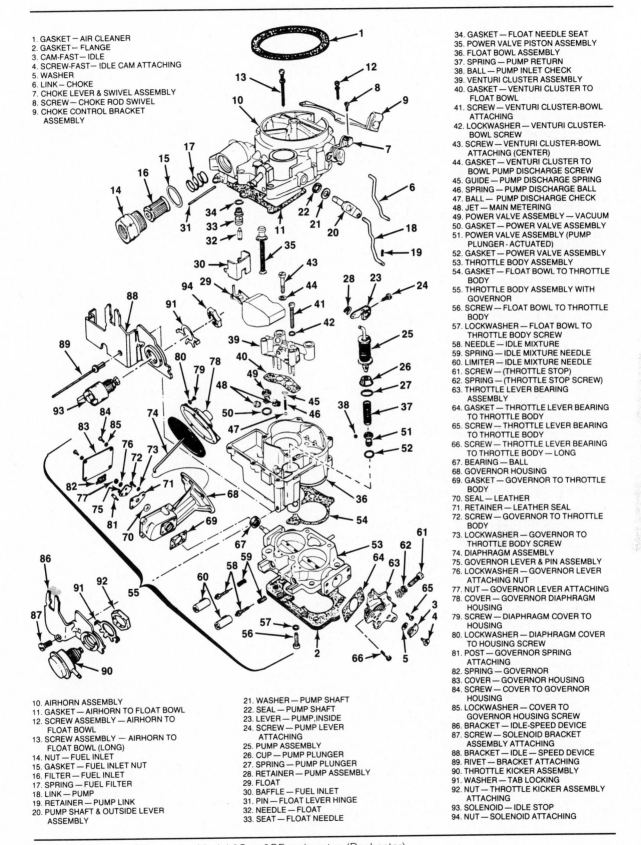

1. GASKET — AIR CLEANER
2. GASKET — FLANGE
3. CAM-FAST — IDLE
4. SCREW-FAST— !DLE CAM ATTACHING
5. WASHER
6. LINK — CHOKE
7. CHOKE LEVER & SWIVEL ASSEMBLY
8. SCREW — CHOKE ROD SWIVEL
9. CHOKE CONTROL BRACKET
   ASSEMBLY

10. AIRHORN ASSEMBLY
11. GASKET — AIRHORN TO FLOAT BOWL
12. SCREW ASSEMBLY — AIRHORN TO
    FLOAT BOWL
13. SCREW ASSEMBLY — AIRHORN TO
    FLOAT BOWL (LONG)
14. NUT — FUEL INLET
15. GASKET — FUEL INLET NUT
16. FILTER — FUEL INLET
17. SPRING — FUEL FILTER
18. LINK — PUMP
19. RETAINER — PUMP LINK
20. PUMP SHAFT & OUTSIDE LEVER
    ASSEMBLY

21. WASHER — PUMP SHAFT
22. SEAL — PUMP SHAFT
23. LEVER — PUMP,INSIDE
24. SCREW — PUMP LEVER
    ATTACHING
25. PUMP ASSEMBLY
26. CUP — PUMP PLUNGER
27. SPRING — PUMP PLUNGER
28. RETAINER — PUMP ASSEMBLY
29. FLOAT
30. BAFFLE — FUEL INLET
31. PIN — FLOAT LEVER HINGE
32. NEEDLE — FLOAT
33. SEAT — FLOAT NEEDLE

34. GASKET — FLOAT NEEDLE SEAT
35. POWER VALVE PISTON ASSEMBLY
36. FLOAT BOWL ASSEMBLY
37. SPRING — PUMP RETURN
38. BALL — PUMP INLET CHECK
39. VENTURI CLUSTER ASSEMBLY
40. GASKET — VENTURI CLUSTER TO
    FLOAT BOWL
41. SCREW — VENTURI CLUSTER-BOWL
    ATTACHING
42. LOCKWASHER — VENTURI CLUSTER-
    BOWL SCREW
43. SCREW — VENTURI CLUSTER-BOWL
    ATTACHING (CENTER)
44. GASKET — VENTURI CLUSTER TO
    BOWL PUMP DISCHARGE SCREW
45. GUIDE — PUMP DISCHARGE SPRING
46. SPRING — PUMP DISCHARGE BALL
47. BALL — PUMP DISCHARGE CHECK
48. JET — MAIN METERING
49. POWER VALVE ASSEMBLY — VACUUM
50. GASKET — POWER VALVE ASSEMBLY
51. POWER VALVE ASSEMBLY (PUMP
    PLUNGER - ACTUATED)
52. GASKET — POWER VALVE ASSEMBLY
53. THROTTLE BODY ASSEMBLY
54. GASKET — FLOAT BOWL TO THROTTLE
    BODY
55. THROTTLE BODY ASSEMBLY WITH
    GOVERNOR
56. SCREW — FLOAT BOWL TO THROTTLE
    BODY
57. LOCKWASHER — FLOAT BOWL TO
    THROTTLE BODY SCREW
58. NEEDLE — IDLE MIXTURE
59. SPRING — IDLE MIXTURE NEEDLE
60. LIMITER — IDLE MIXTURE NEEDLE
61. SCREW — (THROTTLE STOP)
62. SPRING — (THROTTLE STOP SCREW)
63. THROTTLE LEVER BEARING
    ASSEMBLY
64. GASKET — THROTTLE LEVER BEARING
    TO THROTTLE BODY
65. SCREW — THROTTLE LEVER BEARING
    TO THROTTLE BODY
66. SCREW — THROTTLE LEVER BEARING
    TO THROTTLE BODY — LONG
67. BEARING — BALL
68. GOVERNOR HOUSING
69. GASKET — GOVERNOR TO THROTTLE
    BODY
70. SEAL — LEATHER
71. RETAINER — LEATHER SEAL
72. SCREW — GOVERNOR TO THROTTLE
    BODY
73. LOCKWASHER — GOVERNOR TO
    THROTTLE BODY SCREW
74. DIAPHRAGM ASSEMBLY
75. GOVERNOR LEVER & PIN ASSEMBLY
76. LOCKWASHER — GOVERNOR LEVER
    ATTACHING NUT
77. NUT — GOVERNOR LEVER ATTACHING
78. COVER — GOVERNOR DIAPHRAGM
    HOUSING
79. SCREW — DIAPHRAGM COVER TO
    HOUSING
80. LOCKWASHER — DIAPHRAGM COVER
    TO HOUSING SCREW
81. POST — GOVERNOR SPRING
    ATTACHING
82. SPRING — GOVERNOR
83. COVER — GOVERNOR HOUSING
84. SCREW — COVER TO GOVERNOR
    HOUSING
85. LOCKWASHER — COVER TO
    GOVERNOR HOUSING SCREW
86. BRACKET — IDLE-SPEED DEVICE
87. SCREW — SOLENOID BRACKET
    ASSEMBLY ATTACHING
88. BRACKET — IDLE — SPEED DEVICE
89. RIVET — BRACKET ATTACHING
90. THROTTLE KICKER ASSEMBLY
91. WASHER — TAB LOCKING
92. NUT — THROTTLE KICKER ASSEMBLY
    ATTACHING
93. SOLENOID — IDLE STOP
94. NUT — SOLENOID ATTACHING

Figure 12-4. Typical Rochester Model 2G or 2GF carburetor. (Rochester)

and bus applications. It has a manual choke, may be fitted with a governor, and contains an adjustable wide-open throttle mixture control.

Four fuel metering circuits are used: idle, main, accelerator pump, and power enrichment. Each barrel has two venturis and a separate fuel feed, with centrally located metering. A removable venturi cluster contains the main nozzle, idle tubes, mixture passages, air bleeds, and pump jets. The main nozzle and idle tubes are part of the airhorn and are suspended in the fuel bowl main wells. This helps prevent percolation spillover during hot engine operation. Some versions use a hot-idle compensator.

ROCHESTER 2MC, M2MC, M2ME, E2MC, AND E2ME (DUALJET) TWO-BARREL

A single-stage carburetor introduced in 1975, the Dualjet, figure 12-5, uses the design features of the primary side of the Quadrajet four-barrel carburetor. The 2M is the basic Dualjet model designation; the M2M is a modified version used with open-loop fuel systems, and the E2M is a feedback carburetor used with closed-loop fuel systems. A hot air heated cap-type choke is used with "C" designations; an integral cap-type electric choke is used with "E" designations.

The 2MC uses the same casting as the Quadrajet but contains no secondary throttle or system. Each bore of the 2MC uses a separate and independent idle system. All primary circuits are the same as the Quadrajet. A machined pump well is used with some 1985 and later versions for better sealing between the pump cup and bore. Such carburetors have "MW" stamped on the float bowl. If fuel bowl replacement is required, the new part used must have these letters stamped on it.

A revised version of the 2MC was introduced in 1977. Called the M2MC200 series, it retains most of the 2MC features, but does not use the same Quadrajet body casting.

The E2MC and E2ME are called "closed-loop" carburetors because they are used with GM Computer Command Control (CCC) systems and contain a mixture control solenoid which varies the air-fuel ratio as directed by the engine control computer.

ROCHESTER 2SE AND E2SE (VARIJET) TWO-BARREL

The 2SE, figure 12-6, and E2SE, figure 12-7, are 2-stage downdraft carburetors with four basic operating circuits: idle, main, secondary enrichment, and accelerator pump. A single float chamber provides fuel for both barrels. An electrically heated, cap-type choke is controlled by dual vacuum-break diaphragm.

The E2SE is a closed-loop carburetor used with CCC systems. A mixture control solenoid in the airhorn extends into the float bowl and controls air-fuel mixture in the primary barrel as directed by the engine control computer. A throttle position sensor in the float bowl signals the computer of throttle position changes. An air valve and tapered metering rod in a fixed jet control the air-fuel mixture at wide-open throttle.

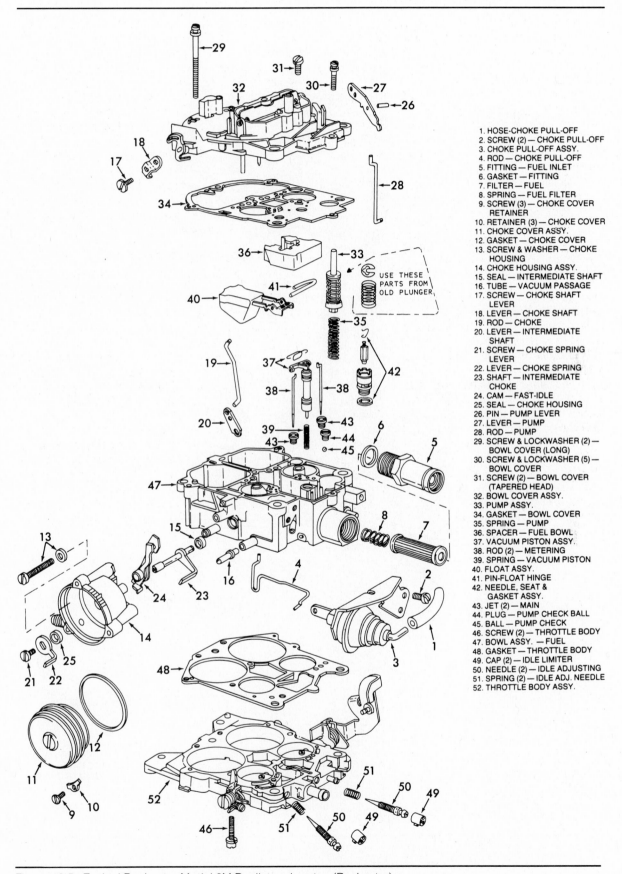

1. HOSE-CHOKE PULL-OFF
2. SCREW (2) — CHOKE PULL-OFF
3. CHOKE PULL-OFF ASSY.
4. ROD — CHOKE PULL-OFF
5. FITTING — FUEL INLET
6. GASKET — FITTING
7. FILTER — FUEL
8. SPRING — FUEL FILTER
9. SCREW (3) — CHOKE COVER
   RETAINER
10. RETAINER (3) — CHOKE COVER
11. CHOKE COVER ASSY.
12. GASKET — CHOKE COVER
13. SCREW & WASHER — CHOKE
    HOUSING
14. CHOKE HOUSING ASSY.
15. SEAL — INTERMEDIATE SHAFT
16. TUBE — VACUUM PASSAGE
17. SCREW — CHOKE SHAFT
    LEVER
18. LEVER — CHOKE SHAFT
19. ROD — CHOKE
20. LEVER — INTERMEDIATE
    SHAFT
21. SCREW — CHOKE SPRING
    LEVER
22. LEVER — CHOKE SPRING
23. SHAFT — INTERMEDIATE
    CHOKE
24. CAM — FAST-IDLE
25. SEAL — CHOKE HOUSING
26. PIN — PUMP LEVER
27. LEVER — PUMP
28. ROD — PUMP
29. SCREW & LOCKWASHER (2) —
    BOWL COVER (LONG)
30. SCREW & LOCKWASHER (5) —
    BOWL COVER
31. SCREW (2) — BOWL COVER
    (TAPERED HEAD)
32. BOWL COVER ASSY.
33. PUMP ASSY.
34. GASKET — BOWL COVER
35. SPRING — PUMP
36. SPACER — FUEL BOWL
37. VACUUM PISTON ASSY.
38. ROD (2) — METERING
39. SPRING — VACUUM PISTON
40. FLOAT ASSY.
41. PIN-FLOAT HINGE
42. NEEDLE, SEAT &
    GASKET ASSY.
43. JET (2) — MAIN
44. PLUG — PUMP CHECK BALL
45. BALL — PUMP CHECK
46. SCREW (2) — THROTTLE BODY
47. BOWL ASSY. — FUEL
48. GASKET — THROTTLE BODY
49. CAP (2) — IDLE LIMITER
50. NEEDLE (2) — IDLE ADJUSTING
51. SPRING (2) — IDLE ADJ. NEEDLE
52. THROTTLE BODY ASSY.

Figure 12-5. Typical Rochester Model 2M Dualjet carburetor. (Rochester)

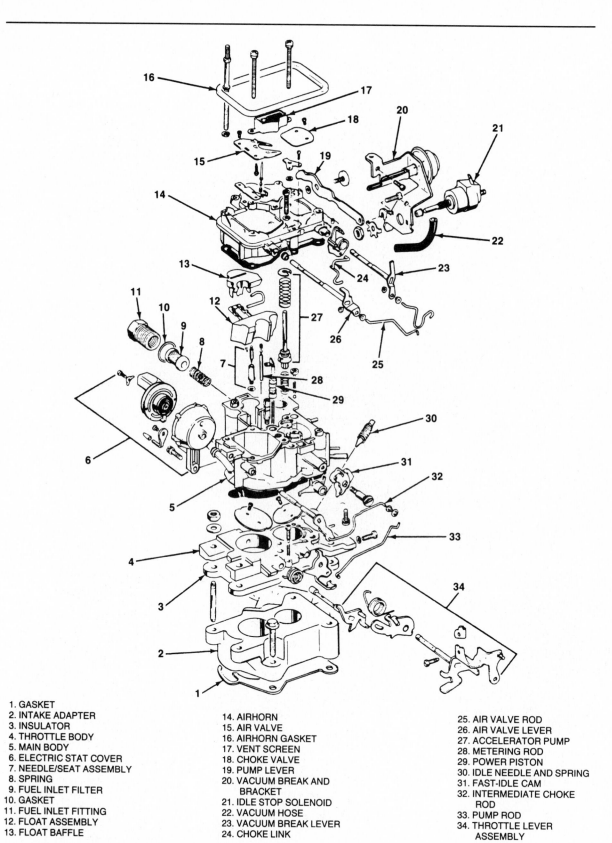

1. GASKET
2. INTAKE ADAPTER
3. INSULATOR
4. THROTTLE BODY
5. MAIN BODY
6. ELECTRIC STAT COVER
7. NEEDLE/SEAT ASSEMBLY
8. SPRING
9. FUEL INLET FILTER
10. GASKET
11. FUEL INLET FITTING
12. FLOAT ASSEMBLY
13. FLOAT BAFFLE

14. AIRHORN
15. AIR VALVE
16. AIRHORN GASKET
17. VENT SCREEN
18. CHOKE VALVE
19. PUMP LEVER
20. VACUUM BREAK AND
    BRACKET
21. IDLE STOP SOLENOID
22. VACUUM HOSE
23. VACUUM BREAK LEVER
24. CHOKE LINK

25. AIR VALVE ROD
26. AIR VALVE LEVER
27. ACCELERATOR PUMP
28. METERING ROD
29. POWER PISTON
30. IDLE NEEDLE AND SPRING
31. FAST-IDLE CAM
32. INTERMEDIATE CHOKE
    ROD
33. PUMP ROD
34. THROTTLE LEVER
    ASSEMBLY

Figure 12-6. Typical Rochester Model 2SE carburetor. (Rochester)

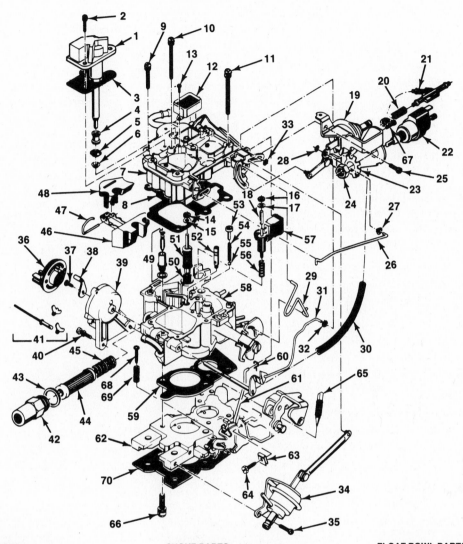

Figure 12-7. Typical Rochester Model E2SE carburetor. (Rochester)

AIRHORN PARTS:
1. MIXTURE CONTROL SOLENOID
2. M/C SOLENOID SCREW (3)
3. M/C SOLENOID GASKET
4. M/C SOLENOID SPACER
5. M/C SOLENOID SEAL
6. M/C SOLENOID SEAL RETAINER
7. AIRHORN ASSEMBLY
8. AIRHORN GASKET
9. AIRHORN SCREW-SHORT (2 OR 3)
10. AIRHORN SCREW-LONG (3)
11. AIRHORN SCREW-LARGE
12. VENT STACK
13. VENT STACK SCREW (2)
14. PUMP PLUNGER SEAL
15. PUMP PLUNGER SEAL RETAINER
16. TPS PLUNGER SEAL
17. TPS PLUNGER SEAL RETAINER
18. TPS PLUNGER (THROTTLE POSITION SENSOR)

CHOKE PARTS:
19. PRIMARY VACUUM BREAK AND BRACKET ASSEMBLY
20. VACUUM BREAK CONNECTING HOSE
21. VACUUM BREAK CONNECTING TEE
22. IDLE-SPEED SOLENOID
23. IDLE-SPEED SOLENOID RETAINER
24. IDLE-SPEED SOLENOID NUT
25. VACUUM BREAK BRACKET ATTACHING SCREW
26. AIR VALVE LINK
27. AIR VALVE LINK BUSHING
28. AIR VALVE LINK RETAINER
29. FAST-IDLE CAM LINK
30. VACUUM BREAK HOSE
31. INTERMEDIATE CHOKE SHAFT/LEVER/ LINK ASSEMBLY
32. INTERMEDIATE CHOKE LINK BUSHING
33. INTERMEDIATE CHOKE LINK RETAINER
34. SECONDARY VACUUM BREAK AND BRACKET ASSEMBLY
35. VACUUM BREAK ATTACHING SCREW (2)
36. CHOKE COVER AND COIL ASSEMBLY
37. CHOKE LEVER ATTACHING SCREW
38. CHOKE LEVER AND CONTACT ASSEMBLY
39. CHOKE HOUSING
40. CHOKE HOUSING ATTACHING SCREW (2)
41. STAT COVER RETAINER KIT

FLOAT BOWL PARTS:
42. FUEL INLET
43. FUEL INLET NUT GASKET
44. FUEL INLET FILTER
45. FUEL FILTER SPRING
46. FLOAT ASSEMBLY
47. FLOAT HINGE PIN
48. FLOAT BOWL INSERT
49. NEEDLE AND SEAT ASSEMBLY
50. PUMP RETURN SPRING
51. PUMP ASSEMBLY
52. METERING JET
53. PUMP SPRING AND CHECK BALL RETAINER
54. PUMP CHECK BALL SPRING
55. PUMP CHECK BALL
56. TPS SPRING
57. TPS (THROTTLE POSITION SENSOR)
58. FLOAT BOWL ASSEMBLY
59. FLOAT BOWL GASKET

THROTTLE BODY PARTS:
60. PUMP ROD CLIP
61. PUMP ROD
62. THROTTLE BODY ASSEMBLY
63. CAM SCREW CLIP
64. FAST-IDLE CAM SCREW
65. IDLE NEEDLE AND SPRING
66. THROTTLE BODY ATTACHING SCREW
67. VACUUM BREAK BRACKET ATTACHING SCREW (NEW)
68. IDLE STOP SCREW
69. IDLE STOP SCREW SPRING
70. INTAKE MANIFOLD GASKET

ROCHESTER 4GC FOUR-BARREL

The Model 4GC, figure 12-8, has an airhorn, a float bowl, and a throttle body. Each primary barrel is aided by a secondary which starts to open at about half-throttle. The primary and secondary sides use independent fuel bowls and twin floats, as well as separate idle and main metering circuits. The secondary throttles are mechanically linked to the primary throttle valves. The primary side also has an accelerator pump, power-enrichment and choke circuits, and an integral choke operated by a thermostatic coil spring. Small venturis are used in the primary side for fuel economy. Large venturis in the secondary side provide more airflow. Some versions have a hot-idle compensator.

ROCHESTER QUADRAJET FOUR-BARREL (MODELS 4M AND M4M OPEN-LOOP VARIATIONS)

The 4M is the basic Quadrajet designation; the M4M is a modified version used with open-loop fuel systems. A "V" designation indicates a well-type or remote choke located on the engine manifold. A hot air heated cap-type choke is used with "C" designations; an integral cap-type electric choke is used with "E" designations. A "D" in the designation indicates the use of a dual-capacity pump; an "F" contains an adjustable wide-open throttle mixture control.

The Quadrajet series is a 2-stage, spread-bore design, figure 12-9, with large secondaries and small primaries. A small central fuel bowl serves both primary and secondary sides. The central bowl also reduces fuel evaporation dur-ing engine shutdown periods. The 3-piece design uses an aluminum throttle body. The primary side contains four basic fuel metering circuits: idle, main, accelerator pump, and power enrichment. The secondary side uses only a main metering system. Both primary and secondary sides use tapered metering rods with an air valve in the secondary side for metering control.

A machined pump well is used with some 1985 and later versions for better sealing between the pump cup and bore. These carburetors have "MW" stamped on the float bowl. If float bowl replacement is required, a new part stamped with these letters must be used.

The wide use of this carburetor by GM, Ford, and Chrysler has resulted in a large number of different versions, such as the M4ME, figure 12-10, M4MED, figure 12-11, and M4MEF, figure 12-12.

ROCHESTER QUADRAJET FOUR-BARREL (MODEL E4M AND CLOSED-LOOP VARIATIONS)

Sometimes called the "Electronic" Quadrajet, the E4M and its numerous variations, figure 12-13, are used with electronic fuel management or engine control systems. The closed-loop carburetor designations are the same as those used with the open-loop Quadrajet.

The basic features of the E4M are the same as those of the M4M, with the addition of a mixture control solenoid to vary the air-fuel mixture ratio on direction from the engine control computer. Models with a dual-capacity pump use a combined mixture control and dual-capacity pump solenoid.

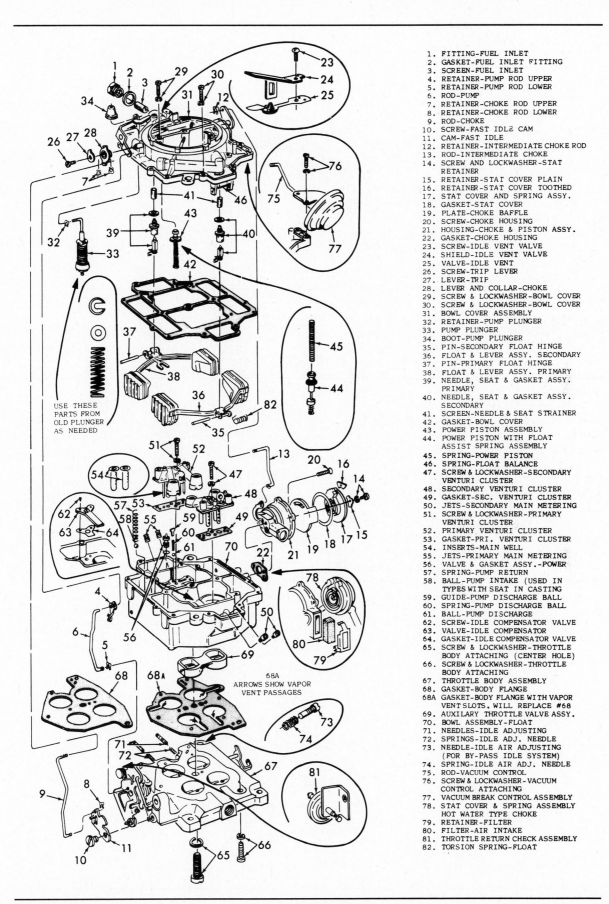

1. FITTING-FUEL INLET
2. GASKET-FUEL INLET FITTING
3. SCREEN-FUEL INLET
4. RETAINER-PUMP ROD UPPER
5. RETAINER-PUMP ROD LOWER
6. ROD-PUMP
7. RETAINER-CHOKE ROD UPPER
8. RETAINER-CHOKE ROD LOWER
9. ROD-CHOKE
10. SCREW-FAST IDLE CAM
11. CAM-FAST IDLE
12. RETAINER-INTERMEDIATE CHOKE ROD
13. ROD-INTERMEDIATE CHOKE
14. SCREW AND LOCKWASHER-STAT RETAINER
15. RETAINER-STAT COVER PLAIN
16. RETAINER-STAT COVER TOOTHED
17. STAT COVER AND SPRING ASSY.
18. GASKET-STAT COVER
19. PLATE-CHOKE BAFFLE
20. SCREW-CHOKE HOUSING
21. HOUSING-CHOKE & PISTON ASSY.
22. GASKET-CHOKE HOUSING
23. SCREW-IDLE VENT VALVE
24. SHIELD-IDLE VENT VALVE
25. VALVE-IDLE VENT
26. SCREW-TRIP LEVER
27. LEVER-TRIP
28. LEVER AND COLLAR-CHOKE
29. SCREW & LOCKWASHER-BOWL COVER
30. SCREW & LOCKWASHER-BOWL COVER
31. BOWL COVER ASSEMBLY
32. RETAINER-PUMP PLUNGER
33. PUMP PLUNGER
34. BOOT-PUMP PLUNGER
35. PIN-SECONDARY FLOAT HINGE
36. FLOAT & LEVER ASSY. SECONDARY
37. PIN-PRIMARY FLOAT HINGE
38. FLOAT & LEVER ASSY. PRIMARY
39. NEEDLE, SEAT & GASKET ASSY. PRIMARY
40. NEEDLE, SEAT & GASKET ASSY. SECONDARY
41. SCREEN-NEEDLE & SEAT STRAINER
42. GASKET-BOWL COVER
43. POWER PISTON ASSEMBLY
44. POWER PISTON WITH FLOAT ASSIST SPRING ASSEMBLY
45. SPRING-POWER PISTON
46. SPRING-FLOAT BALANCE
47. SCREW & LOCKWASHER-SECONDARY VENTURI CLUSTER
48. SECONDARY VENTURI CLUSTER
49. GASKET-SEC. VENTURI CLUSTER
50. JETS-SECONDARY MAIN METERING
51. SCREW & LOCKWASHER-PRIMARY VENTURI CLUSTER
52. PRIMARY VENTURI CLUSTER
53. GASKET-PRI. VENTURI CLUSTER
54. INSERTS-MAIN WELL
55. JETS-PRIMARY MAIN METERING
56. VALVE & GASKET ASSY.-POWER
57. SPRING-PUMP RETURN
58. BALL-PUMP INTAKE (USED IN TYPES WITH SEAT IN CASTING
59. GUIDE-PUMP DISCHARGE BALL
60. SPRING-PUMP DISCHARGE BALL
61. BALL-PUMP DISCHARGE
62. SCREW-IDLE COMPENSATOR VALVE
63. VALVE-IDLE COMPENSATOR
64. GASKET-IDLE COMPENSATOR VALVE
65. SCREW & LOCKWASHER-THROTTLE BODY ATTACHING (CENTER HOLE)
66. SCREW & LOCKWASHER-THROTTLE BODY ATTACHING
67. THROTTLE BODY ASSEMBLY
68. GASKET-BODY FLANGE
68A GASKET-BODY FLANGE WITH VAPOR VENT SLOTS. WILL REPLACE #68
69. AUXILARY THROTTLE VALVE ASSY.
70. BOWL ASSEMBLY-FLOAT
71. NEEDLES-IDLE ADJUSTING
72. SPRINGS-IDLE ADJ. NEEDLE
73. NEEDLE-IDLE AIR ADJUSTING (FOR BY-PASS IDLE SYSTEM)
74. SPRING-IDLE AIR ADJ. NEEDLE
75. ROD-VACUUM CONTROL
76. SCREW & LOCKWASHER-VACUUM CONTROL ATTACHING
77. VACUUM BREAK CONTROL ASSEMBLY
78. STAT COVER & SPRING ASSEMBLY HOT WATER TYPE CHOKE
79. RETAINER-FILTER
80. FILTER-AIR INTAKE
81. THROTTLE RETURN CHECK ASSEMBLY
82. TORSION SPRING-FLOAT

USE THESE PARTS FROM OLD PLUNGER AS NEEDED

68A ARROWS SHOW VAPOR VENT PASSAGES

Figure 12-8. Typical Rochester Model 4GC carburetor. (Rochester)

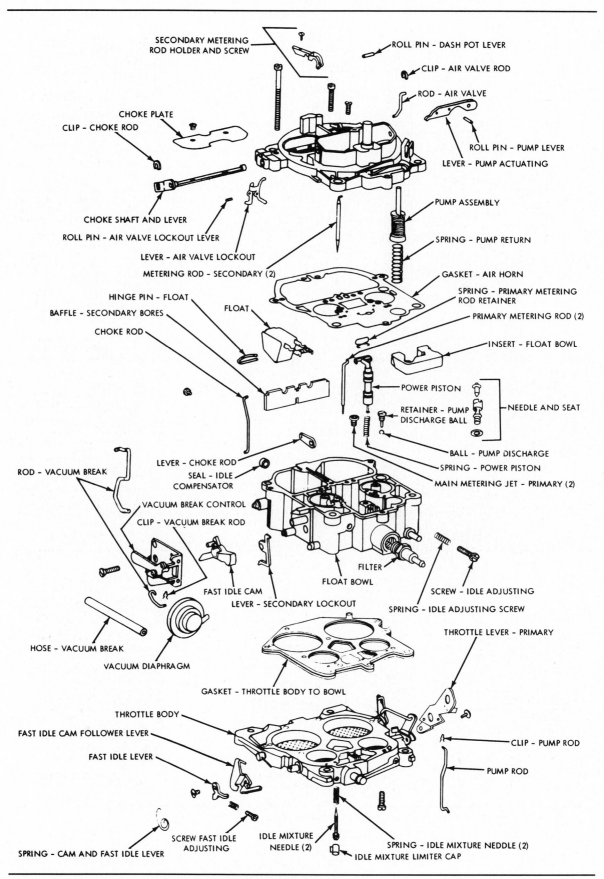

Figure 12-9. Typical Rochester Model 4MV carburetor. (Rochester)

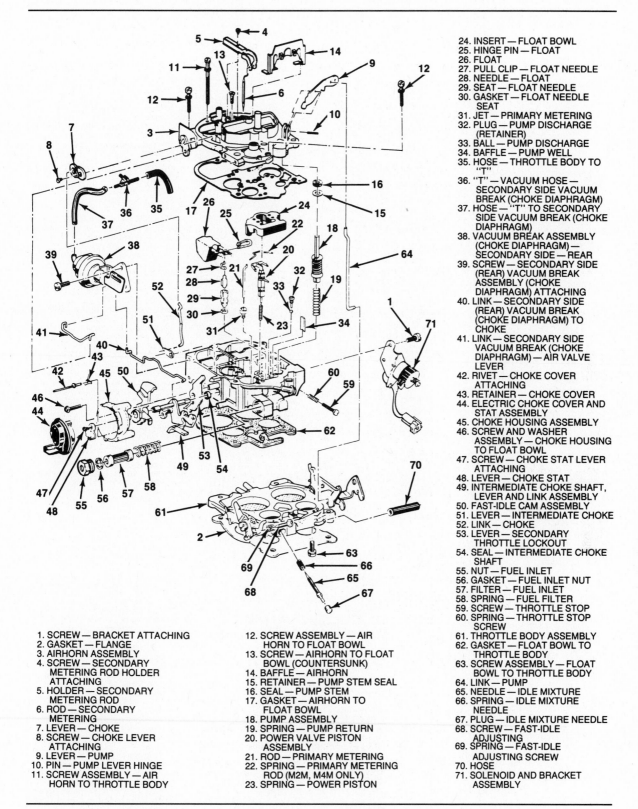

24. INSERT — FLOAT BOWL
25. HINGE PIN — FLOAT
26. FLOAT
27. PULL CLIP — FLOAT NEEDLE
28. NEEDLE — FLOAT
29. SEAT — FLOAT NEEDLE
30. GASKET — FLOAT NEEDLE SEAT
31. JET — PRIMARY METERING
32. PLUG — PUMP DISCHARGE (RETAINER)
33. BALL — PUMP DISCHARGE
34. BAFFLE — PUMP WELL
35. HOSE — THROTTLE BODY TO "T"
36. "T" — VACUUM HOSE — SECONDARY SIDE VACUUM BREAK (CHOKE DIAPHRAGM)
37. HOSE — "T" TO SECONDARY SIDE VACUUM BREAK (CHOKE DIAPHRAGM)
38. VACUUM BREAK ASSEMBLY (CHOKE DIAPHRAGM) — SECONDARY SIDE — REAR
39. SCREW — SECONDARY SIDE (REAR) VACUUM BREAK ASSEMBLY (CHOKE DIAPHRAGM) ATTACHING
40. LINK — SECONDARY SIDE (REAR) VACUUM BREAK (CHOKE DIAPHRAGM) TO CHOKE
41. LINK — SECONDARY SIDE VACUUM BREAK (CHOKE DIAPHRAGM) — AIR VALVE LEVER
42. RIVET — CHOKE COVER ATTACHING
43. RETAINER — CHOKE COVER
44. ELECTRIC CHOKE COVER AND STAT ASSEMBLY
45. CHOKE HOUSING ASSEMBLY
46. SCREW AND WASHER ASSEMBLY — CHOKE HOUSING TO FLOAT BOWL
47. SCREW — CHOKE STAT LEVER ATTACHING
48. LEVER — CHOKE STAT
49. INTERMEDIATE CHOKE SHAFT, LEVER AND LINK ASSEMBLY
50. FAST-IDLE CAM ASSEMBLY
51. LEVER — INTERMEDIATE CHOKE
52. LINK — CHOKE
53. LEVER — SECONDARY THROTTLE LOCKOUT
54. SEAL — INTERMEDIATE CHOKE SHAFT
55. NUT — FUEL INLET
56. GASKET — FUEL INLET NUT
57. FILTER — FUEL INLET
58. SPRING — FUEL FILTER
59. SCREW — THROTTLE STOP
60. SPRING — THROTTLE STOP SCREW
61. THROTTLE BODY ASSEMBLY
62. GASKET — FLOAT BOWL TO THROTTLE BODY
63. SCREW ASSEMBLY — FLOAT BOWL TO THROTTLE BODY
64. LINK — PUMP
65. NEEDLE — IDLE MIXTURE
66. SPRING — IDLE MIXTURE NEEDLE
67. PLUG — IDLE MIXTURE NEEDLE
68. SCREW — FAST-IDLE ADJUSTING
69. SPRING — FAST-IDLE ADJUSTING SCREW
70. HOSE
71. SOLENOID AND BRACKET ASSEMBLY

1. SCREW — BRACKET ATTACHING
2. GASKET — FLANGE
3. AIRHORN ASSEMBLY
4. SCREW — SECONDARY METERING ROD HOLDER ATTACHING
5. HOLDER — SECONDARY METERING ROD
6. ROD — SECONDARY METERING
7. LEVER — CHOKE
8. SCREW — CHOKE LEVER ATTACHING
9. LEVER — PUMP
10. PIN — PUMP LEVER HINGE
11. SCREW ASSEMBLY — AIR HORN TO THROTTLE BODY

12. SCREW ASSEMBLY — AIR HORN TO FLOAT BOWL
13. SCREW — AIRHORN TO FLOAT BOWL (COUNTERSUNK)
14. BAFFLE — AIRHORN
15. RETAINER — PUMP STEM SEAL
16. SEAL — PUMP STEM
17. GASKET — AIRHORN TO FLOAT BOWL
18. PUMP ASSEMBLY
19. SPRING — PUMP RETURN
20. POWER VALVE PISTON ASSEMBLY
21. ROD — PRIMARY METERING
22. SPRING — PRIMARY METERING ROD (M2M, M4M ONLY)
23. SPRING — POWER PISTON

Figure 12-10. Typical Rochester Model M4ME carburetor. (Chrysler)

17. HOSE — PRIMARY SIDE (FRONT) VACUUM BREAK
18. LINK — PRIMARY SIDE VACUUM BREAK — AIR VALVE LEVER
19. RETAINER — PUMP STEM SEAL
20. SEAL — PUMP STEM
21. FLOAT BOWL ASSEMBLY
22. GASKET — AIRHORN TO FLOAT BOWL
23. PUMP ASSEMBLY
24. SPRING — PUMP RETURN
25. POWER VALVE PISTON ASSEMBLY
26. ROD — PRIMARY METERING
27. SPRING — PRIMARY METERING ROD
28. SPRING — POWER PISTON
29. SCREW — SOLENOID CONNECTOR ATTACHING
30. GASKET — SOLENOID CONNECTOR TO AIRHORN
31. INSERT — FLOAT BOWL
32. HINGE PIN — FLOAT
33. FLOAT
34. PULL CLIP — FLOAT NEEDLE
35. NEEDLE — FLOAT
36. SEAT — FLOAT NEEDLE
37. GASKET — FLOAT NEEDLE SEAT
38. VALVE ASSEMBLY — DUAL CAPACITY PUMP
39. GASKET — VALVE ASSEMBLY
40. JET — PRIMARY METERING
41. PLUG — PUMP DISCHARGE (RETAINER)
42. BALL — PUMP DISCHARGE
43. BAFFLE — PUMP WELL
44. SOLENOID — DUAL CAPACITY PUMP
45. HOSE — SECONDARY SIDE (REAR) VACUUM BREAK
46. VACUUM BREAK ASSEMBLY — SECONDARY SIDE (REAR)
47. SCREW — SECONDARY SIDE (REAR) VACUUM BREAK
48. LINK — SECONDARY SIDE (REAR) VACUUM BREAK TO CHOKE
49. LINK — SECONDARY SIDE VACUUM BREAK — AIR VALVE LEVER
50. RIVET — CHOKE COVER ATTACHING
51. RETAINER — CHOKE COVER
52. ELECTRIC CHOKE COVER AND STAT ASSEMBLY
53. CHOKE HOUSING ASSEMBLY
54. SCREW AND WASHER ASSEMBLY — CHOKE HOUSING TO FLOAT BOWL
55. SCREW — CHOKE STAT LEVER ATTACHING
56. LEVER — CHOKE STAT
57. INTERMEDIATE CHOKE SHAFT, LEVER AND LINK ASSEMBLY
58. FAST-IDLE CAM ASSEMBLY
59. LEVER — INTERMEDIATE CHOKE
60. LINK — CHOKE
61. LEVER — SECONDARY THROTTLE LOCKOUT
62. SEAL — INTERMEDIATE CHOKE SHAFT
63. NUT — FUEL INLET
64. GASKET — FUEL INLET NUT
65. FILTER — FUEL INLET
66. SPRING — FUEL FILTER
67. SCREW — THROTTLE STOP
68. SPRING — THROTTLE STOP SCREW
69. THROTTLE BODY ASSEMBLY
70. GASKET — FLOAT BOWL TO THROTTLE BODY
71. SCREW ASSEMBLY — FLOAT BOWL TO THROTTLE BODY
72. LINK — PUMP
73. NEEDLE — IDLE MIXTURE
74. SPRING — IDLE MIXTURE NEEDLE
75. PLUG — IDLE MIXTURE NEEDLE
76. SCREW — FAST-IDLE ADJUSTING
77. SPRING — FAST-IDLE ADJUSTING SCREW
78. SOLENOID AND BRACKET ASSEMBLY
79. SCREW — BRACKET ATTACHING
80. THROTTLE KICKER ASSEMBLY
81. BRACKET — THROTTLE KICKER
82. NUT — THROTTLE KICKER ASSEMBLY ATTACHING
83. WASHER — TAB LOCKING

1. GASKET — AIR CLEANER
2. GASKET — FLANGE
3. AIRHORN ASSEMBLY
4. SCREW — SECONDARY METERING ROD HOLDER ATTACHING
5. HOLDER — SECONDARY METERING ROD
6. ROD — SECONDARY METERING
7. LEVER — CHOKE
8. SCREW — CHOKE LEVER ATTACHING
9. LEVER — PUMP
10. PIN — PUMP LEVER HINGE

11. SCREW ASSEMBLY — AIRHORN TO THROTTLE BODY
12. SCREW ASSEMBLY — AIRHORN TO FLOAT BOWL
13. SCREW — AIRHORN TO FLOAT BOWL (COUNTERSUNK)
14. BAFFLE — AIRHORN
15. VACUUM BREAK ASSEMBLY — PRIMARY SIDE (FRONT)
16. SCREW — PRIMARY SIDE (FRONT) VACUUM BREAK ASSEMBLY ATTACHING

Figure 12-11. Typical Rochester Model M4MED carburetor. (Rochester)

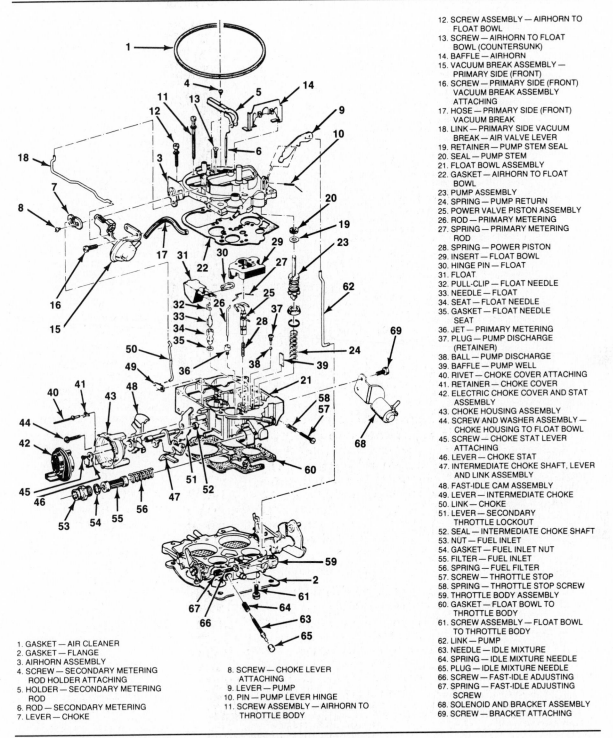

12. SCREW ASSEMBLY — AIRHORN TO FLOAT BOWL
13. SCREW — AIRHORN TO FLOAT BOWL (COUNTERSUNK)
14. BAFFLE — AIRHORN
15. VACUUM BREAK ASSEMBLY — PRIMARY SIDE (FRONT)
16. SCREW — PRIMARY SIDE (FRONT) VACUUM BREAK ASSEMBLY ATTACHING
17. HOSE — PRIMARY SIDE (FRONT) VACUUM BREAK
18. LINK — PRIMARY SIDE VACUUM BREAK — AIR VALVE LEVER
19. RETAINER — PUMP STEM SEAL
20. SEAL — PUMP STEM
21. FLOAT BOWL ASSEMBLY
22. GASKET — AIRHORN TO FLOAT BOWL
23. PUMP ASSEMBLY
24. SPRING — PUMP RETURN
25. POWER VALVE PISTON ASSEMBLY
26. ROD — PRIMARY METERING
27. SPRING — PRIMARY METERING ROD
28. SPRING — POWER PISTON
29. INSERT — FLOAT BOWL
30. HINGE PIN — FLOAT
31. FLOAT
32. PULL-CLIP — FLOAT NEEDLE
33. NEEDLE — FLOAT
34. SEAT — FLOAT NEEDLE
35. GASKET — FLOAT NEEDLE SEAT
36. JET — PRIMARY METERING
37. PLUG — PUMP DISCHARGE (RETAINER)
38. BALL — PUMP DISCHARGE
39. BAFFLE — PUMP WELL
40. RIVET — CHOKE COVER ATTACHING
41. RETAINER — CHOKE COVER
42. ELECTRIC CHOKE COVER AND STAT ASSEMBLY
43. CHOKE HOUSING ASSEMBLY
44. SCREW AND WASHER ASSEMBLY — CHOKE HOUSING TO FLOAT BOWL
45. SCREW — CHOKE STAT LEVER ATTACHING
46. LEVER — CHOKE STAT
47. INTERMEDIATE CHOKE SHAFT, LEVER AND LINK ASSEMBLY
48. FAST-IDLE CAM ASSEMBLY
49. LEVER — INTERMEDIATE CHOKE
50. LINK — CHOKE
51. LEVER — SECONDARY THROTTLE LOCKOUT
52. SEAL — INTERMEDIATE CHOKE SHAFT
53. NUT — FUEL INLET
54. GASKET — FUEL INLET NUT
55. FILTER — FUEL INLET
56. SPRING — FUEL FILTER
57. SCREW — THROTTLE STOP
58. SPRING — THROTTLE STOP SCREW
59. THROTTLE BODY ASSEMBLY
60. GASKET — FLOAT BOWL TO THROTTLE BODY
61. SCREW ASSEMBLY — FLOAT BOWL TO THROTTLE BODY
62. LINK — PUMP
63. NEEDLE — IDLE MIXTURE
64. SPRING — IDLE MIXTURE NEEDLE
65. PLUG — IDLE MIXTURE NEEDLE
66. SCREW — FAST-IDLE ADJUSTING
67. SPRING — FAST-IDLE ADJUSTING SCREW
68. SOLENOID AND BRACKET ASSEMBLY
69. SCREW — BRACKET ATTACHING

1. GASKET — AIR CLEANER
2. GASKET — FLANGE
3. AIRHORN ASSEMBLY
4. SCREW — SECONDARY METERING ROD HOLDER ATTACHING
5. HOLDER — SECONDARY METERING ROD
6. ROD — SECONDARY METERING
7. LEVER — CHOKE

8. SCREW — CHOKE LEVER ATTACHING
9. LEVER — PUMP
10. PIN — PUMP LEVER HINGE
11. SCREW ASSEMBLY — AIRHORN TO THROTTLE BODY

Figure 12-12. Typical Rochester Model M4MEF carburetor. (Rochester)

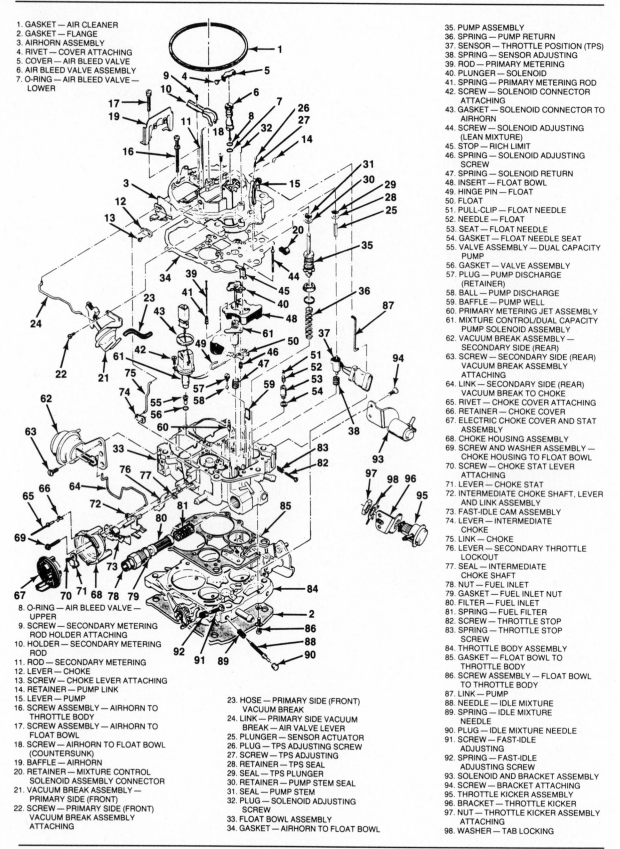

1. GASKET — AIR CLEANER
2. GASKET — FLANGE
3. AIRHORN ASSEMBLY
4. RIVET — COVER ATTACHING
5. COVER — AIR BLEED VALVE
6. AIR BLEED VALVE ASSEMBLY
7. O-RING — AIR BLEED VALVE — LOWER

8. O-RING — AIR BLEED VALVE — UPPER
9. SCREW — SECONDARY METERING ROD HOLDER ATTACHING
10. HOLDER — SECONDARY METERING ROD
11. ROD — SECONDARY METERING
12. LEVER — CHOKE
13. SCREW — CHOKE LEVER ATTACHING
14. RETAINER — PUMP LINK
15. LEVER — PUMP
16. SCREW ASSEMBLY — AIRHORN TO THROTTLE BODY
17. SCREW ASSEMBLY — AIRHORN TO FLOAT BOWL
18. SCREW — AIRHORN TO FLOAT BOWL (COUNTERSUNK)
19. BAFFLE — AIRHORN
20. RETAINER — MIXTURE CONTROL SOLENOID ASSEMBLY CONNECTOR
21. VACUUM BREAK ASSEMBLY — PRIMARY SIDE (FRONT)
22. SCREW — PRIMARY SIDE (FRONT) VACUUM BREAK ASSEMBLY ATTACHING

23. HOSE — PRIMARY SIDE (FRONT) VACUUM BREAK
24. LINK — PRIMARY SIDE VACUUM BREAK — AIR VALVE LEVER
25. PLUNGER — SENSOR ACTUATOR
26. PLUG — TPS ADJUSTING SCREW
27. SCREW — TPS ADJUSTING
28. RETAINER — TPS SEAL
29. SEAL — TPS PLUNGER
30. RETAINER — PUMP STEM SEAL
31. SEAL — PUMP STEM
32. PLUG — SOLENOID ADJUSTING SCREW
33. FLOAT BOWL ASSEMBLY
34. GASKET — AIRHORN TO FLOAT BOWL

35. PUMP ASSEMBLY
36. SPRING — PUMP RETURN
37. SENSOR — THROTTLE POSITION (TPS)
38. SPRING — SENSOR ADJUSTING
39. ROD — PRIMARY METERING
40. PLUNGER — SOLENOID
41. SPRING — PRIMARY METERING ROD
42. SCREW — SOLENOID CONNECTOR ATTACHING
43. GASKET — SOLENOID CONNECTOR TO AIRHORN
44. SCREW — SOLENOID ADJUSTING (LEAN MIXTURE)
45. STOP — RICH LIMIT
46. SPRING — SOLENOID ADJUSTING SCREW
47. SPRING — SOLENOID RETURN
48. INSERT — FLOAT BOWL
49. HINGE PIN — FLOAT
50. FLOAT
51. PULL-CLIP — FLOAT NEEDLE
52. NEEDLE — FLOAT
53. SEAT — FLOAT NEEDLE
54. GASKET — FLOAT NEEDLE SEAT
55. VALVE ASSEMBLY — DUAL CAPACITY PUMP
56. GASKET — VALVE ASSEMBLY
57. PLUG — PUMP DISCHARGE (RETAINER)
58. BALL — PUMP DISCHARGE
59. BAFFLE — PUMP WELL
60. PRIMARY METERING JET ASSEMBLY
61. MIXTURE CONTROL/DUAL CAPACITY PUMP SOLENOID ASSEMBLY
62. VACUUM BREAK ASSEMBLY — SECONDARY SIDE (REAR)
63. SCREW — SECONDARY SIDE (REAR) VACUUM BREAK ASSEMBLY ATTACHING
64. LINK — SECONDARY SIDE (REAR) VACUUM BREAK TO CHOKE
65. RIVET — CHOKE COVER ATTACHING
66. RETAINER — CHOKE COVER
67. ELECTRIC CHOKE COVER AND STAT ASSEMBLY
68. CHOKE HOUSING ASSEMBLY
69. SCREW AND WASHER ASSEMBLY — CHOKE HOUSING TO FLOAT BOWL
70. SCREW — CHOKE STAT LEVER ATTACHING
71. LEVER — CHOKE STAT
72. INTERMEDIATE CHOKE SHAFT, LEVER AND LINK ASSEMBLY
73. FAST-IDLE CAM ASSEMBLY
74. LEVER — INTERMEDIATE CHOKE
75. LINK — CHOKE
76. LEVER — SECONDARY THROTTLE LOCKOUT
77. SEAL — INTERMEDIATE CHOKE SHAFT
78. NUT — FUEL INLET
79. GASKET — FUEL INLET NUT
80. FILTER — FUEL INLET
81. SPRING — FUEL FILTER
82. SCREW — THROTTLE STOP
83. SPRING — THROTTLE STOP SCREW
84. THROTTLE BODY ASSEMBLY
85. GASKET — FLOAT BOWL TO THROTTLE BODY
86. SCREW ASSEMBLY — FLOAT BOWL TO THROTTLE BODY
87. LINK — PUMP
88. NEEDLE — IDLE MIXTURE
89. SPRING — IDLE MIXTURE NEEDLE
90. PLUG — IDLE MIXTURE NEEDLE
91. SCREW — FAST-IDLE ADJUSTING
92. SPRING — FAST-IDLE ADJUSTING SCREW
93. SOLENOID AND BRACKET ASSEMBLY
94. SCREW — BRACKET ATTACHING
95. THROTTLE KICKER ASSEMBLY
96. BRACKET — THROTTLE KICKER
97. NUT — THROTTLE KICKER ASSEMBLY ATTACHING
98. WASHER — TAB LOCKING

Figure 12-13. Typical Rochester Model E4MED carburetor. (Rochester)

PART FOUR

Electronic Engine Management

13

Electronic Fuel Metering Control

The arrival of electronic fuel metering resulted in a gradual modification of the traditional carburetor and how it worked. In more recent years, however, the carburetor has been largely replaced by fuel injection systems that are also electronically controlled.

In this chapter, you will learn about the basic components of an electronic fuel metering system, how the components interact, and how the early systems evolved. We will continue our study of electronic engine management in subsequent chapters dealing with electronic engine controls, fuel injection, supercharging, and turbocharging.

ELECTRONIC CONTROL SYSTEMS

To meet stringent emission control requirements in the early 1970's, automotive engineers began to apply electronic control to basic automotive systems. The use of electronics was first applied to ignition timing and later to fuel metering. Electronic control introduced a degree of precision that electromechanical and vacuum-operated systems could not achieve in matching fuel delivery and ignition timing with engine load and speed requirements. With electronic control came a significant decrease in emission levels, major improvements in driveability, and increased reliability of the systems.

Electronic ignitions were the first to appear. They were followed a few years later by electronic fuel metering systems, which were quickly integrated with the electronic ignitions to form the early engine management systems. By the early 1980's, many automotive systems were under the control of an onboard computer.

To understand how electronic controls function, you must have an understanding of basic electricity and how a computer works.

Electrical Review

Because electronic fuel metering is based on the simple principles of electricity, we will begin with a review of basic electrical theory. You may have studied these principles in an automotive electrical and electronics class. By reviewing them here, and then moving to a study of electronic engine control, you will see that the most complex computer systems are based on fundamental laws of science and engineering.

Electric current can be described by the conventional theory or the electron theory, figure 13-1. Either theory can be used with equal accuracy, as long as it is used consistently. When scientists first began to make discoveries

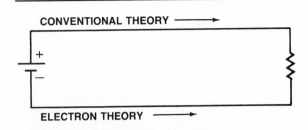

Figure 13-1. Two theories of current flow.

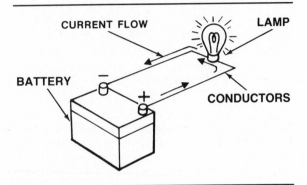

Figure 13-2. A simple circuit.

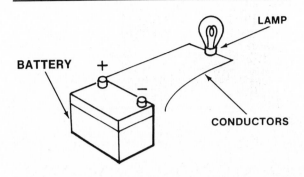

Figure 13-3. An incomplete (open) circuit.

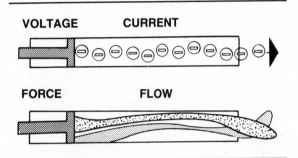

Figure 13-4. Voltage pushes current flow; force pushes water flow.

about electricity, they thought it flowed from positive to negative. This became what we call the **conventional theory of current flow.** In the past, the conventional theory always was used to describe automobile electrical systems. It is still the more common method of describing those systems.

The **electron theory of current flow** states that current moves from negative to positive. This theory generally is used in electronic communications, computers, and all other areas of the electronics industry. In recent years, however, the electron flow theory also has been used to describe some automotive electronic control systems. Our review of current, voltage, and resistance in this chapter is based on the conventional current theory, unless stated otherwise.

Current, voltage, and resistance
An electrical current needs a path along which to flow. This path is called a **circuit,** which means circle. Any break in the circuit prevents current from flowing, since the electrons have nowhere to go.

One of the simplest circuits contains an energy source (battery), conductors (wires), and a load (lamp), figure 13-2. The current flows from the positive side, or terminal, of the battery through the wires and the lamp, lighting the lamp, and back to the negative terminal of the

battery. If one wire is taken off the battery terminal, figure 13-3, the circuit is incomplete. Current will not flow, and the lamp will not light. Breaking the circuit by removing one wire from its terminal is the same as using a switch to break the circuit.

The rate of electric current flow is measured in amperes. Current through a conductor is comparable to water flowing through a pipe, figure 13-4. Water flow is measured counting how many gallons or liters flow past a point within a certain time. When measuring electrical charges, we count electrons instead of gallons or liters. When 6.28 billion electrons pass a

Conventional Theory of Current Flow: The current flow theory which says electricity flows from positive to negative. Also called positive current flow theory.

Electron Theory of Current Flow: The current flow theory which says electricity flows from negative to positive.

Circuit: A circle or unbroken path through which an electric current can flow.

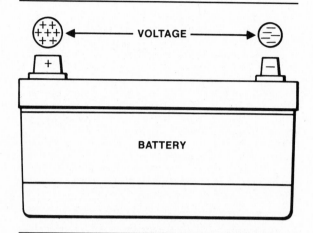

Figure 13-5. Voltage is a potential difference in electromotive force.

point in one second, we say that one **ampere,** or amp, of current is flowing. Remember that current (amperes) flows *through* a circuit.

Current cannot flow unless some force pushes the electrons in one direction. This push is called electromotive force (emf) and is measured in units called **volts.** The force that causes a current to flow through a conductor is called **voltage.** It can be compared to the pressure that moves water through a pipe, figure 13-4. Voltage also is the measurement of a potential difference in force that exists between two points. One point may be negatively charged, and the other may be positively charged, figure 13-5, such as the two terminals of a battery. The strength of the force depends on the strength of the charges at each point.

Voltage can exist even when there is no current, such as when we disconnect one wire from its battery terminal in our simple circuit. Voltage is present at the terminals of the battery in figure 13-5, but no current can flow without a complete circuit.

Voltage is required to force current through a **conductor.** All conductive materials oppose current flow to some extent. This opposition is called **resistance.** However, the resistance of an electrical device is more important than the resistance of a conductor. The lamp in figure 13-2 is an electrical device that has more resistance than the wire conductors. Other electrical devices, such as motors, radios, solenoids, or ignition coils, also offer resistance.

Resistance can be present in places other than the wires and devices of a circuit. A break in the circuit, such as in figure 13-3, creates infinite resistance. Loose or corroded connections also cause resistance to current. Remember that resistance exists *in* a circuit.

Resistance is measure in a unit called an **ohm.** There is an important relationship between volts (electromotive force), amperes (current), and ohms (resistance) that must be understood:

When a force of *one volt* pushes *one ampere* of current through a circuit, the resistance present is *one ohm.* This statement is an expression of Ohm's Law, one of the most important basic electrical rules.

Electricity travels through conductors. The filament of a bulb and the windings of a motor are both conductors. The electrical devices in a circuit, such as bulbs and motors, are also called "loads". These are the devices that do work — produce light, heat, or motion. The amount of resistance present in any part of an electrical circuit depends upon five factors. These five factors are:

1. *The atomic structure of the material* — Any material with few free electrons is a poor conductor, since resistance to current flow will be high. All conductors have some resistance, but the resistance of a good conductor is so small that a fraction of a volt will cause current flow.
2. *The length of the conductor* — Electrons in motion are constantly colliding with the atoms of the conductor. The longer a piece of wire, the farther the flow must travel. The more collisions that occur, the greater the resistance of a conductor.
3. *The cross-sectional area of the conductor* — The thinner a piece of wire is, the higher its resistance will be.
4. *The temperature of the conductor* — In most cases, the higher the temperature of the conducting material, the greater its resistance. That is why alternator regulators are tested at normal operating temperature for accurate readings.
5. *The condition of the conductor* — If a wire is partially cut, it will act almost as if the entire wire were of a smaller diameter, offering a high resistance at the damaged point. Loose or corroded connections have the same effect. High resistance at connections is a major cause of electrical problems.

Every electrical load in a circuit offers some resistance. This means that voltage is reduced as it moves the current through each load. Voltage is electrical energy, and as it moves current through a load, some of the electrical energy is changed to another form of energy, such as light, heat, or motion. The amount of voltage used to move current through each load is called the **voltage drop** across the load. If you measure the voltage drop at every load in a circuit and add the measurements, they will equal the original voltage available. Voltage does not

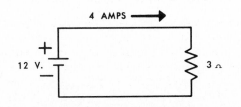

Figure 13-6. A simple series circuit. (Delco-Remy)

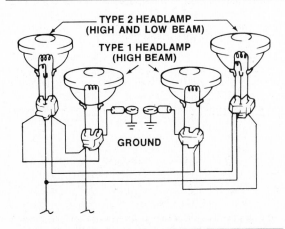

Figure 13-7. The headlamps are wired in parallel with each other in all headlamp circuits.

disappear; the resistance of the load just changes it into a different form of energy.

The resistance of any electrical part (load or conductor) can be measured in three ways:

1. Direct measurement with an ohmmeter, which measures the ohms of resistance offered by the part.
2. Indirect measurement with a voltmeter, which measures the voltage drop through the part.
3. Indirect measurement with an ammeter, which measures the current through the part.

While voltage, current, and resistance are measured in different units, they are directly related. If you know the voltage drop and the current, the resistance and the current, or the resistance and the voltage drop, you can calculate the other factor using Ohm's Law.

Basic circuits

There are three basic circuits:

1. Series
2. Parallel
3. Series-parallel.

In a **series circuit**, the current has only one path to follow. Using conventional current flow theory, you can see that the current in figure 13-6 must flow from the battery, through the resistor, and back to the battery. The circuit must be continuous (have continuity). If one wire is disconnected from the battery (no continuity), the circuit is broken and no current can flow. If electrical loads are wired in series, they must all be switched on and working, or the circuit will be broken and none of them will work.

In a **parallel circuit**, current can follow more than one path to complete the circuit. The points where current paths split and rejoin are called junction points. The separate paths which split and meet at junction points are called branch circuits or shunt circuits. Figure 13-7 shows a parallel circuit.

As the name suggests, a **series-parallel circuit** combines the two types of circuits already discussed. Some of the loads are wired in series, but there are also some loads wired in parallel,

Ampere: The unit for measuring the rate of electrical current flow.

Volt: The unit for measuring the amount of electrical force.

Voltage: The electromotive force that causes current flow. The potential difference in electrical force between two points when one is negatively charged and the other is positively charged.

Conductor: A material that allows easy flow of electricity.

Resistance: Opposition to electrical current flow.

Ohm: The unit for measuring electrical resistance.

Voltage Drop: The amount of voltage required to move current through a load.

Series Circuit: A circuit with only one path for the current to follow.

Parallel Circuit: A circuit with more than one path for the current to follow.

Series-Parallel Circuit: A circuit in which some loads are wired in series and some are wired in parallel.

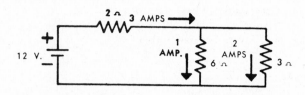

Figure 13-8. A series-parallel circuit. (Delco-Remy)

figure 13-8. The entire headlamp circuit of an automobile is a series-parallel circuit, figure 13-9. The headlamps are in parallel with each other, but the switch is in series with the battery and with each lamp. Both lamps are controlled by the switch, but one lamp still will light if the other is burned out. Most of the circuits in an automobile electrical system are series-parallel.

COMPUTER CONTROL

A computer is simply a machine that receives information which it uses to make a series of decisions and then acts as a result of the decisions made. It cannot think on its own but does everything according to a detailed set of instructions called a program.

Computers use voltage to send and receive information. As we have learned, voltage is electrical pressure and does not flow through circuits. It causes current, which does the real work in an electrical circuit. However, voltage can be used as a signal. A computer converts input information or data into voltage signal combinations that represent number combinations. The number combinations can represent a wide variety of information — temperature, speed, or even words and letters. A computer processes the input voltage signals it receives by computing what they represent and delivering the data in computed or processed form.

The Four Basic Computer Functions

Regardless of the size or use to which it is put, the operation of every computer can be divided into four basic functions, figure 13-10:

1. Input
2. Processing
3. Storage
4. Output.

These basic functions are not unique to computers; they can be found in many noncomputer systems. However, we need to know how the computer handles these functions.

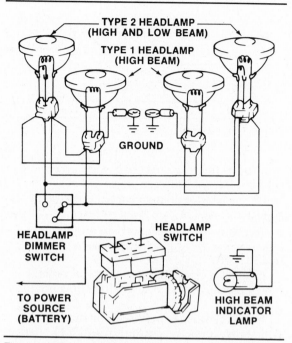

Figure 13-9. A complete headlamp circuit, with all bulbs and switches, is a series-parallel circuit.

1. The computer receives a voltage signal (*input*) from an input device. The device can be as simple as a button or a switch on an instrument panel, or a sensor on an automotive engine. Typical types of automotive sensors are shown in figure 13-11. The keyboard on your personal computer or the programming keyboard of a video cassette recorder are other examples of an input device.

Modern automobiles use various mechanical, electrical, and magnetic sensors to measure factors such as vehicle speed, engine rpm, air pressure, oxygen content of exhaust gas, airflow, and temperature. Each sensor transmits its information in the form of voltage signals. The computer receives these voltage signals, but before it can use them, the signals must undergo a process called **input conditioning**. This process includes amplifying voltage signals that are too small for the computer circuitry to handle. Input conditioners generally are located inside the computer, figure 13-12, but a few sensors have their own input conditioning circuitry.

2. Input voltage signals received by a computer are *processed* through a series of electronic logic circuits maintained in its programmed instructions. These logic circuits change the input voltage signals, or data, into output voltage signals, or commands.

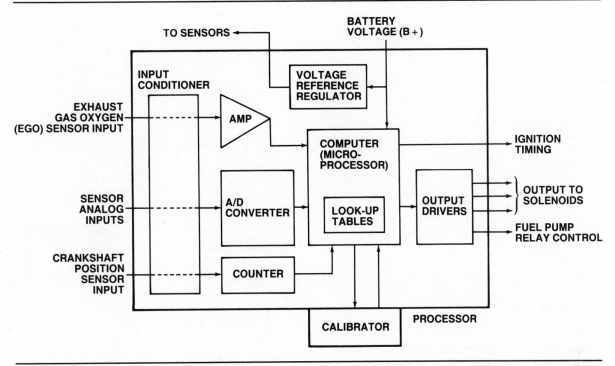

Figure 13-10. The internal components of an engine control computer that do the input, processing, storage, and output functions. (Ford)

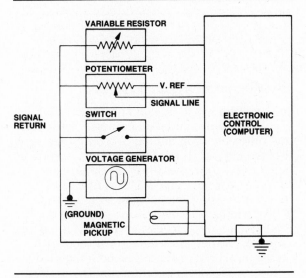

Figure 13-11. Five basic types of sensors send input data to an engine control computer. (Ford)

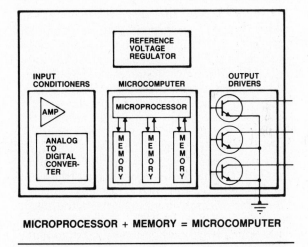

MICROPROCESSOR + MEMORY = MICROCOMPUTER

Figure 13-12. Input conditioners within the computer amplify weak voltage signals and convert analog signals to digital ones. (Ford)

3. The program instructions for a computer are *stored* in electronic memory. Some programs may require that certain input data be stored for later reference or future processing. In others, output commands may be delayed or stored before they are transmitted to devices elsewhere in the system. Computers use a number of different memory devices which we will look at later in this chapter.

4. After the computer has processed the input signals, it sends *output* voltage signals or commands to other devices in the system, such as a

Input Conditioning: The process of amplifying or converting a voltage signal into a form usable by the computer's central processing unit.

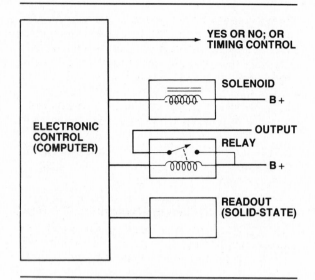

Figure 13-13. When the computer makes a decision, it signals an actuator which translates the voltage signal into mechanical action. (Ford)

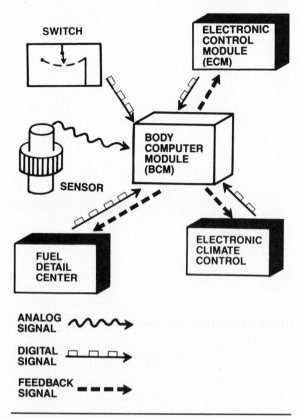

Figure 13-14. Cadillac's BCM accepts inputs from a variety of sources and manages the other onboard computer systems. (General Motors)

system actuator, figure 13-13. An **actuator** is an electrical or mechanical device that does the desired operation, such as adjusting engine idle speed, altering suspension height, or regulating fuel metering.

Computers also can communicate with, and control, each other through their output and input functions. This means that the output signal from one computer system can be the input signal for another computer system. General Motors introduced a body computer module (BCM) on some 1986 models. This acts as a master control unit by managing a network containing all sensors, switches, and other vehicle computers, figure 13-14.

As an example, let's suppose the BCM sends an output signal to disengage the air conditioning compressor clutch. That same output signal can become an input signal to the electronic control module (ECM) that controls engine operation. Based on the signal from the BCM, the ECM signals an actuator to reduce engine speed to account for the decreased load of the compressor. This in turn affects the fuel metering system.

The four basic functions described above are common to all computers, regardless of size or purpose. They also form an organizational pattern to troubleshoot a malfunctioning system. While most input and output devices can be adjusted or repaired, the processing and storage functions can only be replaced.

Analog and Digital Systems

A computer has to be told how to do its job. The instructions and data necessary to do this are called the program. Since a computer cannot read words, the information must be translated into a form the computer can understand — voltage signals. This can be done in two ways, using an analog or a digital system.

An **analog** computer is one in which the voltage signal or processing function is continuously variable, relative to the function being measured or the adjustment required, figure 13-15. Most operating conditions affecting an automobile, such as engine speed, are analog variables. These operating conditions can be measured by sensors. For example, engine speed does not change abruptly from idle to wide-open throttle. It varies in clearly defined, finite steps — 1,500 rpm, 1,501 rpm, 1,502 rpm, etc. — which can be measured. The same is true for temperature, fuel metering, airflow, vehicle speed, and other factors.

If a computer is to measure engine speed changes from 0 rpm through 6,500 rpm, it can

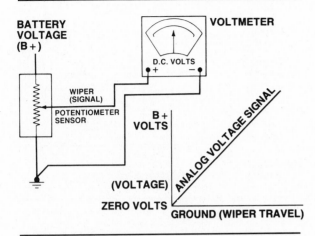

BATTERY
VOLTAGE
(B+)

VOLTMETER

D.C. VOLTS
+ —

WIPER
(SIGNAL)

POTENTIOMETER
SENSOR

B+
VOLTS

ANALOG VOLTAGE SIGNAL

(VOLTAGE)

ZERO VOLTS

GROUND (WIPER TRAVEL)

Figure 13-15. An analog signal is continuously variable. (Ford)

be programmed to respond to an analog voltage that varies from 0 volts at 0 rpm to 6.5 volts at 6,500 rpm. Any analog signal between 0 and 6.5 volts will represent a proportional engine speed between 0 and 6,500 rpm.

Analog computers have several shortcomings, however. They are affected by temperature changes, supply voltage fluctuations, and signal interference. They also are slower in operation, more expensive to manufacture, and more limited in what they can do than digital computers.

Actuator: An electrical or mechanical device that receives an output signal from a computer and does something in response to that signal.

Analog: A voltage signal or processing action that is continuously variable relative to the operation being measured or controlled.

■ Digital Logic

All digital computers handle data bits with three basic logic circuits called logic gates: the NOT, AND, and OR gates. This terminology is used to describe circuit switching functions only — it has nothing to do with their physical construction. Logic gates are called gates because the circuits act as routes or gates for output voltage signals according to different input signal combinations. The thousands of field-effect transistors (FET's) in a microprocessor are logic gates.

A digital computer does its job by switching output voltage on and off according to the input voltage signals. When input voltage enters a logic gate, its transistors can change from a cutoff state (no voltage) to full saturation (voltage). This is equal to an off (low) signal and an on (high) signal. By combining input and output signals in logical combinations, they can be made to equal binary numbers.

The most elementary logic gate is called a NOT gate and inverts the signal. When voltage to its single input terminal is high or on, the output voltage is low or off.

The AND gate has two inputs and one output. Its output is high only if both inputs are high. If one or both inputs are low, output is low.

The OR gate has two or more inputs and one output. It differs from the AND gate in that output is high when one or more outputs are high. OR gate output is low when all inputs are low.

These gates can be combined to produce other logic functions. By placing an inverter or NOT gate after an AND or an OR gate, we can invert the signal and create a NAND (not AND) or a NOR (not OR gate).

INPUT	OUTPUT	INPUTS		OUTPUT	INPUTS		OUTPUT
A	C	A	B	C	A	B	C
1	0	0	0	0	0	0	0
0	1	0	1	0	0	1	1
		1	0	0	1	0	1
		1	1	1	1	1	1

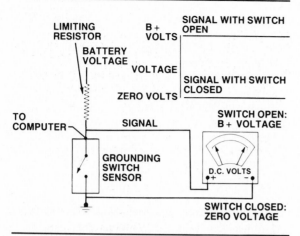

Figure 13-16. A digital signal is a simple on/off, or voltage/no voltage. (Ford)

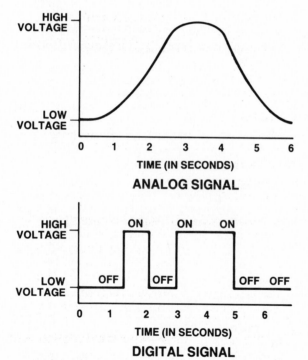

ANALOG SIGNAL

DIGITAL SIGNAL

Figure 13-17. An analog signal takes the form of a sine wave; a digital signal is a square wave. (Ford)

In a **digital** computer, the voltage signal or processing function is a simple high/low, yes/no, on/off. The digital signal voltage is limited to two voltage levels. One is a positive voltage, the other is no voltage, figure 13-16. Since there is no stepped range of voltage or current in between, a digital binary signal is a square wave, figure 13-17.

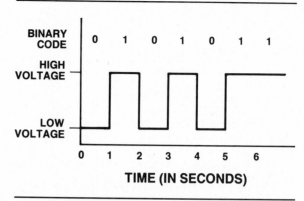

Figure 13-18. Digital computers receive information in a form called binary code. As shown here, a low-voltage signal is represented by 0; a high-voltage signal is 1. (Ford)

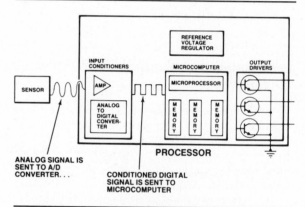

Figure 13-19. An analog signal must be "input conditioned" by a converter in the computer before the computer can deal with it. (Ford)

Using our engine speed example above, suppose that the computer needs to know that engine speed is either above or below a specific level, say 1,800 rpm. Since it doesn't need to know the exact engine speed, but only whether it is above or below 1,800 rpm, the digital signal can be no voltage below 1,800 rpm and any arbitrary voltage when engine speed is above 1,800 rpm. As you can see, a digital signal acts like a simple switch to open and close a circuit, figure 13-16.

An engineer can reverse the switch functions to provide a high input signal below 1,800 rpm and a low (zero voltage) signal above 1,800 rpm. The result at the computer will be the same. The computer will get a simple digital input signal that represents a *change in operating conditions.*

The signal is called "digital" because the on and off signals are processed by the computer

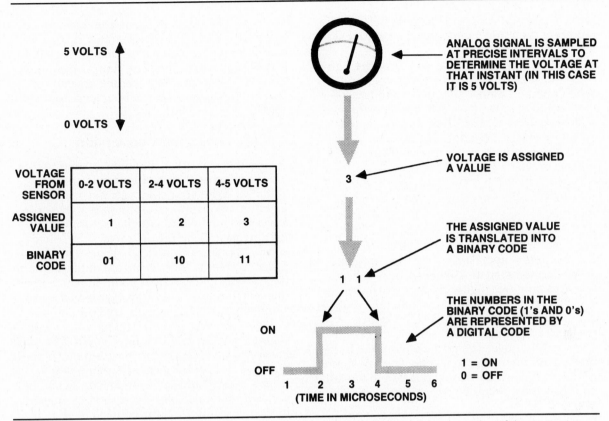

5 VOLTS

0 VOLTS

VOLTAGE FROM SENSOR	0-2 VOLTS	2-4 VOLTS	4-5 VOLTS
ASSIGNED VALUE	1	2	3
BINARY CODE	01	10	11

ANALOG SIGNAL IS SAMPLED AT PRECISE INTERVALS TO DETERMINE THE VOLTAGE AT THAT INSTANT (IN THIS CASE IT IS 5 VOLTS)

VOLTAGE IS ASSIGNED A VALUE

3

THE ASSIGNED VALUE IS TRANSLATED INTO A BINARY CODE

1 1

THE NUMBERS IN THE BINARY CODE (1's AND 0's) ARE REPRESENTED BY A DIGITAL CODE

ON

OFF

1 = ON
0 = OFF

1   2   3   4   5   6
(TIME IN MICROSECONDS)

Figure 13-20. The process of analog-to-digital conversion performed by the converter section of the computer. (Ford)

as the digits or numbers 0 and 1. The number system containing only these two digits is called the **binary** system. Any number or letter from any number system or language alphabet can be translated into a combination of binary 0's and 1's for the digital computer, figure 13-18.

A digital computer changes the analog input signals (voltage) to digital bits (BInary digiTS) of information through an **analog-to-digital (AD)** converter circuit, figure 13-19. The binary digital number, figure 13-20, is used by the computer in its calculations or logic networks. Output signals usually are digital signals that turn system actuators on and off. A digital signal can be changed to an analog output signal through a **digital-to-analog (DA)** converter. This is the opposite of the AD converter circuit that changes analog input signals. More often, however, a digital output signal is made to approximate an analog signal through a variable duty cycle, which you will learn more about later.

The digital computer can process thousands of digital signals per second because its circuits are able to switch voltage signals on and off in billionths of a second.

Binary Numbers

As we have just seen, digital computer switching circuits are characterized by being either off or on. This state can be represented by a 0 (off) or a 1 (on). For a digital computer to understand a command, it must be stated in a binary form — as zeros or ones. Since a binary number system represents all numbers as sequences of

Digital: A 2-level voltage signal or processing function that is either on/off or high/low.

Binary: A mathematical system consisting of only two digits (0 and 1) which allows a digital computer to read and process input voltage signals.

Analog-to-Digital (AD): An electronic conversion process for changing analog voltage signals to digital voltage signals.

Digital-to-Analog (DA): An electronic conversion process for changing digital voltage signals to analog voltage signals.

zeros or ones, off and on voltage signals can represent the 0 and 1 of the binary number system, figure 3-18.

Our decimal (base ten) system uses the numbers 0 through 9 written in a single column. When numbers above 9 are used, another column is added to the left — the ten's position. Therefore, the number 10 is equal to one ten and zero ones. Each additional position to the left multiplies the number by ten, giving us powers of ten:

$1 = 10^0$
$10 = 10^1$
$100 = 10^2$
$1000 = 10^3$

Using the binary (base two) system, we group whole numbers from right to left, just as in the decimal system. However, since the system uses only two digits, the first one must equal either 0 or 1. Then to indicate the value of 2, we must use the second position (the *two's* position) and write it as 10, since it equals one two and zero ones. The next position, the hundred's position in base ten, is the *four's* position in base two. Each additional position to the left multiples the number by two, giving us powers of two:

$1 = 2^0$ (decimal 1)
$10 = 2^1$ (decimal 2)
$100 = 2^2$ (decimal 4)
$1000 = 2^3$ (decimal 8)

In the same way, we can change any decimal number to a binary number:

Decimal	Binary
3	11
4	100
5	101
6	110
7	111
8	1000
9	1001
10	1011

By translating decimals into their binary equivalents, we can tell the computer exactly what we wish it to do. With several thousand transistor circuits arranged in various series and parallel combinations inside the microprocessor, various combinations can switch on and off to equal any binary number in microseconds. You will never have to do this conversion, but understanding it gives you a better insight into how a digital computer works and how it can handle so much information in very short periods of time.

How does the computer know where one binary number (voltage pulse) ends and another

one begins? How does it differentiate between a 01 and a 0011? A clock generator inside the computer provides constant pulses. Each pulse is the length of one bit. The computer monitors these clock pulses while reading or sending data. In this way, it knows how long each voltage pulse should last.

Analog-to-Digital Conversion

We mentioned earlier that most operating conditions which affect an automobile are analog variables. When our computer needs to know whether an operating condition is above or below a specified point, a digital sensor can be used to act as a simple off/on switch. Below the specified point, the switch is open. The computer receives no voltage signal until the condition reaches the specified point, at which time the switch closes. This is an example of a simple digital off/on circuit: off = 0, on = 1.

Let's use engine coolant temperature as an example and specify that the computer needs to know the exact temperature within one degree. Suppose our sensor measures temperature from 0° to 300° and sends an analog signal that varies from 0 to 6 volts. Each 1-volt change in the sensor signal is the equivalent of a 50-degree change in temperature. If 0 volts equals a temperature of 0° and 6 volts equals 300°:

1.0 volt = 50°
0.5 volt = 25°
0.1 volt = 5°
0.02 volt = 1°

In order for the computer to determine temperature within 1 degree, it must react to sensor voltage changes as small as 0.020 volt or 20 millivolts. For example, if the temperature is 125°, the sensor signal will be 2.50 volts. If the temperature rises to 126°, sensor voltage increases to 2.52 volts. In reality, temperature does not pass directly from one degree to another; it passes through many smaller increments, as does voltage as it changes from 2.50 to 2.52 volts. Our digital computer, however, processes only signals equaling 1-degree changes in temperature. To do so, the computer sends the signal through analog-to-digital (AD) conversion circuits, where the analog sensor voltage is converted to a series of 0.020-volt changes for each degree. This is called "digitizing" an analog signal.

The analog-to-digital conversion process brings us back to binary numbers. Transistors can be designed to switch on and off at different voltage levels or with differing combinations of voltage signals. In the computer we are

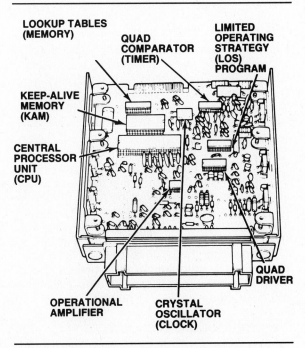

Figure 13-21. Some of the basic components of an engine control computer, housed in a metal box for protection. (Ford)

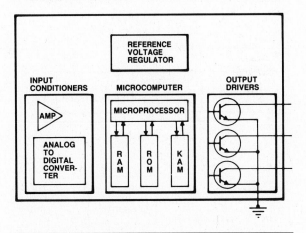

Figure 13-22. Three different types of memory are used by an engine control computer. (Ford)

discussing, transistor groups must switch from off to on at 20-millivolt increments. The input signal is created by varying the transistor combinations that are on or off. Since the computer can read the various voltage signal combinations as binary numbers, it performs its calculations. It does so almost instantly because the current travels through the miniature circuits at almost the speed of light.

PARTS OF A COMPUTER

We have dealt with the functions, logic, and software used by a computer. The software consists of the programs and logic functions stored in the computer's circuitry. The hardware is the mechanical and electronic parts of computer. Figure 13-21 shows the basic structure of a computer.

Central Processing Unit (CPU)

As mentioned earlier, the microprocessor is the **central processing unit**, or **CPU**, of a computer. Since it does the essential mathematical operations and logic decisions that make up its processing function, the CPU can be considered the heart of a computer. Some computers use more than one microprocessor, called a coprocessor.

Computer Memory

The computer storage or memory function is provided by other integrated circuit (IC) devices. These simply store the computer operating program, system sensor input data, and system actuator output data for use by the CPU. Automobile computers use three different types of memory for their storage functions, figure 13-22:

- Read-only memory (ROM) or programmable read-only memory (PROM)
- Random-access memory (RAM)
- Keep-alive memory (KAM).

Permanent memory is called **read-only memory (ROM)** because the central computer can read the contents of the memory but cannot change the information contained within it. Data stored in ROM is retained even when power to the computer is turned off. The computer control program and specific vehicle data are stored in ROM so they will not be lost when power to the computer is interrupted. ROM containing the control program is built into the computer. The specific vehicle data is located in

Central Processing Unit (CPU): The processing and calculating portion of a computer.

Read-Only Memory (ROM): The permanent part of a computer's memory storage function. ROM can be read but not changed, and is retained when power is shut off to the computer.

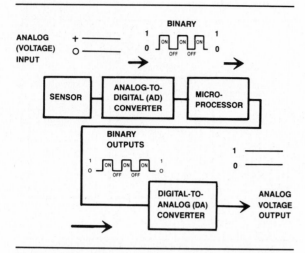

Figure 13-23. A/D and D/A converters interface the microprocessor with its input and output devices. (Chrysler)

a separate ROM chip called a **programmable read-only memory (PROM)**. The PROM is used to individualize a single computer for use in various models.

Temporary memory is called **random-access memory (RAM)** because the central computer can both read information from it and write new information into it as dictated by the computer program. However, data contained in RAM is lost whenever power to the computer is interrupted. Depending upon the computer design, RAM can provide both short- or long-term memory. Short-term memory is lost every time the ignition switch is turned off; long-term memory is retained until the computer power supply is completely disconnected. System trouble codes and diagnostic test results are common items stored in RAM.

Keep-alive memory (KAM) shares characteristics of ROM and RAM. Like RAM, data can be written into keep-alive memory. It also can be read and erased, but like ROM, it is not lost when the ignition is turned off. However, like long-term memory, KAM is erased whenever the power supply to the computer is disconnected. KAM is used primarily in conjunction with adaptive strategies, which we will study in the next chapter.

Input and Output Circuits

A computer is not directly connected to every input or output device. The signals are received and sent by other IC devices, many of which provide the computer with parallel connections. This allows it to receive several input signals while it is sending several output signals.

Converter Circuits

The computer must have circuits to convert input data into a form with which it can work. The analog signals which we have discussed must be digitized, or changed to digital signals. This conversion is done by separate IC devices called analog-to-digital (AD) converter circuits, figure 13-19.

The computer's output signals also must be converted into a form which the output device can recognize and upon which it can act. Since some of the output devices are analog, the digital signals must be changed to analog signals. This conversion is done by digital-to-analog (DA) converter circuits. Because the circuits which perform these functions "interface" the CPU with the input and output devices, they are sometimes called the input/output (I/O) interface, figure 13-23.

Control Module Locations

The onboard automotive computer may be called an electronic control unit, module, or assembly, depending upon the carmaker and the computer application. The computer hardware is all mounted on one or more circuit boards and installed in a metal case, figure 13-21, to help shield it from electromagnetic interference (EMI). The wiring harnesses which link the computer to sensors and actuators connect to multipin connectors or edge connectors on the circuit boards.

Onboard computers range from single-function units that control a single operation to multifunction units that manage all of the separate (but linked) electronic systems in the vehicle. They vary in size from a small module to a notebook-sized box. Chrysler's early computers were attached to the air cleaner housing in the engine compartment. The computers used with Chrysler 4-cylinder engines are 2-piece units, with the power module installed between the battery and the fenderwell in the engine compartment, and the logic module behind a kick panel in the passenger compartment. Most other computers are installed in the passenger compartment either under the instrument panel or in a side kick panel where they can be shielded from physical damage caused by temperature extremes, dirt and vibration, or from interference from the high currents and voltages of various underhood systems, figure 13-24.

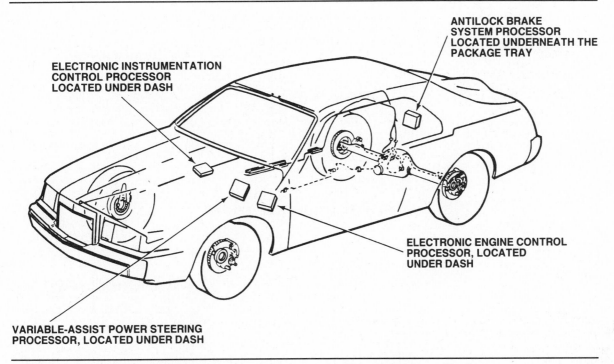

ELECTRONIC INSTRUMENTATION
CONTROL PROCESSOR
LOCATED UNDER DASH

ANTILOCK BRAKE
SYSTEM PROCESSOR
LOCATED UNDERNEATH THE
PACKAGE TRAY

ELECTRONIC ENGINE CONTROL
PROCESSOR, LOCATED
UNDER DASH

VARIABLE-ASSIST POWER STEERING
PROCESSOR, LOCATED UNDER DASH

Figure 13-24. Most onboard computers are located in the passenger compartment. (Ford)

FUEL CONTROL SYSTEM OPERATING MODES

A computer-controlled fuel metering system can be selective; depending upon the computer program, it may have different operating modes. The onboard computer does not have to respond to data from all of its sensors, nor does it have to respond to the data in the same way each time. Under specified conditions, it may ignore sensor input. Or, it may respond in different ways to the same input signal, based on inputs from other sensors. Most current control systems, figure 13-25, have two operating modes: open loop and closed loop. We touched on these briefly in Chapter 8. The most common application of these modes is in fuel-metering feedback control, although there are other open- and closed-loop functions. Air conditioning automatic temperature control is an example. Control logic programmed into the computer determines the choice of operating mode according to engine operating conditions.

Open-Loop Control

Open-loop control means that the onboard computer works according to established conditions in its program. It gives the orders and the output actuators carry them out. The computer ignores sensor feedback signals as long as the established conditions exist.

For example, the computer is programmed to provide a specific amount of fuel and spark timing when the engine is first started. Since these factors are predetermined in the program (regardless of other factors), the computer will ignore signals (feedback) from the exhaust gas oxygen (EGO) sensor until coolant temperature reaches the predetermined level, figure 13-26.

Programmable Read-Only Memory (PROM): An integrated circuit chip installed in a computer which contains appropriate operating instructions and database information for a particular application.

Random-Access Memory (RAM): Temporary short-term or long-term computer memory that can be read and changed, but is lost whenever power is shut off to the computer.

Keep-Alive Memory (KAM): A form of long-term RAM used mostly with adaptive strategies. Requires a separate power supply circuit to maintain voltage when the ignition is off.

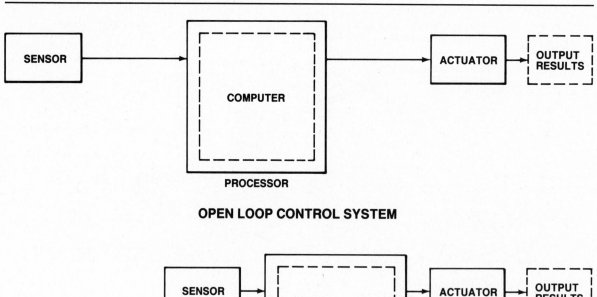

OPEN LOOP CONTROL SYSTEM

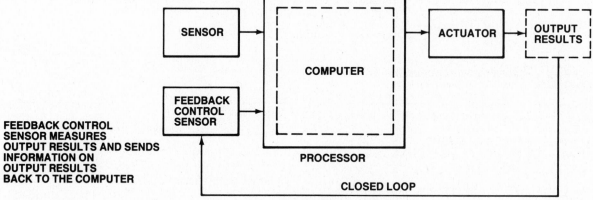

**FEEDBACK CONTROL
SENSOR MEASURES
OUTPUT RESULTS AND SENDS
INFORMATION ON
OUTPUT RESULTS
BACK TO THE COMPUTER**

CLOSED LOOP

CLOSED LOOP CONTROL SYSTEM

Figure 13-25. The two control modes of a computer-controlled engine system. (Ford)

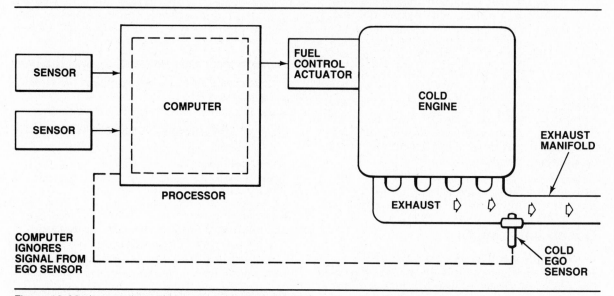

Figure 13-26. In open loop, the computer ignores the EGO sensor signal and operates on a predetermined program. (Ford)

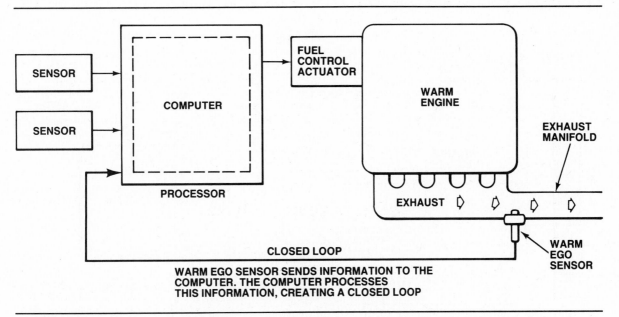

Figure 13-27. In closed loop, the computer accepts the EGO sensor signal and readjusts the air-fuel mixture accordingly. (Ford)

Closed-Loop Control

Once certain conditions (such as coolant temperature) have been met, the system goes into closed loop. The computer now reads and responds to signals from *all* of its sensors. When the engine is first started (open loop), the computer ignores input from the EGO sensor. As soon as the coolant temperature reaches a predetermined level (specified in the computer program), the computer accepts the sensor input and adjusts the fuel and spark timing accordingly. We say that the computer is responding to a "feedback" signal; that is, the sensor is telling the computer that there is an error factor in its operation that must be corrected, figure 13-27.

SENSORS AND ACTUATORS

All sensors and actuators are **transducers**, or devices which change one form of energy into another:

- Sensors convert light, temperature, motion, pressure, heat and other types of energy into voltage signals.
- Actuators convert voltage signals into mechanical energy or work.

Sensors and actuators used with automobile computer systems often do the same work that mechanical transducers such as vacuum diaphragms do on older vehicles without computer systems.

Sensors

Engine control system sensors fall into several basic categories including switches, timers, resistors, transformers, and generators. Except for generators, all automotive sensors are resistive devices. This means that they cannot create a voltage, but only can modify a voltage applied to them. The voltage applied is controlled by the computer and is called the **reference voltage**.

The computer sends this reference voltage to a sensor, and receives a different voltage back, figure 13-28. The returning signal is determined by the changing sensor resistance. The computer interprets the altered return voltage as a sign of specific changes in the engine operating condition and adjusts engine operation accordingly.

Transducer: A device that changes one form of energy into another.

Reference Voltage: A constant voltage signal (below battery voltage) applied to a sensor by the computer. The sensor alters the voltage according to engine operating conditions and returns it as a variable input signal to the computer, which adjusts system operation accordingly.

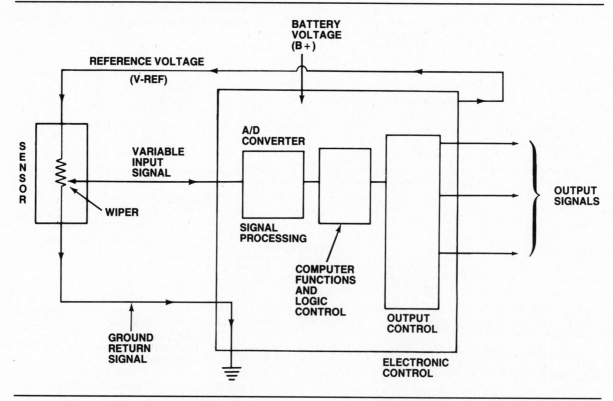

Figure 13-28. The computer reference voltage is sent out to a sensor and returns changed. This tells the computer how to readjust engine operation. (Ford)

Either a 5-volt or a 9-volt signal generally is used as the reference voltage. It must always be less than minimum battery voltage so that it can be maintained at a constant level at all times (even when battery power is low) to prevent faulty input signals from the sensors.

Characteristics and features

Automotive sensors function in a severe environment. For this reason, they must be designed for long-term, dependable operation while providing reliable signals. A sensor must have certain characteristics, or operating features, for it to operate properly. These characteristics affect the selection of a particular sensor for a given function, and establish the specifications for troubleshooting and service. The important characteristics are:

● *Repeatability* — the sensor must function consistently. This means that a temperature switch must open and close at the design points thousands of times without deviation. If the sensor produces a voltage in proportion to the condition being measured, it must do so throughout its operating range.

● *Accuracy* — the sensor must work within the tolerances or limits designed into it. Our temperature switch may close at 195° ± 1°, or it may close at 195° ± 10°. The tolerances depend on

how the sensor is used, but once established, the sensor must work consistently. These tolerances are used to design sensor test specifications for troubleshooting.

● *Operating range* — an operating or dynamic range within which it must function is established for the sensor. A digital sensor has only one or two switching points. Since the operating range of an analog sensor is wider, it must be proportional. Signals outside the operating range are ignored by the computer.

● *Linearity* — this refers to sensor accuracy throughout its dynamic range. Within this range, an analog sensor must be as consistently proportional as possible to the measured value. While sensor linearity is most accurate near the center of its dynamic range, no sensor has perfect linearity and computer programs rely on memory data to compensate for this.

In the following paragraphs, you will learn about the various sensors used in an automobile. These include:

● Switches and timers
● Potentiometers
● Thermistors
● Piezoresistive devices
● Transformers
● Signal generators.

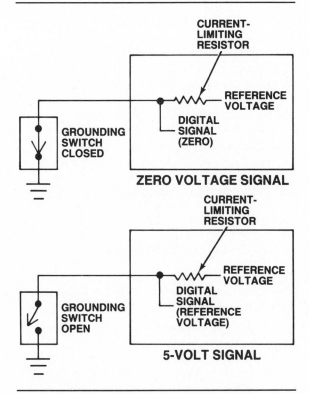

ZERO VOLTAGE SIGNAL

5-VOLT SIGNAL

Figure 13-29. How a grounding switch operates. (Ford)

Switches and timers

The simplest form of sensor is a switch. A switch signals either a totally on, or a totally off, condition. A switch can signal the computer in one of two ways: through reference voltage, or by grounding a signal.

In the first way, full reference voltage is returned to the computer when the switch is closed. When the switch is open, no return voltage signal is sent. Not all switches relay reference voltage back to the computer; some send a battery voltage signal directly to the computer when the condition they are monitoring is met.

In the second way, a grounding switch is used in series with a fixed, current-limiting resistor and operates just the opposite of the first way of signalling. When the switch is closed, no voltage signal is sent. When the switch is open, reference voltage returns to the computer, figure 13-29.

A common use for a switch sensor is to signal the computer when a high-load accessory, such as an air conditioning compressor or rear window defogger, is turned on or off. The computer uses the switch signal to adjust the idle speed to compensate for the added or reduced load. Coolant temperature switches that are closed when the engine has reached a certain temperature are also used in some systems.

■ **The Rise And Fall Of Gaskets In A Tube**

In the 1970's, carmakers replaced many traditional cork gasket installations with room-temperature-vulcanizing (RTV) silicone sealants. RTV sealants quickly gained an unchallenged reign as "gaskets in a tube" for installing such parts as water pumps, valve covers, oil pans, transmission and differential covers, and other components.

Cork gaskets had been the industry standard for decades in these applications. Cork, however, dries out with age and loses its shape. Parts departments often found themselves with an inventory of unusable, overage gaskets. Moreover, car dealers and garages had to stock a variety of gaskets for different car models. RTV sealant in a tube seemed to be a revolutionary breakthrough.

When properly applied, RTV is an excellent sealer, but mating surfaces must be perfectly clean of oil and grease because oil dissolves the RTV sealant. Additionally, RTV residue in bolt holes can cause a hydraulic effect that affects torque when a bolt is installed. These are minor problems, however, and were easily mastered by professional technicians. But, RTV also has longer term disadvantages that took time for the service industry to discover. For example, it:

● Has a short shelf life of about one year
● Does not cure properly when the shelf life has expired
● Will spoil if the cap is left off the tube because moisture in the air causes it to cure
● Is expensive to manufacture and stock for long periods.

Despite these disadvantages, carmakers used and specified RTV sealants and created a virtual depression in the cork gasket industry. In the mid-1980's, however, General Motor discovered that RTV sealants can cause long-term problems that had been unforeseen a decade earlier.

Although RTV sealant cures sufficiently to provide a firm seal in 24 to 48 hours, it can require as long as one year after installation to cure *completely*. Final curing time depends on where it is used in a vehicle and how thickly it is applied. During the prolonged curing time, RTV sealant gives off acidic fumes that can corrode electrical connections and sensitive electronic parts.

As solid-state electronic components increased in use during the 1980's, GM found that RTV sealant can contribute to failure of these sensitive devices. As goes GM, so goes the entire auto industry, and carmakers are again providing cork-based gaskets for applications where RTV sealant had been used. The gasket makers, meanwhile, have learned to bond cork to both sides of thin metal to manufacture gaskets that meet the needs of modern vehicles and eliminate the shrinkage and deterioration problems that plagued gaskets of a generation ago.

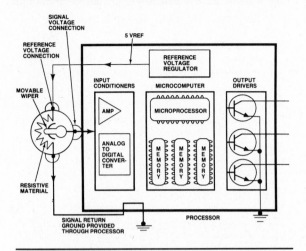

Figure 13-30. The basic components of a potentiometer and how it interacts with the computer. (Ford)

When combined with a switch, a timer can delay a signal for a specific and predetermined time. Timers prevent the computer from having to compensate for momentary conditions that do not significantly affect engine operation. The timer may be built into the computer, or it can be part of the switch itself.

Potentiometers
A potentiometer is a variable-resistance sensor with three terminals. One end of the resistor receives reference voltage, while the other end is grounded. The third terminal is attached to a movable contact that slides across the resistor to vary its resistance. Depending on whether the contact is near the supply end or the ground end of the resistor, return voltage will be high or low, figure 13-30.

Throttle position sensors are among the most common potentiometer-type sensors. The computer uses their input to determine the amount of throttle opening and the rate of change. Vane-type intake airflow meters use a potentiometer to signal the amount of air entering the engine. EGR valve flow sensors use a potentiometer to signal the valve position which the computer can use to interpret the amount of flow.

Thermistors
A **thermistor** is a solid-state variable resistor whose resistance changes with temperature. The resistance of any resistor changes as temperature changes, but the resistance variations across the operating range of a thermistor make it very accurate as an analog temperature sensor.

Thermistors are classified in two groups: **positive temperature coefficient (PTC) resistors** and **negative temperature coefficient (NTC) resistors**. These names simply mean that:
- The resistance of a PTC thermistor increases as temperature increases.
- The resistance of an NTC thermistor decreases as temperature increases.

Both kinds are used in automobile systems, but the NTC thermistor is more common. Heat can come from an external source or from current through the resistor. Externally heated NTC thermistors are the most common analog sensors for engine coolant temperature and intake air temperature.

The computer applies the reference voltage to one sensor terminal and receives the input (return) signal from the other, figure 13-31. As the sensor warms up, resistance decreases, and signal voltage increases.

Piezoresistive sensors
A **piezoelectric** crystal develops voltage across its surfaces when pressure is applied to it, figure 13-32. Similar crystals change their resistance when pressure is applied to them. This feature makes **piezoresistive** sensors ideal for analog pressure measurement.

When used as an engine detonation sensor, a piezoresistive device senses vibration and converts the degree of vibration into an electrical signal that tells the computer the extent of engine knock present.

When used as a barometric or manifold pressure sensor, the piezoelectric device changes the frequency of its output signal rather than the voltage. A pressure-sensing capacitor contains a sealed vacuum reference input on one side of a diaphragm, with barometric or manifold pressure on the other side. Any change in pressure results in a change in capacitance.

Transformers
A transformer sensor has input and output windings with a movable core in between them, figure 13-33. The electric coupling between the two cores varies with the core position. Reference voltage is applied to the input winding. The signal voltage generated in the output winding is the same as that in the input winding when the core is centered. As the core moves away from the center position, the return signal voltage changes. Transformer sensors are used as one type of manifold pressure sensor.

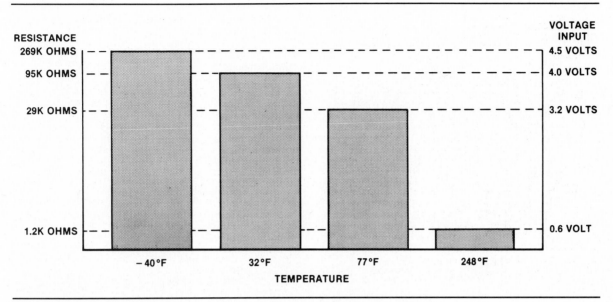

Figure 13-31. A thermistor requires a reference voltage from the computer. (Ford)

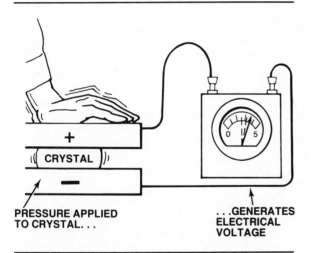

PRESSURE APPLIED TO CRYSTAL...   ...GENERATES ELECTRICAL VOLTAGE

Figure 13-32. Sensors containing a certain type of quartz generate a voltage when pressure or force is applied. (Ford)

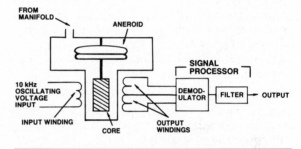

Figure 13-33. A transformer sensor creates a voltage differential between two output windings.

Signal generators

Signal generator sensors do not depend on a reference voltage. They generate an output signal that is sent to the computer. This output signal may be a varying voltage, a varying frequency, or a combination of the two. There are several types of generator sensors, including magnetic pulse generators, Hall-effect switches, and galvanic batteries.

Magnetic pulse generators operate in a manner similar to the pickup coil and reluctor used in many electronic ignition systems. Hall-effect switches are used in some newer versions. These sensors commonly are used to provide

Thermistor (Thermal Resistor): A resistor especially built to change its resistance as the temperature changes.

Positive Temperature Coefficient (PTC) Resistor: A thermistor whose resistance decreases as the temperature increases.

Negative Temperature Coefficient (NTC) Resistor: A thermistor whose resistance decreases as the temperature increases.

Piezoelectric: Voltage caused by physical pressure applied to the faces of certain crystals.

Piezoresistive: A sensor whose resistance varies in relation to pressure or force applied to it. A piezoresistive sensor receives a constant reference voltage and returns a variable signal in relation to its varying resistance.

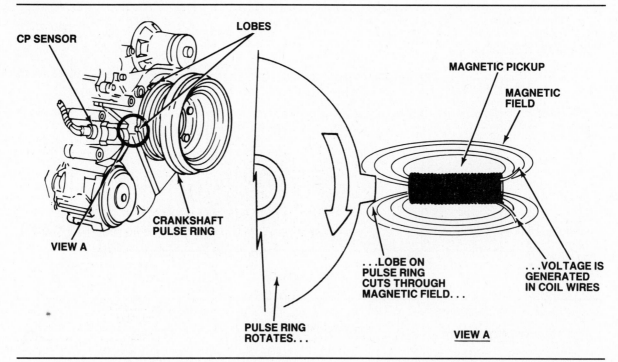

Figure 13-34. A crankshaft position (CP) sensor uses a magnetic pickup similar to those used in breakerless distributors. (Ford)

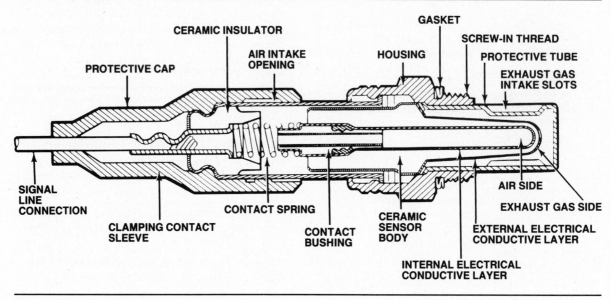

Figure 13-35. The components of a typical EGO sensor. (Ford)

an ignition trigger signal or crankshaft or camshaft position information. They may be located in the distributor or mounted in the block to respond to a tooth or cutout on the crankshaft reluctor, harmonic damper, or flywheel, figure 13-34. Some engines use separate sensors to signal crankshaft and camshaft position.

A **galvanic battery** is a generator sensor that produces a voltage by comparing the oxygen level in the ambient air to that in the engine exhaust. The only galvanic battery on modern cars is the EGO sensor, figure 13-35, used to help control the air-fuel mixture.

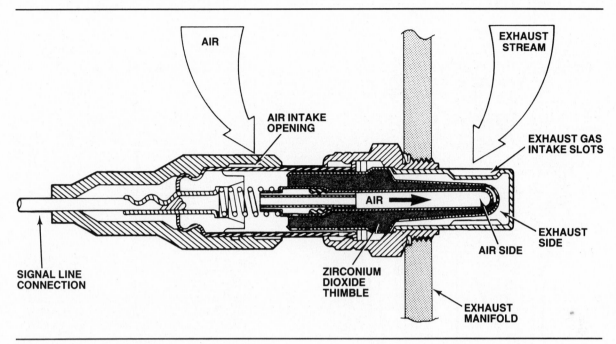

Figure 13-36. How the EGO sensor works. (Ford)

Exhaust Gas Oxygen Sensors and Fuel Metering Control

The exhaust gas oxygen (EGO) sensor is one of the most important sensors on a car. It is usually installed in the exhaust manifold, although in some vehicles it may be located downstream from the manifold in the headpipe (but before the catalytic converter). This places it directly in the path of the exhaust gas stream where it can monitor both the exhaust gas and ambient air. The sensor tip contains a thimble made of zirconium dioxide (ZrO_2), an electrically conductive material capable of generating a small voltage in the presence of oxygen.

Exhaust gases from the engine pass through the end of the sensor installed in the manifold where they contact the outer side of the thimble. Atmospheric air enters through the other end of the sensor and contacts the inner side of the thimble, figure 13-36. The inner and outer surfaces of the thimble are plated with platinum. The inner surface is a negative electrode; the outer surface is a positive electrode.

The atmosphere contains a relatively constant 21 percent oxygen. Rich exhaust gases contain virtually no oxygen. Exhaust from a lean mixture combustion, or from a misfire, contains more uncombined oxygen (still far less than the atmosphere, however).

Negatively charged oxygen ions are drawn to the thimble, where they collect on both the inner and outer surfaces, figure 13-37. Because

the oxygen present in the atmosphere exceeds that in exhaust gases, the air side of the thimble draws more negative oxygen ions than the exhaust side. The difference between the two sides creates an electrical potential. When the concentration of oxygen on the exhaust gas side of the thimble is low, a high voltage (0.60 to 1.00 volt) is generated between the electrodes. As the oxygen concentration on the exhaust side increases, the voltage generated drops (0.00 to 0.40 volt).

This voltage signal is sent to the computer, where it passes through the input conditioner for amplification. The computer interprets the high-voltage signal as a rich air-fuel ratio, and a low-voltage signal as a lean air-fuel ratio. Based on the EGO signal, the computer will either lean or richen the mixture as required to maintain as close to a 14.7 air-fuel ratio as possible. The EGO sensor is therefore the key sensor of an electronically controlled fuel-metering system.

An EGO sensor does not send a voltage signal until its tip reaches a temperature of about 572°F (300°C). EGO sensors provide their fastest response to mixture changes at about 1472°F

Galvanic Battery: A direct current voltage source, generated by the chemical action of an electrolyte.

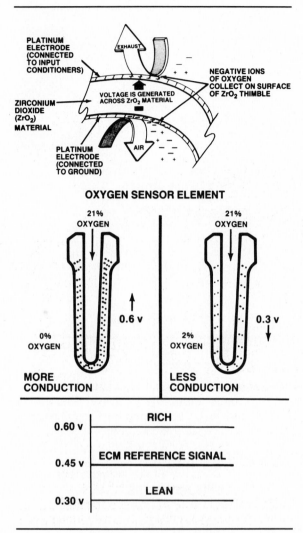

Figure 13-37. The difference in negative ion collection between the outside and inside of the EGO sensor creates a voltage potential. (Ford)

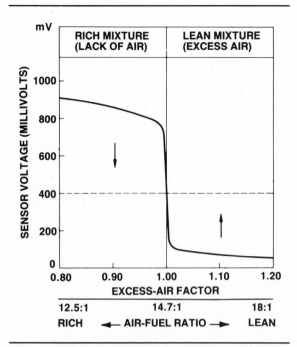

Figure 13-38. The EGO sensor provides its fastest response at the stoichiometric air-fuel ratio of 14.7:1.

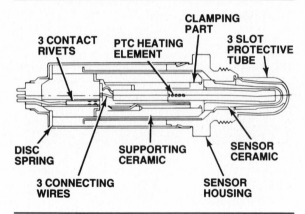

Figure 13-39. The components of a typical HEGO sensor. (Ford)

(800°C). This is the primary reason for open-loop fuel control on a cold engine.

Figure 13-38 shows the operating range of an EGO sensor at 1472°F (800°C). Sensor voltage changes fastest at an air-fuel ratio of 14.7 at this temperature.

Another important point about an EGO sensor is that *it measures oxygen: it does not measure air-fuel ratio*. If the engine misfires, no oxygen is consumed in combustion. There is a large amount of oxygen in the unburned exhaust mixture, and the sensor will deliver a false "lean mixture" signal. This is one reason why computer control of ignition timing and EGR is essential for effective fuel metering control.

Unlike resistive sensors, an EGO sensor (as well as any other generator sensor) does not require a reference voltage. The computer,

however, uses an internal reference voltage as a comparison for the sensor signal. Because the EGO sensor signal ranges from 0.1 to 0.3 volt (100 to 300 millivolts) with a lean mixture to 0.6 or 0.9 volt (600 to 900 millivolts) with a rich signal, the computer uses an internal reference of 0.45 volt (450 millivolts) as a reference, figure 13-37. The internal reference voltage also is the basis for fuel-metering signals during open-loop operation.

To make the system more responsive, car-makers went first to the concept of installing a separate EGO sensor in each manifold of a V-type engine. While this arrangement is still

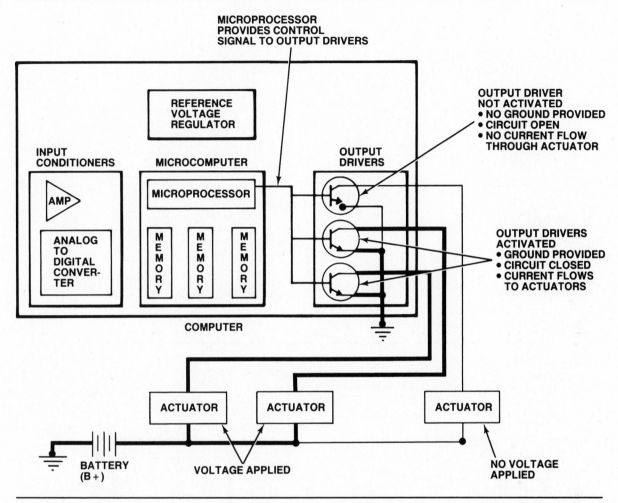

MICROPROCESSOR PROVIDES CONTROL SIGNAL TO OUTPUT DRIVERS

REFERENCE VOLTAGE REGULATOR

INPUT CONDITIONERS

MICROCOMPUTER

OUTPUT DRIVERS

AMP

MICROPROCESSOR

ANALOG TO DIGITAL CONVERTER

MEMORY **MEMORY** **MEMORY**

COMPUTER

OUTPUT DRIVER NOT ACTIVATED
- **NO GROUND PROVIDED**
- **CIRCUIT OPEN**
- **NO CURRENT FLOW THROUGH ACTUATOR**

OUTPUT DRIVERS ACTIVATED
- **GROUND PROVIDED**
- **CIRCUIT CLOSED**
- **CURRENT FLOWS TO ACTUATORS**

ACTUATOR   **ACTUATOR**   **ACTUATOR**

BATTERY (B+)   **VOLTAGE APPLIED**   **NO VOLTAGE APPLIED**

Figure 13-40. The output drivers receive a voltage signal from the microprocessor. This tells the drivers to open or close the ground circuit of the actuators they control. (Ford)

used, the heated exhaust gas oxygen (HEGO) sensor, figure 13-39, is the most recent development.

An HEGO sensor is constructed and operates the same as an EGO sensor, but contains a built-in heater powered by the vehicle battery whenever the ignition is in the run position. A third wire to the sensor delivers battery current (1 ampere or less) to the sensor electrode. This helps warm the sensor to operating temperature more quickly and permits the sensor to operate at a lower exhaust gas temperature (approximately 392°F or 200°C). The heating element also keeps the sensor from cooling off when exhaust temperature drops, such as during prolonged idling in cold weather. HEGO sensors often are used on turbocharged engines where the sensor is installed downstream from the turbocharger, which absorbs much of the heat in the exhaust.

■ **The Key Number**

Do you have a lucky number? Perhaps a set of numbers that are important to your business or your life? Everyone remembers their own birthday by the numbers of the day, month, and year. Most people remember their social security number, their car license number(s), or their driver's license number. Here's a key number for any automobile technician to remember:

14.7

That's the number that represents *both* atmospheric pressure at sea level (14.7 psi) and the stoichiometric air-fuel ratio: 14.7 parts of air (by weight) to each part of gasoline (by weight).

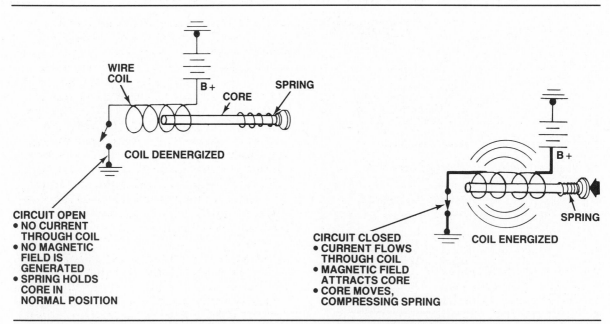

Figure 13-41. How a solenoid works. (Ford)

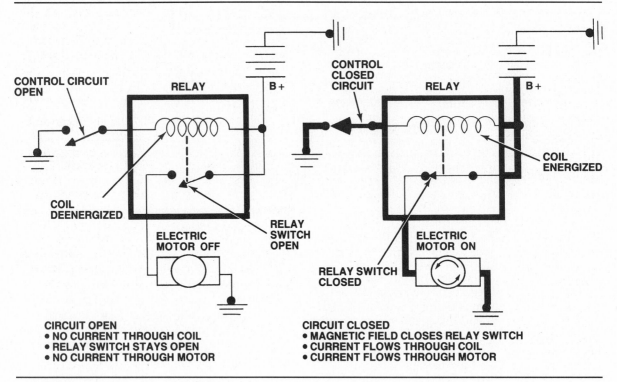

Figure 13-42. How a relay works. (Ford)

All EGO sensors work on the principles just discussed, but they are not all built the same. They may have one, two, or three wires that connect to the vehicle wiring harness. Early model sensors had two wires and were grounded through the computer or to some point on the chassis or engine. Later sensors have a single wire and ground through their outer shell to the exhaust pipe or manifold. Single- and double-wire EGO sensors are not interchangeable (the three-wire sensor is an HEGO).

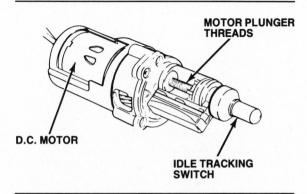

MOTOR PLUNGER THREADS

D.C. MOTOR

IDLE TRACKING SWITCH

Figure 13-43. The stepper motor is a DC motor which moves in specific increments. (Ford)

Some sensors have a silicone boot to protect the sensor and to provide a vent for ambient air circulation. The positioning of a boot (when used) is important. If the boot is seated too far down on the sensor body, it can block the air vent, resulting in an inaccurate signal to the computer. The silicone boot has been abandoned on some late-model engines because it was thought that the silicone material gives off fumes that corrode electrical connections and terminals.

Actuators

The computer receives the sensor inputs and does the necessary calculations to determine which engine systems must be adjusted to meet the demands of the moment. Then it sends electrical control signals to one or more output drivers, which in turn operate the control devices or actuators by completing the actuator ground circuits. This process is illustrated in figure 13-40.

Some of the computer's output signals, such as those that regulate ignition timing, control engine operation directly. However, an actuator is required whenever the output must regulate a mechanical device. An actuator converts the computer's electrical signal into a mechanical action.

Most engine actuators either are solenoids or relays, although stepper motors also are used in some cases.

Solenoids and relays

Solenoids and relays both operate on the principle of electro-magnetism. These two simple mechanical devices allow the computer to control almost any automotive system. A solenoid essentially is a digital actuator; it is either on or off. Battery voltage is applied to one terminal of the solenoid and the computer opens and closes the ground circuit attached to the other terminal. A solenoid contains a coil winding

around a spring-loaded metallic plunger, figure 13-41. When the switch is closed and current flows through the windings, the magnetic field of the coil attracts the movable plunger. This pulls it against spring pressure into the center of the coil toward the plate. Once current is shut off, the magnetic field collapses and spring pressure moves the plunger out of the coil.

In most applications, the solenoid is energized for varying periods of time determined by the computer program. When energized, a solenoid may extend a plunger to control engine speed. Other types of solenoids regulate vacuum flow to various emission-related systems such as air injection, vapor canister purge, and EGR.

A relay is a switch that uses electromagnetism to move internal contacts, allowing a small electrical current to control a large current. To do this, it contains a control circuit and a power circuit, figure 13-42. The small flow of current through the relay coil moves an armature to open or close a set of contact points. This is called the control circuit because it controls the flow of a much larger current through a separate circuit called the power circuit. The computer controls the operation of a relay through its output drivers, which open and close the relay control circuit.

Stepper motors

A stepper motor, figure 13-43, also is a digital actuator. Stepper motors are d.c. motors that move in fixed increments from deenergized (no voltage) to fully energized (full voltage). A stepper motor can have as many as 120 discrete steps of motion which allow it to serve as an analog output operated by a digital signal.

The most common uses for stepper motors are as idle speed controls. On carbureted engines, the motor often acts directly on the throttle linkage, but in most fuel injection systems, it controls an idle air bypass built into the throttle body.

Pulse Width and Duty Cycle

Solenoids are more precisely controlled on some late-model cars with a procedure called pulse width modulation (PWM). This is the same technique used to control carburetor mixture control solenoids, as we learned in Chapter 8. With this technique, the solenoid is continuously cycled on and off a fixed number of times per second. The solenoid is on (energized) for part of each cycle, and off (deenergized) for the remainder of that cycle. The percentage of the total cycle time that the solenoid is energized is

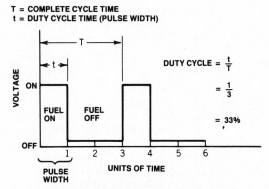

T = COMPLETE CYCLE TIME
t = DUTY CYCLE TIME (PULSE WIDTH)

DUTY CYCLE $= \frac{t}{T}$

$= \frac{1}{3}$

$= 33\%$

A. SHORT DUTY CYCLE (PULSE WIDTH), MINIMUM FUEL INJECTION

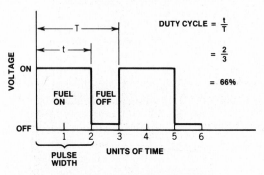

DUTY CYCLE $= \frac{t}{T}$

$= \frac{2}{3}$

$= 66\%$

B. LONG DUTY CYCLE (PULSE WIDTH), MAXIMUM FUEL INJECTION

Figure 13-44. The GM electronic fuel control (EFC) system.

called its duty cycle. The duty cycle is determined by a timed voltage pulse from the computer. The computer varies, or modulates, this pulse width to establish the duty cycle and achieve the desired solenoid output.

A solenoid can operate at any number of cycles per second: 10, 20, 30, 60, or whatever the engineer chooses to design. Each complete cycle lasts the same amount of time, but duty cycle can vary as a percentage of each cycle. Pulse width varies along with the duty cycle because it is the actual *time* that the solenoid is energized. Figure 13-44 shows two different pulse widths and duty cycles for the same complete cycle time. The system computer calculates the necessary pulse width and duty cycle from information provided by system sensors. Modern digital computers operate fast enough to change pulse width in fractions of a second to maintain precise fuel metering.

Pulse width modulation allows a digital output signal to provide varied or analog control of a mechanical device. It also is used to control fuel injectors and carburetor mixture control solenoids.

EARLY FUEL MANAGEMENT SYSTEMS

As you learned at the beginning of this chapter, the first application of electronic control to an automotive system was to control ignition timing. Chrysler Corporation receives historical credit as the first automaker to equip its engines with a breakerless ignition in 1972. However, the first computer-controlled fuel metering system made its appearance on European vehicles. In this section, we'll take a brief historical look at the development of electronic fuel management systems.

Bosch Lambda-Sond

Volvo offered the first electronically controlled feedback fuel system on 1977 models sold in the United States. This application was produced by the Robert Bosch company, a pioneer European manufacturer of fuel injection systems and electronic components.

The Volvo system combined the use of a 3-way catalytic converter (TWC) and EGO sensor with K-Jetronic fuel injection, figure 13-45. Sensor input was processed by the electronic control module, which then signaled a timing valve in the K-Jetronic system to vary injection control pressure and regulate the amount of fuel delivered by the continuous injection system. Saab and other European carmakers quickly followed with their own versions.

Ford and GM Feedback Fuel Control Systems

Ford and GM both introduced 3-way catalyst and feedback carburetor systems on some 1978 4-cylinder engines sold in California. Ford's system was called the "electronic control feedback carburetor with a three-way catalyst and conventional oxidation catalyst (TWC-COC)", figure 13-46. GM'S system was called "Phase II" emission control or "Electronic Fuel Control (EFC)", figure 13-47. Both systems controlled only fuel metering.

In both systems, the signal voltage from the EGO sensor was sent to a control module which in turn activated or deactivated a solenoid-operated vacuum valve. The vacuum valve or regulator controlled the flow of engine vacuum to a diaphragm in the carburetor. The diaphragm controlled the position of a fuel metering valve, which effectively varied the air-fuel ratio as desired.

The Ford system used an analog computer which was replaced by a digital computer on

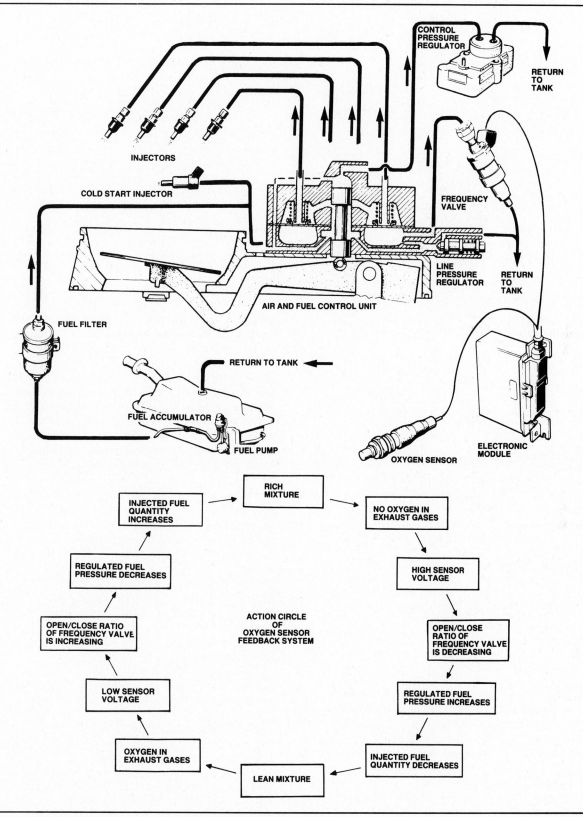

Figure 13-45. The Volvo Lambda-Sond system. (Volvo)

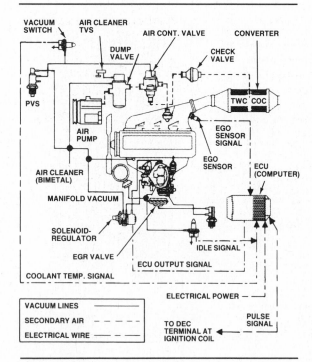

Figure 13-46. Ford's Feedback Electronic Engine Control system.

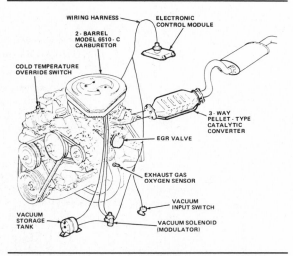

Figure 13-47. The GM electronic fuel control (EFC) system.

1980 models when the system was redesignated as the microprocessor control unit (MCU) system. It remained unchanged on many Ford engines during the early 1980's. The GM system was replaced on 1980 front-wheel drive vehicles by a fully integrated computer-controlled catalytic converter (C-4) system, which evolved into the computer command control (CCC or C-3) system on 1981 models.

SUMMARY

There are two theories of electric current flow. The conventional current theory says that current flows from positive to negative. The electron current theory says that current flows from negative to positive. Either theory can be used to describe current in a circuit.

A circuit is a complete path from an electrical energy source, through wires and electrical loads, and back to the source. A complete circuit is necessary for current to flow. Current flows *through* a circuit. Voltage is *applied to* or *impressed on* a circuit. Resistance, measured in ohms, opposes current flow and is *contained within* the circuit.

There are three kinds of electrical circuits: series, parallel, and series-parallel. In a series circuit, there is only one current path. In a parallel circuit, there are two or more current

paths. A series-parallel circuit has several current paths in parallel with each other, but in series with other parts of the circuit.

Every computer has four main functions: input, processing, storage, and output. Computers can operate on analog or digital signals. An analog signal is infinitely variable. A digital signal is an on/off or high/low signal. In automobile applications, most variable measurements produce analog signals which must be changed to digital signals for computer processing.

Digital computers use the binary system in which on/off or high/low voltage signals are represented by combinations of 0's and 1's.

Onboard computers are used in automotive fuel metering control systems. The control system regulates the operation of the vehicle's fuel system and operates in open loop or closed loop modes. In an open loop mode, the system does not respond to an output feedback signal. In a closed loop mode, the computer responds to the feedback signal and adjusts the output value accordingly.

Computer input is provided by sensors. Sensors can be switches, timers, potentiometers, piezoresistors, transformers, or generators. Computer output is sent to actuators, which transduce, or convert, the electrical signal to mechanical action. Most actuators are solenoids or relays, but some are stepper motors. The first feedback fuel metering system was Volvo's Lambda-Sond system in 1977. Ford and GM followed in 1978 with their own versions, which quickly evolved into integrated engine management systems.

Review Questions

Choose the single most correct answer.
Compare your answers with the correct answers on page 414.

1. The electromotive force that moves electrons from one point to another is called:
   a. Resistance
   b. Voltage
   c. Amperes
   d. Ohms

2. The rate of current flow in a circuit is measured in:
   a. Ohms
   b. Volts
   c. Amperes
   d. Watts

3. Resistance is measured in:
   a. Ohms
   b. Amperes
   c. Watts
   d. Voltage drop

4. Mechanic A says that a parallel circuit contains several current paths.
   Mechanic B says that a series-parallel circuit contains several current paths.
   Who is right?
   a. A only
   b. B only
   c. Both A and B
   d. Neither A nor B

5. Mechanic A says that an engine computer receives input information from its actuators, processes data, stores data, and sends output information to its sensors.
   Mechanic B says that most late-model automotive computers are based on analog microprocessors.
   Who is correct?
   a. A only
   b. B only
   c. Both A and B
   d. Neither A nor B

6. The operational program for a specific engine and vehicle is stored in the computer's:
   a. Logic module
   b. Programmable read-only memory (PROM)
   c. Random-access memory (RAM)
   d. Keep-alive memory (KAM)

7. An exhaust gas oxygen (EGO) sensor is an example of a:
   a. Resistor
   b. Potentiometer
   c. Generator
   d. Solenoid

8. An engine detonation sensor uses a:
   a. Piezoresistive crystal
   b. Voltage divider pickup
   c. Potentiometer
   d. Thermistor

9. The binary system used by a digital computer consists of:
   a. 10 numbers
   b. 5 numbers
   c. 3 numbers
   d. 2 numbers

10. The computer can read but not change the information stored in:
    a. ROM
    b. RAM
    c. KAM
    d. None of these

11. Mechanic A says that analog input data must be digitized by an A/D converter.
    Mechanic B says that output data must be changed to analog signals by a D/A converter.
    Who is right?
    a. A only
    b. B only
    c. Both A and B
    d. Neither A nor B

12. An onboard computer can do all of the following except:
    a. Ignore sensor input under certain conditions
    b. Respond in different ways to the same input
    c. Accept an input signal from another computer
    d. Ignores its program instructions under certain conditions

13. When the onboard computer is in open loop operation, it:
    a. Controls fuel metering to a predetermined value
    b. Ignores the temperature sensor signals
    c. Responds to the EGO sensor signal
    d. All of the above

14. The reference value sent to a sensor by the computer must be:
    a. Above battery voltage
    b. Exactly the same as battery voltage
    c. Less than minimum battery voltage
    d. Either a or c

15. The simplest digital sensor is a:
    a. Solenoid
    b. Switch
    c. Timer
    d. Relay

16. A variable resistance sensor is called a:
    a. Potentiometer
    b. Thermistor
    c. Transformer
    d. Generator

17. The percentage of time a solenoid is energized relative to total cycle time is called the:
    a. Pulse width modulation (PWM)
    b. Frequency
    c. Duty cycle
    d. KAM

18. A stepper motor:
    a. Is either on or off
    b. Is one form of solenoid
    c. Is used to operate the EGR valve
    d. Has discrete steps of movement

19. Most onboard computers work with:
    a. Binary numbers
    b. Voltage signals
    c. Both a and b
    d. Neither a nor b

20. Mechanic A says that throttle position sensors are potentiometers.
    Mechanic B says that throttle position sensors are analog devices.
    Who is right?
    a. A only
    b. B only
    c. Both A and B
    d. Neither A nor B

14

Electronic Engine Control Systems

Electronic Engine Control Systems

The computers used to control various electrical systems, including engine operation, may be called modules, assemblies, or electronic control units, as you learned in Chapter 13. Some are single-function devices that control a single system. Others are multiple-function devices that regulate more than one system. A few even act as master units, supervising a network of computer-controlled systems. These computers use input signals from various sensors to control a given system through a series of actuators, figure 14-1, and the output signal from one computer can act as an input signal to another computer.

COMPUTER FUNCTIONS — A REVIEW

As you learned in previous chapters, every computer does four basic functions. Electronic engine control systems provide some of the best examples of these functions:

1. *Input* — Variable voltage signals provided by sensors are the computer's input data. After conditioning the signals, the computer compares them to programmed information and makes a decision.
2. *Processing* — Processing begins with the receipt of input signals and continues as the computer evaluates and compares multiple signals and makes decisions for output commands.
3. *Storage* — A computer stores its own program, or operating instructions, in its memory. It also stores a basic set of data about vehicle design, such as weight, engine, transmission, and accessory combinations. The read-only memory (ROM) and random access memory (RAM) allow the computer to store this programmed data, as well as input and output signals for later reference.
4. *Output* — After receiving and processing input data, the computer sends output voltage signals to various actuators in the engine control system. The actuators are electromechanical or electronic devices that control fuel metering, ignition timing, emission control operation, and other engine operations.

Open- and Closed-Loop Operation

Every engine control system has two basic operating modes: open loop and closed loop. In open-loop operation, however, the computer does not respond to a feedback error signal from an actuator or from a sensor that measures output results. The computer simply *assumes* that its output signals achieve the desired

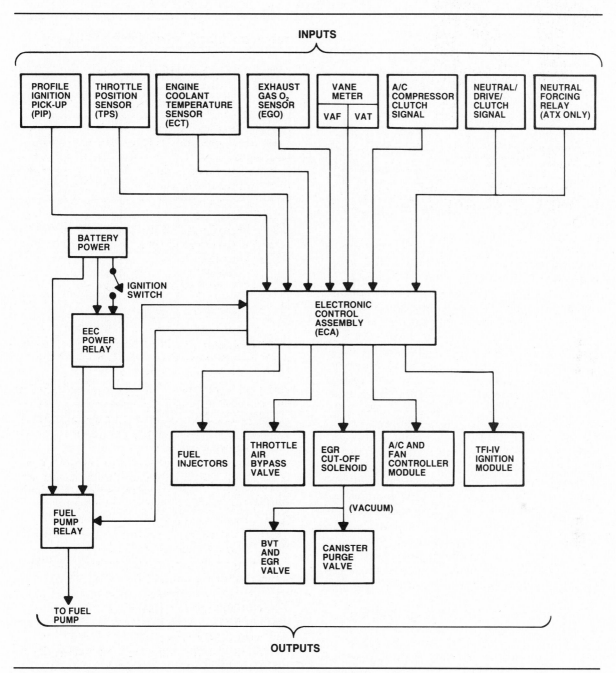

INPUTS

| PROFILE IGNITION PICK-UP (PIP) | THROTTLE POSITION SENSOR (TPS) | ENGINE COOLANT TEMPERATURE SENSOR (ECT) | EXHAUST GAS O₂ SENSOR (EGO) | VANE METER | A/C COMPRESSOR CLUTCH SIGNAL | NEUTRAL/ DRIVE/ CLUTCH SIGNAL | NEUTRAL FORCING RELAY (ATX ONLY) |

Figure 14-1. Typical engine control system inputs and outputs. (Ford)

performance. Nothing tells it differently unless it gets an input signal that is radically out of limits.

In closed-loop operation, a computer receives, and responds to, a signal from a sensor that measures output results. The principal feedback signal in an engine control system is the exhaust gas oxygen (EGO) sensor. The EGO sensor measures oxygen in the exhaust and sends a corresponding voltage signal to the computer. The computer interprets this signal

as a measurement of the air-fuel ratio. The EGO sensor signal closes the loop. The computer no longer has to assume that its output signals achieve the desired results. It measures the results and adjusts the signals if the results are out of the desired limits. Figure 14-2 is a simple block diagram of closed-loop control system operation.

Although the EGO sensor is the principal closed-loop feedback sensor, engine control computers receive, or infer, feedback signals

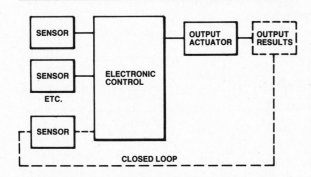

Figure 14-2. Typical open- and closed-loop block diagram. (Ford)

from other sensors. A detonation sensor, for example, sends a signal when the combined results of engine temperature, ignition timing, and air-fuel ratio cause engine pinging. Although a detonation sensor signal is not a feedback from a single output, it measures the results of combined output signals and allows the computer to correct its signals accordingly. Exhaust gas recirculation (EGR) sensors provide similar feedback signals that indicate EGR valve position.

Idle speed control is another example of a feedback signal. If idle speed changes from a programmed value, an ignition (tachometer) signal informs the computer. The computer then directs an actuator (a solenoid or a stepper motor) to adjust airflow or throttle position and return the engine to the desired idle speed.

Essentially, a computer operating in closed loop is constantly "retuning" the engine to keep it within programmed limits. This retuning compensates for changes in operating factors such as temperature, speed, load, and altitude.

The distinctions between open-loop and closed-loop operation are fundamental to any engine control system. The basic difference is simple: does or does not the computer respond to a feedback error signal. The specific applications can become quite subtle and complex, however. Figure 14-3 is a chart that summarizes seven basic engine operating modes and how the computer responds to each. The following paragraphs summarize open- and closed-loop computer operations.

Open-loop operating mode

When a vehicle is first started, the control system is in an open-loop mode. This means that:

- The sensors provide information to the computer in the form of voltage signals
- The computer compares the signals to its stored program and makes a decision

ENGINE OPERATING MODES

Engine Operating Mode	Air/Fuel Ratio	Engine Temperature	Exhaust Gas Sensor Input	Air/Fuel Temperature
Engine Crank	Fixed 2:1 to 12:1	Cold to cool	None	Cold to cool
Engine Warm-up	Fixed 2:1 to 15:1	Warming	None until engine warm-up	Warming
Open Loop	Fixed 2:1 to 15:1	Cold or warm	May signal, but ignored by processor	Cold or warm
Closed Loop	14.7:1 Depends on exhaust gas sensor input	Warm	Signalling	Warm
Hard Acceleration	Variable rich mixture, depends on driver demands	Warm	Signals, but ignored by processor	Warm
Deceleration	Variable lean mixture	Warm	Signals, but ignored by processor	Warm
Idle	Rich or lean depends on calibration	Warm	Signal, may be ignored (depends on calibration)	Warm

Figure 14-3. Ford computers are programmed for 7 different engine operating modes. Sensor input determines the mode selected. (Ford)

- The computer transmits a voltage signal to its output drivers to implement the decision
- The output drivers react to the signal by opening or closing the ground circuit of one or more output actuators
- The actuators operate in a specific manner without providing any feedback to the computer.

This unidirectional operating sequence is shown by the solid lines in figure 14-2.

For an engine computer to determine whether the air-fuel mixture is correct, it requires an EGO sensor feedback signal. If the computer ignores this signal, it relies on its stored program and signals from other sensors to make its decision. In this case, suppose the engine coolant sensor tells the computer that the engine is cold. The throttle position sensor also signals that engine speed is increasing. Based on this information, the computer tells the fuel control actuator to enrich the mixture. It does this under certain specified conditions:

- During a cold start or hot restart
- Under low vacuum conditions
- At wide-open throttle or under full load, regardless of engine speed
- During idle or deceleration conditions (some systems).

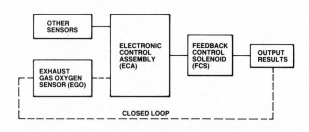

Figure 14-4. Closed-loop fuel metering block diagram. (Ford)

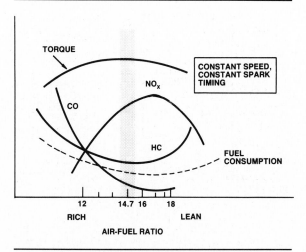

Figure 14-5. Changes in air-fuel ratio produce these fuel consumption, emission, and torque curves.

Closed-loop operating mode

Once the computer switches system operation into a closed-loop mode, it responds to feedback signals provided by sensors or actuators. These feedback signals tell the computer whether the output is insufficient, optimum, or excessive. In other words, the feedback signals regulate the output control. This cycle is summarized by the dotted line in figure 14-4.

In our fuel metering example above, the computer will respond to the EGO sensor signal. If the sensor measures an excessively rich mixture, it signals the computer of this. The computer then directs the fuel control actuator to lean the mixture. If the actuator leans the mixture too much, the EGO sensor informs the computer, which then directs the fuel control actuator to again enrich the mixture. This is an on-going process that occurs many times per second.

AIR-FUEL RATIO, TIMING, AND EGR EFFECTS ON OPERATION

This is a good time to consider the effects of air-fuel ratio, timing and exhaust gas recirculation (EGR) operation on overall engine operation. Almost all late-model engine systems recirculate a measured amount of exhaust gas into the intake air-fuel mixture. Exhaust gas recirculation was introduced in the early 1970's to reduce NO_x emissions by diluting the intake air-fuel charge (displacing oxygen) and lowering combustion temperatures. EGR also is an effective detonation control method, which allows spark timing to be maintained at optimum advance for performance and economy. The following explanations will help you to understand why only a computer can control air-fuel ratios, spark timing, and EGR for maximum engine efficiency.

Air-Fuel Ratio — A Review

Assuming fixed timing and engine speed, variations in air-fuel ratio have a dramatic effect on pollutants, figure 14-5. When the ratio is richer than 14.7 to 1, hydrocarbon (HC) and carbon monoxide (CO) emissions are high, as would be expected with the increase in fuel consumption. However, oxides of nitrogen (NO_x) emissions are low. Torque is greatest at ratios between 12 and 16. Above 16, torque decreases, as does NO_x, while HC increases. When the ratio is very lean, HC emissions also increase, as does fuel consumption. This is caused by the engine misfiring from the lean mixture and passing unburned fuel through the cylinders. The result is a reduction in both power and torque.

Stoichiometric ratio

The stoichiometric or ideal air-fuel ratio is 14.7 to 1. This ratio gives the most efficient combination of air and fuel during combustion. This ratio is a compromise between maximum power and economy, and it delivers the optimum combination of performance and mileage.

Emission control also is optimum at this ratio, if a 3-way oxidation-reduction catalytic converter is used, figure 14-6. As the mixture gets richer, HC and CO conversion efficiency falls off. With leaner mixtures, NO_x conversion efficiency also falls off. As figure 14-6 shows, the conversion efficiency range is very narrow — between 14.65 and 14.75 to 1. A fuel system without feedback control cannot maintain this narrow range.

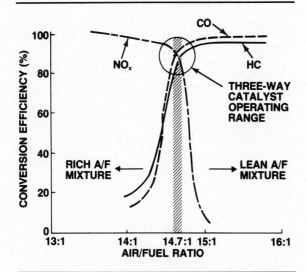

Figure 14-6. Three-way catalysts work properly only in a narrow air-fuel ratio range.

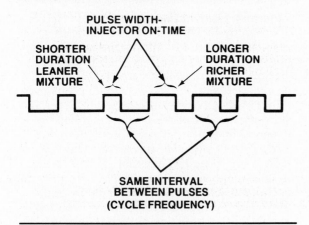

Figure 14-7. Fuel injector pulse width determines the air-fuel ratio.

Ratio control with carburetors and fuel injection

Two types of fuel control actuators are used with carburetors:

● A solenoid or stepper motor mounted on or in the carburetor to directly control the fuel-metering rods or the air bleeds, or both
● A remote-mounted, solenoid-actuated vacuum valve to regulate vacuum diaphragms that control the fuel-metering rods and air bleeds.

The computer sends a pulsed voltage signal to the control device, varying the ratio of on-time to off-time according to the signals received from the EGO sensor. As the percentage of on-time is increased or decreased, the mixture becomes leaner or richer.

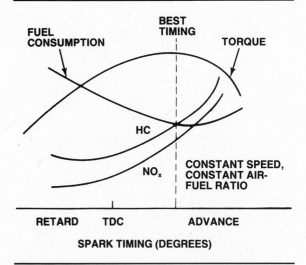

Figure 14-8. Changes in ignition timing produce these fuel consumption, emission, and torque curves.

With fuel injection systems, however, the computer controls the ratio by switching one or more fuel injectors on and off. Engine speed determines the switching rate, and the computer varies the length of time the injectors remain open (pulse width) to establish the air-fuel ratio, figure 14-7. As the computer receives data from its inputs, it increases the pulse width to supply more fuel for situations such as cold running, heavy loads, or fast acceleration. In a similar manner, it shortens the pulse width to lean the mixture for situations such as idling, cruising, or decelerating.

Ignition Timing

Assuming a fixed air-fuel ratio and engine speed, variations in ignition timing also have a dramatic effect on fuel consumption and pollutants, figure 14-8. When timing is at top dead center (tdc) or slightly retarded, emissions are low and fuel consumption is high. As timing is advanced, fuel consumption drops off but emissions increase. Engine computers are programmed to calculate the best timing for any combination of air-fuel ratio and engine speed without detonation problems.

Exhaust Gas Recirculation

Exhaust gas recirculation is the most efficient way to reduce NO_x emissions without adversely affecting fuel economy, driveability, and HC emission control. The recirculation of exhaust gases lowers the combustion temperature, so NO_x emissions drop off sharply when EGR is introduced into the air-fuel mixture. However, excessive reliance on EGR leads to an

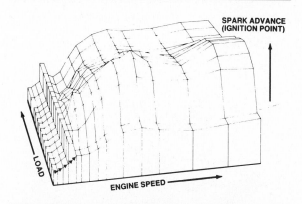

Figure 14-9. A typical spark advance "map" produced by the engine mapping process. (Bosch)

Figure 14-10. A typical GM PROM.

increase in both HC emissions and fuel consumption. Again, the engine computer is programmed to calculate the percentage of EGR that delivers the best compromise between NO_x control, HC emissions, and fuel economy without detonation problems.

Computer Integration

One of the primary values of a computer is its ability to integrate the operation of two or more individual and relatively uncomplicated systems to form a larger and more complex system. For example, we know that centrifugal and vacuum advance mechanisms can control spark timing relative to engine speed and load. We also know that fuel metering through a carburetor is controlled by airflow, and that manifold or ported vacuum can manage basic EGR flow. Integrating such independent systems through a computer provides faster, more precise regulation of each system, and allows the computer to calculate the effect of changing several variable factors at the same time.

Engine Mapping

Every computer needs instructions to do its job. These instructions are called a computer program. The program for an engine control computer consists of several elements:

● The mathematical instructions that tell the computer how to process, or "compute", the information it receives.
● The information that pertains to *fixed* vehicle values, such as vehicle weight, the number of cylinders, engine compression ratio, transmission type and gear ratios, firing order, and emission control devices.
● The data that pertains to *variable* vehicle values such as engine rpm, car speed, coolant

temperature, intake airflow, fuel flow, ignition timing, and others.

Since the mathematical instructions and vehicle valves are constant values, they are fixed and are easily placed into computer memory. To place the variable values into memory, it is necessary to simulate the vehicle and its system in operation. Carmakers use a large mainframe computer to calculate all of the possible variable conditions for any given system. This process of system simulation is called **engine mapping** and provides the control program for the individual onboard computer.

By operating a vehicle on a dynamometer and manually adjusting the variable factors such as speed, load, and spark timing, it is possible to determine the optimum output settings for the best driveability, economy, and emission control. Engine mapping creates a 3-dimensional performance graph, figure 14-9, which applies to a given vehicle and powertrain combination.

The vehicle information mapped in this manner is stored along with the mathematical instructions in a computer chip called a programmable read-only memory (PROM), figure 14-10, which is installed in the central computer for that particular car model. The computer used the PROM as its memory; it compares the input from various sensors to the data in memory, and adjusts the systems under its control accordingly.

Engine Mapping: Vehicle operation simulation procedure used to tailor the onboard computer program to a specific engine/powertrain combination. This program is stored in a PROM or calibration assembly.

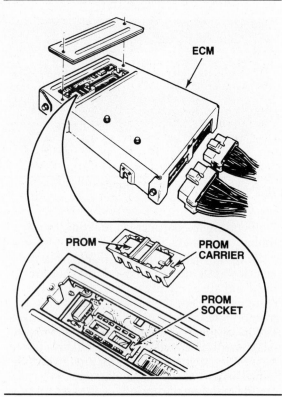

Figure 14-11. A GM engine control module (ECM) with a replaceable PROM. (GM)

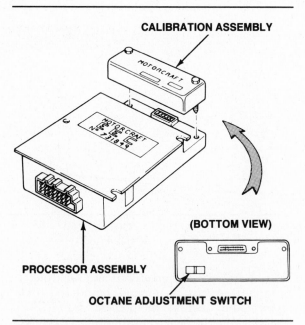

Figure 14-12. A Ford computer with a replaceable calibration assembly. (Ford)

Mapping allows a carmaker to use one basic computer for all models; the unique PROM individualizes the computer for a particular model. Also, if a driveability problem can be resolved by a change in the program, the carmaker can release a revised PROM to supersede the earlier part. Some PROMs are made so that they can be erased when exposed to ultraviolet light and reprogrammed. These are called erasable programmable read-only memories (EPROMs).

Most carmakers use a single PROM which plugs into the computer, figure 14-11. Some Ford computers use a larger "calibration module" that contains the system PROM, figure 14-12. If the onboard computer must be replaced, the PROM or calibration unit is removed from the defective unit and installed in the replacement computer.

Adaptive Memory, Integrator, and Block Learn

Recent engine control systems can be programmed to learn from their own experience. This is called **adaptive memory**. It allows the computer to adjust its memory for computing open-loop operation. Once the system is operating in closed loop, the computer compares its open-loop calculated air-fuel ratios against the average limit cycle values in closed loop. If there is a substantial difference, the computer corrects its memory so that it can match closed-loop control as closely as possible when it is in open-loop. This data is stored in keep-alive memory (KAM) so that it can be used the next time the vehicle is started, providing more accurate air-fuel ratio control. KAM is a specialized form of random access memory (RAM). It also allows a computer to adapt its program to deal with long-term changes in engine operation resulting from wear.

The ECM used on GM fuel-injected engines contains a pair of functions called **integrator** and **block learn**. These are responsible for making minor adjustments to the air-fuel ratio of a fuel-injected engine, similar to the mixture control solenoid dwell on a carbureted engine. Integrator and block learn represent injector on-time.

The ECM program contains a base fuel calculation. When the EGO sensor tells the ECM to richen the air-fuel mixture, the ECM adds fuel to the base calculation. If the mixture is to be leaned, the ECM subtracts fuel from the base calculation.

This information can be retrieved from the ECM using a scan tool, or scanner. The tool connects to the ECM serial data transmission line through either the assembly line communication link (ALCL) or the assembly line diagnostic link (ALDL) connector. Acting as a

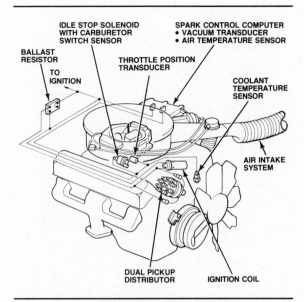

IDLE STOP SOLENOID WITH CARBURETOR SWITCH SENSOR

SPARK CONTROL COMPUTER
• **VACUUM TRANSDUCER**
• **AIR TEMPERATURE SENSOR**

BALLAST RESISTOR

THROTTLE POSITION TRANSDUCER

TO IGNITION

COOLANT TEMPERATURE SENSOR

AIR INTAKE SYSTEM

DUAL PICKUP DISTRIBUTOR

IGNITION COIL

Figure 14-13. A Chrysler Lean-Burn system.

bystander, the SCAN tool monitors the inputs and outputs as received and sent by the ECM.

The scan tool reads the base fuel calculation as the number 128. If the scan tool reads a higher number, the ECM is adding fuel to the mixture. If the number is less than 128, the ECM is subtracting fuel. This corrective action taken by the ECM is the integrator function; it is effective only on a short-term basis. Block learn, or long-term correction, only changes if the integrator sees a condition which remains for a predetermined length of time.

In summary, adaptive memory allows the engine computer to alter its program over the life of the vehicle. The computer can adapt its program to such long-term variables as:

- Engine wear
- Changes in fuel quality
- Changes in regular driving habits
- Changes in environmental conditions.

The integrator capability allows the computer to make short-term — minute-by-minute — corrections in fuel metering. Such corrections may be necessary, for example, when a car is driven from low altitude across a high mountain pass and back to low altitude in an hour or two.

Block-learn represents the long-term effects of integrator corrections. As such, it complements adaptive memory. If the computer continually must overcompensate fuel metering to maintain the stoichiometric ratio, it "learns" the correction and adapts its memory to make the correction factor part of its basic program. The period necessary for block-learn to become part of basic memory may be 8 hours, 40 hours,

80 hours, or some other period determined by the system engineers.

PARTIAL-FUNCTION CONTROL SYSTEMS

Early electronic engine control systems of the late 1970's were partial-function systems. They regulated fuel metering or ignition timing, but not both. They were, however, the starting point for the full-function systems used on today's vehicles. You have been introduced to some of these systems in chapter 13, and the last section of this chapter has histories of system development by major domestic and foreign carmakers. We will use some of these same systems as examples of partial-function engine controls in the following paragraphs.

Partial-function systems have one or more of the following features. Late-model, full-function systems all have:

- A 3-way catalytic converter and an EGO sensor for stoichiometric air-fuel ratio control
- Electronic feedback control of fuel metering in the carburetor or fuel injection system
- Open- and closed-loop operating modes
- Electronic spark timing control in place of traditional centrifugal and vacuum spark advance
- Electronic control of air injection switching, EGR, and vapor canister purging
- Electronic control of transmission shifting, torque converter lockup, and accessory operation.

Ignition Timing Control

Chrysler's electronic lean burn (ELB) system, figure 14-13, was introduced in 1976. It was an

---

Adaptive Memory: A feature of computer memory that allows the microprocessor to adjust its memory for computing open-loop operation, based on changes in engine operation.

Integrator: The ability of the computer to make short-term — minute-by-minute — corrections in fuel metering.

Block-Learn: The long-term effects of integrator corrections. As such, block-learn complements adaptive memory. If, for example, the computer continually must overcompensate fuel metering to maintain the stoichiometric ratio, it "learns" the correction and adapts its memory to make the correction factor part of its basic program.

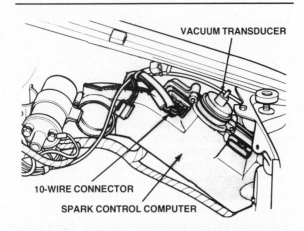

Figure 14-14. The Chrysler 4-cylinder spark control computer.

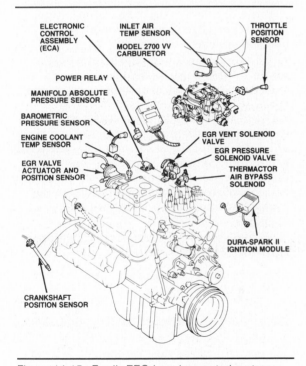

Figure 14-15. Ford's EEC-I engine control system.

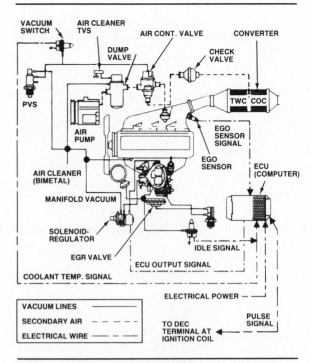

Figure 14-16. Ford's Feedback Electronic Engine Control system.

early example of a partial-function ignition timing control system. Although the name sounds like it was a fuel control system, ELB simply used a carburetor calibrated for lean air-fuel ratios. No feedback, or variable, fuel control was used. The system computer regulated ignition timing to allow the engine to operate smoothly on lean ratios. The first ELB systems used a distributor with centrifugal advance but no vacuum advance. In 1977, centrifugal advance was eliminated, and timing was under full electronic control. Chrysler's ELB system continued through 1978 on V-8 and 4-cylinder engines, and in 1979 was renamed electronic spark control (ESC), figure 14-14.

Ford's first electronic engine control system (EEC-I), figure 14-15, was introduced in 1978 and controlled spark timing, EGR, and air injection. It did not include feedback fuel metering control. Ignition timing signals were provided by a crankshaft position sensor. On 1978 models, the sensor was installed in the rear of the engine block; in 1979, it was moved to the front of the crankshaft, behind the vibration damper.

General Motors' first partial-function timing control system was the microprocessor sensing and automatic regulation (MISAR) system used on 1977 Oldsmobile Toronados. The system used a crankshaft position sensor, a standard high energy ignition (HEI) distributor, and an electronic control module (the system computer) to control spark advance. In 1978, the system was renamed electronic spark timing (EST).

Fuel Metering Control Only

In 1979, Chrysler modified its ELB, or electronic spark control system, to work with feedback carburetors and EGO sensors. This system was used on some 6-cylinder engines and was close to a full-function engine control system.

In Chapter 13 you read about Ford's earliest feedback control (FBC) system used on some 1978 2.3-liter, 4-cylinder engines, figure 14-16.

Used with a 3-way catalytic converter (TWC-COC) and an EGO sensor, this system controlled only fuel metering through the Holley 6500 carburetor. It did not control ignition timing or other engine functions.

Ford changed the system computer from an analog to a digital processor on 1980 models and reidentified the system as the microprocessor control unit (MCU) system. The first versions controlled fuel metering in a way similar to the earlier FBC or TWC-COC systems. Later versions in the mid-1980's controlled canister purging, idle speed, and detonation spark

timing. Some MCU systems include simple self-diagnostic capabilities.

Ford's EEC-II system appeared on some 1979 V-8's and controlled air injection switching and vapor canister purging, along with feedback fuel metering. Fully electronic spark timing control was not included. Ford's EEC-III, introduced in 1980, also was based on 3-way catalytic converters and EGO sensors. It was used with feedback carburetors and with throttle body fuel injectors (called central fuel injection — CFI — by Ford). EEC-III was the first Ford system to have a self-test program.

■ Drive By Wire

You have learned that cable-operated throttle linkage is a standard design, used on many vehicles for several decades. It's a pretty obvious design choice when you think about it. Connect the accelerator pedal to the carburetor or fuel injection throttle body with a flexible cable. On many cars, it's a lot simpler than an arrangement of levers, cranks, and springs. Cable-operated throttles are simple and straightforward devices, but they are not what engineers refer to when they talk about "drive-by-wire" systems today.

Current electronic engine control systems use various actuators to control fuel metering, EGR operation, and idle speed. Imagine, for a moment, the possibility of replacing a throttle lever or cable with an electronic control system. Then the driver could control the engine through electrical wires, without a direct mechanical connection to the throttle.

Such systems are not a scientific pipedream. Developmental systems are in operation today. Here is how they generally work:

● The accelerator pedal is connected to a position sensor, usually a potentiometer. This is similar to the airflow sensors or throttle position sensors used with engine control systems.
● The accelerator sensor sends a variable (analog) voltage signal to the engine control computer to indicate pedal position.
● The computer evaluates this signal along with signals from other sensors that indicate engine and vehicle speed, crankshaft position, temperature, ignition timing, air-fuel metering, and other operating variables.
● The computer sends an output signal to a servomotor connected to the throttle valve.
● The servomotor opens and closes the throttle valve in direct relation to the driver's movement of the accelerator pedal.

It's all quite practical, and such a system can be built with the technology available since the late 1970's.

Besides applications to cars, drive-by-wire systems have major advantages for construction equipment and other specialized machinery where the driver is in a remote and mechanically awkward position in relation to the engine. For such equipment, electronic throttle control can be much more precise and economical than a complex arrangement of cables and levers.

Even though drive-by-wire systems can be built with today's technology, one final electronic development is necessary to "fine tune" them for automotive applications. That is the production of an economical and reliable engine torque sensor. Engine torque — pulling power — is one operating condition that a driver's brain and right foot can sense more accurately and more quickly than a simple electronic device can.

Torque sensors exist today, and engineers are working to make them practical, cheap, and reliable for automotive use. When torque sensors are added to engine control systems, driving by wire will be commonplace in the twenty-first century.

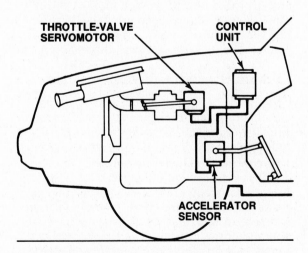

An accelerator sensor, an electronic control unit (computer), and a throttle servomotor are the major parts of a drive-by-wire system.

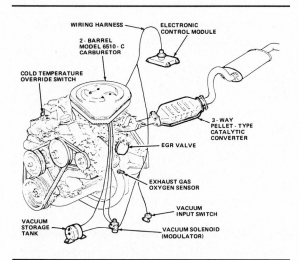

Figure 14-17. The GM electronic fuel control (EFC) system.

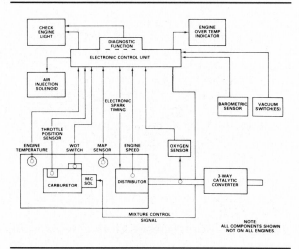

Figure 14-18. The GM C-4 system.

Chapter 13 of this manual also introduced you to General Motors' electronic fuel control (EFC) system used on some 1978-79 151-cid (2.5-liter) engines, figure 14-17. The GM EFC system was quite similar to Ford's early FBC system. It used a 3-way converter, an EGO sensor, and a Holley 6510-C feedback carburetor. The early GM and Ford systems both controlled the carburetor through a vacuum solenoid that operated diaphragms to regulate metering rods and air bleeds.

In mid-1979, GM introduced the computer-controlled catalytic converter (C-4) system on the 1980 X-cars and some Buick V-6 engines, figure 14-18. The first C-4 systems were partial-function systems that controlled fuel metering and provided limited ignition timing control. The C-4 system developed into the GM full-function computer command control (CCC or C-3) system, described later in this chapter.

Limited (Partial) Function Systems

GM introduced a minimum function control system on certain 1982 Chevette and Pontiac T-1000 4-cylinder engines. It differs from the full-function systems in four major ways:

1. It controls fuel metering only.
2. It has only seven trouble codes.
3. A coolant temperature switch closes when a predetermined engine coolant temperature is reached. At this time, the ECM switches system operation from open to closed loop, if the EGO sensor is at operating temperature and a programmed time interval has elapsed after the engine is started. Once the coolant switch closes, there is no effect on system operation if it opens due to a malfunction. The temperature

switch was replaced by a coolant sensor on 1983 models.
4. Since the ECM does not provide a fixed dwell during open-loop operation, the range of dwell control is limited. Open-loop time consists of four regions, or areas. Dwell control varies according to which region the ECM is in at that time. For example, the system may be in open loop, but if dwell were measure, it would give a reading as if the system were in closed loop.
5. Memory is short-term, as is the case with some early C-4 systems. Any trouble codes set in memory are lost each time the ignition is turned off. With short-term memory, trouble codes must be located before shutting the engine off.

FULL-FUNCTION CONTROL SYSTEMS

As we have seen, the earliest control systems affected a single system: ignition timing or fuel metering. Full-function systems control two or more engine functions. In doing so, they share many components and operational principles.

The early full-function control systems had one or more of the following characteristics; late-model systems have all of them.

• The computer controls timing electronically instead of relying on distributor vacuum and centrifugal advance mechanisms.
• The computer controls air-fuel ratio as close as possible to the stoichiometric value (14.7) through an EGO sensor and a 3-way catalytic converter.
• The computer controls fuel metering by operating a carburetor mixture control solenoid or by pulsing fuel injectors according to data received from various sensors.

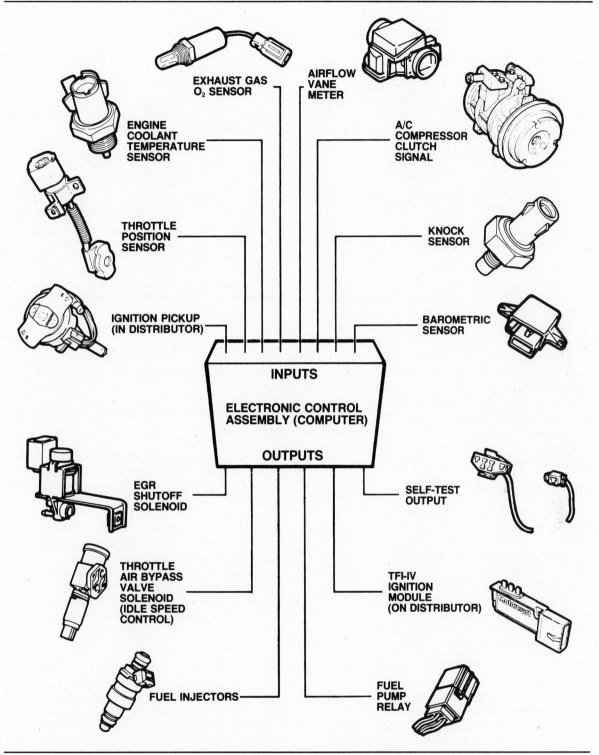

Figure 14-19. Typical components of a late-model Ford electronic engine control (EEC-IV) system.

● Engine operation is divided into open- and closed-loop operational modes. A separate "limp-in mode", or "limited operational strategy", is provided when a serious system malfunction occurs, allowing the vehicle to be driven in for service.

● The computer can set a number of trouble codes and may have a self-diagnostic capability.

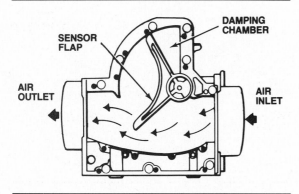

Figure 14-20. A vane-type airflow sensor.

CONTROL SYSTEM DEVELOPMENT

Engine systems have developed considerably since they first appeared a decade ago. Today's computer control systems, however, are undergoing an even more rapid development into a highly sophisticated electronic system composed of many microprocessors (computers) that eventually will manage all operational and customer convenience systems in a vehicle. The body computer module (BCM) concept which appeared on some 1986 GM luxury cars is the first generation of these "total control" systems, and their development should be more rapid than the progress we've seen with engine control systems.

COMMON COMPONENTS

As we have seen, all engine control systems use a computer, a series of sensors, and various actuators. The sensors feed data to the computer in the form of voltage signals. The computer processes the sensor data according to its internal program and then signals the actuators to exercise the desired control over the subsystems that require adjustment.

System Sensors

As we learned in Chapter 13, a sensor is an input device used to change temperature, motion, light, pressure, and other forms of energy into voltage signals that a computer can read. Input sensors tell the computer what is happening in several areas of vehicle operation at any given moment. Typical sensors used as computer inputs are shown at the top of figure 14-19.

These basic types of sensors were dealt with at length in Chapter 13. For our purposes here, a quick review of their outstanding characteristics should refresh your memory before we look at specific sensor applications.

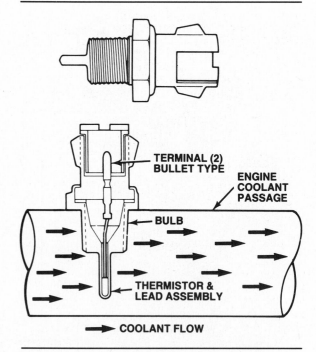

Figure 14-21. An engine coolant temperature sensor is installed in a coolant passage. (Ford)

- *Switch* — the simplest form of sensor, it signals an on or off condition.
- *Timer* — used to delay a signal for a predetermined length of time to prevent the computer from compensating for momentary conditions that do not significantly affect engine operation.
- *Resistive sensors* — may be potentiometers, thermistors, or piezoresistive devices.
- *Transformer* — contains a movable core that varies its position between input and output windings to produce a voltage signal.
- *Generator* — may be a magnetic pulse generator, a Hall-effect switch, or a galvanic battery. These sensors do not require a reference voltage but generate their own signal voltage.

Airflow sensors
Some fuel injection systems use a vane-type airflow sensor, figure 14-20, positioned between the air filter and the intake manifold. The airflow sensor monitors the volume of air entering the intake manifold. A thermistor is used to sense air temperature and is part of the vane. Since the angular position of the vane is proportional to airflow, a potentiometer connected to the vane sends a voltage signal proportional to intake air volume to the computer.

Manifold pressure, vacuum, and barometric pressure sensors
These sensors keep the computer informed about air volume and engine load, allowing it

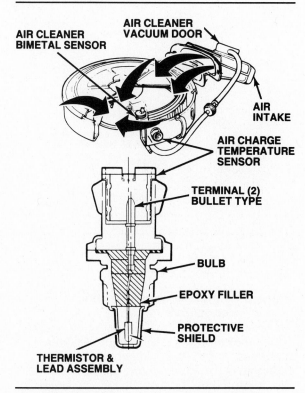

Figure 14-22. An air charge temperature (ACT) sensor may be installed in the air cleaner or intake manifold. (Ford)

to adjust fuel metering accordingly. Their input also is used by the computer to adjust timing and EGR flow relative to load. They may be a piezoresistive device, a transformer, or a potentiometer operated by an aneroid bellows or vacuum diaphragm.

Temperature sensors

Two types of temperature sensors are used: engine coolant and **air charge temperature (ACT)**. If the computer only is interested in whether coolant or air temperature is above a stated point, a simple bimetal switch is used. However, when the computer requires information about a temperature range, a thermistor is used.

The thermistor-type sensor used to track coolant temperature is threaded into a passage where its sensing bulb is immersed in engine coolant, figure 14-21.

An air charge temperature sensor is similar in construction to the coolant temperature sensor, but provides a faster response time to air temperature changes. It may be located in the air cleaner, figure 14-22, to measure only air temperature, or in the intake manifold where it measures the air-fuel mixture temperature.

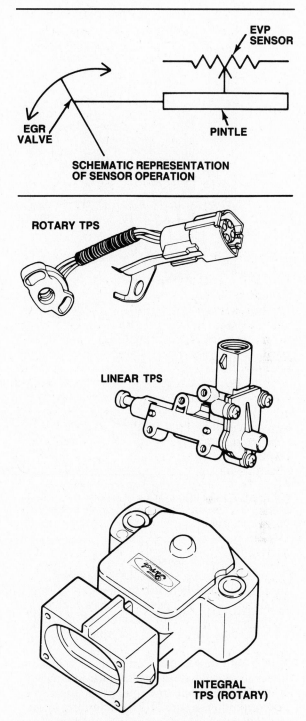

Figure 14-23. Typical throttle position sensors and their operation. (Ford)

Air Charge Temperature (ACT) Sensor: A thermistor used to measure intake air temperature or air-fuel mixture temperature.

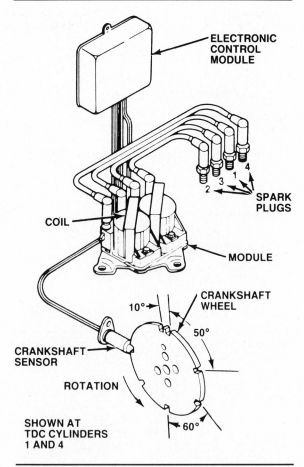

Figure 14-24. Rotation of a crankshaft timing disc through a sensor field tells the computer the cylinder position and provides a triggering signal to fire the proper coil. (GM)

Throttle position sensors
A throttle position sensor (TPS) may be a simple on/off switch, used to indicate wide-open throttle or idle position with a high or low voltage. The carburetor switch used on some Chrysler carburetors is an example of this type of TPS.

A potentiometer also can be used to indicate the exact position and speed of throttle movement. The TPS may be a rotary or a linear potentiometer, figure 14-23, depending upon its application. A rotary throttle position sensor is used on fuel injection assemblies. A linear TPS generally is used with carburetors. The two types differ primarily in how they work, but both send the computer an analog signal proportional to the angle of the throttle plate opening. The rotary potentiometer moves on an axis with the throttle shaft; the linear potentiometer uses a plunger that rides on a throttle shaft cam.

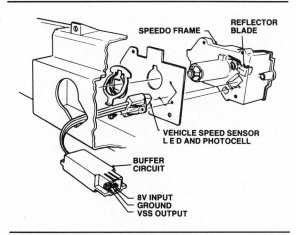

Figure 14-25. An LED and a photocell in the speedometer act as a speed sensor. (GM)

Ignition timing, crankshaft position, and engine speed sensors
These sensors send analog and digital signals which the computer uses to control timing, fuel metering, and EGR. A Hall-effect switch is mounted on or in the engine block. A timing disc mounted on the harmonic balancer or cast as part of the crankshaft, figure 14-24, passes through the Hall-effect switch field. This is used as a signal to inform the computer of cylinder position and firing order.

Vehicle speed sensors
Vehicle speed information necessary to control torque converter lockup is provided to the computer by a pulse generator or optical sensor. The most common type now in use has a reflective blade attached to the speedometer assembly, figure 14-25. As the blade spins, it passes through a light emitting diode (LED) beam. Each time the blade cuts through the LED beam, it reflects light back to a phototransistor. This creates a low-power signal which is amplified and sent to the ECM.

EGR sensors
A sliding-contact potentiometer may be connected to the top of the EGR valve stem to inform the computer of EGR flow rate, figure 14-26. This information is used by the computer to control timing, fuel metering and EGR valve operation.

Air conditioning sensors
An air conditioning compressor adds to the engine load when the compressor clutch is engaged. To allow the computer to make the necessary adjustments to compensate for the increased load, a simple on/off switch is used to tell the computer whether the compressor clutch is engaged or disengaged.

Figure 14-31. A mixture control solenoid used in Rochester carburetors.

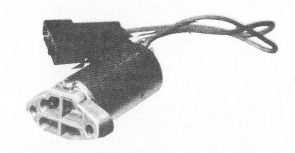

Figure 14-32. A pulse solenoid controls variable air bleeds in some Carter carburetors. (Carter)

figure 14-31, is an integral part of the carburetor. It operates both a metering rod in the main jet and a rod that controls an idle air bleed passage. Other systems use a solenoid that controls vacuum to a diaphragm installed in the carburetor. The vacuum diaphragm controls the operation of the metering rods and air bleeds.

Some Carter carburetors use a pulse solenoid with a variable duty cycle to control the air bleeds, figure 14-32.

EGR Actuators

Engine control computer systems that manage EGR flow use solenoids to regulate the amount of vacuum applied to the EGR valve. Although system designs differ, figure 14-33 shows a typical system, as used in Ford's EEC-II.

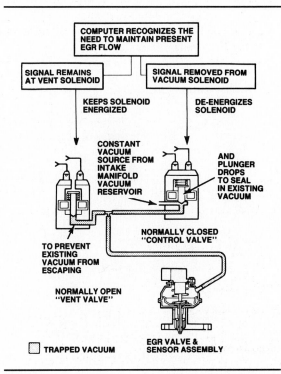

Figure 14-33. A typical computer-controlled EGR system. (Ford)

Other System Actuators

In addition to controlling ignition timing and fuel metering, the newer integrated electronic engine systems control several other functions:
- Torque converter lockup or engine shifting: solenoid valves in the transmission or transmission or transaxle hydraulic circuits respond to computer signals based on vehicle speed and engine load sensors
- Air injection switching: one or more solenoids operate a valve in the vacuum line to the air switching or air control valve, figure 14-34
- Vapor canister purge: a solenoid installed in the canister-to-carburetor or manifold valve line opens and closes as directed by the computer.

HISTORY OF ENGINE CONTROL SYSTEMS

Electronic engine controls appeared on the automotive scene with the 1977 models. The early control systems regulated only a single function: either ignition timing or fuel metering. However, they were rapidly expanded to control both systems, as well as numerous other engine functions. Most late-model engine

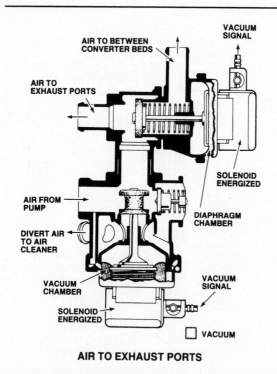

AIR TO BETWEEN
CONVERTER BEDS

VACUUM
SIGNAL

AIR TO
EXHAUST PORTS

SOLENOID
ENERGIZED

AIR FROM
PUMP

DIAPHRAGM
CHAMBER

DIVERT AIR
TO AIR
CLEANER

VACUUM
CHAMBER

VACUUM
SIGNAL

SOLENOID
ENERGIZED

☐ VACUUM

AIR TO EXHAUST PORTS

Figure 14-34. The Delco air switching valve uses two solenoids. (Delco-Remy)

control systems have a self-diagnostic capability, and display trouble codes for troubleshooting, and do the following:
- Open- and closed-loop operation
- Electronic ignition timing control
- Fuel metering control
- Stoichiometric air-fuel ratio control
- EGR flow control
- Air injection switching
- Vapor canister purging
- Automatic transmission or transaxle torque converter lockup.

Bosch Lambda and Motronic Systems

The Robert Bosch Company pioneered fuel injection and electronic controls used in European vehicles. The Volvo Lambda-Sond system (manufactured by Bosch) was combined with K-Jetronic fuel injection on 1977 Volvo models sold in the U.S. This was the first electronically controlled fuel metering system using a 3-way catalytic converter and EGO sensor. (This system was also described in Chapter 13 and illustrated in figure 13-45.)

An input signal from the EGO sensor results in an output signal to the K-Jetronic timing valve. This valve varies the injection control pressure and regulates the fuel supplied by the continuous injection nozzles. The original Lambda-Sond system controls only fuel metering.

The Bosch Motronic or digital motor electronics (DME) system added ignition timing and electronic spark control to the fuel metering control of the Lambda-Sond system. In addition to the 3-way converter and EGO sensor, the DME system receives input signals from crankshaft speed and position sensors, a magnetic pulse generator in the distributor, and the L-Jetronic sensors. This allows the computer to:
- Adjust injector pulse width for air-fuel ratio control
- Adjust ignition timing for combined speed and load conditions
- Shut off injection completely during closed-throttle deceleration.

Chrysler

Chrysler introduced its electronic lean-burn (ELB) spark timing control system in 1976 on some 6.6L (400-cid) V-8 engines, figure 14-13. The system is based on a special carburetor that provides air-fuel ratios as lean as 18:1 and a modified electronic ignition controlled by an analog spark control computer attached to the air cleaner housing. Two printed circuit boards inside the computer contain the spark control circuitry. The program schedule module receives the sensor inputs and interprets them for the ignition control module, which directs the spark timing output. The 1976 ELB distributor has dual ignition pickups and a centrifugal advance mechanism. The distributor secondary components are similar to those used with the basic Chrysler electronic ignition. A dual ballast resistor controls primary current and protects the spark control computer from voltage spikes.

In 1977, ELB became available on all Chrysler V-8 engines and the centrifugal advance mechanism was eliminated. A second-generation ELB design was used on 5.2L (318-cid) V-8's. The start pickup in the distributor was dropped, and the computer was redesigned so all the circuitry fit on a single board.

The second-generation system was adopted on all V-8's in 1978, and a new ELB version was introduced on the Omni and Horizon 4-cylinder engine. The 4-cylinder system uses a Hall-effect distributor instead of a magnetic pickup. It has variable dwell to control primary current and does not use a ballast resistor. The 4-cylinder spark control computer is mounted on the left front fender, figure 14-14.

With the introduction of 3-way converters, the ELB system was modified in 1979 to work with revised carburetors that provided air-fuel ratios closer to 14.7. The protection circuitry was integrated into the computer, and the

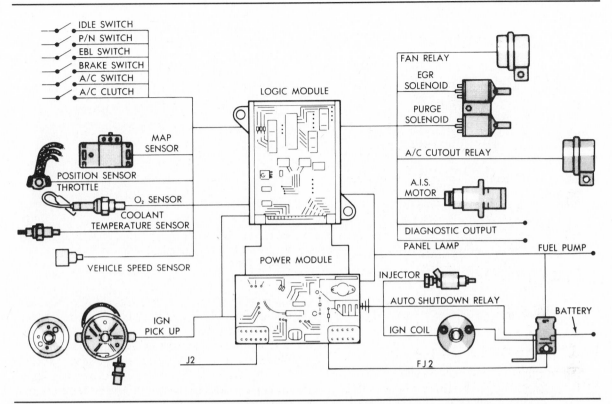

Figure 14-35. Chrysler's modular engine control system (single-point injection version shown).

dual ballast resistor was replaced by a single 1.2-ohm resistor to control primary current only. This third-generation system was renamed electronic spark control (ESC) and appeared on some 6-cylinder inline engines with an EGO sensor and a feedback carburetor.

The 1980 model year was a transitional one for Chrysler. All California engines received ESC with feedback fuel control. The 5.9L (360-cid) V-8 and Canadian 5.2L (318-cid) 4-barrel V-8's continued to use ESC without the feedback system. All other engines reverted to basic electronic ignition with mechanical and vacuum advance mechanisms. Detonation sensors were introduced on some 1980 ESC systems, but the biggest change was the switch from an analog to a digital computer. Models with the digital computer in 1980 can be identified by their return to dual ignition pickups in the distributor. Systems with digital computers also eliminate the ballast resistor completely. Since 1981, all of Chrysler's domestic, carbureted, 4-cylinder engines and all 6-cylinder and V-8 powerplants have feedback fuel control and digital ESC systems without ballast resistors.

Chrysler introduced the modular control system (MCS), figure 14-35, in late 1983 on throttle-body fuel-injected 4-cylinder engines. In 1984, its use was expanded to turbocharged port-injected engines as well. The modular control system regulates vehicle functions using two separate modules whose functions are similar to the two circuit boards in the original ELB computer. The logic module handles all of the low-current tasks within the system, including receiving the inputs and making control decisions. A replaceable PROM is mounted in the logic module housing, and a self-test program is provided to aid in system diagnosis. The logic module is mounted inside the car to avoid underhood electrical interference.

The power module handles the high-current tasks and is located in the left front fender. It looks similar to the spark control computer used in 4-cylinder ESC systems. The power module contains the regulated power supply for the entire control system, along with the switching controls for the ignition coil, fuel injectors, and auto-shutdown relay (ASD). The ASD supplies power to the coil, the fuel pump relay, and the power module when it detects a distributor cranking signal.

All of these Chrysler computer control systems are designed with an emergency "limp-in" mode. In case of a system failure, the computer reverts to a fixed set of operating values. This allows the vehicle to be driven to a shop for repair. If the failure is in the start pickup or the coil triggering circuitry, however, the engine will not start.

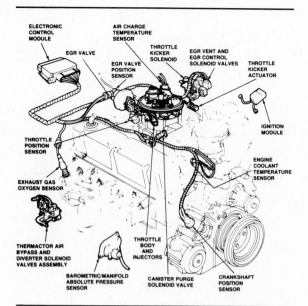

Figure 14-36. EEC-III/CFI system components.

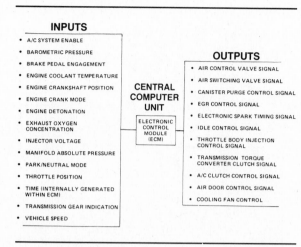

Figure 14-37. Typical CCC system sensors (inputs) and controls (outputs).

Ford

Ford introduced its feedback electronic engine control system, figure 14-16, on some 1978 2.3-liter, 4-cylinder engines. The system contains a 3-way catalyst and conventional oxidation catalyst (TWC-COC) converter, an EGO sensor, a vacuum control solenoid, and an analog computer. Its control was limited to fuel metering. In 1980, a digital computer replaced the analog unit and the system was renamed the microprocessor control unit (MCU) fuel feedback system. The major change in early applications is the addition of self-diagnostics, but later designs have expanded capabilities including control over idle speedup, canister purge, and detonation spark control. They might be considered complete engine control systems except that they lack continuous spark timing control.

Ford also introduced its first generation electronic engine control (EEC-I) system, figure 14-15, on the 1978 Lincoln Versailles. This system controls spark timing, EGR flow, and air injection. A digital microprocessor electronic control assembly (ECA) installed in the passenger compartment receives signals from various sensors. It then determines the best spark timing, EGR flow rate, and air injection operation and sends signals to the appropriate control devices.

All 1978-79 California EEC-I systems use a variation of the blue-grommet Dura-Spark II ignition. The 1979 Federal EEC-I system has a yellow-grommet dual-mode Dura-Spark II module. Although these modules appear simi-

lar to their non-EEC counterparts, they are controlled through the ECA and cannot be tested with the same procedures.

The ignition switching signal on 1978 EEC-I systems is provided by a sensor at the rear of the engine block that detects four raised ridges on a magnetic pulse ring mounted to the end of the crankshaft. In 1979, the pickup and pulse ring were moved to the front of the engine immediately behind the vibration damper. This design is used on the later EEC-II and EEC-III systems as well.

Late in 1979, Ford's second generation EEC-II system appeared on some 5.8L (351-cid) V-8 engines. EEC-II added electronic controls for vapor canister purging and air injection switching. In addition, dual 3-way converters are used with a feedback carburetor for precise air-fuel mixture control.

Ford's third generation EEC-III system appeared in 1980 and is available in two versions through 1984. EEC-III/FBC incorporates a feedback carburetor similar to the EEC-II system; EEC-III/CFI, figure 14-36, has a throttle-body-type central fuel injection system. All EEC-III systems use the Dura-Spark III ignition module. EEC-III is the first Ford computer engine control system to have a self-test program.

Ford's EEC-IV, figure 14-19, was introduced in 1983 and incorporates the thick-film integrated (TFI) ignition system. The 2-microchip EEC-IV microprocessor is much more powerful than the 4- or 5-microchip ECA's used with earlier EEC systems. EEC-IV has both increased memory and the ability to handle almost one million computations per second. Unlike earlier EEC systems, however, the EEC-IV's calibration assembly is located inside the ECA and cannot be replaced separately. All EEC-IV systems have an improved self-test capability with

BASIC COMPONENT LOCATIONS

1 INJECTOR	13 ELECTRONIC CONTROL UNIT (ECU)
2 THROTTLE POSITION SENSOR	14 SOLENOID-TO-EVAP CANISTER CONTROL
3 PRESSURE REGULATOR	15 STARTER MOTOR RELAY
4 IDLE SPEED CONTROL MOTOR	16 FUEL PUMP RELAY
5 SOLENOID-TO-EGR VALVE	17 FUEL PUMP
6 EGR VALVE	18 IGNITION CONTROL MODULE
7 MANIFOLD AIR/FUEL TEMPERATURE SENSOR	19 IN-LINE FUEL FILTER
8 O_2 - SENSOR	20 AIR CONDITIONER ON
9 SPEED SENSOR	21 TRANSAXLE NEUTRAL/PARK SWITCH
10 IGNITION SWITCH	22 CLOSED-THROTTLE (IDLE) SWITCH
11 POWER RELAY	23 WIDE-OPEN THROTTLE (WOT) SWITCH
12 MAP SENSOR	24 TEMPERATURE SENSOR (COOLANT)

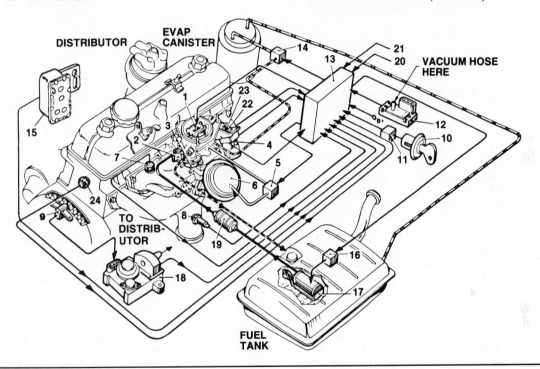

Figure 14-38. This AMC-Renault system is used to control fuel metering, ignition timing, emission devices, and transmission shift points. (AMC-Renault)

trouble codes that are stored for readout at a later date.

All Ford EEC systems have a limited operating strategy (LOS) mode in case of a failure within the system. The exact nature of the LOS varies from one system to another, but generally, the timing is fixed at 10°, and other ECA outputs are rendered inoperable.

General Motors

The first GM spark timing control was offered on 1977-78 Oldsmobile Toronados. The 1977 system was called microprocessed sensing and automatic regulation (MISAR). MISAR is a basic spark timing system only. A rotating disc

and stationary sensor on the front of the engine replace the pickup coil and trigger wheel of the distributor. Except for this change, the 1977 MISAR system uses a standard HEI distributor with a basic 4-terminal ignition module.

The MISAR system was modified in 1978 and renamed electronic spark timing (EST). The crankshaft-mounted disc and stationary sensor were dropped, and a conventional pickup coil and trigger wheel were again fitted in the HEI distributor. The ignition module, however, is a special 3-terminal design that is not interchangeable with any other HEI system.

In mid-1979, GM introduced the computer-controlled catalytic converter (C-4) system, figure 14-18. At first, the C-4 system was purely a fuel control system used with 3-way converters. But in 1980, Buick V-6 engines with C-4 were also fitted with an electronic spark timing (EST) system. Although this is the same name applied to the MISAR system just discussed, the two systems are not the same. C-4 with EST has a single electronic control module (ECM) that regulates both fuel delivery and spark timing. It was GM's first complete computer engine control system. The C-4 system was further upgraded in 1981 with EST in almost all applications, and additional control capabilities were added. The expanded system, figure 14-37, was renamed computer command control (CCC or C-3).

In 1986, GM began to update the CCC system through the introduction of a new ECM on certain vehicles. The new ECM is smaller than previous models, but has more functional capabilities. It operates at twice the speed of previous ECM's, and has fewer IC chips and internal connections. The new ECM draws less current with the ignition off, provides more diagnostic functions, and operates reliably on battery voltage as low as 6.3 volts.

Service procedures for the new ECM allow repair and reprogramming by replacing several different integrated circuits in the computer. All C-4 and C-3 systems have self-diagnostic capabilities. The newer the system, the more comprehensive the diagnostic capabilities are. Several GM cars provide diagnostic readouts accessible through instrument panel displays.

American Motors

AMC uses the GM C-4 and C-3 systems on its Pontiac-built, 4-cylinder engines and its Chevrolet-built V-6's. AMC also has used a computerized emission control (CEC) system on 4- and 6-cylinder engines of its own manufacture since 1980.

AMC-Renault uses a fully integrated electronic engine control system made by Renix, a partnership between Renault and Bendix. The Renix system, figure 14-38, uses an array of sensors to control fuel metering, ignition timing, automatic transmission shifting, and emission controls.

Control System Development

As you have learned in this chapter, engine systems have developed considerably since they first appeared a decade ago. Today's computer control systems, however, are undergoing an even more rapid development into a highly sophisticated electronic system composed of many microprocessors (computers) that will eventually manage all operational and convenience systems in a vehicle. The body computer module (BCM) concept which appeared on some 1986 GM luxury cars is the first generation of these "total control" systems, and their development should be more rapid than the progress we've seen with engine control systems.

When a particular vehicle arrives for service, you will have to know the system's components, understand its operation, and use the correct procedures and specifications. As described in Chapter 1 of the *Shop Manual*, this information is available in service manuals available from the carmaker or independent publishers such as Chek-Chart. It is impossible for the individual technicians to carry all of this information in their heads.

However, you should not forget that the laws of electricity, physics, and chemistry have no manufacturer's trademark on them. They remain the same, and all of the systems we have studied operate on the same principles.

SUMMARY

All computers must perform four basic functions: input, processing, storage, and output. Engine control computers use various sensors to receive input data. This data is compared to lookup tables in the computer's memory. Some data may be stored in memory for future use. The computer output takes the form of voltage signals to its actuators.

A control system operates in an open loop-mode until the EGO sensor is warm enough, then the computer switches into the closed-loop mode. In open loop, the computer ignores feedback signals and functions with a predetermined set of values. Once the system switches into closed loop, the computer acts on the feedback signals and is constantly "retuning" the engine while it is running. The most recent computers have the ability to adapt their operating strategies to account for a number of conditional changes, including the wear that results from engine operation.

System sensors generally measure analog variables. Their voltage signals are digitized by the computer, which compares the signals to its program and sends an output signal to the actuators. Actuators change the computer voltage signal into electromechanical motion. Sensors are generally switches, resistors, transformers, or generators. Actuators are usually solenoids and stepper motors.

Computer-controlled engine systems began as a way of managing fuel metering for better mileage and emission control. Manufacturers used electronic ignitions with the fuel metering systems to form the basis for an engine management system. Chrysler's lean-burn, the Bosch DME system, Ford's EEC-I, and GM's C-4 system are examples of early engine control systems. The most recent systems, such as GM's CCC and Ford's EEC-IV systems, have self-diagnostic capabilities, control many more functions, and are far more efficient and powerful than their predecessors.

Review Questions

Choose the single most correct answer.
Compare your answers with the correct answers on page 414.

1. The first domestic engine control system was Chrysler's Electronic Lean-Burn. It controls:
   a. Ignition timing
   b. Fuel metering
   c. Both a and b
   d. Neither a nor b

2. A recent Ford engine control system is called:
   a. Electronic Feedback Fuel Control (EFC)
   b. Microprocessor Control Unit (MCU)
   c. Electronic Engine Control-IV (EEC-IV)
   d. Computerized Emission Control (CEC)

3. NO_x emissions may be reduced by:
   a. Enriching the fuel mixture
   b. Lowering the engine's compression ratio
   c. Recirculating exhaust gases
   d. Retarding spark timing

4. An engine coolant temperature sensor:
   a. Receives reference voltage from the computer
   b. Is a potentiometer
   c. Provides the computer with a digital signal
   d. Contains a piezoelectric crystal

5. Mechanic A says that an exhaust gas oxygen (EGO) sensor with two wires is grounded to the exhaust manifold.
   Mechanic B says that the computer ignores an EGO sensor in closed-loop operation.
   Who is right?
   a. A only
   b. B only
   c. Both A and B
   d. Neither A nor B

6. Mechanic A says that a throttle position sensor (TPS) may be either a rotary or a linear potentiometer.
   Mechanic B says that a TPS can be a simple on/off switch.
   Who is right?
   a. A only
   b. B only
   c. Both A and B
   d. Neither A nor B

7. Mechanic A says that the output signal from one computer can act as an input signal for another computer.
   Mechanic B says that late-model computers have self-diagnostic capabilities and can display trouble codes.
   Who is right?
   a. A only
   b. B only
   c. Both A and B
   d. Neither A nor B

8. Mechanic A says that excessive exhaust gas recirculation will increase HC emissions and fuel consumption.
   Mechanic B says that emissions are low and fuel consumption is high when timing is advanced.
   Who is right?
   a. A only
   b. B only
   c. Both A and B
   d. Neither A nor B

9. The first electrically controlled fuel metering system with a 3-way converter and EGO sensor was the:
   a. C-4 system
   b. Lambda-Sond system
   c. Ford FEEC system
   d. None of these

10. A GM "minimum function" system:
    a. May use a coolant temperature switch
    b. Has long-term memory
    c. Provides fixed dwell
    d. Controls only ignition timing

11. Which of the following is *not* true of a fuel-injected engine?
    a. The computer controls the air-fuel ratio by switching injectors on or off
    b. The pulse width is increased to supply more fuel
    c. To lean the mixture, the computer opens the air bleeds
    d. Engine speed determines the injector switching rate

12. What is the most efficient way to reduce NO_x without adversely affecting fuel economy, driveability, and HC emissions?
    a. Engine mapping
    b. Adaptive memory
    c. Spark timing delay
    d. Exhaust gas recirculation

13. The first GM ignition timing control was called:
    a. MISAR
    b. HEI
    c. EST
    d. TWC

14. Ford's EEC-II added what element to the earlier spark timing, EGR flow, and Thermactor air injection?
    a. Feedback carburetor
    b. Electronic controls for vapor canister purging
    c. Central fuel injection with TBI
    d. Electronic fault codes

15. A self-test program was part of which version of Ford's electronic engine controls?
    a. EEC-I
    b. EEC-II
    c. EEC-III
    d. EEC-IV

16. Which of the following is *not* true of a late-model full-function control system?
    a. The computer controls timing electronically
    b. The computer changes the air-fuel ratio within the range of 14:1 to 17:1
    c. Fuel metering is controlled with a carburetor MC solenoid or by pulsing fuel injectors
    d. The car can be driven in a limited operational strategy

Chapter

15

Gasoline Fuel Injection Systems

A carburetor is a mechanical device that is neither totally accurate nor particularly fast in responding to changing engine requirements. Adding electronic feedback mixture control improves a carburetor's fuel metering capabilities under some circumstances, but most of the work is still done mechanically by the many jets, passages, and air bleeds. Adding feedback controls and other emission-related devices in recent years has resulted in very complex carburetors that are extremely expensive to repair or replace.

The intake manifold also is a mechanical device that, when teamed with a carburetor, results in less than ideal air-fuel control. Because of a carburetor's limitations, the manifold must locate it centrally over (V-engines), or next to (inline engines), the intake ports while remaining within the space limitations under the hood. The manifold runners have to be kept as short as possible to minimize fuel delivery lag, and there cannot be any low points where fuel might puddle. These restrictions severely limit the amount of manifold tuning possible, and even the best designs still have problems with fuel condensing on cold manifold walls.

The solution to the problems posed by a carbureted fuel system is **electronic fuel injection (EFI)**. EFI provides precise mixture control over all speed ranges and under all operating conditions. Its fuel delivery components are simpler and often less expensive than a feedback carburetor. Some designs allow a wider range of manifold designs. Equally important is that EFI offers the potential of highly reliable electronic control.

In this chapter, we will discuss:

● The advantages of fuel injection over carburetion
● The differences among various fuel injection systems
● The fundamentals of electronic fuel injection
● The subsystems and components of typical fuel injection systems now in use.

FUEL INJECTION OPERATING REQUIREMENTS

The major difference between a carbureted and an injected fuel system is the method of fuel delivery. In a carbureted system, the carburetor mixes air and fuel. In a fuel injection system, one or more injectors meter the fuel into the intake air stream.

An electronic fuel injection system uses the same principle of pressure differential used in a carbureted system, but in a slightly different way. As you learned in Chapter 8, it is the difference in pressure between the inside and the

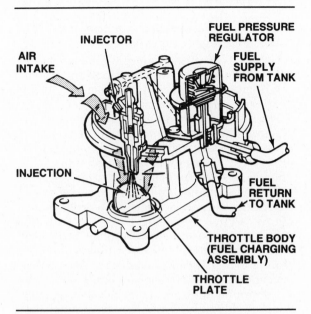

Figure 15-1. Air and fuel combine in a TBI unit at about the same point as in a carburetor. (Ford)

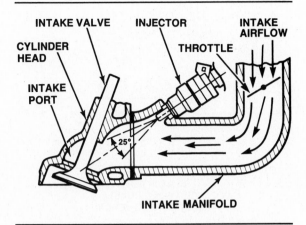

Figure 15-2. In a multipoint system, the air and fuel combine at the intake valve. (GM)

outside of the engine that forces fuel out of the carburetor fuel bowl. In EFI systems, the airflow or air pressure sensor determines the pressure difference and informs the computer. The computer evaluates the air sensor's input along with that of other sensors to decide how much fuel is required for engine operation, then controls injector operation to provide the correct fuel quantity. The computer makes it possible for the fuel injectors to do all the functions of the main systems used in a carburetor. However, throttle response in an injected system is more rapid than in a carbureted system because the fuel is under pressure at all times. The computer calculates the changing pressure and opens the injector much quicker than a carburetor can react under similar conditions.

Carburetors must break up liquid gasoline into a fine mist, change the liquid into a vapor, and distribute the vapor evenly to the cylinders. Fuel injectors do the same. A fuel injector delivers atomized fuel into the air stream where it is instantly vaporized. In **throttle body injection (TBI)** systems, this occurs above the throttle at about the same point as in a carbureted system, figure 15-1. With **multipoint injection** systems, the injectors are mounted in the intake manifold near the cylinder head and injection occurs as close as possible to the intake valve, figure 15-2. We will look at both systems in greater detail later in the chapter.

To review basic engine air-fuel requirements:
1. When an engine starts, low airflow and manifold vacuum, combined with a cold engine and poor fuel vaporization, require a rich air-fuel ratio.
2. At idle, low airflow and high manifold vacuum, combined with low carburetor vacuum and poor vaporization, still require a slightly rich air-fuel ratio.
3. At low speed, the air-fuel ratio becomes progressively leaner as engine speed, airflow, and carburetor vacuum increase.
4. At cruising speed, air-fuel ratios become the leanest for best economy with light load and high, constant vacuum and airflow.
5. For extra power such as acceleration or heavy load operation, the engine needs a richer air-fuel ratio. This requirement is combined

Electronic Fuel Injection (EFI): A computer-controlled fuel injection system which gives precise mixture control and almost instant response to all operating conditions at all speed ranges.

Throttle Body Injection (TBI): A fuel injection system in which one or two injectors are installed in a carburetor-like throttle body mounted on a conventional intake manifold. Fuel is sprayed at a constant pressure above the throttle plate to mix with the incoming air charge.

Multipoint (Port) Injection: A fuel injection system in which individual injectors are installed in the intake manifold at a point close to the intake valve. Air passing through the manifold mixes with the injector spray just as the intake valve opens.

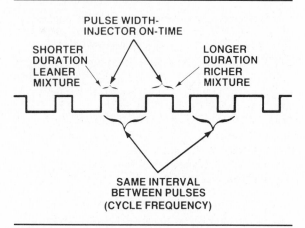

Figure 15-3. Fuel injector pulse width determines the air-fuel ratio.

with low vacuum and airflow on acceleration or with low vacuum and high airflow at wide-open throttle.

A carburetor satisfies all of these requirements as its systems respond to changes in air pressure and airflow. A fuel injection system does exactly the same things. All of the sensors and other devices in an injection system respond to the same operating conditions that a carburetor does, but an injection system responds faster and more precisely for a better combination of performance, economy, and emission control.

ADVANTAGES OF FUEL INJECTION

Fuel injection is not a new development. The first gasoline fuel injection systems can be traced back to 1912 and the Robert Bosch Company in Germany. Bosch's early work in fuel injection was restricted to aircraft applications, but it was transferred to automotive use in the 1930's. A decade later, two Americans named Hilborn and Enderle developed injection systems for use on racing engines. Chevrolet, Pontiac, and Chrysler offered mechanical fuel injection systems during the late 1950's and early 1960's. Some imported vehicles have had fuel injection since 1968. Among domestic manufacturers, Cadillac made electronic fuel injection standard on its 1976 Seville model and other automakers soon followed suit.

What really made fuel injection a practical alternative to the carburetor was the development of reliable solid-state components during the 1970's. Automakers were quick to apply these advances in electronics to fuel injection. The result was a more efficient and dependable system of fuel delivery.

Electronic fuel injection systems offer several major advantages over carburetors:

- Injectors can precisely match fuel delivery to engine requirements under all load and speed conditions. This reduces fuel consumption with no loss of engine performance.
- Since intake air and fuel are mixed at the engine port, keeping a uniform mixture temperature is not as difficult with multipoint fuel injection systems as it is with carburetors. There is no need for manifold heat valves.
- The manifold in a multipoint injection system carries only air, so there is no problem of the air and fuel separating.
- Exhaust emissions are lowered by maintaining a precise air-fuel ratio according to engine requirements. The improved air-fuel flow in an injection system also helps to reduce emissions.
- Continuing improvements in electronic fuel injection design have allowed some automakers to eliminate other emission control systems such as heated intake air and air pumps.

AIR-FUEL MIXTURE CONTROL

Almost all of the electronic fuel injection systems used on domestic vehicles in recent years are integrated into complete electronic engine control systems with exhaust gas oxygen (EGO) feedback mixture control. The electronic controls of a fuel-injected engine are similar to those of an engine with a feedback carburetor. The main difference is that instead of regulating a vacuum solenoid, a stepper motor, or a mixture control solenoid, the computer switches one or more solenoid-operated fuel injectors on and off. Engine rpm determines the rate of switching, and the computer varies the length of time the injectors remain open (pulse width) to establish the air-fuel ratio, figure 15-3. Injector pulse width and duty cycle are explained in more detail later in this chapter.

This variable pulse width takes the place of all the different carburetor metering circuits. As the computer receives data from its inputs, it can lengthen the pulse width to supply additional fuel for cold running (choke), heavy loads (power enrichment), fast acceleration (accelerator pump), or several other situations. Similarly, it can shorten the pulse width to lean the mixture at idle (idle circuit), under cruise (main circuit), or during deceleration. On such systems, the EGO sensor provides fine tuning of the mixture under most operating conditions.

By weight, fuel makes up only about one-fifteenth of the air-fuel mixture. The computer is mainly concerned with regulating the fuel delivery, but to do this accurately, it must know the amount of air entering the engine. Because airflow and air pressure cannot act directly on the fuel as occurs in a carburetor, fuel injection systems use one of three basic kinds of intake air sensors:

1. A manifold pressure sensor
2. An airflow sensor
3. An air mass sensor.

Manifold pressure, airflow velocity, and the molecular mass of the intake air are all directly related to the weight of intake air. (Remember that air-fuel ratios are expressions of air and fuel weights by volume.) The computer has a set of values (similar to a graph) stored in memory. These values relate manifold pressure, airflow, or air mass to engine speed and the required fuel entering for any combination of conditions. Once the computer knows the amount of air entering the engine, as well as other conditions, it adjusts injector pulse width to achieve the required air-fuel ratio.

Some injection systems use combinations of the three basic kinds of sensors listed above. You will learn more about these sensors later in this chapter's section on air control systems.

TYPES OF FUEL INJECTION SYSTEMS

There are three general ways to categorize modern fuel injection systems:
- Mechanical or electronic injection
- Throttle body or multipoint (port or manifold) injection
- Continuous or intermittent injection.

However, the distinctions between the various types of fuel injection systems are not simple and clear-cut. All mechanical systems make use of some electronic components, and electronic systems may share some of the features of multipoint and TBI in a single system. One example is Chrysler's continuous-flow TBI system introduced in 1981 and used exclusively on V-8 engines in the Imperial. Although a technological deadend, it breaks the rules of operation we're about to discuss. It has two injectors mounted in the throttle body. They respond to varying pressure from a control pump and deliver fuel continuously. A unique airflow sensor in the air cleaner inlet measures the volume of air passing into the system. The entire operation is controlled electronically.

Mechanical or Electronic Injection

A mechanical fuel injection system delivers gasoline by using its pressure to open the injector valve. Mechanical injection systems generally are continuously operating. In other words, they inject fuel constantly while the engine is operating. The Bosch K-Jetronic is a typical example of a mechanical injection system.

An electronic fuel injection EFI system generally uses one or more solenoid-operated injectors to spray fuel in timed pulses, either into the intake manifold or near the intake port. EFI systems operate intermittently. All domestic manufacturers use some form of EFI.

Throttle Body or Multipoint Injection

The most common type of EFI system is throttle body injection (TBI), figure 15-4. This design is something of a halfway measure between feedback carburetion and multipoint injection. It generally is classified as a single-point, pulse-time, modulated injection system. TBI injection, as its name implies, has one or two injectors in a carburetor-like TBI assembly that mounts on a traditional intake manifold. Fuel is kept at a constant pressure by a regulator built into the throttle body, figure 15-1.

■ **Why Vapor Lock?**

When gasoline vapors form in the fuel system, vapor lock occurs. This is the partial or complete stoppage of fuel flow to the carburetor. Partial vapor lock will lean the air-fuel mixture and reduce both the top speed and the power of an engine. Complete vapor lock will cause the engine to stall, and make restarting impossible until the fuel system has cooled.

Four factors usually cause vapor lock.

1. High gasoline temperature and pressure in the fuel system.
2. Vapor-forming characteristics of a particular gasoline.
3. The fuel system's inability to minimize vapors.
4. Poor engine operating conditions, such as overheating.

Vapor may form anywhere in the fuel system, but the critical temperature point is the fuel pump.

Engineers have improved fuel pumps and fuel systems to make today's cars less likely to have vapor lock. Oil companies have succeeded in reducing the vapor-locking tendencies of gasoline by adjusting its volatility according to weather requirements. But vapor lock may still occur during long periods of idle (such as in heavy rush-hour traffic) or when the car's fuel system is not properly maintained. Periodically inspect the fuel system and correct all air leaks and defects to prevent vapor lock with today's cars.

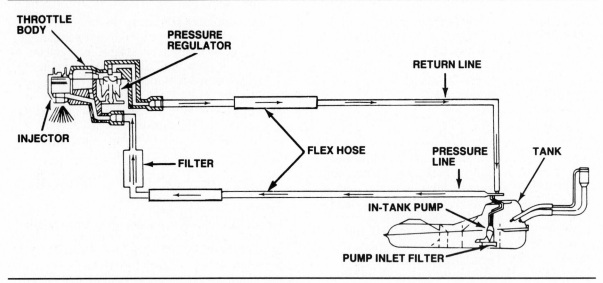

Figure 15-4. A typical throttle body injection system. (GM)

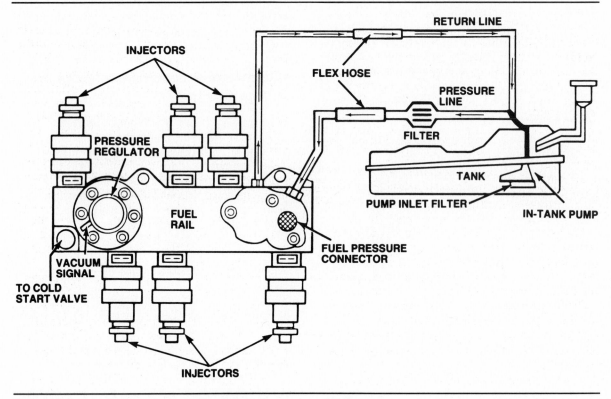

Figure 15-5. A typical 6-cylinder multipoint injection system. (GM)

The computer controls injector pulsing in one of two ways: synchronized or nonsynchronized. If the system uses a synchronized mode, the injector pulses once for each distributor reference pulse. When dual injectors are used in a synchronized system, the injectors pulse alternately. In a nonsynchronized system, the injectors are pulsed once during a given period (which varies according to calibration) completely independent of distributor reference pulses.

A TBI system has certain advantages: it provides improved fuel metering over a carburetor, it is easier to service, and it is less expensive to

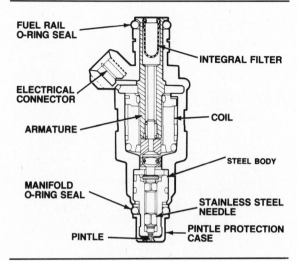

Figure 15-6. A multipoint (port) electronic fuel injector.

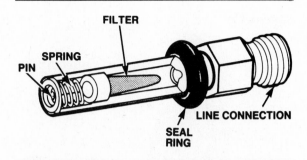

Figure 15-7. A Bosch K-Jetronic mechanical fuel injector. (Bosch)

manufacture. Its disadvantages are primarily related to the manifold: fuel distribution is unequal and a cold manifold still causes fuel to condense and puddle. To compensate for this placement, some systems use two differently calibrated injectors. This results in a different amount of fuel being sprayed by each injector. Also, a TBI unit, like a carburetor, must be mounted above the combustion chamber level. This generally prevents the use of tuned intake manifold designs.

Multipoint, or port, injection, figure 15-5, is older than TBI and because of its many advantages, will probably be the system of choice for all but economy-class vehicles in the future. Multipoint systems have one injector for each engine cylinder. The injectors are mounted in the intake manifold near the cylinder head where they can inject fuel as close as possible to the intake valve, figure 15-2.

The advantages of this design also are related to characteristics of intake manifolds:
● Fuel distribution is equal to all cylinders because each cylinder has its own injector, figure 15-6.
● The fuel is injected almost directly into the combustion chamber, so there is no chance for it to condense on the walls of a cold intake manifold.
● Because the manifold does not have to carry fuel or properly position a carburetor or TBI unit, it can be shaped and sized to tune the intake airflow to achieve specific engine performance characteristics.

The primary disadvantage of multipoint injection is the higher cost of individual injectors and other parts, as well as the computer software to control their operation.

Continuous or Intermittent Injection

Continuous injection systems constantly inject fuel whenever the engine is running. The most common example used on production vehicles is the Bosch K-Jetronic mechanical continuous injection system. (Many carmakers refer to K-Jetronic simply as CIS.)

The individual injectors, figure 15-7, operate on the opposing forces of fuel pressure and the spring-loaded valve in the injector tip. When inlet fuel pressure reaches about 45 psi (310 kPa), it overcomes spring pressure and forces the injector open. Each injector then delivers fuel continuously to the port near the intake valve. Fuel collects or "waits" at the valve to mix with incoming air. When the intake valve opens, the air-fuel mixture enters the cylinder.

In the K-Jetronic system, fuel pressure controls injector opening and the amount of fuel that is injected. Fuel pressure varies continually during different engine operating conditions. Under heavy load, for example, fuel pressure may reach 70 psi (483 kPa). This forces the injector farther open to admit more fuel.

The main fuel distributor and regulator controls fuel pressure in response to airflow and preset regulator pressure. Injector spring force and fuel pressure cause the injector valve to vibrate. This constantly varies the amount of fuel in relation to engine requirements and helps to atomize the fuel as it is injected.

When the engine is shut off, fuel pressure drops below injector tip spring pressure, and the injectors close. Residual fuel pressure is retained in the lines to ensure a ready fuel supply when the engine is restarted.

Continuous Injection System: A fuel injection system in which fuel is injected constantly whenever the engine is running. Bosch K-Jetronic and CIS systems are typical examples.

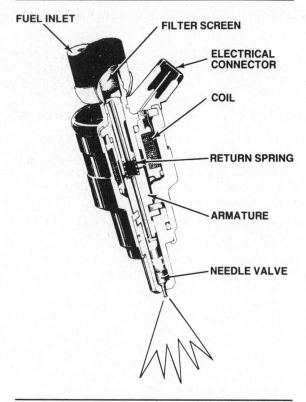

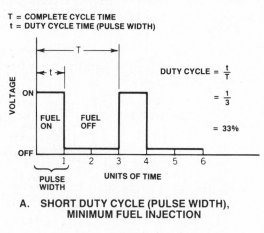

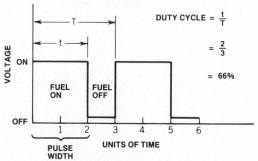

Figure 15-8. Solenoid actions intermittently open the EFI nozzles. (Volvo)

A. SHORT DUTY CYCLE (PULSE WIDTH), MINIMUM FUEL INJECTION

B. LONG DUTY CYCLE (PULSE WIDTH), MAXIMUM FUEL INJECTION

Figure 15-9. Comparison of pulse width and duty cycle for the same cycle time.

Electronic Injection Nozzle

To understand fuel injection as part of a complete engine control system, you must understand the operation and control of electronic injection nozzles. An electronic injection nozzle is simply a specialized solenoid, figure 15-8. It has an armature and a needle or ball valve. A spring holds the needle or ball closed against the valve seat, and the armature opens the valve when it receives a current pulse from the system computer. When the solenoid is energized, it unseats the valve to inject fuel.

The injector always opens the same distance, and the fuel pressure is maintained at a constant value by the pressure regulator. The amount of fuel delivered by the injector depends on the amount of time that the nozzle is open. This is the injector pulse width: the time in milliseconds that the nozzle is open.

The system computer varies pulse width to supply the amount of fuel that an engine needs at a specific moment. A long pulse width delivers more fuel; a short pulse width delivers less fuel.

Injector pulse width relates to another important concept that you will use to test injection systems. This is the injector duty cycle.

The duty cycle relates to any intermittently operating device. Ignition dwell, for example, is really the ignition duty cycle. The complete operating cycle of any solenoid-operated device is the entire time from *on* to *off* to *back on* again. The duty cycle is the percentage of on-time to total cycle time.

A solenoid can operate at any number of cycles per second: 10, 20, 30, 60, or whatever the engineer chooses to design. Each complete cycle lasts the same amount of time, but duty cycle can vary as a percentage of each cycle. Pulse width varies along with the duty cycle because it is the actual *time* that the solenoid is energized. Figure 15-9 shows two different pulse widths and duty cycles for the same complete cycle time. The system computer calculates the necessary pulse width and duty cycle from information provided by system sensors. Modern digital computers operate fast enough to change injector pulse width in fractions of a second to maintain precise fuel metering.

Various Sequence Combinations

Electronic fuel injection systems use a solenoid-operated injector, figure 15-6, to spray atomized fuel in timed pulses (intermittently) into the manifold or near the intake valve. Injectors may be sequenced and fired in one of several ways, as we will see shortly, but their duty cycle and pulse width are determined and controlled by the engine computer.

Multipoint systems contain an injector for each cylinder but they do not all fire the injectors in the same way. Domestic systems use one of four ways to trigger the injectors:

- Grouped single-fire
- Grouped double-fire
- Simultaneous double-fire
- Sequential.

In the first three of these word combinations, the first words refer to the way in which the injectors are connected electrically within the system. Some systems fire, or trigger, the injectors in groups; others fire them all together. The second words (hyphenated) in each combination refer to how many firings of each injector are used to make up the fuel charge for one combustion stroke.

Grouped single-fire

Injectors in this system are split into two groups. The groups are fired alternately with one group firing each engine revolution. Only one injector pulse is used for each combustion stroke. Early Cadillac V-8 systems use this design and while it works reasonably well, it is not as precise as newer designs. Since only two injectors can be fired relatively close to the point where the intake valve is about to open, the fuel charge for the remaining six cylinders must stand in the intake manifold for varying times. Since a new charge is released only once every two crankshaft revolutions, it is necessary to wait this long before any change can be made in the air-fuel mixture.

Grouped double-fire

This system again splits the injectors into two equal-sized groups. The groups fire alternately, but each group fires once each revolution. Two injector pulses make up each intake charge, which means that the air-fuel mixture remains in the manifold for a shorter time and that mixture changes can be made sooner than with a single-fire system.

Simultaneous double-fire

This design fires all of the injectors at the same time once every engine revolution. Many multipoint injection systems on 4-cylinder engines

use this pattern of injector firing. It is easier for engineers to program and it can make relatively quick adjustments in the air-fuel ratio, but it still requires the intake charge to wait in the manifold for varying lengths of time.

Sequential

Sequential firing of the injectors according to engine firing order is the most accurate and desirable method of regulating multipoint injection. However, it is also the most complex and expensive to design and manufacture. In this system, the injectors are controlled individually. Each cylinder receives one charge every two revolutions just before the intake valve opens. This means that the mixture is never static in the intake manifold and mixture adjustments can be made almost instantaneously between the firing of one injector and the next. A camshaft sensor signal or a special distributor reference pulse informs the computer when the No. 1 cylinder is on the compression stroke. If the sensor fails or the distributor reference pulse is interrupted, some injection systems shut down, while others revert to pulsing the injectors simultaneously.

COMMON SUBSYSTEMS AND COMPONENTS

Regardless of the type, all fuel injection systems have three basic subsystems:

- Fuel delivery
- Air control
- Electronic control with auxiliary sensors or actuators.

Fuel Delivery System

The fuel delivery system consists of an electric fuel pump, a filter, a pressure regulator, one or more injectors, and the necessary connecting fuel distribution lines.

Fuel pump

An electric fuel pump provides constant and uniform fuel pressure at the injectors. Most systems use a positive-displacement vane, turbine, or roller pump, figure 15-10. Fuel-injected vehicles generally use one of the following pump applications:

- High pressure, in-tank
- Low pressure, in-tank
- Low pressure, in-tank and high pressure, inline.

In-tank fuel pumps may be separate units, but often are combined with the fuel gauge

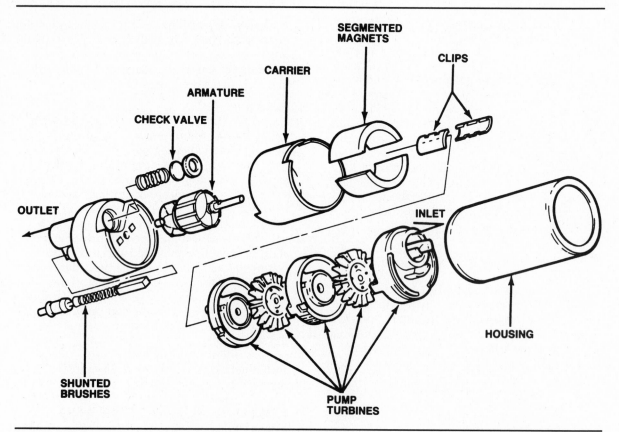

Figure 15-10. An in-tank turbine-type electric pump provides continuous low pressure (above 9 psi or 62 kPa). (GM)

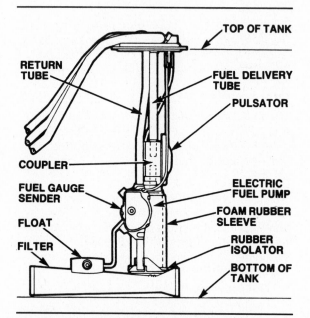

Figure 15-11. This in-tank roller vane fuel pump is combined with the fuel gauge sender unit. (GM)

sender assembly, figure 15-11. The pump supplies more fuel than the system requires through an inline filter to the TBI unit or fuel rail, with the pressure regulator controlling volume and pressure. An internal relief valve protects the pump from excessive pressure if the filter or fuel lines become restricted.

Fuel filter
Clean fuel is extremely important in a fuel injection system because of the small orifice in the injector tip through which fuel must pass. Most injection systems use three forms of filters. The first is a fuel strainer or filter of woven plastic attached to the fuel pump inlet, figure 15-11. This prevents contamination from entering the fuel line and separates water from the fuel. The filter is self-cleaning and requires no maintenance. If a fuel restriction occurs at this point, the tank contains too much sediment or moisture and should be removed and cleaned.

Additional filtration is provided by a high-capacity inline filter, figure 15-12, to remove contamination larger than 10 to 20 microns (0.0025 to 0.0050 inch). This filter generally is mounted on a frame rail under the vehicle, but sometimes it may be mounted in the engine compartment.

A — CLAMP-TYPE

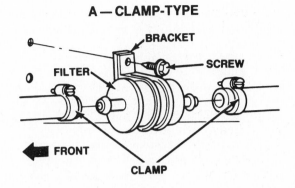

B — O-RING TYPE

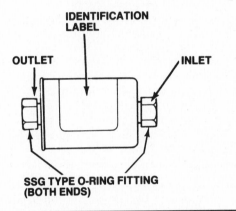

Figure 15-12. Fuel injection systems use large disposable filter canisters to ensure proper filtration. When threaded connections are used instead of clamps, O-ring seals are required. (GM)

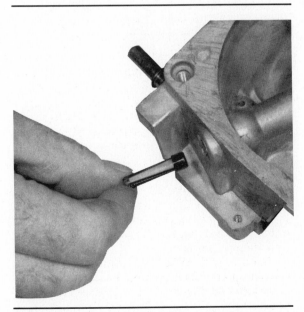

Figure 15-13. A filter screen may be installed in the fuel inlet of some throttle bodies.

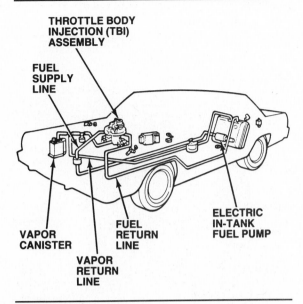

Figure 15-14. A typical fuel injection system showing fuel line routing. (Ford)

The final line of defense against fuel contamination is in the fuel injector or TBI unit itself. Each injector contains its own inlet filter screen, figure 15-6, to prevent contamination from reaching the tip. Some TBI units have a similar filter screen installed in the fuel inlet fitting, figure 15-13.

Fuel lines
A supply line carries fuel from the pump through the filter to the TBI unit or fuel rail. A return line carries excess fuel back to the tank. This allows fuel to be continuously circulated from the tank to the injectors and back to the tank, figure 15-14. The system maintains constant fuel pressure and volume while minimizing fuel heating and vapor lock. Most injection system fuel lines are made of steel tubing, although Ford vehicles use nylon fuel line tubing. Flexible hose may be used in some low-pressure return lines.

Pressure regulator
The pressure regulator and fuel pump work together to maintain the required constant pressure at the injector tips. The regulator consists of a spring-loaded diaphragm-operated valve in a metal housing. Two types of pressure regulators are used in injection systems.

All fuel pressure regulators on current EFI systems are installed on the return (downstream) side of the injectors. Downstream regulation minimizes fuel pressure pulsations caused by pressure drop across the injectors as the nozzles open. It also ensures positive fuel

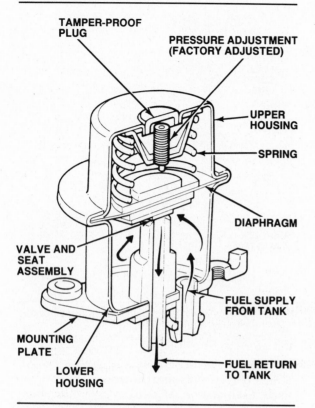

TAMPER-PROOF PLUG

PRESSURE ADJUSTMENT (FACTORY ADJUSTED)

UPPER HOUSING

SPRING

DIAPHRAGM

VALVE AND SEAT ASSEMBLY

MOUNTING PLATE

LOWER HOUSING

FUEL SUPPLY FROM TANK

FUEL RETURN TO TANK

Figure 15-15. The pressure regulator commonly used with a TBI system functions by fuel pressure alone. (Ford)

pressure at the injectors at all times and holds residual pressure in the lines when the engine is off.

The pressure regulator used with a TBI system is built into or on the TBI unit. With this type, figure 15-15, fuel pressure must overcome spring pressure on the spring-loaded diaphragm to uncover the return line to the tank. This happens when system pressure exceeds operating requirements. The spring side of the diaphragm is exposed to inlet air pressure, or manifold vacuum, inside the TBI unit. Because it is close to the injector tip, the regulator senses essentially the same air pressure as the injector.

The pressure regulator used in a multipoint injection system, figure 15-16, has an intake manifold connection on the regulator vacuum chamber. This allows fuel pressure to be modulated by a combination of spring pressure and manifold vacuum acting on the diaphragm.

In both systems, the regulator shuts off the return line when the fuel pump is not running. This maintains pressure at the injectors for easy restarting and reduced vapor lock.

Mechanical injection systems without electronic control operate on the differential between fuel pump pressure and the control pressure developed by the fuel distributor. The

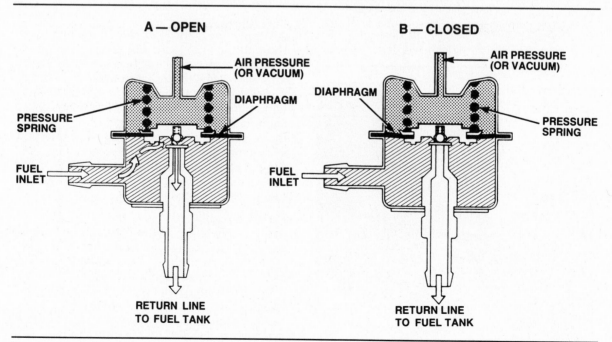

A — OPEN

B — CLOSED

AIR PRESSURE (OR VACUUM)

DIAPHRAGM

PRESSURE SPRING

FUEL INLET

RETURN LINE TO FUEL TANK

Figure 15-16. The pressure regulator commonly used with a multipoint system works on a combination of fuel pressure and manifold vacuum or air pressure. (GM)

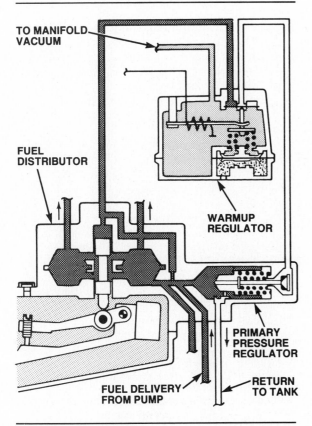

Figure 15-17. Injection pressure is controlled in a Bosch K-Jetronic system by a main pressure regulator and a separate warmup regulator. (Bosch)

pressure regulator is located on the control pressure side of the fuel distributor and connected to the fuel return line, figure 15-17. Spring force and control pressure operate on one side of the regulator with pump pressure on the other side to modulate the return line opening. Bosch K-Jetronic systems vary control pressure for different operation requirements with a primary regulator and a warmup regulator.

In later mechanical injection system designs, such as the CIS electronic system used on the Volkswagen Golf, a differential pressure regulator replaces the control pressure regulator. This device is a part of the fuel distributor, figure 15-18, and is operated by the computer. A separate spring-loaded, diaphragm-operated pressure regulator governs system pressure. This eliminates the need for a pressure relief valve in the fuel distributor.

Multipoint injection systems generally operate with pressures at the injector of about 30 to 55 psi (207 to 379 kPa), while TBI systems work with injector pressures of about 10 to 20 psi (69 to 138 kPa). The difference in system pressures results from the difference in the systems operation. Remember that an injection system requires only enough pressure to move the fuel through the injector and help in atomizing it.

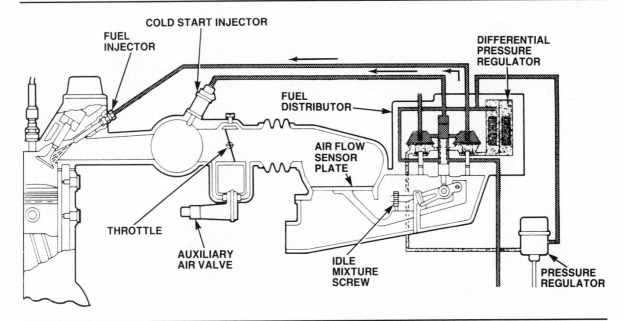

Figure 15-18. A differential pressure regulator is used instead of a control pressure regulator on some CIS late-model electronic systems. (Volkswagen)

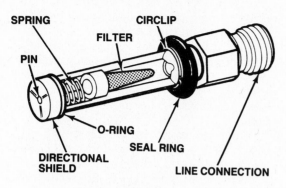

A — CUTAWAY VIEW

SPRING
CIRCLIP
FILTER
PIN
O-RING
SEAL RING
DIRECTIONAL SHIELD
LINE CONNECTION

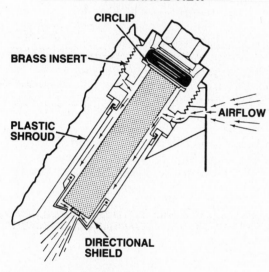

B — EXTERNAL VIEW

CIRCLIP
BRASS INSERT
PLASTIC SHROUD
AIRFLOW
DIRECTIONAL SHIELD

Figure 15-19. Air-shrouded injectors are similar to the K-Jetronic design, but flow air past the nozzle to improve atomization. (Volkswagen)

Since injectors in a TBI system inject the fuel into the airflow at the manifold inlet (above the throttle), there is more time for atomization in the manifold before the air-fuel charge reaches the intake valve. This allows TBI injectors to work at lower pressures than injectors used in a multipoint system.

Mechanical injection nozzles

The most common type of mechanical injection is the Bosch K-Jetronic system. Two types of injectors are used. Non-electronic systems use an injector with a spring-loaded tip to seal the nozzle against fuel pressure, figure 15-7. When pressure is greater than a specified level, it forces the nozzle open. The amount of fuel that enters the manifold is controlled by the nozzle

opening, which in turn depends on fuel pressure. The mixture control unit controls injection pressure by a combination of control pressure and pump pressure. K-Jetronic nozzles inject fuel constantly as long as the pump is operating. Bosch K-Jetronic nozzles are described in detail in the Continuous Injection section, earlier in this chapter.

Electronically controlled K-Jetronic systems use an air-shrouded injector. These injectors work in the same way as those used with non-electronic systems, but design changes have made them more efficient. Air flows in through a cylinder head passage, passes between the injector and plastic shroud surrounding it, and exits close to the injector tip. This improves atomization of the fuel as it leaves the injector, which reduces fuel condensation in the manifold. A second-generation air-shrouded injector, figure 15-19, uses a circlip to retain the seal ring. A separate plastic injector shroud and fluted air directional shield on the injector tip improve airflow around the injector tip. These changes further improve fuel atomization.

Electronic injection nozzles

EFI systems use a solenoid-operated injector. Figure 15-20 shows typical TBI injectors; figure 15-21 shows the two types of multipoint injectors now in use. This electromagnetic device contains an armature and a spring-loaded needle valve or ball valve assembly. When the computer energizes the solenoid, voltage is applied to the solenoid coil until the current reaches a specified reference level (usually about 4 amperes). This permits a quick pull-in of the armature during turn-on. The armature is pulled off its seat against spring force, allowing fuel to flow through a 10-micron screen to the spray nozzle, where it is sprayed in a conical pattern. When current reaches the reference level, it is regulated at a specified value (usually about 1 ampere) until the injector is turned off. The low energy level during the holding state prevents overheating of the solenoid coil. The injector opens the same amount each time it is energized, so the amount of fuel injected depends on the length of time the injector remains open. The injector duty cycle and the operating pulse width are explained earlier in this chapter.

Most injectors are manufactured by Bosch (or made under Bosch license) and use a needle valve or pintle design, figure 15-21. This design allows deposits to gather on the sides of the pintle, causing the injector to malfunction. To overcome this problem, Rochester Products introduced the Multec injector for both TBI and multipoint systems on some 1987 automobiles.

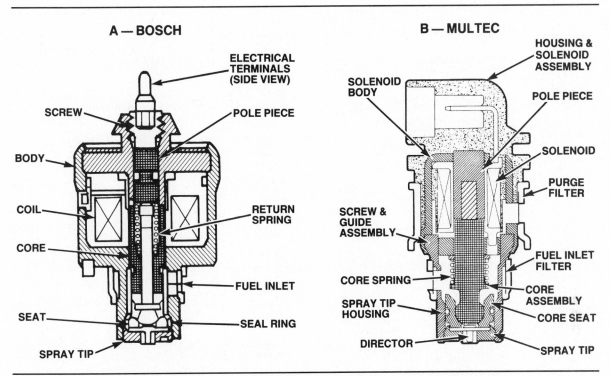

Figure 15-20. Bosch (left) and Rochester Multec (right) injectors used in a TBI unit. (GM)

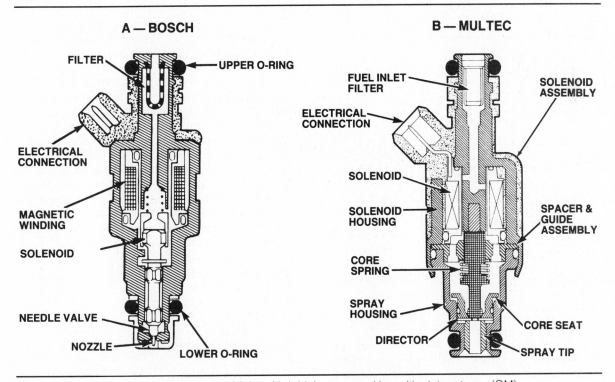

Figure 15-21. Bosch (left) and Rochester Multec (right) injectors used in multipoint systems. (GM)

The Multec injector design uses a stainless steel ball and seat valve with a director plate for fuel control. The director plate gives a more precise spray pattern. Coupled with the near-mirror finish on the ball and seat assembly, Multec injectors are not as prone to clogging as the pintle valve design.

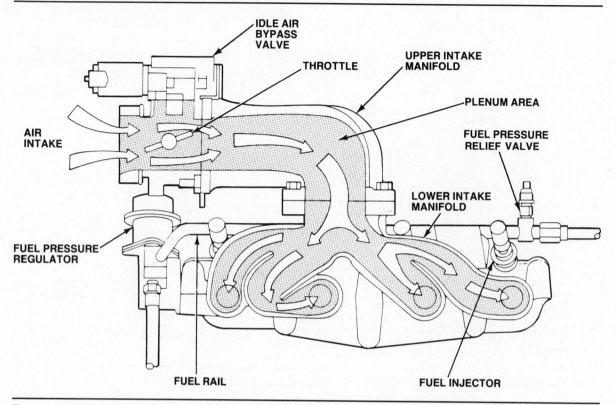

IDLE AIR
BYPASS
VALVE

THROTTLE

UPPER INTAKE
MANIFOLD

PLENUM AREA

AIR
INTAKE

FUEL PRESSURE
RELIEF VALVE

LOWER INTAKE
MANIFOLD

FUEL PRESSURE
REGULATOR

FUEL RAIL

FUEL INJECTOR

Figure 15-22. The horizontal throttle used in a multipoint system controls airflow only. (Ford)

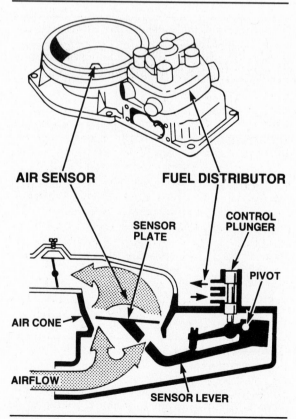

AIR SENSOR

FUEL DISTRIBUTOR

SENSOR
PLATE

CONTROL
PLUNGER

PIVOT

AIR CONE

AIRFLOW

SENSOR LEVER

Figure 15-23. The Bosch K-Jetronic airflow sensor. (Bosch)

Air Control System

As you learned in Chapter 8, the difference between fuel bowl air pressure and carburetor barrel air pressure controls fuel metering in a carbureted system. A fuel injection system needs the same kind of control. However, since the fuel is not introduced to the airflow until it passes through the injectors, there must be some way to measure intake air volume and meter the fuel accordingly. All fuel injection systems use a throttle to control the volume of air. They also must measure air volume with one of the following:

- An airflow meter or sensor
- A manifold or barometric pressure sensor
- An air mass sensor.

Throttle

The throttle works exactly the same in both a carbureted and a fuel-injected system. It is connected to the accelerator linkage, and regulates the amount of air taken into the engine. With a TBI system, the throttle is mounted between the injectors and the manifold, figure 15-1. Since the fuel is introduced to the air above the throttle plate, the throttle regulates the amount of air-fuel mixture that enters the manifold. On multipoint systems, the throttle is mounted horizontally in the air intake throttle body on

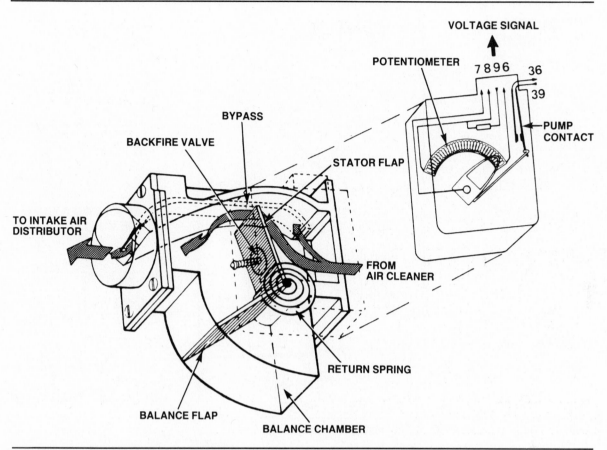

Figure 15-24. The Bosch L-Jetronic and many other airflow systems use a movable vane sensor connected to a potentiometer. (Bosch)

the air inlet side of the injectors and controls only airflow, figure 15-22. A sensor determines the intake air volume. It can do this in one of three ways: by measuring airflow speed, manifold pressure, or the molecular mass of the air drawn into the engine. All of these methods directly relates to the amount of fuel required from the injectors.

Airflow sensors

Multipoint injection systems that use mass airflow for fuel adjustment have a movable plate or vane called an airflow sensor which is deflected by intake airflow, figure 15-23. In a Bosch K-Jetronic system, the movement of the sensor plate moves two counterweighted levers which operate the fuel distributor control plunger. The sensor plate deflects the control plunger proportionate to the amount of air flowing through the system. This causes the control plunger to regulate fuel pump pressure to each cylinder's differential pressure valve. Control pressure exerted by the fuel on the opposite end of the plunger acts to balance airflow sensor force. The combined pressures found at

the differential pressure valve will control injection relative to airflow.

The airflow sensor used in Bosch L-Jetronic, Ford, and most Japanese electronic multipoint systems is a movable vane connected to a potentiometer, figure 15-24. The vane is mounted on a pivot pin and is deflected by intake airflow proportionate to air velocity. As the vane moves, it also moves the potentiometer. This causes a change in the reference voltage applied to it by the computer. For example, if the reference voltage is 5 volts, the potentiometer's return signal to the computer will vary from a zero voltage signal (no airflow) to almost a 5-volt signal (maximum airflow), figure 15-25. In this way, the potentiometer provides the information the computer needs to vary the injector pulse width proportionate to airflow.

Manifold and barometric (absolute) pressure sensors

As you have already learned, oxygen content and barometric pressure change with differences in altitude. Because of this, the computer must be able to compensate for changes in the

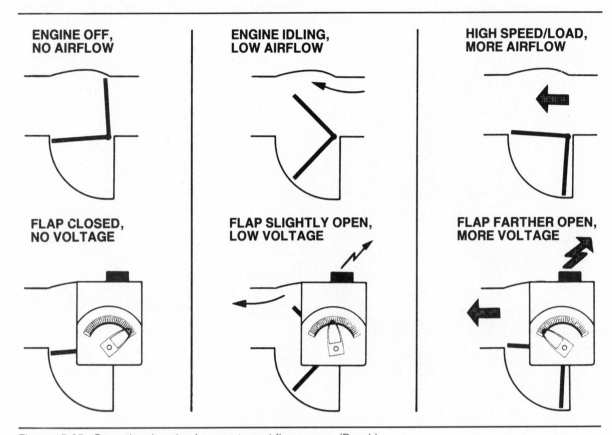

ENGINE OFF, NO AIRFLOW	ENGINE IDLING, LOW AIRFLOW	HIGH SPEED/LOAD, MORE AIRFLOW
FLAP CLOSED, NO VOLTAGE	FLAP SLIGHTLY OPEN, LOW VOLTAGE	FLAP FARTHER OPEN, MORE VOLTAGE

Figure 15-25. Operational cycle of a vane-type airflow sensor. (Bosch)

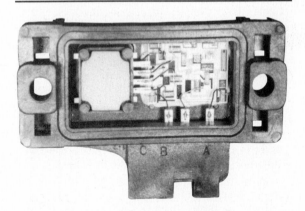

Figure 15-26. The signal conditioning circuitry of a GM MAP sensor.

flow of fuel entering the engine. To provide the computer with this information, a fuel injection system may use one of three kinds of pressure sensors:

● Manifold absolute pressure (MAP)
● Barometric absolute pressure (BAP)
● Barometric and manifold absolute pressure (BMAP)

The MAP sensor may be a ceramic capacitor diaphragm, an aneroid bellows, or a piezo-resistive crystal. It has a sealed vacuum reference input on one side; the other side is connected (vented) to the intake manifold. The sensor housing also contains signal conditioning circuitry, figure 15-26. Pressure changes in the manifold cause the sensor to deflect, varying its return signal to the computer.

A BAP sensor is similar in design, but senses changes in barometric absolute pressure (atmospheric air pressure). It is vented directly to the atmosphere. The BMAP sensor is actually a combination of a BAP and MAP sensor in the same housing. The BMAP sensor has individual circuits to measure barometric and manifold pressure. It sends separate signals to the computer for comparison to determine the operating manifold pressure and required amount of fuel. Pressure sensors generally are remotely mounted, figure 15-27, and connected to the intake manifold by tubing.

Air mass sensor
There are two types of air mass sensors. One contains a platinum wire; the other uses a metal foil sensing element, figure 15-28. Both sensors operate in essentially the same way. A resistor wire or screen is installed in the path of intake airflow and is heated to a constant temperature

Figure 15-27. Many GM MAP sensors are mounted on the side of the air cleaner housing. They also may be mounted on the cowl or fenderwell.

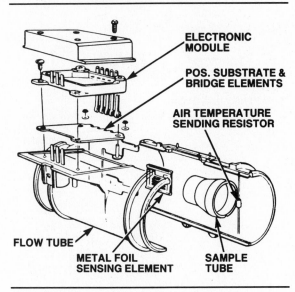

Figure 15-28. Components of a metal foil air mass sensor. (GM)

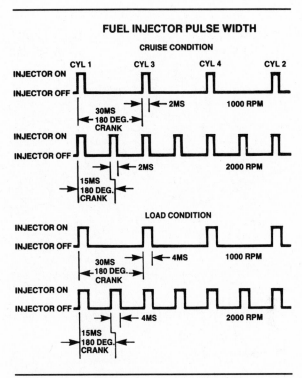

Figure 15-29. Fuel flow timing and pulse width. (Chrysler)

by electric current provided by the computer. Air flowing past the wire or screen cools it. The degree of cooling varies with air velocity and temperature. These factors combine to indicate the mass volume of air entering the engine. As the wire or screen cools, more current is required to maintain the specified temperature. As the wire or screen heats up, less current is required.

The operating principles can be summarized as follows:

● More intake air volume = cooler sensor, more current

● Less intake air volume = warmer sensor, less current

The computer constantly monitors the change in current and translates it into a voltage signal which is used to determine injector pulse width.

Electronic Control System with Auxiliary Sensors and Actuators

You may have noticed that to this point, we have not mentioned chokes, power valves, accelerator pumps, or idle speed and mixture controls in our discussion of fuel injection. Earlier in this chapter, however, you learned that a fuel injection system does the same jobs that a carburetor does. The enrichment and mixture control functions of a carburetor are done by the devices just mentioned. Now you'll learn how a fuel injection system does the same thing.

Computer control

An electronic fuel injection system uses engine sensors and auxiliary metering devices to perform the required enrichment and mixture control functions. Sensors provide input data to the engine computer. After processing the data, the computer sends output signals to actuators. In an EFI system, the actuators are solenoid-operated fuel injectors.

You should remember from the section on electronic fuel injectors that the injector opens the same amount each time it is energized, so the amount of fuel injected depends on the length of time the injector is energized. As you learned earlier this time duration is called pulse width, and is measured in milliseconds. The

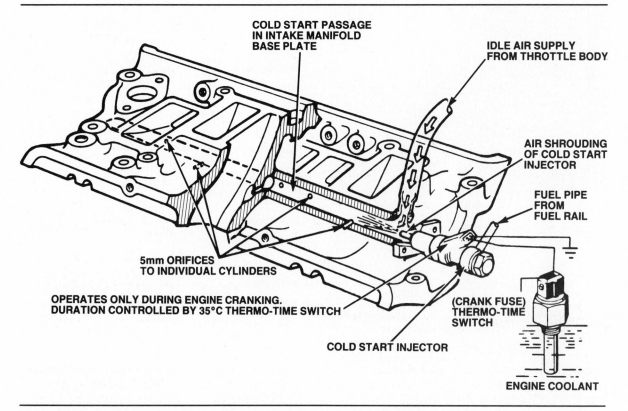

Figure 15-30. A cold-start injector provides extra fuel during cranking. In this design, it mixes with air and flows through small orifices into individual cylinders. (GM)

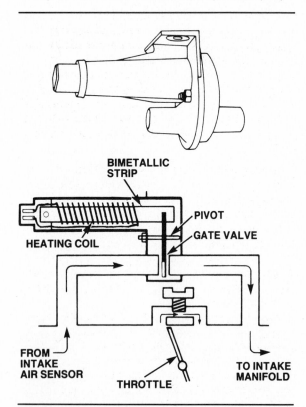

Figure 15-31. The auxiliary air regulator controls the air intake through a gate valve operated by a heated bimetallic strip. (Bosch)

computer varies the injector pulse width according to the amount of fuel required under any operating condition, figure 15-29. Pulse width also may be called duty cycle. Each on/off sequence of an injector is a *full* cycle. The pulse width, or duty cycle, is a variable portion, or percentage, of that full cycle.

Cold starting
A cold engine requires a richer air-fuel mixture for starting. Multipoint systems often use a separate cold-start injector to provide additional fuel during the crank mode. During engine cranking, the individual injectors send fuel into the cylinder ports. At the same time, the cold-start injector provides the additional fuel by injecting it into a central area of the manifold or into a separate passage in the manifold, figure 15-30. In either location, the extra fuel is distributed more or less equally to all cylinders. The colder the engine, the longer the cold-start injector remains on.

Cold-start injector duration is regulated by a thermal-time switch in the thermostat housing. This switch grounds the cold-start injector circuit when the engine is cranked at a coolant temperature below a specified value. A heating element inside the switch also is activated during cranking and starts to heat a bimetallic strip

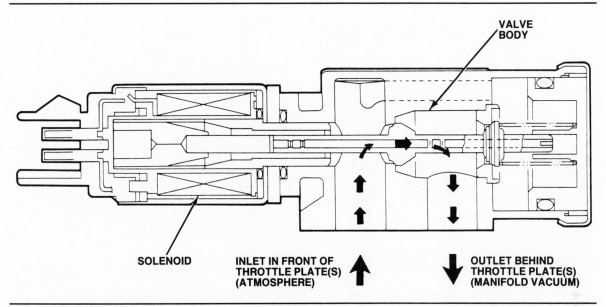

Figure 15-32. A solenoid-operated air bypass valve is used with most domestic multipoint systems to provide idle air flow around the throttle plate. (Ford)

in the ground circuit. Once the bimetallic strip reaches the specified temperature, it opens the ground circuit and shuts off power to the cold-start injector.

Idle control

Multipoint injection systems use an auxiliary air bypass, figure 15-22, or auxiliary air regulator, figure 15-31, to do the same job as the air bypass circuit in a carburetor. This air bypass or regulator provides needed additional airflow by opening an intake air passage to let more air into the engine. The system is calibrated to maintain engine idle speed at a specified value regardless of engine temperature.

Most domestic multipoint injection systems use a computer-controlled, solenoid-operated valve, figure 15-32, to regulate the airflow. The Bosch auxiliary air regulator used in several kinds of systems, figure 15-31, contains a heated bimetallic strip. When the engine is first started, air passes through the gate valve and current is applied to the bimetallic strip. As the engine warms up, the bimetallic strip starts to deflect and gradually closes off the gate valve passage to shut off the additional airflow. By the time the engine has fully warmed up, the gate valve is completely closed and the idle by-pass in the throttle chamber determines engine speed.

Idle airflow in a TBI system travels through a passage around the throttle and is controlled by a stepper motor. Idle fuel metering on a Bosch K-Jetronic system is controlled by a mechanical adjustment. Electronic fuel injection systems may control idle fuel metering either by a pre-determined program in the computer or from signals provided by a manifold pressure sensor.

Acceleration and load enrichment

Airflow and manifold pressure change rapidly when an engine needs more fuel for additional power. With a mechanical injection system, these changes affect fuel pressure, causing the metering assembly to increase the volume of fuel flow. In an electronic injection system, signals from the airflow or pressure sensors to the computer result in an increase in injector pulse width.

Engine speed and crankshaft position

Engine speed is used by mechanical injection systems to operate the cold-start and cranking-enrichment subsystems. Electronic injection systems use engine speed and crankshaft position to time the injection pulses and determine the pulse width. Speed and position signals can be provided either by a magnetic sensor in the engine or by Hall-effect switches or pulse generators in the distributor.

Engine-mounted sensors generally are positioned in the bell housing. They react to the changes in magnetic reluctance caused by flywheel grooves or teeth, figure 15-33. On GM engines with a distributorless ignition system, the sensor is in the engine block and uses a special reluctor or wheel on the crankshaft to

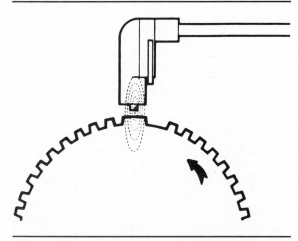

Figure 15-33. A typical crankshaft sensor which reads crankshaft position from the flywheel teeth. (AMC/Renault)

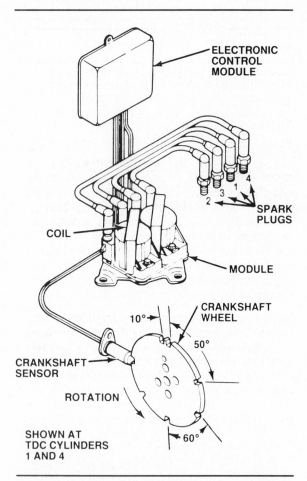

Figure 15-34. Distributorless ignitions use a sensor and notched wheel or reluctor on the crankshaft to inform the computer of crankshaft position. (GM)

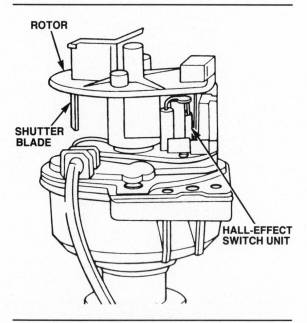

Figure 15-35. A Hall-effect switch in the distributor also can be used to determine crankshaft position. (Chrysler)

generate the signal, figure 15-34. Distributor signals generally are provided by the magnetic pulse generator or Hall-effect switch, figure 15-35, although a separate signal generator or switching device can be used.

As mentioned previously, sequential multi-point systems require a number 1 cylinder signal to time the injector firing order. This signal can be provided by the same sensor that indicates crankshaft position or engine speed. Some engines use a separate sensor for the sequential timing signal, but it works on the same principles as other position or speed sensors. Without the number 1 cylinder signal, a sequential injection system will not run; or it will evert to a programmed pulsed firing sequence that allows the car to be driven in for repair.

Throttle position
A sensor monitors throttle position for the computer. Throttle position sensors (TPS) may be a simple on/off switch, or a potentiometer that sends a variable resistance signal to the computer, figure 15-36. By interpreting the signal, the computer determines when the engine is at:

● Closed throttle (during idle and deceleration)
● part throttle (normal operation)
● Wide-open throttle (acceleration and full-power operation).

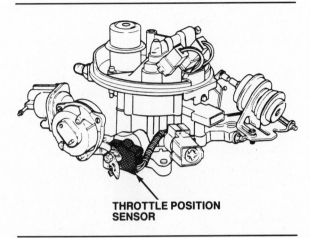

Figure 15-36. A typical throttle position sensor. (Ford)

Proper TPS adjustment is critical to ensure that fuel enrichment and shutoff occur at the appropriate times. Otherwise, driveability problems such as stalling, surging, and hesitation will occur. Some throttle position sensors can be adjusted; others are nonadjustable and must be replaced if out of specifications.

Air and coolant temperature

The computer must be kept aware of intake air temperature and engine coolant temperature. Intake air temperature is used as a density corrector for calculating fuel flow and proportioning cold enrichment fuel flow. Coolant temperature is used to determine the correct air-fuel ratio according to the engine's operating temperature. This information is provided to the computer by coolant and air temperature sensors.

Most temperature sensors are thermistors or variable resistors, figure 15-37. With this type of sensor, resistance decreases as the sensor is heated. The reference voltage applied by the computer is thus altered according to temperature, resulting in a varying return voltage to the computer from the sensor. This data is integrated with data from other sensors by the computer, and used to change the injector pulse width or fuel pressure.

SPECIFIC SYSTEMS

Fuel injection systems have almost replaced the carburetor. The summary descriptions that follow provide the basic features of these systems and can help you to recognize the different devices, and how they do similar tasks.

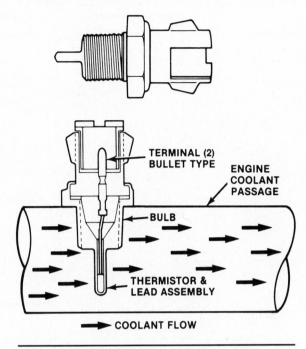

Figure 15-37. Thermistors are used to measure engine coolant or intake air temperatures. (Ford)

Bosch K-Jetronic System

The Bosch K-Jetronic system was used on many European vehicles in the early 1970's and now is the most common example of a production mechanical (continuous) port injection system, figure 15-38. The system also is a continuous injection system (CIS). Air volume is measured by sensing airflow; fuel pressure is regulated relative to air volume. Fuel pressure also activates the individual injectors.

Bosch D-Jetronic System

This electronically controlled injection system, figure 15-39, was used on 1968-73 Volkswagens and other European vehicles. It uses solenoid-operated injectors which operate in intermittent pulses; fuel is measured relative to engine speed and manifold air pressure.

Bosch L-Jetronic System

This system was introduced in 1974 and uses individual solenoid-operated injectors that operate intermittently. It is similar to the mechanical K-Jetronic system in that fuel metering is controlled relative to airflow, but the airflow sensor is connected to an electronic module which also regulates fuel metering relative to

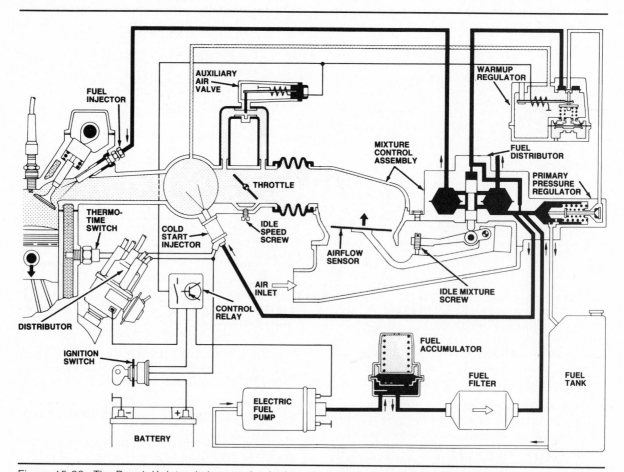

Figure 15-38. The Bosch K-Jetronic is a mechanical continuous injection system (CIS). (Bosch)

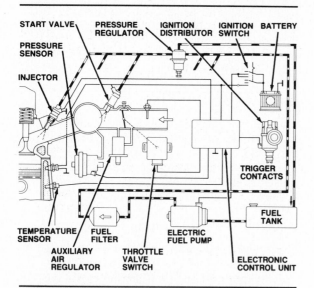

Figure 15-39. An electronic intermittent injection system, the Bosch D-Jetronic is controlled by manifold pressure. (Bosch)

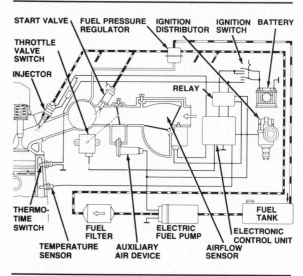

Figure 15-40. The Bosch L-Jetronic is controlled by engine speed and airflow measurement. (Bosch)

temperature and engine speed, figure 15-40. The L-Jetronic system sometimes is called airflow-controlled (AFC) injection.

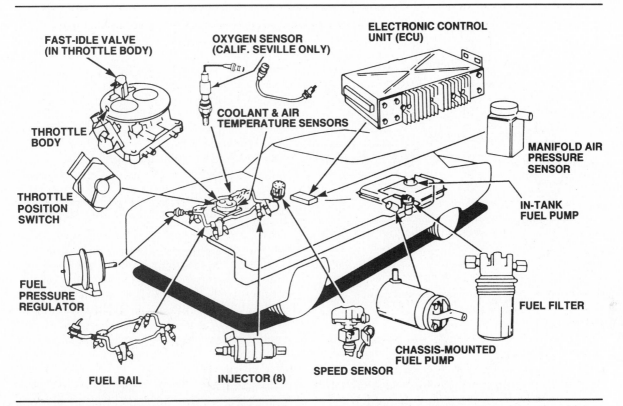

Figure 15-41. The Bendix EFI system was the first modern multi-point injection system used on domestic vehicles. (Cadillac)

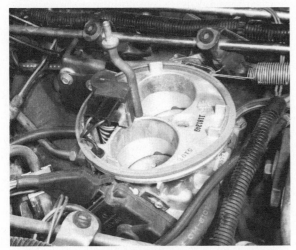

Figure 15-42. The Bendix EFI system on Cadillac Sevilles used a carburetor-like throttle body for air intake and a fuel rail with individual injectors.

Cadillac EFI System

This EFI system by Bendix, figure 15-41, was used on the 1976 + 1979 Cadillac Seville. It was the first electronically controlled multipoint injection system to be used on domestic vehicles. Fuel is supplied to individual injectors in the intake manifold through a fuel rail, with air intake controlled by a throttle body which looks much like a carburetor, figure 15-42. An electronic control unit (ECU) controls fuel delivery according to the various sensor inputs shown in figure 15-41. The injectors were pulsed intermittently in two banks. Feedback control of fuel metering with an EGO sensor was used in later versions of this system.

Late-Model Multipoint Injection Systems

The multipoint injection systems used on domestic and many Japanese vehicles are based on Bosch designs and contain many components manufactured by or under license from Bosch. All have individual injectors installed in some form of fuel rail and mounted in the intake manifold.

General Motors

GM has used three multipoint injection systems since 1984:

1. *Port fuel injection (PFI)* is a simultaneous double-fire system in which all injectors fire once during each engine revolution. In other words, two injections of fuel are mixed with incoming air for each combustion cycle. PFI

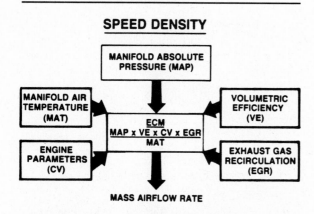

Figure 15-43. A GM speed-density fuel injection system takes several inputs into consideration to determine mass airflow rate. (GM)

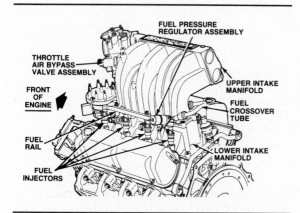

Figure 15-44. A typical Ford SEFI system. (Ford)

systems are used on 4-cylinder and small V-6 engines *without* the distributorless ignition system. On some 1984-85 engines, this system was called multipoint fuel injection (MFI).

2. *Sequential fuel injection (SFI)* is a port injection system like PFI and TPI, but the injectors operate in firing order sequence. SFI systems are used on all GM engines with a distributorless ignition system. This type of ignition system requires sequential fuel injection so that the "waste-spark" that fires in the cylinder on an exhaust stroke does not ignite any air-fuel mixture.

Most GM multipoint fuel injection systems use a heated-film, mass-airflow sensor, figure 15-28. This consists of a screen to break up the airflow, a ceramic resistor which measures the incoming air temperature, the heated film, and an electronic module. The sensor is mounted in the intake air duct between the air cleaner and air intake throttle body. The ECM of the CCC system heat the sensor to a constant 167°F

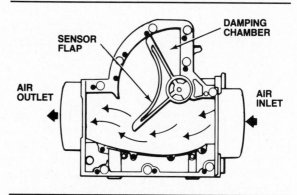

Figure 15-45. A vane-type airflow sensor. (Ford)

(75°C) above ambient air temperature. As incoming air cools the sensor, the ECM provides additional current to maintain the sensor temperature. As the intake airflow changes for different driving conditions, the ECM translates the varying current requirements into measurements of intake air mass. The ECM then regulates injector pulse width to provide the required air-fuel ratio.

Some GM multipoint fuel injection systems use a speed-density air measurement system instead of the mass airflow. A speed-density system uses manifold absolute pressure and temperature along with an engine mapping program in the computer to calculate mass airflow rate, figure 15-43. This type of system is sensitive to engine and EGR variations.

Many 1987 and later GM multiport systems on V-6 and V-8 engines have dual, or "split", fuel rails. The injectors for each bank of the engine have their own fuel distribution rail. As a further refinement, some of these split-rail systems have "biased" injectors. The computer controls each bank differently, and the injectors for right and left banks may have different flow rates for identical duty cycles and pulse widths. For example, with a pulse width of 5 milliseconds, the left bank injectors may flow 1.5 cc, and the right bank injectors may flow 1.6 cc. Biased injectors are color coded for identification. Most engines with split-rail injection also have two EGO sensors so that the engine computer can control fuel metering independently for each bank of the engine.

Ford

Ford refers to its multipoint injection systems as Electronic Fuel Injection (EFI). Based on the Bosch L-Jetronic design, Ford EFI systems for 4-cylinder engine use simultaneous double-fire injector control. Ford introduced its first multipoint EFI system on 1.6-liter engines in 1983

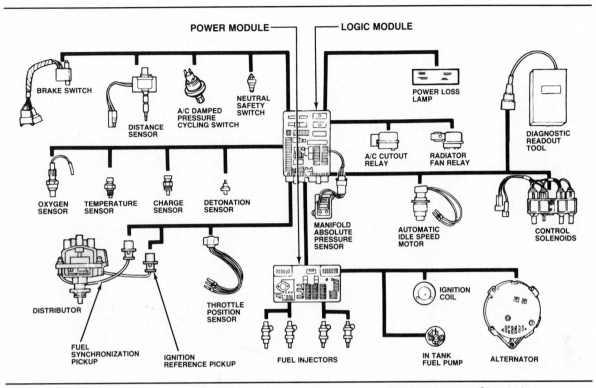

POWER MODULE ─── ──── LOGIC MODULE

BRAKE SWITCH

A/C DAMPED
PRESSURE
CYCLING SWITCH

DISTANCE
SENSOR

NEUTRAL
SAFETY
SWITCH

POWER LOSS
LAMP

DIAGNOSTIC
READOUT
TOOL

A/C CUTOUT
RELAY

RADIATOR
FAN RELAY

OXYGEN
SENSOR

TEMPERATURE
SENSOR

CHARGE
SENSOR

DETONATION
SENSOR

MANIFOLD
ABSOLUTE
PRESSURE
SENSOR

AUTOMATIC
IDLE SPEED
MOTOR

CONTROL
SOLENOIDS

DISTRIBUTOR

THROTTLE
POSITION
SENSOR

IGNITION
COIL

FUEL
SYNCHRONIZATION
PICKUP

IGNITION
REFERENCE PICKUP

FUEL INJECTORS

IN TANK
FUEL PUMP

ALTERNATOR

Figure 15-46. The Chrysler grouped double-fire injection system used on 2.2-liter engines. (Chrysler)

Escort, Lynx, and EXP/LN7 models. The same basic system continues in use on some later 1.6- and 1.9-liter engines. Ford uses a similar multipoint EFI system on 1984 and later turbocharged 1.6- and 2.3-liter engines.

Two new EFI systems were introduced in 1986. Taurus and Sable models with 3.0-liter V-6 engines use a multipoint simultaneous injection system similar to the ones on 4-cylinder engines. All 1986 5.0-liter V-8 engines use a new sequential electronic fuel injection (SEFI) system, figure 15-44. In the SEFI system, the EEC-IV computer operates the injectors in firing order sequence. Each injector fires once every two crankshaft revolutions as the intake valve opens.

Ford EFI systems on 4-cylinder engines use a vane-type airflow meter, figure 15-45, to measure intake airflow. Similar to the Bosch L-Jetronic design, the airflow meter is installed ahead of the throttle body in the air intake system. The vane moves a potentiometer sensor that sends a voltage signal to the EEC computer in proportion to intake airflow.

Chrysler

A grouped double-fire multipoint injection system, figure 15-46, is used on some 1984 and later 2.2-liter 4-cylinder engines. This system is similar to a Bosch D-Jetronic system and uses Bosch-designed fuel injectors in combination with Chrysler-engineered electronics. The pressure regulator on the fuel rail maintains a nominal 55 psi (379 kPa) of fuel pressure at the injector nozzles.

AMC-Renault

The multipoint injection system used on AMC-Renault vehicles manufactured in the United States is a variation on the Bosch L-Jetronic system, figure 15-47. This is an electronic intermittent injection system controlled by airflow measurement and engine speed.

Imports

The major Japanese automakers use multipoint injection systems derived from the Bosch L-Jetronic or AFC system. Figure 15-48 shows common components used in both Nissan and Toyota EFI systems. Nissan introduced its system in 1975, with Toyota following in 1979. These early Japanese systems control only injector duration. Systems introduced in 1981 (Nissan) and 1983 (Toyota) are fully integrated into the engine management system, controlling idle speed, EGR, and ignition timing in relation to fuel injection.

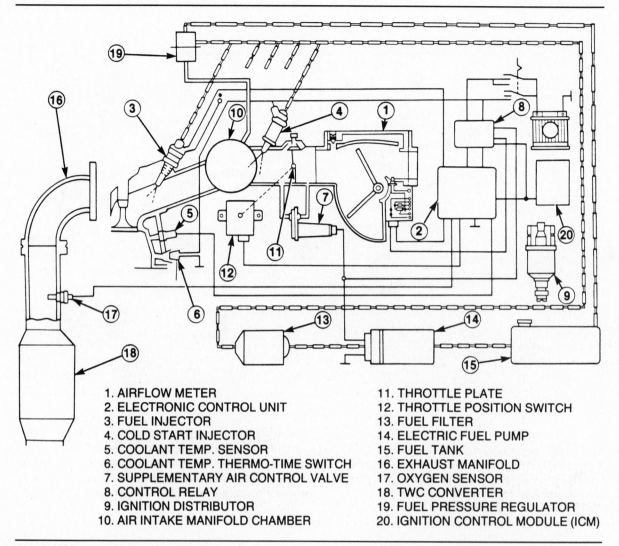

1. AIRFLOW METER
2. ELECTRONIC CONTROL UNIT
3. FUEL INJECTOR
4. COLD START INJECTOR
5. COOLANT TEMP. SENSOR
6. COOLANT TEMP. THERMO-TIME SWITCH
7. SUPPLEMENTARY AIR CONTROL VALVE
8. CONTROL RELAY
9. IGNITION DISTRIBUTOR
10. AIR INTAKE MANIFOLD CHAMBER

11. THROTTLE PLATE
12. THROTTLE POSITION SWITCH
13. FUEL FILTER
14. ELECTRIC FUEL PUMP
15. FUEL TANK
16. EXHAUST MANIFOLD
17. OXYGEN SENSOR
18. TWC CONVERTER
19. FUEL PRESSURE REGULATOR
20. IGNITION CONTROL MODULE (ICM)

Figure 15-47. A variation of the Bosch D-Jetronic system is used on American-built AMC-Renault vehicles. (AMC-Renault)

Late-Model Throttle Body Injection Systems

TBI systems made their first appearance on domestic engines in the early 1980's and now account for about half of the EFI systems in use, although multiport injection seems to be gaining in the number of applications.

With the exception of the Chrysler V-8 system introduced on 1981 Imperials, all TBI systems use similar components. Single or dual solenoid-operated injectors are positioned in a throttle body assembly that looks much like a carburetor base, figure 15-49. A MAP sensor measures intake air volume and the computer controls injection volume proportionately.

General Motors

All GM TBI systems since 1980 are manufactured by Rochester and have one or two injectors on the throttle body. There are several basic models:

• Models 100, 200 and 220 are 2-barrel, 2-injector assemblies. Cadillac has used these types of systems since 1980, although Cadillac has not used a Rochester model number for the system. In 1980, the system was called digital electronic fuel injection (DEFI). In 1981, the name was shortened to digital fuel injection (DFI). The Model 220 was introduced in 1985. Some Model 220 TBI units use injectors which are calibrated to flow fuel at a different rate in each throttle bore.

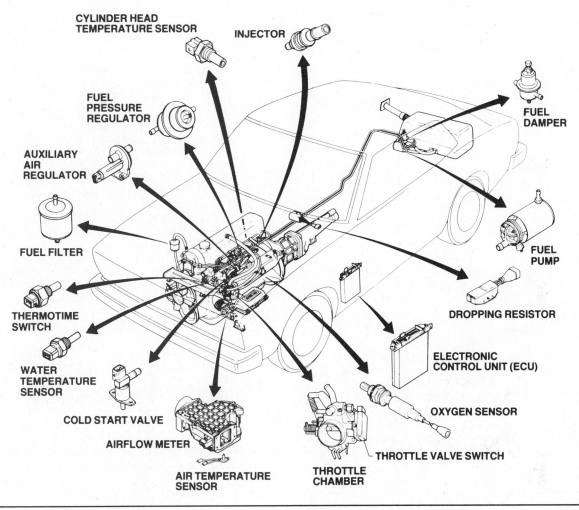

Figure 15-48. Common components used in almost all Nissan and Toyota fuel injection systems.

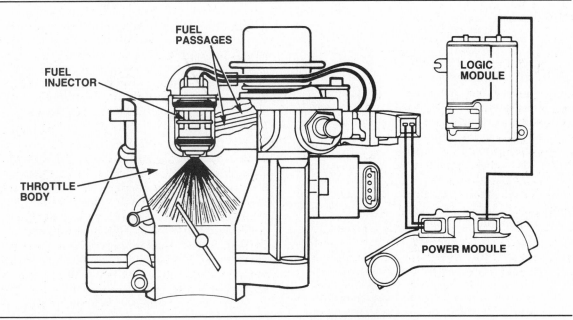

Figure 15-49. The throttle body assembly in a TBI system looks much like a carburetor, but contains only an injector, idle air motor, and pressure regulator. (Chrysler)

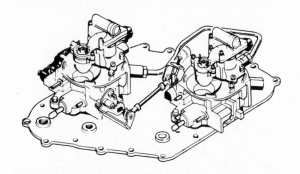

Figure 15-50. The Model 400 cross-fire injection (CFI) assembly consists of two TBI units mounted on a common manifold. (Chevrolet)

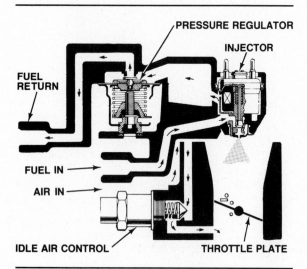

Figure 15-51. Throttle body injection fuel pressure regulation. (GM)

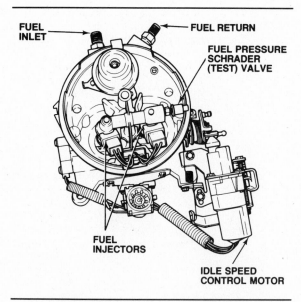

Figure 15-52. A Ford CFI fuel charging (throttle body) assembly. (Ford)

• Models 300, 500 and 700 are 1-barrel, 1-injector systems. The Model 300 was used on 1.8- and 2.5-liter 4-cylinder engines, and the Model 500 was used on 2.0-liter 4-cylinder engines. With the exception of minor fuel and airflow differences, these two units are the same. The Model 700 was introduced on 1987 4-cylinder engines as a replacement for both the 300 and 500. It uses a Multec low-pressure injector, and has a redesigned pressure regulator that permits replacement of the diaphragm assembly instead of the entire regulator unit.
• Model 400 is the unique crossfire assembly of two 1-barrel, 1-injector throttle bodies mounted on a single intake manifold, figure 15-50. This system is used on some 1982-83 5.0-liter V-8 engines in Camaros and Firebirds, as well as 1982-84 5.7-liter V-8 Corvette engines. The ECM of the CCC system is reprogrammed to handle the two units, but operation is otherwise the same as other GM TBI systems. Like

the Chrysler V-8 TBI system, the GM crossfire fuel injection (CFI) system proved to be a technological deadend.

All GM TBI systems have a fuel pressure regulator built into the throttle body, figure 15-51. The regulator senses intake manifold vacuum (or air pressure drop across the injector) at the same point as the injector nozzle tip. The regulator controls fuel pressure by opening and closing the fuel return line. All regulators are serviced as an assembly except for the Model 700, which has a replaceable diaphragm.

Ford

Ford calls all of its throttle body injection systems central fuel injection (CFI) because the fuel is injected at a central point. The first such system was a twin-injector design, figure 15-52, introduced in 1980 and used on some 5.0-liter V-8 engines through 1985. The twin CFI system was adapted to the 3.8-liter V-6 in 1984. Fuel pressure in this system is regulated between 35 and 45 psi (241 and 310 kPa). All 1980-83 and 1984 California V-8 twin CFI systems are controlled by the EEC-III system. Later V-8 and all V-6 CFI systems are regulated the newer EEC-IV system.

A single injector CFI system was introduced in 1985 for the 2.3-liter 4-cylinder engine used in Tempo and Topaz models. This same design is used on the 2.5-liter engine in Taurus and Sable models. Unlike the twin CFI system, the single injector version operated at a lower fuel pressure of approximately 14.5 psi (100 kPa).

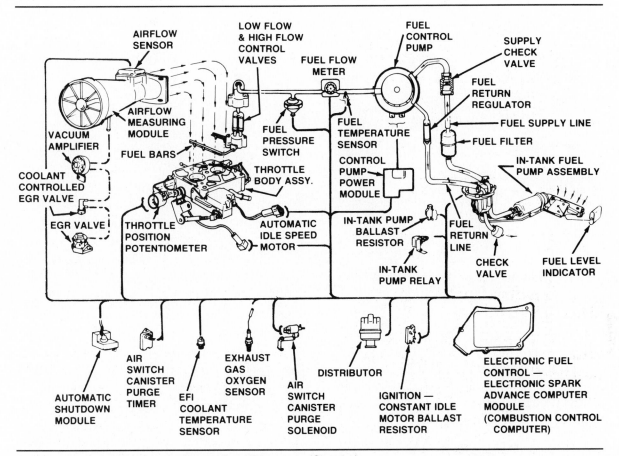

Figure 15-53. The Chrysler continuous-flow TBI system. (Chrysler)

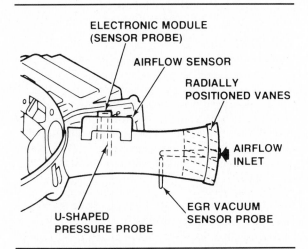

Figure 15-54. Chrysler's continuous-flow TBI airflow sensor. (Chrysler)

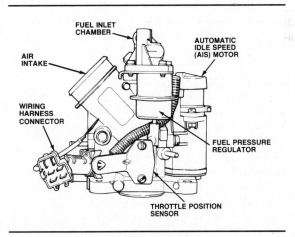

Figure 15-55. The Chrysler-Holly-Bendix high-pressure TBI unit. (Chrysler)

Chrysler

Chrysler's first TBI system was a continuous flow system used on its V-8 engine in 1981-83 Imperials, figure 15-53. A special sensor in the air cleaner snorkel measures air intake volume, figure 15-54. Radial vanes swirl the air and a

U-shaped pressure probe in the vortex of the swirling air determines the amount of air entering the engine.

Most 1981 and later nonturbocharged Chrysler-built 4-cylinder engines use a single injector TBI system. The original system used through 1985 had a Holley-designed throttle body, figure 15-55, with an injector supplied by

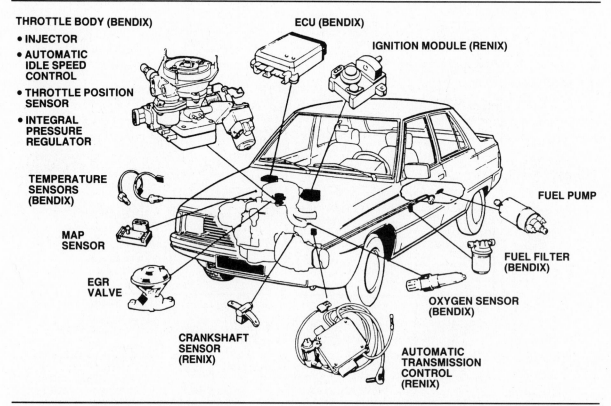

THROTTLE BODY (BENDIX)
- INJECTOR
- AUTOMATIC IDLE SPEED CONTROL
- THROTTLE POSITION SENSOR
- INTEGRAL PRESSURE REGULATOR

TEMPERATURE SENSORS (BENDIX)

MAP SENSOR

EGR VALVE

CRANKSHAFT SENSOR (RENIX)

ECU (BENDIX)

IGNITION MODULE (RENIX)

FUEL PUMP

FUEL FILTER (BENDIX)

OXYGEN SENSOR (BENDIX)

AUTOMATIC TRANSMISSION CONTROL (RENIX)

Figure 15-56. The major components of AMC-Renault's TBI and electronic engine management system. (AMC-Renault)

Bendix. A pressure regulator on the throttle body maintains a nominal 36 psi (248 kPa) of fuel pressure at the injector nozzle. Chrysler refers to this system as a "high-pressure" EFI system, but in the context of other injection systems, it is a medium-pressure system.

The TBI system was revised in 1986, with the newer design using a low-pressure injector designed by Bosch, figure 15-49. The throttle body includes a pressure regulator that maintains fuel pressure at 14.5 to 15 psi (100 to 103 kPa).

AMC-Renault

The TBI system used on AMC-Renault vehicles manufactured in the United States is a combination of Bendix and Renix components, figure 15-56, which work about the same as the domestic TBI systems described earlier. The injection system is fully integrated into the engine management system.

Imports

Most import vehicles equipped with TBI use systems that are quite similar in design and function to those used on domestic vehicles. The AMC/Renault design is a typical example.

Turbocharged Japanese vehicles sold by Mitsubishi and Chrysler Corporation in the United States use a TBI system which measures intake airflow rate based on Karman's vortex theory (a well-known theory dealing with aerodynamics and ultrasound waves), figure 15-57. This system is considerably more complex than the other TBI systems which we have seen.

The airflow sensor and intake air temperature sensor are located inside the filter element in the air cleaner housing, figure 15-58. Filtered air passes through the airflow sensor, which contains a stabilizer plate and generating rod. The plate and rod vibrate according to the rate of airflow, setting up ultrasonic wave patterns which are transformed into a voltage signal by the sensor and sent to the computer indicating flow rate. At the same time, the air temperature sensor signals the computer concerning the temperature of the airflow.

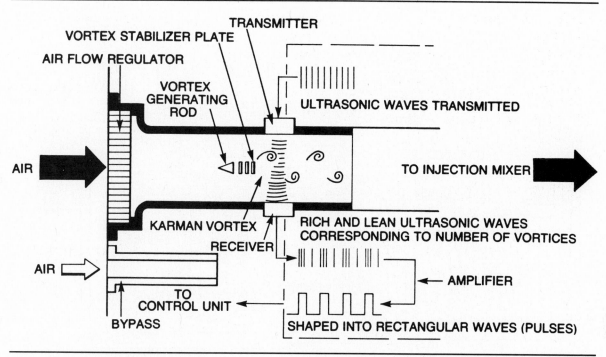

Figure 15-57. Operation of the Mitsubishi-Chrysler TBI system which measures airflow rate according to Karman's vortex theory. (Chrysler)

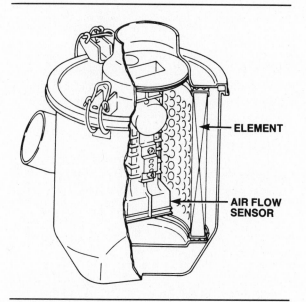

Figure 15-58. The airflow sensor is mounted inside the air cleaner filter element. (Chrysler)

Two injectors installed in the inlet port of the injection mixer or air intake throttle body operate in an alternating sequence. Injector operation is similar to that of other solenoid-operated injectors, but a swirl nozzle design is used to atomize the fuel and "jiggle" the fuel spray, figure 15-59. A schematic of the complete system is shown in figure 15-60.

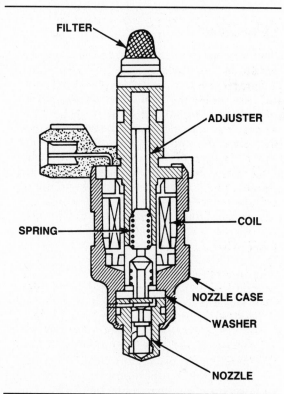

Figure 15-59. The Mitsubishi-Chrysler TBI injector uses a swirl nozzle for better fuel atomization, but operates like any other solenoid-actuated injector. (Chrysler)

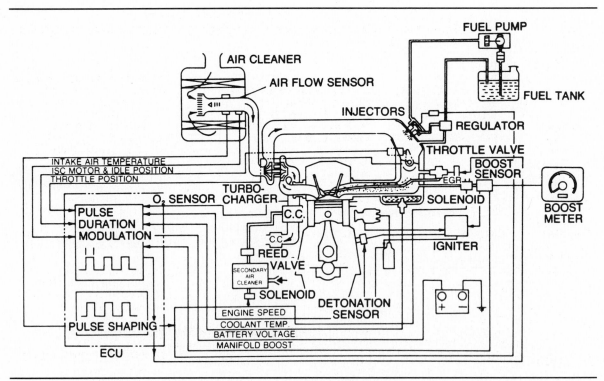

Figure 15-60. The Mitsubishi-Chrysler ECI system components. (Chrysler)

SUMMARY

Fuel injection systems must do the same tasks as carburetors. However, they do so in a slightly different way. Two types of injection systems are used: throttle body injection (TBI) or multipoint (port) injection.

The TBI system uses a carburetor-like throttle body containing one or two injectors. This throttle body is mounted on the intake manifold in the same position as a carburetor. Fuel is injected above the throttle plate and mixed with incoming air.

In a multipoint system, individual injectors are installed in the intake manifold at a point close to the intake valve where they inject the fuel to mix with the air as the valve opens. Intake air passes through an air intake throttle body containing a butterfly valve or throttle plate and travels through the intake manifold where it meets the incoming fuel charge at the intake valve. The difference in injector location permits more advance manifold designs in multipoint systems to aid in fuel distribution.

Fuel injection has numerous advantages over carburetion. It can match fuel delivery to engine requirements under all load and speed conditions, maintain an even mixture temperature, provide more efficient fuel distribution,

and reduce emissions while improving driveability.

Electronic fuel injection (EFI) is far more common than the earlier mechanical injection represented by the Bosch K-Jetronic. EFI integrates the injection system into a complete engine management system which includes control of EGR, ignition timing, canister purging, and various other functions.

The EFI computer uses various sensors to gather data on engine operation. Intake air volume is determined by measuring airflow speed, manifold pressure, or the mass of the air drawn into the engine. This data is added to information about throttle position, idle speed, engine coolant and air temperature, crankshaft position, and other operating conditions. After processing the data, the computer signals the solenoid-operated injectors when to open and close. Injectors may be fired in various combinations, but all use the principle of pulse width modulation, in which the computer varies the percentage of on-time in each full on/off cycle according to engine requirements. In this way, the amount of fuel injected can be varied instantly to accommodate changing conditions.

Review Questions

Choose the single most correct answer.
Compare your answers with the correct answers on page 414.

1. Fuel injection systems can lower emissions:
   a. By matching the air-fuel ratio to engine requirements
   b. Only at high speeds
   c. By using the intake manifold to vaporize fuel
   d. By matching engine speed to load conditions

2. All of the following are common kinds of gasoline fuel injection systems *except:*
   a. Multipoint injection
   b. Throttle body injection
   c. Direct cylinder injection
   d. Continuous injection

3. Mechanic A says that Bosch K-Jetronic systems are mechanical injection systems that operate on differential fuel pressure.
   Mechanic B says that Bosch K-Jetronic systems are intermittent injection systems.
   Who is right?
   a. A only
   b. B only
   c. Both A and B
   d. Neither A nor B

4. Bosch L-Jetronic and similar systems measure airflow with a:
   a. BMAP sensor
   b. Hot-wire air mass sensor
   c. Vane-type airflow sensor
   d. Air charge temperature sensor

5. Mechanic A says that TBI systems are electronic, intermittent injection systems.
   Mechanic B says that TBI systems require fuel pressure regulators.
   Who is right?
   a. A only
   b. B only
   c. Both A and B
   d. Neither A nor B

6. Mechanic A says that most TBI systems measure airflow and volume manifold absolute pressure, barometric absolute pressure or barometric and manifold absolute pressure sensors.
   Mechanic B says that most Ford multipoint injection systems use a heated-film air mass sensor.
   Who is right?
   a. A only
   b. B only
   c. Both A and B
   d. Neither A nor B

7. Mechanic A says that fuel pressure regulators always are on the upstream (inlet) side of the injectors in an EFI system.
   Mechanic B says that mechanical injection systems do not require a pressure regulator.
   Who is right?
   a. A only
   b. B only
   c. Both A and B
   d. Neither A nor B

8. Mechanic A says that the throttle in a multipoint system is located between the injectors and the manifold.
   Mechanic B says that the throttle in an EFI system operates the same as a throttle on a carbureted engine.
   Who is right?
   a. A only
   b. B only
   c. Both A and B
   d. Neither A nor B

9. Mechanical fuel injectors are operated:
   a. Continuously by fuel pressure
   b. Intermittently by the computer
   c. Alternately by air pressure
   d. Sequentially by firing order

10. Modern fuel injection systems are based on work begun by:
    a. Rochester Products Division of GM
    b. Ford Motor Company
    c. Hitachi
    d. Robert Bosch GmbH

11. A cold-start injector is a:
    a. Temperature-controlled detonation device
    b. Solenoid-operated auxiliary injector
    c. Mechanical intermittent injector
    d. None of the above

12. Multec injectors used in a multipoint injection system differ from those produced by Robert Bosch in that they use a:
    a. Pintle valve
    b. Swirl nozzle
    c. Pin and spring
    d. Ball-seat valve with director plate

13. All of the following are filter locations in a fuel injection system *except:*
    a. In the pressure regulator
    b. At the fuel pump
    c. In the fuel line
    d. In the fuel injectors

14. Mechanic A says that an EFI system can use an on/off throttle position switch.
    Mechanic B says that a throttle position sensor is a variable resistor.
    Who is right?
    a. A only
    b. B only
    c. Both A and B
    d. Neither A nor B

15. A hot-wire, or heated-film, air mass sensor measures:
    a. Airflow velocity entering the engine
    b. Barometric absolute pressure
    c. Air temperature in a turbocharged system
    d. Molecular mass of air entering the engine

Chapter

16

Supercharging and Turbocharging

Engines with carburetors rely on atmospheric pressure to push an air-fuel mixture into the combustion chamber vacuum created by the downstroke of a piston. The mixture is then compressed before ignition to increase the force of the burning, expanding gases. The greater the mixture compression, the greater the power resulting from combustion. In this chapter, we will study ways to increase mixture compression.

ENGINE COMPRESSION VS. SUPERCHARGING

A high compression ratio is one way in which mixture compression can be increased. In the late 1960's, compression ratios on high-performance car engines reached 11 to 1. During the 1970's, emission control requirements brought compression ratios down to the range of 8 or 8.5 to 1. This was necessary because higher-compression engines tend to emit too much NO_x. Lower compression ratios also were required due to the greatly reduced lead content in the gasoline. However, the use of electronic engine management systems in the 1980's has allowed compression ratios in the 9 to 10 range.

High compression ratios have two major benefits:

• Volumetric efficiency is improved because the piston displaces a larger percentage of the total cylinder volume on each intake stroke.
• Since temperature increases as pressure increases, combustion is more complete and so thermal efficiency is higher with a high-compression engine.

There are, however, major disadvantages with high compression ratios:

• Since the compression ratio remains the same throughout the engine's operating range, temperature and pressure also are high and cause increased emissions during deceleration, idle, and part-throttle operation.
• A high compression ratio causes increased NO_x emissions.
• High compression ratios require the use of effective antiknock additives. While lead is the most effective additive, it produces harmful emissions and destroys catalytic converters.

Another way to achieve this increase in mixture compression is called **supercharging**. This is a method by which a denser air-fuel charge can be packed into the engine's cylinders. Since the density of the air-fuel charge is greater, so is its weight — and power is directly related to the weight of an air-fuel charge consumed within a given time period. The result is much

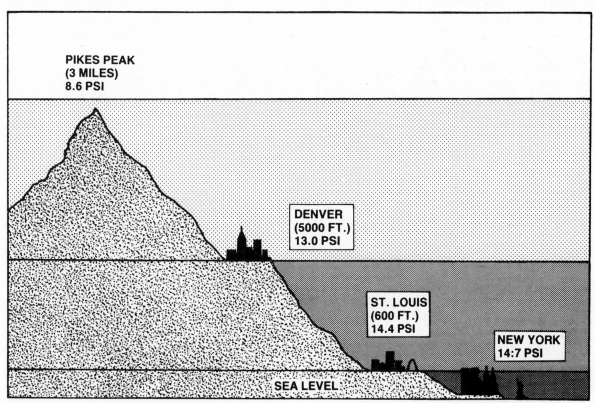

Figure 16-1. Air is a substance and has weight, which causes atmospheric pressure. Atmospheric pressure decreases with increases in altitude. (Ford)

the same as a higher compression ratio, but the effect can be controlled during idle and deceleration to avoid high emissions.

Supercharging Principles

To this point, we have dealt with what we call **normally aspirated** engines. This term describes how an automotive engine breathes, which you studied in Part 1 of this manual. As you recall, air is drawn into a normally aspirated engine by atmospheric pressure forcing it into the low-pressure area of the intake manifold. The low pressure or vacuum in the manifold is created by the reciprocating motion of the pistons and is controlled by the action of the valves and throttle. Basically, this works well, but the engine's volumetric efficiency is considerably less than 100 percent. Volumetric efficiency, as we have seen, is a percentage measurement of the volume of air drawn into a running engine compared to the maximum amount this engine could draw in (total displacement volume). While a stock engine averages approximately 70 to 80 percent volumetric efficiency, racing engines with the proper intake and exhaust tuning can exceed 100 percent volumetric efficiency at their power curve peak.

The volumetric efficiency of any normally aspirated engine is related to the density of the air drawn into it. Since atmospheric pressure and air density are greatest at (or below) sea level, both pressure and density normally decrease as altitude above sea level increases. For example, atmospheric pressure at sea level is about 14.7 psi (101 kPa); at higher elevations, atmospheric pressure may be only 8 or 9 psi (55 or 62 kPa), figure 16-1. Therefore, the volumetric efficiency of any engine will be greater at sea level than at an altitude above sea level.

By increasing the pressure on the air drawn into the engine above that of atmospheric pressure, it is possible to force more air into a cylinder during the intake stroke. Increasing the engine's air intake in this manner means that

Supercharging: Use of an air pump to deliver an air-fuel mixture to the engine cylinders at a pressure greater than atmospheric pressure.

Normally Aspirated: An engine that uses normal vacuum to draw in its air-fuel mixture. Not supercharged.

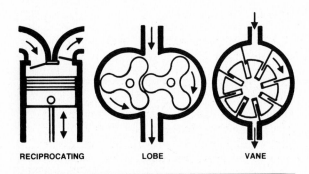

RECIPROCATING　　LOBE　　VANE

Figure 16-2. These are three common positive-displacement pump designs.

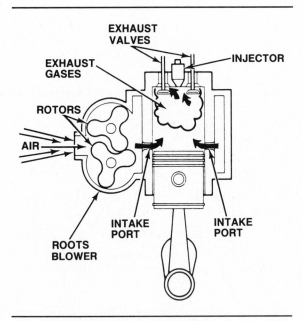

Figure 16-3. The Roots blower is the most commonly used positive displacement pump used as a supercharger. (Ford)

more fuel can be mixed with the air while still maintaining the same air-fuel ratio. The combination of more air and fuel entering an engine during its intake stroke means greater potential energy during combustion and more power as a result of combustion.

A supercharger pressurizes air to greater than atmospheric pressure. This amount of pressurization above atmospheric pressure is called **boost**. Boost can be measured in the same way as atmospheric pressure, which we studied in Chapter 8. Normal sea-level atmospheric pressure is 14.7 psi (101 kPa). Some superchargers can almost double that pressure, providing about 12 psi (83 kPa) boost *above* atmospheric pressure. At sea level, that would make a total intake pressure of 26.7 psi (184 kPa).

Atmospheric pressure drops as altitude increases, but boost pressure remains the same. For example, in Denver, where the atmospheric pressure is 13 psi (90 kPa), figure 16-1, our supercharger described above would provide a boost of 12 psi (83 kPa). This equals a total intake pressure of 25 psi (172 kPa). Superchargers also can be designed to provide increased boost at higher altitudes to make up for the loss of atmospheric pressure. When used on racing engines, superchargers can be designed to double or even triple the horsepower of a normally aspirated engine.

One of the major advantages of using a supercharger with a normally aspirated engine is that it can increase the air-fuel charge density to provide high-compression pressure when power is required, but it allows the engine to run on lower pressures when the additional power is not required.

Superchargers

In basic concept, a supercharger is nothing more than an air pump mechanically driven by the engine itself. Gears, shafts, chains, or belts from the crankshaft turn the pump. This means that the air pump or supercharger pumps air in direct relation to engine speed.

There are two general types of supercharger pumps:
- Positive displacement
- Centrifugal.

Positive displacement pumps include the reciprocating, the lobe (Roots), and the vane designs, figure 16-2. These are called positive displacement pumps because every revolution pumps the same volume of air. They operate by taking in a large column of air, compressing it into a small area and then forcing it through an outlet at high pressure. Some of these designs are very inefficient, and the efficient designs have been too expensive for passenger car use in the past. They are more often found on diesel trucks or racing engines.

The lobe-type, or Roots blower, figure 16-3, is the most common positive displacement supercharger in use. It was first developed in 1864 as a device to separate wheat from chaff but was applied to automotive engines around the turn of the century. Roots blowers are used on many high-performance and racing engines, but also are found on 2-stroke diesel engines, where they improve air intake and exhaust scavenging instead of acting as primary superchargers. The vane-type and axial-flow (Latham) superchargers are not commonly used.

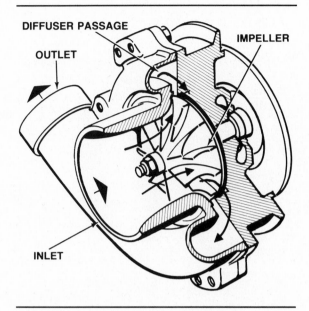

Figure 16-4. The centrifugal pump design accelerates and then slows the air in order to compress it.

The centrifugal pump, figure 16-4, is the most commonly used and most efficient variable-displacement supercharger. As the impeller turns at high speed, air pulled into the center of the impeller is speeded up and then thrown outward from the blades by centrifugal force. Air moved to the perimeter of the pump housing is forced through an outlet. Since centrifugal force increases as speed increases, a centrifugal supercharger will draw in more air and create a higher boost pressure as its speed increases.

Specific Supercharger Designs

The mechanical drive methods that operate a supercharger all have inherent limitations. They require up to 20 percent of the engine's power even when the supercharger is not in use. They also need a high overdrive ratio to obtain a speed great enough to provide the desired boost. Since the drive belt or gear speed must be quite high, the mechanical components are subjected to heavy wear.

However, the advent of small 4-cylinder engines has led automotive engineers to take a second look at the supercharger as a viable means of improving low-speed torque while raising power output. Volkswagen and Toyota are manufacturing supercharged engines, and domestic manufacturers are preparing to bring out their own versions.

Eaton

The Eaton Corporation has been developing a supercharger since 1977. Its design is based on the traditional Roots-type and is integrated into an engine management system with fuel injection, figure 16-5. Intake airflow is the same as in any mass airflow injection system until it travels past the throttle plate. Instead of being sent directly to the manifold, the airflow is directed into the supercharger, where the air is flung from one series of lobes to another (which has the effect of compressing it). The air exits the supercharger, then it passes through an intercooler to lower the heat from the turbulence created by the lobes (increasing density). It is then directed into the intake manifold, where it is mixed with the fuel charge from the injectors.

Boost: A measure of the amount of air pressurization, above atmospheric, that a supercharger can deliver.

■ Switching The Emphasis Turbos

Some drivers may think of turbochargers as exotic trapping for souped-up street cars and racers. That's natural when you consider that automotive engineers normally bring a turbo into play high up on the engine's performance curve to deliver maximum power at top rpm. The result is additional peak power that has won many races, and further convinced the general driving public that turbine-operated superchargers are not for them.

The reappearance of turbocharging as a factory option approached the subject from the opposite viewpoint. Engineers wanted 6-cylinder fuel economy with V-8 performance and the turbocharger was the way to get it. Since its appearance on 1978 domestic engines, the turbocharger has been used for low- and medium-speed passing, and acceleration from about 1,200 rpm up. As engine speed climbs, the wastegate begins to open, preventing an overload of the engine. This means that the turbocharger will be used for only about five percent of the time the engine is operating.

The performance-oriented 1980's have brought turbocharging into favor with consumers. When used with today's 4-cylinder engines, turbochargers provide the power to make them perform like V-6 and small V-8 engines, while retaining good fuel economy. By the mid-1980's, 7 percent of the 4-cylinder engines produced by domestic manufacturers were turbocharged. With the upswing in turbocharger popularity, it faces a challenge from the supercharger, which is about to debut on the domestic scene.

ROOTS SUPERCHARGER

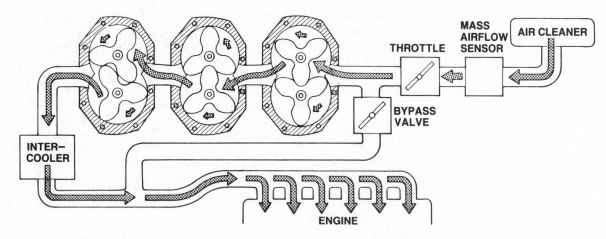

Figure 16-5. The Eaton supercharger system integrates a Roots blower with fuel injection and an engine management system.

VOLKSWAGEN G-LADER

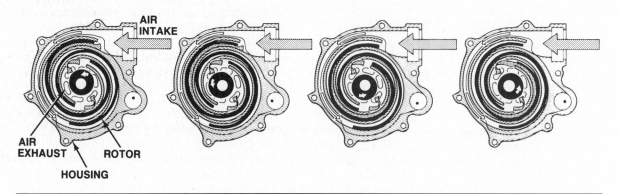

Figure 16-6. The G-Lader supercharger design compresses air by squeezing it through a spiral, then forcing it through outlets in the center of the unit.

A bypass valve routes intake air directly to the manifold when supercharging is not required. The entire system is controlled by the engine computer, which turns the boost on and off as required by operating conditions. Since it operates on pressure pulses instead of a constant flow of air, the Roots blower has a characteristic high-pitched whine which some find objectionable. Eaton has minimized this problem by acoustically tuning the entire intake system. At this time, the Eaton supercharger application has been extensively tested, but has not yet appeared on any production vehicles.

Toyota

Toyota has equipped a 1.6-liter 4-valve, 4-cylinder engine used in its MR2 sports car with a 2-lobe Roots-type supercharger and intercooler. The Toyota application uses an electromagnetic clutch (like that used on an air conditioning compressor) to provide computer control over engagement and disengagement according to engine requirements. While the Eaton supercharger spins freely when it is not on boost, Toyota's computer control reduces any loss of engine efficiency during partial load.

Volkswagen

The supercharger used on some European Volkswagen models is based on a less-traditional design. Called the **G-Lader**, this spiral-channel unit is derived from a French design patented in 1905, figure 16-6. Concentric

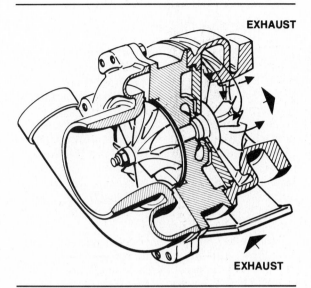

Figure 16-7. A turbine wheel is turned by the expansion of gases against its blades.

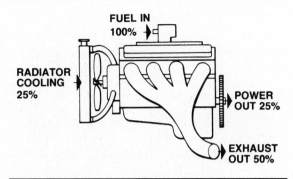

Figure 16-8. A turbine uses some of the heat energy that normally would be wasted.

spiral ramps in both sides of a rotor mesh with similar ramps cast in the split casing. The rotor moves around an eccentric shaft instead of spinning on its axis, as in most other supercharger designs. Air drawn into the casing is squeezed through the spiral. This squeezing action compresses the air, which is then forced through a cluster of ports in the center of the casing and into the engine. The G-Lader supercharger is claimed to have several advantages over a Roots-type blower:

● Intake noise is lower because airflow is fairly constant instead of intermittent
● Wear is minimal because contact between the spiral and housing is slight
● It is considerably more efficient.

Aftermarket applications
Superchargers on domestic vehicles date from the 1906 Chadwick and were popular on various American and European luxury cars during the 1920's and 1930's, as well as on pre-WWII race cars. They started to lose popularity after the war and, in recent years, were found primarily on street machines built by enthusiasts. The majority of superchargers or "blowers" in use are variations on the Roots design.

TURBOCHARGERS

The major disadvantage of a supercharger is its reliance on engine power to drive the unit. In some installations, as much as 20 percent of the engine's power is used by a mechanical supercharger. However, by connecting a centrifugal supercharger to a turbine drive wheel and installing it in the exhaust path, that engine horsepower is regained to perform other work and the combustion heat energy lost in the engine exhaust (as much as 40 to 50 percent) can be harnessed to do useful work. That is the concept of a **turbocharger**.

A turbocharger turbine, figure 16-7, looks much like the centrifugal pump shown in figure 16-4. Hot exhaust gases flow from the combustion chamber to the turbine wheel. The gases are heated and expanded as they leave the engine. It is not the speed or force of the exhaust gases that forces the turbine wheel to turn, as is commonly thought, but the expansion of the hot gases against the turbine wheel's blades.

The turbine's main advantage over a mechanically driven supercharger is that the turbine does not drain power from the engine. In a normally aspirated engine, about half of the heat energy contained in the fuel goes out the exhaust system, figure 16-8. Another 25 percent is lost through radiator cooling. Only about 25 percent is actually converted to mechanical power. A mechanically driven pump uses some of this mechanical output, but a turbine gets its energy from the exhaust gases, converting more of the fuel's heat energy into mechanical energy.

G-Lader: A type of supercharger pump which compresses air by squeezing it through an internal spiral, then forcing it through ports into the engine.

Turbocharger: A supercharging device that uses exhaust gases to turn a turbine that forces extra air-fuel mixture into the cylinders.

Figure 16-9. This cutaway of a typical turbocharger used by Ford shows how simple it is.

Turbocharger Design

Most automotive turbocharger designs use a centrifugal pump because of cost, size, reliability, and efficiency. The modern turbocharger, figure 16-9, is both simple and compact, with few moving parts. But since those moving parts work at very high speeds and under extreme heat, the turbocharger must be manufactured to very exacting and precise tolerances. A turbocharger consists of two chambers connected by a center housing. The two chambers contain a turbine wheel and a compressor wheel connected by a shaft which passes through the center housing, figure 16-10.

To take full advantage of the exhaust heat which provides the rotating force, a turbocharger must be positioned as close as possible to the exhaust manifold. This allows the hot exhaust to pass directly into the unit with a minimum of heat loss. As exhaust gas enters the

■ The Next Generation Of Turbochargers

Integrating the turbocharger into electronic engine management systems has brought it a well-deserved reputation for reliability and performance. To this point, all production turbocharger applications have been based on the centrifugal pump. The next generation of turbochargers will see numerous design variations, the first of which is Nissan's N2-VN or variable nozzle (scroll) turbo.

A movable curved flap in the turbo housing changes the throat area to vary its output according to engine requirements. The flap can move 27 degrees in stepless increments. Flap position is determined by a

vacuum-operated diaphragm controlled by a pressure-controlled modulator and the engine computer.

With the flap closed, the flow of exhaust gas which enters the turbine housing is speeded up, increasing its pressure and rotating the turbine rapidly at low-speed, low-load conditions. At high-speed operation, the flap opens fully and the reduction in exhaust resistance aids in filling the cylinders.

The design provides a normal boost up to 15 psi (103 kPa). However, if operating conditions will permit it without engine damage, the computer will allow boost up to 17 psi (117 kPa) for short periods, such as overtaking another vehicle.

VARIABLE NOZZLE TURBOCHARGER

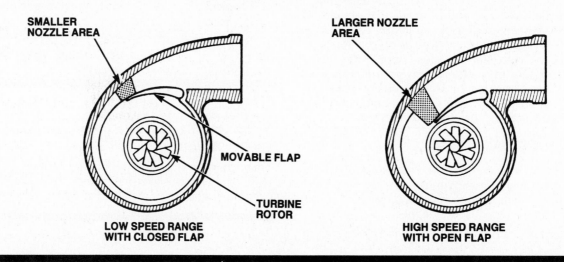

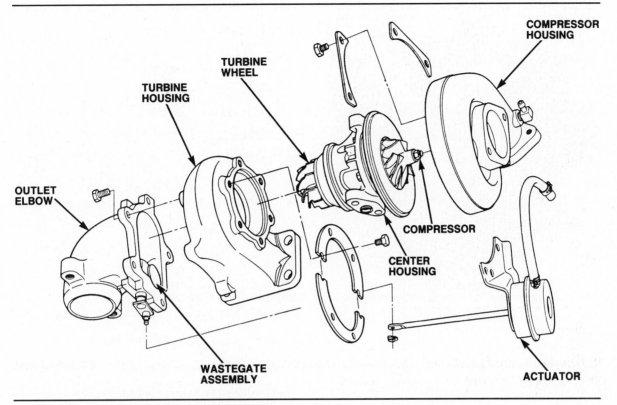

Figure 16-10. The components of a typical turbocharger. (Ford)

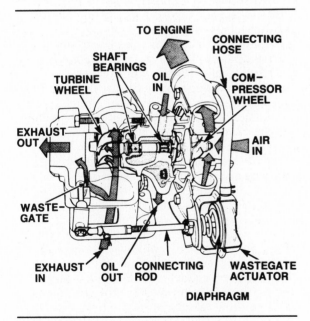

Figure 16-11. Basic operational cycle of a turbocharger. (Ford)

turbocharger, it rotates the turbine wheel blades, figure 16-11. The turbine wheel and compressor wheel are on the same shaft so that they turn at the same speed. Rotation of the compressor wheel draws air in through a cen-

tral inlet and centrifugal force pumps it through an outlet at the edge of the housing. A pair of bearings in the center housing supports the turbine and compressor wheel shaft, figure 16-11, and is lubricated by engine oil.

To minimize possible leakage around their blades, both wheels must operate with extremely close clearances. Leakage around the turbine blades causes a dissipation of the heat energy required for compressor rotation. Leakage around the compressor blades prevents the turbocharger from developing its full boost pressure.

At low engine speeds, both exhaust heat and pressure are low and the turbine runs at a low speed (approximately 1,000 rpm). Since the compressor does not turn fast enough to develop boost pressure, air simply passes through it and the engine works like any normally aspirated engine. As engine speed or load increases, both exhaust heat and flow increase, causing the turbine and compressor wheels to rotate faster. Since there is little rotating resistance on the turbocharger shaft, the turbine and compressor wheels will accelerate as the exhaust heat energy increases. With an engine running at full power, the typical turbocharger will rotate at speeds between 100,000 and 150,000 rpm.

Figure 16-12. Water-cooled turbochargers induct engine coolant into passages in the center housing (arrow) to cool the center bearings.

At this point, you can recognize one of the turbocharger's main disadvantages — there is a time lag between increasing engine speed and the turbocharger's ability to overcome inertia and spin up to speed as the exhaust gas flow increases. This is called **turbo lag**.

Similarly, engine deceleration from full power to idle requires only a second or two because of its internal friction, pumping resistance and drive train load. But the turbocharger has no such load on its shaft, and is already turning many times faster than the engine at top speed. Thus, it may take a minute or more after the engine has returned to idle speed before the turbocharger also has returned to idle.

This is another disadvantage of the turbocharger. If the engine is decelerated to idle and then shut off, engine lubrication stops flowing to the center housing bearings while the turbocharger is still spinning. The oil in the center housing is then subjected to extreme heat and will gradually "coke" or oxidize.

All clearances within a turbocharger are extremely close and the bearings are no exception. Radial clearance is generally kept between 0.003" and 0.006" (0.08 mm and 0.15 mm) and axial clearance is maintained between 0.001" and 0.003" (0.025 mm and 0.08 mm). Since the bearings must maintain these very critical clearances at extreme speeds, constant lubrication with clean oil is very important. Even a small particle of contamination entering the intake or exhaust housing can damage the turbocharger.

Liquid-cooled turbochargers

As we have seen, the turbocharger is a simple but extremely precise device in which heat plays a critical role. When properly maintained, a turbocharger also is a trouble-free device. To prevent problems with the turbocharger, three conditions must be met:

1. There must be constant and proper lubrication of the turbocharger shaft and bearings with clean engine oil.
2. Dirt particles and other contamination must be kept out of the intake and exhaust housings.
3. Whenever a basic engine bearing has been damaged, the turbocharger must be flushed with clean engine oil. Conversely, if the turbocharger is damaged, the engine oil should be flushed and the oil filter replaced as part of the repair procedure.

Many automakers have turned to liquid cooling to prevent the turbocharger center bearings from being damaged by too much heat. In a liquid-cooled turbo design, figure 16-12, engine coolant is circulated through passages provided in the center housing to draw off the excess heat. This allows the bearings to run cooler and minimizes the probability of oil coking when the engine is shut down.

Turbocharger size and response time

We have already mentioned turbo lag as one of the main disadvantages of a turbocharger. Turbo lag is caused by the inertia inherent in the turbocharger's motive power. Like any material, moving exhaust gas has inertia. When full power is applied to an engine, there is a brief delay before the flow of exhaust gas develops enough energy to rapidly accelerate the turbine wheel. Inertia also is present in the turbine and compressor wheels, as well as the intake airflow. Since the turbocharger is a variable-displacement pump, it cannot supply an adequate amount of boost at low speed.

Turbocharger response time is directly related to the size of the turbine and compressor wheels. Small wheels accelerate rapidly; large wheels accelerate slowly. While the small wheels would seem to have an advantage over larger ones, they may not have enough airflow capacity for an engine. To minimize turbo lag, the intake and exhaust breathing capacities of an engine must be matched to the exhaust and intake airflow capabilities of the turbocharger.

Another factor influencing turbocharger response time is the location of the turbocharger on the engine. The most efficient setup is to place the turbine close to the exhaust manifold and the compressor outlet close to the intake manifold. Figure 16-13 shows this placement, as well as the operation cycle of a typical turbocharger used with a carbureted engine.

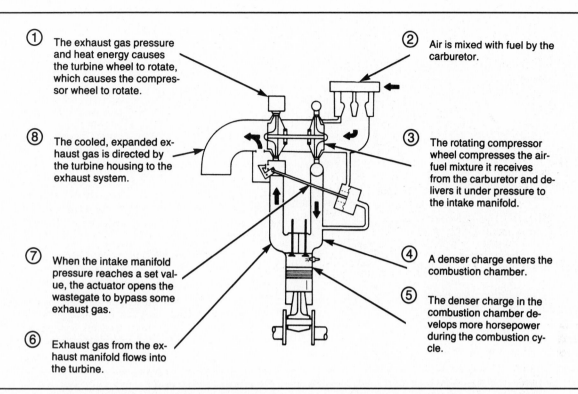

① The exhaust gas pressure and heat energy causes the turbine wheel to rotate, which causes the compressor wheel to rotate.

② Air is mixed with fuel by the carburetor.

③ The rotating compressor wheel compresses the air-fuel mixture it receives from the carburetor and delivers it under pressure to the intake manifold.

④ A denser charge enters the combustion chamber.

⑤ The denser charge in the combustion chamber develops more horsepower during the combustion cycle.

⑥ Exhaust gas from the exhaust manifold flows into the turbine.

⑦ When the intake manifold pressure reaches a set value, the actuator opens the wastegate to bypass some exhaust gas.

⑧ The cooled, expanded exhaust gas is directed by the turbine housing to the exhaust system.

Figure 16-13. Turbo lag is reduced by locating the turbine and compressor close to the exhaust and intake manifolds respectively. This is the operational cycle of a turbocharger used with a carbureted system. (Ford)

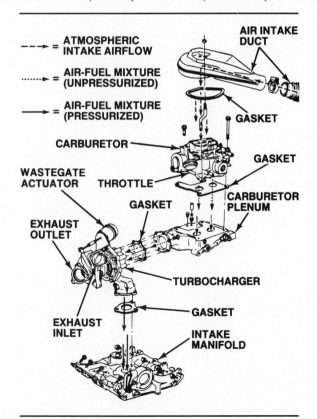

Figure 16-14. The turbocharger compresses the air-fuel mixture when the carburetor is installed at the compressor inlet. (Buick)

Turbocharger Installation

Turbochargers can be installed either on the air intake side (upstream) or on the exhaust side (downstream) of the carburetor or injectors. In an upstream installation, the turbocharger compresses and delivers a denser air charge to the carburetor or injectors. When installed downstream, the air-fuel mixture is compressed and delivered to the cylinders. Both types of installations are used with modern turbocharger applications.

Downstream installations

Turbochargers are usually installed downstream with carbureted engines (this is called a draw-through turbocharger). The carburetor generally is located on the intake side of the turbocharger, although it is occasionally positioned on the outlet side. Positioning the carburetor at the compressor inlet allows the turbocharger to increase airflow and pressure drop through the carburetor. This provides the required air-fuel mixture, which is compressed

Turbo Lag: The time interval required for a turbocharger to overcome inertia and spin up to speed.

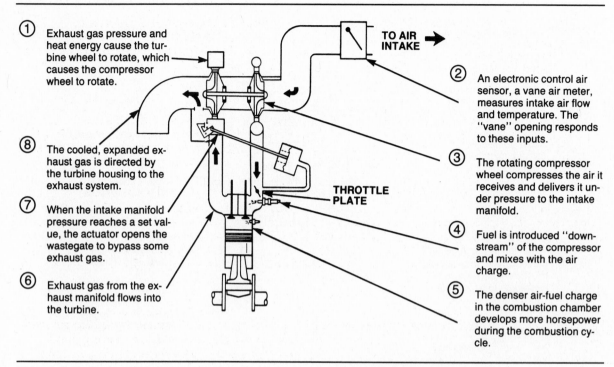

① Exhaust gas pressure and heat energy cause the turbine wheel to rotate, which causes the compressor wheel to rotate.

⑧ The cooled, expanded exhaust gas is directed by the turbine housing to the exhaust system.

⑦ When the intake manifold pressure reaches a set value, the actuator opens the wastegate to bypass some exhaust gas.

⑥ Exhaust gas from the exhaust manifold flows into the turbine.

TO AIR INTAKE ➡

② An electronic control air sensor, a vane air meter, measures intake air flow and temperature. The "vane" opening responds to these inputs.

③ The rotating compressor wheel compresses the air it receives and delivers it under pressure to the intake manifold.

④ Fuel is introduced "downstream" of the compressor and mixes with the air charge.

⑤ The denser air-fuel charge in the combustion chamber develops more horsepower during the combustion cycle.

THROTTLE PLATE

Figure 16-15. The operational cycle of a turbocharger used with a fuel-injected system. (Ford)

and sent to the cylinders through the manifold, figure 16-14.

There is a distinct advantage to positioning the carburetor at the compressor inlet. In this location, the carburetor does not have to be modified to accept and withstand boost pressure. This position of the carburetor simplifies air-fuel ratio control. Locating the carburetor at the turbocharger outlet, on the other hand, means that it must be calibrated to meter fuel correctly under both atmospheric and above-atmospheric conditions. In addition, a carburetor positioned at the turbocharger outlet must also be pressurized to withstand boost pressure without leaking fuel. The fuel system must also move the fuel at a higher pressure to overcome the boost pressure present at the carburetor. These factors all make air-fuel ratio control difficult when the carburetor is located at the turbocharger outlet.

At the same time, there are some minor drawbacks that must be overcome with carburetor placement at the compressor inlet:

• The carburetor is positioned away from the intake manifold, which may cause a slight amount of hesitation when the throttle is opened rapidly.
• Throttle and choke linkages are more complex.
• Preheating the air-fuel mixture on a cold engine is more difficult.

• A special seal must be installed between the center housing and compressor to prevent the air-fuel mixture from entering the engine lubrication system.
• Fuel separation from the compressed mixture may occur before it reaches the manifold.
These disadvantages are not as difficult to deal with as are the problems inherent with a pressurized carburetor. The drawbacks of carburetor placement at the compressor inlet can be overcome with careful manifold design and the proper choice of carburetor and turbocharger.

Upstream installations
Turbochargers are installed upstream with fuel-injected engines (this is called a blow-through turbocharger). The unit is positioned at the manifold air intake and compresses only intake air, figure 16-15. This reduces fuel delivery time and increases the amount of turbine energy available. Most turbocharged engines with fuel injection use a multipoint system, with an individual injector at each cylinder. Just before the compressed air charge enters the cylinder, fuel is injected into it, as shown in step 4, figure 16-15. The throttle plate installed between the turbocharger and injectors regulates airflow.

Mitsubishi uses a throttle body injection (TBI) system on some engines in which the turbocharger delivers compressed air to the injectors in the TBI unit. The throttle plate, however, is located between the injectors and

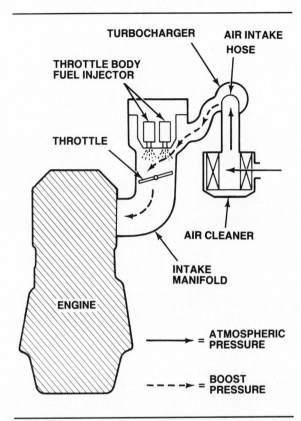

Figure 16-16. The throttle is placed downstream of the turbocharger and fuel injectors in this Mitsubishi TBI system. (Mitsubishi)

the intake manifold, figure 16-16. In this system, the throttle plate regulates the intake volume of compressed air-fuel mixture.

Engine Emissions

Turbochargers have just about the same effect as an increased compression ratio. Why, then, can turbochargers be used on emission-controlled engines when increased compression ratios cannot be used?

The advantage of a turbocharger is the way in which it controls compression. An engine built with a certain compression ratio always will have that ratio. As mentioned earlier, high-compression engines tend to create high exhaust emissions during idle, deceleration, and choked operation. Turbocharger boost can be controlled so that the engine runs as a normally aspirated unit during these operating modes. Under acceleration or heavy load conditions, turbocharger boost is applied to the intake manifold. The resulting increase in compression improves the engine's volumetric efficiency because the percentage of the total cylinder volume which the piston displaces on the intake stroke increases.

Most turbocharged engines now in production use essentially the same emission control devices as do nonturbocharged versions of the same engines.

TURBOCHARGER CONTROLS

You cannot simply bolt a turbocharger onto an engine and automatically pick up free power. As we have seen, a turbocharger increases both compression pressures and loads on the moving parts of an engine. When automakers turned to the turbocharger in the late 1970's, engines required stronger pistons, new piston ring designs, strengthened crankshafts, and other internal modifications in order to survive the stress of greater loads and pressures. Current engine designs, however, generally are strong enough to withstand these forces without major modifications.

Even so, the operation of a turbocharger must be controlled to avoid high exhaust emissions and detonation. The maximum boost pressure of a turbocharger also must be controlled, or the higher compression and power potential could damage the engine. Three factors, then, must be controlled because of the high pressures and temperatures created by a turbocharger. These are:

1. Boost pressure
2. Air-fuel mixture temperature
3. Detonation.

These factors are interrelated. Higher pressures (boost) increase the temperature of the air-fuel mixture, which raises combustion temperatures. Higher combustion temperatures and pressures may combine to cause detonation. The following paragraphs explain how carmakers control these factors and what effects the controls have on turbocharger operation.

Boost Control

Boost increases as turbocharger rotation increases. Boost must be held below the maximum that a given engine can withstand without causing detonation or serious engine damage. Boost control systems either can limit the amount of exhaust gas reaching the turbine, or vent off some of the compressed air mixture before it reaches the combustion chamber. The mechanisms involved are wastegates and blow-off valves.

Wastegates
Turbochargers installed on production engines during the 1970's were limited to 6 or 8 psi (41 or 55 kPa) boost. However, steady advances in electronic engine management systems, turbo-

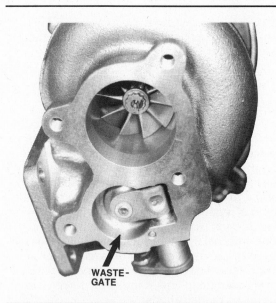

Figure 16-17. A typical wastegate assembly in a turbocharger turbine housing.

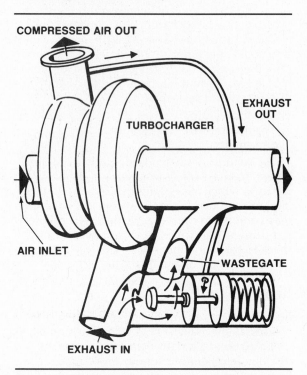

Figure 16-18. The wastegate reacts to intake manifold pressure and controls the amount of exhaust gas reaching the turbine wheel.

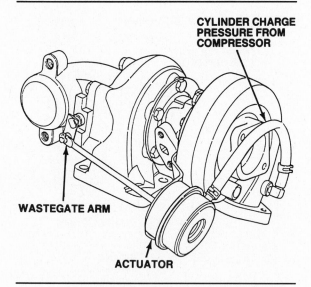

Figure 16-19. Wastegate operation is controlled by linkage connected to a vacuum diaphragm. (Ford)

chargers and engine design now allow many turbocharged engines to operate at boost pressures ranging between 10 and 14 psi (69 and 97 kPa). The most efficient way of controlling boost pressure is with a **wastegate** to control the flow of exhaust gas to the turbine. This is a

bypass valve at the exhaust inlet to the turbine, figure 16-17. The wastegate can allow all of the exhaust into the turbine, or it can route the exhaust directly to the exhaust system, figure 16-18.

Wastegates generally are controlled by an actuator diaphragm linkage, figure 16-19. The diaphragm is exposed to intake manifold pressure. When this pressure reaches a certain level, the diaphragm moves far enough to open the wastegate, which routes exhaust gases directly to the muffler, bypassing the turbine. Since wastegates are constantly exposed to corrosive exhaust gases, they must be made of corrosion resistant alloys.

By bleeding off exhaust gases from the turbine, the wastegate limits the speed of the exhaust turbine and thus controls maximum boost pressure. It also controls detonation to the extent that it holds down boost pressure.

A pressure line connects the actuator to the compressor outlet, figure 16-19. Pressure rises at this point during deceleration because the compressor is working against the closed throttle of an engine that requires little airflow. This causes the actuator to open the wastegate and eliminate overboost during a closed-throttle condition.

Wastegate boost control operates the same on all turbochargers using the system, but the operation of some wastegates can be adapted to adjust the boost level for either low-octane or high-octane fuel. Ford does this on its high-performance engines through a boost control module which interfaces with the engine management computer. The driver indicates the

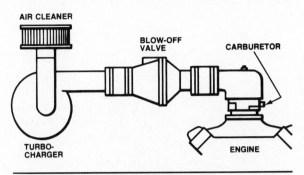

Figure 16-20. A blowoff valve can open to relieve intake manifold pressure.

type of fuel being used with an octane switch on the instrument panel. The switch automatically recalibrates the EEC-IV microprocessor for the fuel selected. The microprocessor then controls a solenoid which routes pressure between the actuator line and the compressor inlet. This allows the wastegate to maintain a lower boost level with low-octane fuel or a higher boost level with high-octane fuel.

Blowoff valve

A **blowoff valve**, or pressure control valve, affects the flow of compressed air between the turbocharger and the engine, figure 16-20. The valve can be operated in different ways according to what it is meant to control.

If the valve is to control the maximum boost pressure, then it is a simple spring-loaded unit. When boost pressure reaches a certain level, the spring tension is overcome and some of the boost pressure is allowed to escape. If the turbocharger is compressing only air, as shown in figure 16-20, then the air can be vented to the atmosphere. If the turbocharger is compressing the combustible air-fuel mixture, then the mixture must be vented back into the turbocharger inlet.

If the blowoff valve is meant to control exhaust emissions, it is operated by intake manifold vacuum. During idle, closed-throttle deceleration, and choked operation, vacuum in the intake manifold causes a diaphragm to move and open the valve. The engine operates as a normally-aspirated unit, avoiding the excessive emissions that the turbocharger compression would cause. Some blowoff valves are operated by vacuum routed to one side of the diaphragm and boost pressure routed to the other.

Mixture Cooling

You have learned that a turbocharger increases the pressure of intake air. As the pressure of air

increases, its temperature also increases. This means that the temperature of the air-fuel mixture provided in a turbocharged engine is higher than that of a normally-aspirated engine.

This higher air-fuel mixture temperature has two unwanted effects: the air-fuel charge density is reduced and it tends to detonate more easily. Detonation is a serious problem with any engine, but even more so with one that is turbocharged. Detonation combined with the high pressures involved in a turbocharged engine can bend a piston rod, burn pistons, or quickly cause other internal damage.

Two methods are used to cool the compressed mixture before it reaches the combustion chamber: water injection systems and intercoolers. The major purpose of each system is to control detonation.

Water injection

The use of **water injection** to cool the intake mixture of a turbocharged engine has been confined mainly to aftermarket or add-on turbochargers, rather than production installations available from carmakers. Water injection can be used most easily on carbureted engines when the turbocharger is located downstream from the carburetor. At maximum boost, a fine spray of water is injected into the carburetor inlet by an electric or vacuum-operated pump. The water has no effect on the combustion process, but since it vaporizes with the gasoline, it tends to cool the intake charge and leaves the engine as water vapor in the exhaust.

Water injection can be highly effective, but it requires the use of precision pumps and valves. Also, water injection requires that the driver maintain a supply of water, something that many drivers would likely forget. Since it is not a driver-proof system, carmakers generally have selected intercooling as a better alternative.

Wastegate: A diaphragm-actuated bypass valve used to limit turbocharger boost pressure by limiting the speed of the exhaust turbine.

Blowoff Valve: A spring-loaded valve that opens when boost pressure overcomes the spring tension.

Water Injection: A method of lowering the air-fuel mixture temperature by injecting a fine spray of water which evaporates as it cools the intake charge.

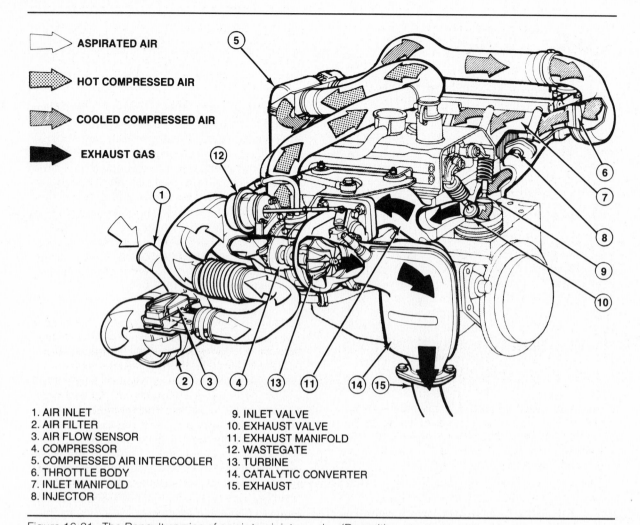

ASPIRATED AIR

HOT COMPRESSED AIR

COOLED COMPRESSED AIR

EXHAUST GAS

1. AIR INLET
2. AIR FILTER
3. AIR FLOW SENSOR
4. COMPRESSOR
5. COMPRESSED AIR INTERCOOLER
6. THROTTLE BODY
7. INLET MANIFOLD
8. INJECTOR
9. INLET VALVE
10. EXHAUST VALVE
11. EXHAUST MANIFOLD
12. WASTEGATE
13. TURBINE
14. CATALYTIC CONVERTER
15. EXHAUST

Figure 16-21. The Renault version of an air-to-air intercooler. (Renault)

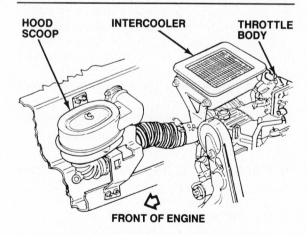

HOOD SCOOP   INTERCOOLER   THROTTLE BODY

FRONT OF ENGINE

Figure 16-22. Ford's high-performance engines use an air-to-air intercooler that draws air from a hood scoop. (Ford)

Intercoolers

An **intercooler** is nothing more than a heat exchanger, figure 16-21. By passing the compressed air from the turbo through an intercooler on its way to the manifold, up to 60 percent of the heat can be removed by air-to-air transfer. Since the cooler air charge is much denser, it contains more oxygen, which results in greater combustion efficiency. Air-to-liquid heat exchangers have also been used as intercoolers, but they are heavier and more complicated than the air-to-air type.

Intercoolers work best on fuel-injected systems or with carbureted systems in which the carburetor is downstream from the turbocharger. With only air passing through the intercooler, there is no possibility of fuel separation from the mixture as it cools.

Among domestic manufacturers, Ford has led the way in air-to-air intercooler use. For maximum efficiency, Ford mounts the intercooler on top of the engine, figure 16-22. The

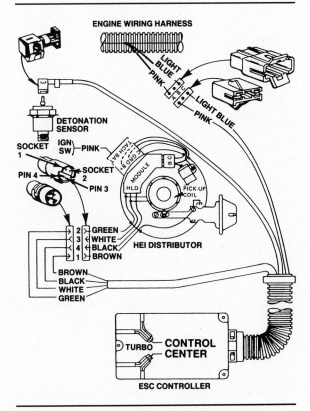

Figure 16-23. A detonation sensor signals the electronic module to retard timing when engine vibrations or knocking occurs. (Buick)

intercooler inlet connects directly to the turbocharger; its outlet connects to the air intake throttle body. The top of the intercooler connects to a hood scoop. This scoop gathers ambient air from outside the engine and directs it into the intercooler, whose air fins remove the heat from the air compressed by the turbocharger.

Spark Timing and Detonation

Excessive heating of the air-fuel mixture can cause an explosion in the combustion chamber known as detonation. This wastes power and can damage the engine, as we learned in Chapter 3. Since a turbocharger compresses and

therefore heats the air-fuel mixture, detonation is a common problem. Various control methods are used to minimize or prevent detonation.

Retarding the spark timing can control detonation by lowering the peak temperature of combustion. Spark control systems can be mechanically or electronically operated.

Mechanical operation

Turbocharged engines with breaker-point ignitions use a distributor vacuum advance unit where manifold pressure acts on a diaphragm to retard ignition timing under high boost conditions. The retard diaphragm is connected to the breaker plate to move it in the opposite direction. This method is similar to the use of a dual-diaphragm distributor to retard timing with high manifold vacuum during idle and deceleration.

Electronic operation

Electronic detonation sensing offers the precise control lacking in a mechanical system. A detonation sensor (piezoelectric crystal) mounted in the engine block or intake manifold senses vibration caused by detonation or engine knock. The detonation sensor works in different ways depending on the system design. When vibration is sensed, the sensor either independently generates a voltage signal to the computer, or voltage signals constantly sent out by the computer are altered by the sensor and then returned to the computer, figure 16-23. The computer retards ignition timing in increments of 2- or 4-degrees until either a maximum retard setting is reached or the detonation stops. When the sensor no longer picks up knocking vibrations, its signal notifies the computer, which gradually restores spark advance to the amount required for optimum engine operation.

Intercooler: An air-to-air or air-to-liquid heat exchanger used to lower the temperature of the air-fuel mixture by removing heat from the intake air charge.

SUMMARY

Supercharging and turbocharging are both proven ways in which mixture compression can be achieved to increase the power resulting from combustion without increasing exhaust emissions. A supercharger is an air pump mechanically driven by the engine in direct relationship to engine speed. Supercharger pumps generally are a positive displacement design in which every revolution pumps the same volume of air. The most common positive displacement supercharger is the lobe-type, or Roots blower. Mechanically driven superchargers may use up to 20 percent of the engine's power even when not in use. The mechanical components involved also are subject to excessive wear.

The Eaton and Toyota superchargers are based on the traditional Roots-type design in which lobes or rotors literally fling air into the intake manifold. Volkswagen uses a spiral-channel design called the G-Lader in which concentric spiral ramps in both sides of a rotor mesh with similar ramps cast in the split casing. The rotor moves around an eccentric shaft instead of spinning on its axis, compressing the air by squeezing it through the spiral.

A turbocharger is an exhaust turbine-drive centrifugal pump that also increases intake air above atmospheric pressure. Except for its motive power, it does essentially the same job as a supercharger in increasing the volumetric efficiency of an engine and developing more power from the air-fuel mixture.

While a turbocharger is a simple device, its clearances are extremely close. Turbochargers operate at speeds up to 150,000 rpm. One major disadvantage of a turbocharger is "turbo lag", the time interval between increasing engine speed and the turbocharger's ability to overcome inertia and spin up to speed. Turbocharger life can be extended by circulating engine coolant through passages provided in the center housing to draw off the excess heat.

Turbochargers can be installed either on the air intake side (upstream) or on the exhaust side (downstream) of the carburetor or injectors. In an upstream installation, the turbocharger compresses and delivers a denser charge of air to the carburetor or injectors. When installed downstream, the air-fuel mixture is compressed and delivered to the cylinders.

Boost pressure, air-fuel mixture temperature, and detonation must be controlled to prevent high exhaust emissions and possible engine damage. Boost pressure is controlled by a waste gate or bypass valve in the exhaust inlet to the turbine. This limits the speed of the exhaust turbine and controls maximum boost pressure. A blowoff valve also can be used. This is a simple spring-loaded unit. When boost pressure reaches a certain level, the spring tension is overcome and some of the boost pressure is allowed to escape.

Mixture temperature can be controlled by injecting a fine spray of water into the intake charge or by use of an intercooler. This is a heat exchanger which removes heat from the air charge before it mixes with the fuel.

Detonation can be controlled mechanically or electronically. Current systems all use electronic control to retard spark timing and lower the peak temperature of combustion. Electronic control is precise. A detonation sensor mounted in the engine block or intake manifold signals the computer whenever it senses vibration caused by detonation or engine knock. The computer gradually retards ignition timing until detonation disappears, then advances timing to its most efficient point.

Review Questions

Choose the single most correct answer.
Compare your answers with the correct answers on page 414.

1. Supercharging delivers the air-fuel mixture to the cylinder at:
   a. Lower than atmospheric pressure
   b. Atmospheric pressure
   c. Higher than atmospheric pressure
   d. Three times the atmospheric pressure

2. Which is *not* true of superchargers?
   a. They can operate at very low rpm
   b. They are mechanically driven
   c. There are two types
   d. Mechanical superchargers consume a lot of engine power

3. Positive displacement pumps:
   a. Contain impellers
   b. Pump the same volume of air each revolution
   c. Increase air pressure by decelerating it
   d. Are highly efficient

4. A turbine wheel turns because of:
   a. Centrifugal force
   b. Exhaust gas speed
   c. Manifold vacuum
   d. Expansion of hot exhaust gases

5. Of the heat energy contained in gasoline:
   a. 50 percent is converted to mechanical power
   b. 50 percent is lost to cooling
   c. 25 percent is converted to engine power
   d. 50 percent goes out the exhaust system

6. Despite high compression pressures achieved, turbochargers can be used on emission-controlled engines because:
   a. They have fixed compression ratios
   b. Turbocharger boost can be varied to meet engine needs
   c. They are placed between the air cleaner and the carburetor
   d. They are placed between the carburetor and the engine

7. Which is *not* a method of controlling a turbocharger system?
   a. Changing the amount of boost
   b. Cooling the compressed mixture
   c. Readjusting the carburetor idle
   d. Altering spark timing

8. The waste gate controls boost by:
   a. Controlling flow of exhaust gas
   b. Controlling compressed air
   c. Controlling the air-fuel mixture
   d. Controlling exhaust emissions

9. Retarding spark time will control detonation by:
   a. Cooling the compressed mixture
   b. Cooling the exhaust manifold
   c. Reducing compression pressure
   d. Lowering peak temperature of combustion

10. Mechanic A says that turbochargers are constant-displacement pumps.
    Mechanic B says that Roots blowers are variable-displacement pumps.
    Who is right?
    a. A only
    b. B only
    c. Both A and B
    d. Neither A nor B

11. Boost control can be limited by a wastegate or by:
    a. An intercooler
    b. A blowoff valve
    c. A detonation sensor
    d. Manifold vacuum

12. Engine exhaust is used to drive a turbocharger:
    a. Turbine
    b. Wastegate
    c. Compressor
    d. Intercooler

13. The momentary hesitation between throttle opening and boost delivery by the turbocharger is called:
    a. Underboost lag
    b. Turbo lag
    c. Ignition lag
    d. Compression lag

14. Mechanic A says that a turbocharger installed upstream from the fuel source compresses only air.
    Mechanic B says that a turbocharger installed downstream from the fuel source compresses air and fuel.
    Who is right?
    a. A only
    b. B only
    c. Both A and B
    d. Neither A nor B

PART FIVE

Emission Control Systems

17

Positive Crankcase Ventilation

The problem of crankcase ventilation has existed since the beginning of the automobile. No piston ring, new or old, can provide a perfect seal between the piston and the cylinder wall. When an engine is running, the pressure of combustion forces the piston downward. This same pressure also forces gases and unburned fuel from the combustion chamber past the piston rings and into the crankcase. These gases are called crankcase vapors, or blowby, as you learned in Chapter 6.

Under perfect conditions, combustion of an engine's air-fuel mixture would completely consume all of the air and the fuel. It would leave only harmless byproducts, such as water vapor and carbon dioxide. However, combustion seldom is perfect and usually is incomplete. The byproducts of incomplete combustion include carbon monoxide (CO), hydrocarbons (HC) and oxides of nitrogen (NO_x).

These combustion byproducts, particularly unburned hydrocarbons, form blowby, figure 17-1. The crankcase must be ventilated to remove these vapors and gases. However, the crankcase on modern engines cannot be ventilated directly to the atmosphere, because the hydrocarbon vapors add to air pollution. Positive crankcase ventilation (PCV) systems were developed to ventilate the crankcase and recirculate the vapors to the engine's induction system.

In this chapter, you will learn:

● The reasons for crankcase ventilation
● The difference between the open and closed ventilation systems
● The components of the modern PCV system
● How the PCV system ventilates the crankcase without polluting the air
● What happens when the PCV system does not work correctly.

DRAFT TUBE VENTILATION

Blowby has three undesirable features:
1. It destroys the lubricating qualities of engine oil.
2. It causes sludge and varnish to form.
3. It helps cause formation of corrosive acids, which can damage engine parts.

After trying various methods of ventilating the engine crankcase, carmakers at first settled on the road draft tube, figure 17-2. This is nothing more than a tube connected to the engine crankcase that allows vapors to pass into the air. Fresh air to ventilate the crankcase enters through a vented oil filler cap. This air passes into the crankcase where it mixes with the vapors. When the car is moving, a vacuum

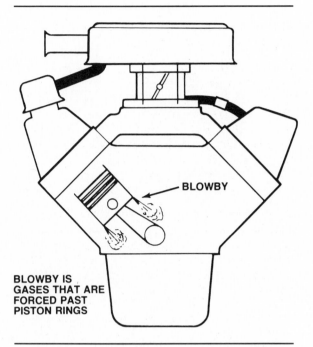

Figure 17-1. Piston rings do not provide a perfect seal. Combustion gases blow by the rings into the crankcase.

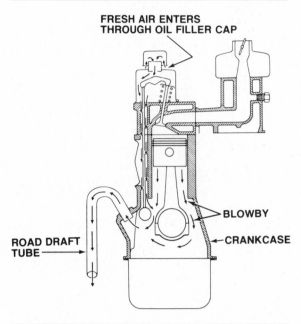

Figure 17-2. The road draft tube ventilates the crankcase to the atmosphere when the car is moving.

is created by the airflow flowing past the road draft tube. This vacuum draws the crankcase vapors out into the atmosphere.

The road draft tube has three major short-comings:

1. It works best only when there is a pressure difference between the oil filler cap and the draft tube. This pressure depends upon car movement. When the car is moving slower than about 25 mph (40 kph), there is not enough vacuum to remove the vapors from the crankcase.

2. It passes the crankcase vapors directly into the atmosphere causing air pollution.

3. At higher vehicle speeds, too much crankcase ventilation will increase engine oil consumption because oil droplets are drawn out through the road draft tube.

POSITIVE CRANKCASE VENTILATION SYSTEMS

The drawbacks of the road draft tube were eliminated when the controlled crankcase ventilation system was introduced. Controlled, or positive, crankcase ventilation relies on intake manifold vacuum to draw the vapors from the crankcase up into the intake manifold. This results in a positive movement of the air through the crankcase whenever the engine is running.

■ It Wasn't Always As Simple

The early PCV systems caused a good deal of grief and engine troubles for automakers. Many garages, even franchised dealers, ignored the PCV systems on 1963-64 cars. They required a lot of care and cleaning, and they clogged quickly when ignored. Contaminants remained in the crankcase, and sludge and moisture formed. This clogged oil lines and prevented adequate engine lubrication. The result was disaster for the engine, and major overhauls on engines still under warranty were often required.

The situation reached a crisis point for one major manufacturer, who stopped using PCV on its cars from the spring of 1964 until early in 1965. Auto engineers were frustrated by the problems PCV systems were creating. The systems had been designed to be simple and require only a minimum amount of service. But mechanics in the field completely ignored the emission control device, and engines began to fail.

These problems resulted in a crash project by the automakers. While engineers worked overtime developing a "better" PCV system, manufacturers started a program to educate dealers, servicemen, and car owners. The so-called "self-cleaning" PCV valve was developed and began appearing on mid-1965 models. This second-generation PCV system is practically the same one in use today.

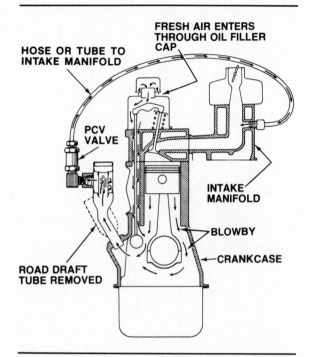

Figure 17-3. In a Type 1 open PCV system, fresh air enters through the oil filler cap. Crankcase vapors are returned to the intake manifold through a PCV valve and hose or tube.

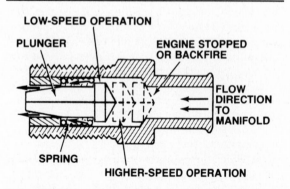

Figure 17-4. The PCV valve plunger responds to spring force and manifold vacuum to regulate the airflow rate.

The vapors are then returned to the combustion chambers, where they are burned. These systems may be classified as open or closed PCV systems, depending upon their design.

Open PCV Systems

Positive crankcase ventilation, as we know it today, received its first major use on 1961 California cars. Open systems were installed on many 1963 models nationwide. Open crankcase ventilation systems can be divided into three types.

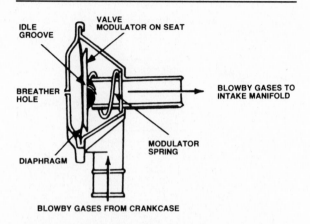

Figure 17-5. When the engine is at idle, the Type 2 PCV valve is closed by crankcase vacuum. Vapors flow through the idle groove at about 3 cubic feet (85 liters) per minute.

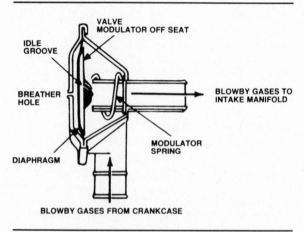

Figure 17-6. When the engine is at cruising speed, the Type 2 PCV valve is opened by crankcase pressure. The flow rate depends on the amount of blowby created by the engine.

Type 1 system operation

A hose connects the crankcase with the intake manifold, figure 17-3. When the engine is running, fresh air is drawn into the crankcase through the vented oil filler cap. This air mixes with the crankcase vapors, travels to the intake manifold where it is drawn into the engine cylinders. Airflow from the crankcase is metered through a PCV valve, figure 17-4, that contains a spring-oriented plunger to control the rate of airflow through the engine.

Type 1 open PCV systems were used as original equipment on many domestic passenger car engines through the mid-1960's.

Type 2 system operation

A Type 2 system is similar to a Type 1 system, but uses a special PCV valve and oil filler cap.

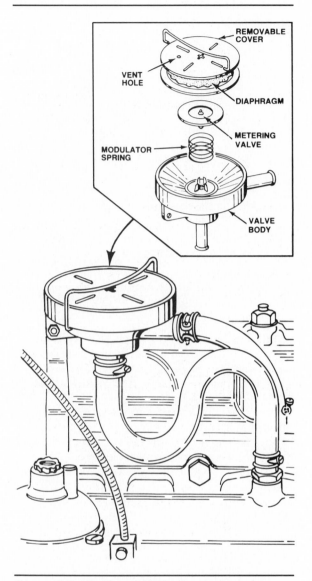

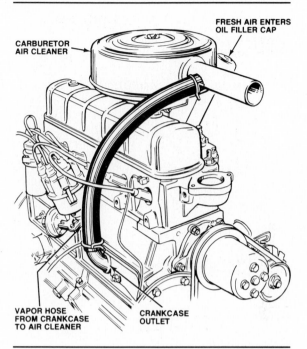

Figure 17-8. A Type 3 open PCV system has no PCV valve.

Type 3 system operation

In a Type 3 system, the crankcase is connected to the air cleaner by a hose, figure 17-8. Vapors are drawn from the crankcase by a slight suction in the air cleaner. No control valve is used. Fresh air enters through the oil filler cap, mixes with the crankcase vapors, and is drawn to the air cleaner. From the air cleaner, the vapors are then drawn through the carburetor and into the cylinders.

■ Check PCV Valve For Fuel Dilution

Winter driving, with its short trips and cold starts, can cause fuel dilution of the engine oil. This will not only thin out the oil, but the gasoline vapors fed to the intake manifold through the PCV system will produce a rich idle mixture. If you are working on a car with a rich idle mixture and you suspect fuel dilution of the oil, you can check it quickly.

Remove the PCV valve from the valve cover and let it draw in fresh air with the engine running. If the idle mixture becomes normal or leaner than normal, the engine oil probably has too much fuel dilution. Drive the car at highway speeds for up to half an hour to purge fuel vapors from the crankcase. Then readjust the idle mixture. As an alternative, drain the crankcase and refill with fresh oil.

Figure 17-7. The Type 2 PCV valve used on some imported cars could be disassembled for cleaning.

The diaphragm-type PCV valve regulates the airflow according to crankcase vacuum. When a vacuum exists in the crankcase, the valve is closed, figure 17-5. When the crankcase is under pressure, figure 17-6, the valve opens. The oil filler cap has an orifice large enough to allow the right amount of air to enter but small enough to maintain crankcase vacuum. The orifice must remain open at all times for the system to work correctly.

Some imported cars, particularly British models in the mid-1960's, had Type 2 PCV systems as original equipment. These were called Smith's PCV valves, figure 17-7. They worked like the one we just described and could be opened for cleaning.

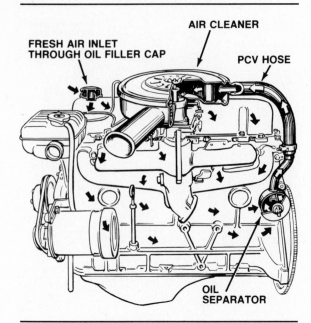

Figure 17-9. Ford Type 3 PCV systems had oil separators at the crankcase outlet.

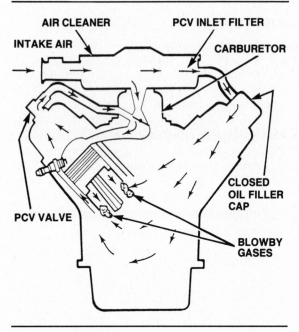

Figure 17-10. Type 4 sealed PCV system operation.

Because blowby reaches the carburetor *before* the rich air-fuel vaporization, this system tends to richen the air-fuel mixture. Carburetors used with a Type 3 system must be adjusted to make up for this richer mixture. Some imported cars use a "sealed" Type 3 system in which the oil filler cap is not vented.

Type 3 PCV systems were used as original equipment on some Ford and American Motors

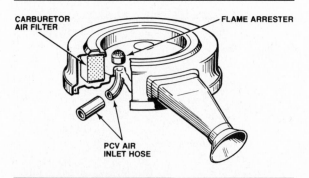

Figure 17-11. When PCV inlet air is drawn from the clean side of the air cleaner, a flame arrestor is required in case of a backfire.

cars in the early 1960's. The systems used on Ford 6-cylinder engines had oil separators at the crankcase outlets to minimize the amount of oil drawn through the PCV hose to the air cleaner, figure 17-9.

Limited efficiency of the open PCV system
Open PCV systems only partly control crankcase emissions. Manifold vacuum decreases considerably under heavy load or acceleration, causing crankcase pressures to build. This forces some of the vapors into the atmosphere through the vented oil filler cap. Crankcase vapors also pass through the vented cap into the air if the system becomes clogged.

Open PCV systems do provide some benefits:
1. They promote longer engine life by removing most harmful vapors from the crankcase.
2. They reduce the amount of crankcase vapors which pollute the air.

Closed PCV Systems

Closed, or Type 4, PCV systems were required on all new California cars in 1964. They were standard nationwide by 1968 and are still used today on all new domestically built cars, and all cars imported into the United States.

In a closed PCV system, figure 17-10, the oil filler cap is not vented to the atmosphere. Air for the crankcase is drawn through a hose from the air cleaner to one of the valve covers or to a crankcase inlet below the intake manifold. The dipstick also is sealed to prevent air from leaking into the crankcase.

Crankcase ventilation air may come from either the clean side (inside) or the dirty side (outside) of the carburetor air filter. When air is drawn from the clean side (using the air cleaner filter as a PCV filter), a flame arrester, figure 17-11, is used. This wire screen is in the PCV air intake line, either at the air cleaner or at the

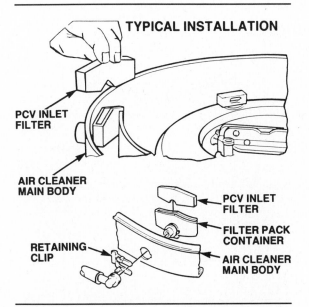

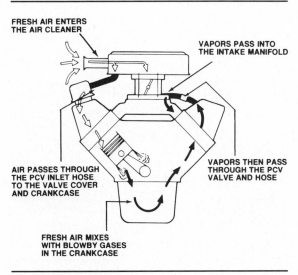

Figure 17-14. Closed PCV system operation under normal conditions.

Figure 17-12. The PCV inlet filter on many vehicles is located in the air cleaner housing and is serviced as shown.

valve cover. Its purpose is to prevent a crankcase explosion if the engine backfires.

When air is drawn from the dirty side of the air cleaner, a separate PCV air filter, or crankcase ventilation filter, is used. This can be located in the air cleaner, figure 17-12, in the oil filler cap, figure 17-13, or in the inlet air hose where it connects to the valve cover. A flame arrester is not required in this case, since the air cleaner filter does that.

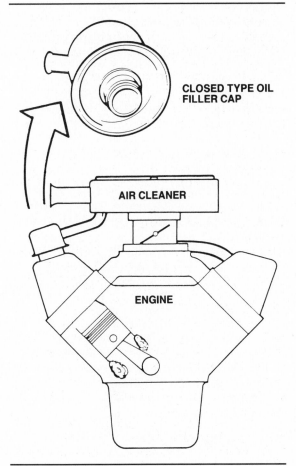

■ PCV System Service

When a PCV system becomes restricted or clogged, the cause is usually an engine problem or the lack of proper maintenance. For example, scored cylinder walls or badly worn rings and pistons will allow too much blowby. Start-and-stop driving requires more frequent maintenance and causes PCV problems more quickly than highway driving, as will any condition allowing raw fuel to reach the crankcase. Using the wrong grade of oil, or not changing the crankcase oil at periodic intervals will also cause the ventilation system to clog.

When a PCV system begins to clog, the engine tends to stall, idle roughly, or overheat. As ventilation becomes more restricted, burned plugs or valves, bearing failure, or scuffed pistons can result. Also look for an oil-soaked distributor or points, or leaking out around valve covers or other gaskets. Do not overlook the PCV system while troubleshooting. A partly or completely clogged PCV valve, or one of the incorrect capacity, may well be the cause of poor engine performance.

Figure 17-13. On other engines, the PCV inlet filter may be located in the oil filler cap.

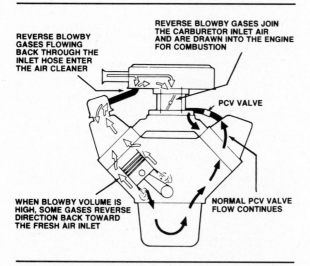

Figure 17-15. Closed PCV system operation under heavy load.

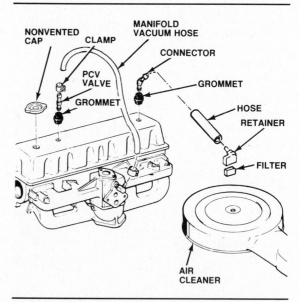

Figure 17-16. Late-model closed PCV systems generally share these common components. (AMC)

Type 4 system operation

Under normal conditions, fresh air from the air cleaner passes through the inlet hose to the crankcase, figure 17-14. The fresh air mixes with the crankcase vapors and passes through a PCV valve before being drawn into the intake manifold. Vapors that back up under certain conditions cannot escape from the closed system. If manifold vacuum drops or if the system becomes clogged, extra crankcase vapors will reverse their direction. In the closed crankcase ventilation system, these excess vapors flow back to the air cleaner, figure 17-15, instead of passing out of the engine and into the atmosphere. Once in the air cleaner, they mix with incoming air and pass through the carburetor, throttle body, or intake manifold (port injection) to be burned in the combustion chamber. This makes the closed system almost 100 percent effective in controlling crankcase emissions.

Closed PCV system efficiency

Closed crankcase ventilation provides three benefits:

1. It promotes longer engine life by removing harmful vapors from the crankcase.
2. It eliminates crankcase vapors that pollute the air.
3. It increases fuel economy by recirculating all unburned blowback to the intake manifold.

When air flows freely, the PCV system will function properly, as long as the PCV valve is not clogged. Modern engine design includes the air and vapor flow as a calibrated portion of the air-fuel mixture. In fact, some engines receive as much as 30 percent of their idle air through the PCV system. For this reason, a flow problem in the PCV system will result in driveability problems.

A PCV system that is not properly vented, or scavenged, will result in oil dilution, formation of sludge, and oil deposits collecting in the air cleaner. Unlike driveability problems, these are not immediately noticed and their effect is long-term, resulting in premature engine wear and inefficient air cleaner operation.

ORIGINAL EQUIPMENT CLOSED PCV SYSTEMS

All new cars sold in the United States since 1968 have a Type 4 closed PCV system, figure 17-16. The design of closed PCV systems is essentially the same, regardless of manufacturer. All use a PCV valve or calibrated orifice, an air inlet filter, and connecting hoses. An oil/vapor or oil/water separator is used in some systems. The oil/vapor separator lets oil vapors condense and drain back into the crankcase, figure 17-17. The oil/water separator accumulates moisture and prevents it from freezing during cold engine starts, figure 17-18. Separators generally are used with turbocharged and fuel-injected engines. The location of each of these components may vary from one engine to another, but all work the same.

Air Inlet Filter

PCV air inlet filters usually are installed in a retainer inside the air cleaner, figure 17-12, in the oil filter cap, figure 17-13, or in the

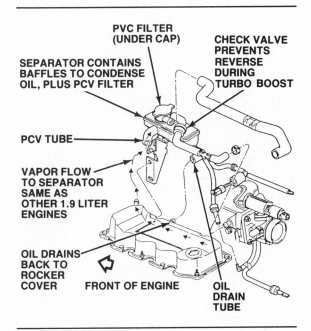

PVC FILTER
(UNDER CAP)

SEPARATOR CONTAINS
BAFFLES TO CONDENSE
OIL, PLUS PCV FILTER

CHECK VALVE
PREVENTS
REVERSE
DURING
TURBO BOOST

PCV TUBE

VAPOR FLOW
TO SEPARATOR
SAME AS
OTHER 1.9 LITER
ENGINES

OIL DRAINS
BACK TO
ROCKER
COVER

FRONT OF ENGINE

OIL
DRAIN
TUBE

Figure 17-17. Turbocharger pressure increases blowby and creates a large quantity of oil vapor. The oil/vapor separator allows the vapors to condense and drain back into the crankcase. (Ford)

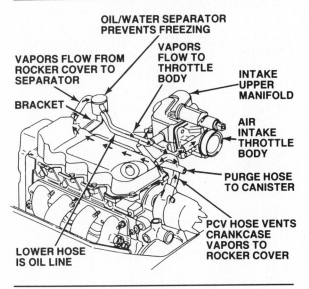

OIL/WATER SEPARATOR
PREVENTS FREEZING

VAPORS FLOW FROM
ROCKER COVER TO
SEPARATOR

VAPORS
FLOW TO
THROTTLE
BODY

INTAKE
UPPER
MANIFOLD

BRACKET

AIR
INTAKE
THROTTLE
BODY

PURGE HOSE
TO CANISTER

PCV HOSE VENTS
CRANKCASE
VAPORS TO
ROCKER COVER

LOWER HOSE
IS OIL LINE

Figure 17-18. An oil/water separator may be used on some fuel injected engines to accumulate moisture and prevent it from freezing during cold engine starts. (Ford)

oil/vapor separator, figure 17-17. They are made of wire gauze, wire mesh, or polyurethane foam. Some wire gauze or wire mesh filters can be cleaned and reused. Others must be replaced, and like the polyurethane foam type, usually are replaced at the same time as the air cleaner filter element.

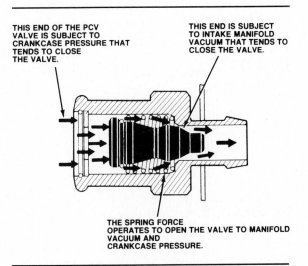

THIS END OF THE PCV
VALVE IS SUBJECT TO
CRANKCASE PRESSURE THAT
TENDS TO CLOSE
THE VALVE.

THIS END IS SUBJECT
TO INTAKE MANIFOLD
VACUUM THAT TENDS TO
CLOSE THE VALVE.

THE SPRING FORCE
OPERATES TO OPEN THE VALVE TO MANIFOLD
VACUUM AND
CRANKCASE PRESSURE.

Figure 17-19. Spring force, crankcase pressure, and manifold vacuum work together in regulating PCV valve flow rate.

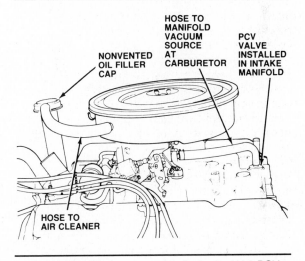

HOSE TO
MANIFOLD
VACUUM
SOURCE
AT
CARBURETOR

PCV
VALVE
INSTALLED
IN INTAKE
MANIFOLD

NONVENTED
OIL FILLER
CAP

HOSE TO
AIR CLEANER

Figure 17-20. The components of a typical V-8 PCV system. (AMC)

Connecting Hoses

The closed PCV system uses two connecting hoses. Fresh air travels from the air cleaner to the engine through the air inlet hose. Crankcase vapors travel from the engine to the intake manifold through the manifold vacuum hose. PCV hoses are made of special materials that resist oil vapors. Heater or fuel line hose should *not* be used as a substitute.

PCV Valve

This one-way valve has a spring-operated plunger, figure 17-19, to control valve flow rate. Flow rate is set for each engine and a valve for a

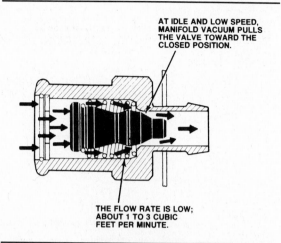

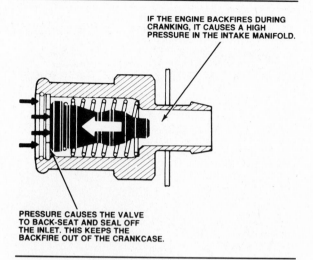

AT IDLE AND LOW SPEED, MANIFOLD VACUUM PULLS THE VALVE TOWARD THE CLOSED POSITION.

THE FLOW RATE IS LOW; ABOUT 1 TO 3 CUBIC FEET PER MINUTE.

Figure 17-21. PCV valve airflow during cruising and light-load operation.

IF THE ENGINE BACKFIRES DURING CRANKING, IT CAUSES A HIGH PRESSURE IN THE INTAKE MANIFOLD.

PRESSURE CAUSES THE VALVE TO BACK-SEAT AND SEAL OFF THE INLET. THIS KEEPS THE BACKFIRE OUT OF THE CRANKCASE.

Figure 17-23. PCV valve operation in case of a backfire.

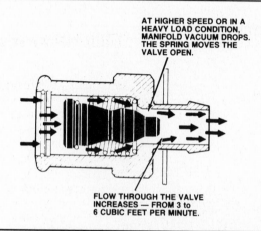

AT HIGHER SPEED OR IN A HEAVY LOAD CONDITION, MANIFOLD VACUUM DROPS. THE SPRING MOVES THE VALVE OPEN.

FLOW THROUGH THE VALVE INCREASES — FROM 3 to 6 CUBIC FEET PER MINUTE.

Figure 17-22. PCV valve airflow during acceleration and heavy-load operation.

different engine should not be substituted. This setting is determined by the size of the plunger and the holes inside the valve. PCV valves usually are located in the valve cover, figure 17-16, or in the intake manifold, figure 17-20.

The PCV valve regulates the airflow through the crankcase under all driving conditions and speeds. When manifold vacuum is high (cruising and light-load operation), the PCV valve restricts the airflow, figure 17-21, to maintain a balanced air-fuel ratio. It also prevents high intake manifold vacuum from pulling oil out of the crankcase and into the intake manifold.

Under high speed or heavy loads, the valve opens and allows maximum airflow, figure 17-22. If the engine backfires, the valve will close instantly, figure 17-23, to prevent a crankcase explosion.

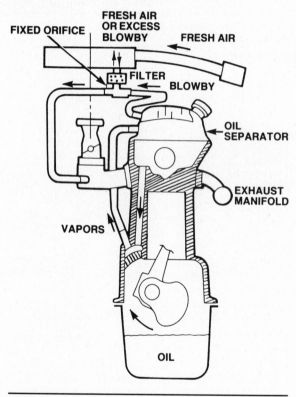

Figure 17-24. The orifice flow control system used on 1981-85 Ford 1.6L/1.9L Escort engines does not use fresh air scavenging of the crankcase.

Orifice-Controlled Systems

The closed PCV system used on 4-cylinder engines in some domestic and imported cars contains a calibrated orifice instead of a PCV valve. This often is called an orifice-controlled system.

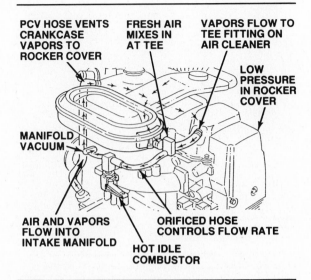

Figure 17-25. Normal operation of the 1981-85 Ford 1.6L/1.9L Escort orifice flow control system.

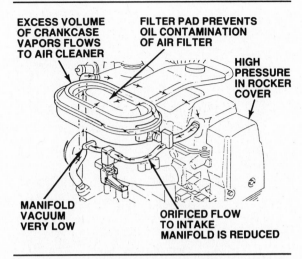

Figure 17-26. Operation of the 1981-85 Ford 1.6L/1.9L Escort orifice flow control system under heavy acceleration or high speed conditions.

The orifice may be located in the valve cover or intake manifold, or in a hose connected between the valve cover, air cleaner, and intake manifold.

While most orifice flow control systems work the same as a PCV valve system, they may not use fresh air scavenging of the crankcase. The carbureted 1981-85 Ford 1.6L/1.9L Escort engine is a good example of this design, figure 17-24. Crankcase vapors are drawn into the intake manifold in calibrated amounts depending on manifold pressure and the orifice size. If vapor availability is low, as during idle, air is drawn in with the vapors. During off-idle operation, excess vapors are sent to the air cleaner. Normal operation of this particular orifice flow control system is shown in figure 17-25; figure 17-26 shows the system under heavy acceleration or high-speed operation.

A dual orifice valve is used on carbureted 1986 and later Ford 1.9L Escort engines to increase PCV flow during off-idle engine operation. At idle, PCV flow is controlled by a 0.050-inch (1.3-mm) orifice. As the engine moves off-idle, spark port vacuum pulls a spring-loaded valve off its seat, allowing PCV flow to pass through a 0.090-inch (2.3-mm) orifice.

■ Retrofit PCV Systems

Several retrofit devices are made and sold to convert an open-type PCV system to the closed-type. These devices are about the same as the factory-installed closed PCV systems, and are offered in two versions: one for the crankcase and one for the air cleaner. The kit components are used to install PCV systems on older engines that did not have them.

Why would anyone want to install a retrofit device? The State of California provides one answer to the question. Used cars sold in California must have a PCV system approved by the state. Those 1955-67 cars originally sold outside California with open PCV systems can be legally driven by their original owner. However, when ownership is transferred within the state (sold as used cars), their open PCV system must be converted to a closed-type system to meet state requirements before they can be properly registered and licensed.

Retrofit kits are made for specific engines and each kit contains detailed instructions for installation and service. Once installed, these kits are serviced the same way as the factory-installed, closed PCV systems.

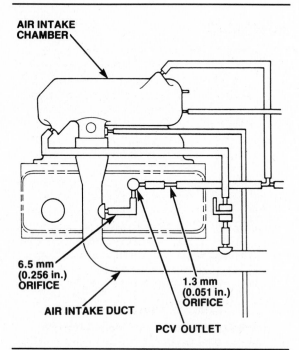

Figure 17-27. The AMC/Renault dual-orifice PCV system used with multiport injection. (AMC/Renault)

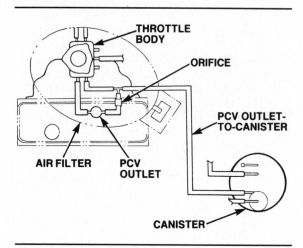

Figure 17-28. The AMC/Renault single orifice PCV system used with throttle body injection. (AMC/Renault)

Other variations of the orifice flow control system have been used over the years, primarily by foreign manufacturers. AMC/Renault 4-cylinder engines with multipoint fuel injection use a 2-orifice system, figure 17-27. One orifice is installed in the air intake chamber circuit; the other orifice is located in the air intake duct circuit. PCV flow passes through one or the other of these orifices according to engine operating conditions.

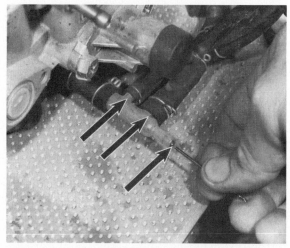

Figure 17-29. The dual-return crankcase ventilation system on Honda CVCC engines connects to the carburetor with a joint containing orifices at the points shown. They can be cleaned with an appropriate size drill bit.

Figure 17-30. The breather hose should be removed from Honda CVCC condensation chambers to drain accumulated fluids.

On AMC/Renault 4-cylinder engines with throttle body fuel injection, a single orifice is located in a vacuum line that tees into the canister purge-to-throttle body line and connects to the valve cover, figure 17-28.

Another common example is the "dual-return" crankcase ventilation system found on Honda CVCC engines. In this system, blowby travels up through the engine, where it passes through a liquid/vapor separator in the top of the valve cover. The vapors then are directed through a breather hose into a plastic condensation chamber installed on the underside of the air cleaner housing. Separation occurs again

Figure 17-31. Combustion by-products which form harmful sludge and varnish are trapped in the Dual Return condensation chamber.

inside the chamber. The lighter vapors are directed into the carburetor through a series of fixed orifices, figure 17-29, according to the amount of manifold vacuum present.

Condensation settles in the chamber, with solids remaining at the bottom and liquids passing into a drain tube. Early models had a slit in the tube that allowed accumulated oil and water to drain when the tube was squeezed enough to open the slit. Later models have a regular hose which must be removed from the chamber and inverted to drain accumulated fluid, figure 17-30. The condensation chamber also must be removed from the air cleaner housing periodically and cleaned in solvent to remove sludge and varnish, figure 17-31.

SUMMARY

Pressure in the engine cylinders forces combustion gases past the pistons. These gases, called blowby, settle in the engine crankcase where they contaminate the lubricating oil and create harmful acids. Ventilation is necessary to remove the vapors from the crankcase. The draft tube system was used until the 1960's but it doesn't work well at low speeds. It also allows the vapors to pollute the air.

The recirculation of crankcase vapors to the intake manifold is called positive crankcase ventilation. Four types of positive crankcase ventilation (PCV) systems have been used. Types 1, 2, and 3 are called open systems because vapors are forced into the atmosphere whenever manifold vacuum is low, or if the system is clogged. Type 4 is called a closed system because vapors cannot escape into the air under any conditions.

A malfunctioning PCV system can cause driveability problems and create harmful acids and sludge that affect engine lubrication and result in premature wear.

PCV systems use a fixed or variable metering device to regulate airflow. The variable metering device, or PCV valve, is standard on most modern PCV systems. It contains a spring-operated plunger and reacts to manifold vacuum. It is important that the correct PCV valve be installed in an engine. Differences in PCV valve idle airflow can greatly influence idle smoothness and vehicle driveability when incorrect valves are used.

The fixed metering device, or orifice, is used on some domestic and foreign 4-cylinder engines. Honda's dual-return system is a variation of this design.

Review Questions
Choose the single most correct answer.
Compare your answers with the correct answers on page 414.

1. In a PCV system, crankcase vapors are recycled to the:
   a. Exhaust system
   b. Road draft tube
   c. Oil-filler breather cap
   d. Intake manifold

2. Which of these effects of blowby is not harmful to the engine?
   a. It destroys lubricating qualities of the engine oil
   b. It causes air pollution
   c. It causes sludge and varnish to form
   d. It causes formation of corrosive acids

3. Road draft tubes:
   a. Draw fresh air from the carburetor
   b. Work off intake manifold vacuum
   c. Are efficient at low speeds
   d. None of the above

4. Controlled crankcase ventilation was introduced in California in:
   a. 1959
   b. 1960
   c. 1961
   d. 1962

5. The illustration shows a:
   a. Type 1 PCV system
   b. Type 2 PCV system
   c. Type 3 PCV system
   d. Type 4 PCV system

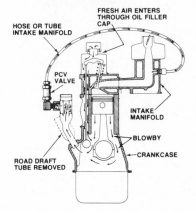

HOSE OR TUBE
INTAKE MANIFOLD

FRESH AIR ENTERS
THROUGH OIL FILLER
CAP

PCV
VALVE

INTAKE
MANIFOLD

BLOWBY

CRANKCASE

ROAD DRAFT
TUBE REMOVED

6. Type 2 PCV systems have:
   a. A closed oil filler cap
   b. An open oil filler cap
   c. An oil filler cap with a preset orifice
   d. None of the above

7. In Type 3 PCV systems:
   a. The crankcase is connected to the intake manifold
   b. The PCV valve is a plunger type
   c. The PCV valve is a diaphragm type
   d. There is no PCV valve

8. The Type 3 PCV system:
   a. Is vented to the intake manifold
   b. Tends to make the air fuel mixture leaner
   c. Has no effect on fuel mixture
   d. Tends to make the fuel mixture richer

9. Closed (Type 4) PCV systems became standard nationwide in which year?
   a. 1964
   b. 1966
   c. 1967
   d. 1968

10. A separate flame arrester is used in a closed PCV system when inlet air is drawn from the:
    a. Clean side of the carburetor filter
    b. Dirty side of the carburetor filter
    c. The intake manifold
    d. The oil filler cap on a valve cover

11. The advantages of a closed PCV system are:
    a. It promotes longer engine life
    b. It eliminates crankcase vapors almost entirely
    c. It increases fuel economy
    d. All of the above

12. Which is not part of a Type 4 PCV system?
    a. A PCV valve
    b. A vented oil filler cap
    c. An air inlet filter
    d. A manifold vacuum hose

13. PCV system hoses are made from:
    a. Heater hose material
    b. Low-temperature-resistant nylon
    c. Oil-resistant rubber material
    d. Fuel-resistant neoprene

14. The PCV valve operates in which of the following ways?
    a. Restricts airflow when intake manifold vacuum is high
    b. Increases airflow when intake manifold vacuum is low
    c. Acts as a check valve in case of carburetor backfire
    d. All of the above

15. Most orifice-controlled systems work the same as PCV valve *except*:
    a. They are open systems
    b. They may not use fresh air scavenging of the crankcase
    c. They cannot be used with fuel-injected engines
    d. All of the above

Chapter

18

Air Injection

Carmakers have various names for their air injection systems, but all do about the same job. They provide additional air to the exhaust manifold where it mixes with the hot exhaust leaving the engine. This helps the oxidation, or burning reaction, necessary to lower HC and CO emissions.

In this chapter, you will learn:

● The reasons for, and principles of, air injection

● The components used in a typical pump air injection or pulse air injection system

● The changes which created second-generation systems used with catalytic converters.

THE NEED FOR AIR INJECTION

During the early days of emission control, **air injection** was an easy way to meet the required standards. Also known as the air pump system, figure 18-1, air injection was one of the first add-on devices used to help oxidize HC and CO exhaust emissions. By 1966, Chrysler was the only domestic automaker not using an air pump system on at least some cars. Early air injection systems contained many hoses and tubes placed across the engine. The external connecting lines made it difficult to work on the engine, and hose failure due to engine heat was common.

While the air injection system was an add-on device, it actually modified the basic process of combustion in the engine. Combustion, as you know, is an oxidation reaction, but usually an incomplete reaction. By providing air to the exhaust system as soon as the hot exhaust gases leave the cylinder, the injection system makes possible the continued oxidation of any HC and CO remaining in the exhaust. As a result, the HC and CO combine with O_2 to form H_2O vapor and CO_2.

Chrysler, however, chose to modify the design of its engines instead of adding air injection. The results proved that emission standards could be met without the add-on pump system. The other automakers profited from Chrysler's experience, and so domestic cars of the later 1960's relied more on engine modifications and less on air injection. The use of air injection decreased on most engines until 1972. To meet the more stringent emission standards that year, even Chrysler was forced to install an air injection system on some engines.

As emission standards became even more strict during the 1970's, automakers became more dependent on air injection. When the catalytic converter was introduced in 1975, it

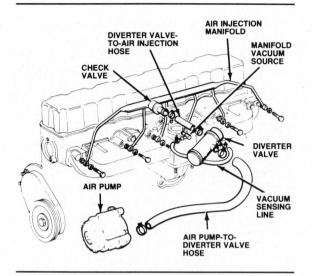

Figure 18-1. The basic components of an air injection system.

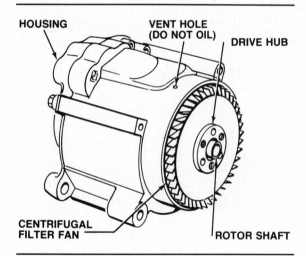

Figure 18-2. Late-model 2-vane air pumps have an external centrifugal filter fan mounted on the front of the housing.

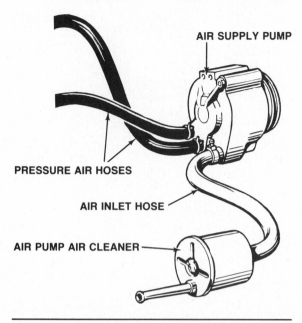

Figure 18-3. Older 3-vane pumps had separate inlet air cleaners.

BASIC SYSTEM DESIGN AND OPERATION

Manufacturers use the following names for their air injection systems:

● American Motors — Air Guard
● Chrysler — Air Injection System
● Ford — Thermactor or Managed Thermactor Air System
● General Motors — Air Injection Reactor (AIR).

Regardless of its name, the basic air injection system is a relatively simple system, generally consisting of:

1. A belt-driven air pump with inlet air filter
2. One or more air distribution manifolds and nozzles
3. Antibackfire valve
4. One or more check valves
5. Connecting hoses.

Air Injection Pump

The air pump, figure 18-2, is normally mounted at the front of the engine and is driven by a belt from the crankshaft pulley. The pump pulls fresh air in through an external filter and pumps it under slight pressure to each exhaust port through connecting hoses. Adding this extra air to the hot HC and CO emissions in the exhaust manifold causes oxidation to take place. This helps change the HC and CO into H_2O and CO_2.

was hoped that the air pump system could be abandoned for good. However, the additional air provided by the injection system was necessary to increase the catalytic action on many engines. With these changes, the air injection system seemed a permanent part of emission control for many engines.

However, the development of sophisticated electronic fuel management and engine control systems may have turned the tide again. A number of the new and smaller engines equipped with electronic fuel injection now can meet emission standards without air injection. If the trend continues, air injection systems may become less important in the future.

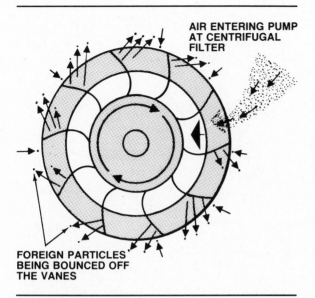

Figure 18-4. Dust and dirt are removed from the inlet air by centrifugal force.

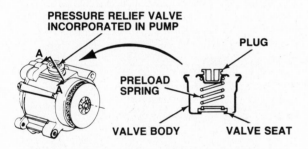

Figure 18-5. Some air pumps have built-in pressure relief valves. (AC-Delco)

The Saginaw Division of General Motors makes most air pumps used on domestic cars. Early Saginaw pumps (1966-67) were a 3-vane design, figure 18-3, and could be rebuilt if necessary. A Saginaw 2-vane design replaced the 3-vane pump on 1968 and later models. The 2-vane pumps cannot be rebuilt but must be replaced if they fail. Older 3-vane pumps still in use usually are replaced with new 2-vane pumps if they fail.

The main difference between the two pumps is the way they filter the intake air. The 3-vane pump draws its fresh air supply through a separate air filter or from the clean side of the air cleaner. The 2-vane pump uses an impeller-type, centrifugal air filter fan mounted on the air pump rotor shaft. This is not a true filter, but cleans the air entering the pump by centrifugal force, figure 18-4. The relatively heavy

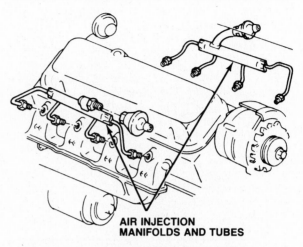

Figure 18-6. External air manifolds are used with many air injection systems. (Chevrolet)

dust particles in the air are forced in the opposite direction to the inlet air flow. The lighter air is then drawn into the pump by the impeller-type fan.

To prevent pump pressure from becoming too high, many pumps use a pressure relief valve, figure 18-5, which opens at high engine speed. Some late-model pumps use a replaceable plastic plug to control the pressure setting of the relief valve. This pressure setting can be changed by putting in a plug with a different pressure setting. Pumps without a pressure relief valve use a diverter valve, explained later in this section.

Air Distribution Manifolds and Nozzles

In early air injection systems, air is delivered to the engine's exhaust system in one of two ways:

1. An external air manifold, figure 18-6, distributes the air through injection tubes to the exhaust port near each exhaust valve. This method is used primarily on smaller engines.

Air Injection: A way of reducing exhaust emissions by injecting air into each of the exhaust ports of an engine. It mixes with the hot exhaust and oxidizes the HC and CO to form H_2O and CO_2.

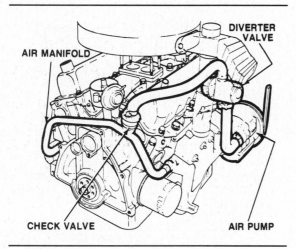

Figure 18-7. Some engines have air distributor passages built into the cylinder heads. (Ford)

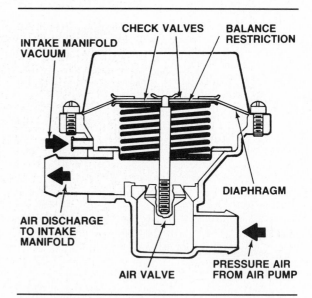

Figure 18-9. The gulp valve was used on early air injection systems and still can be found on some late-model vehicles.

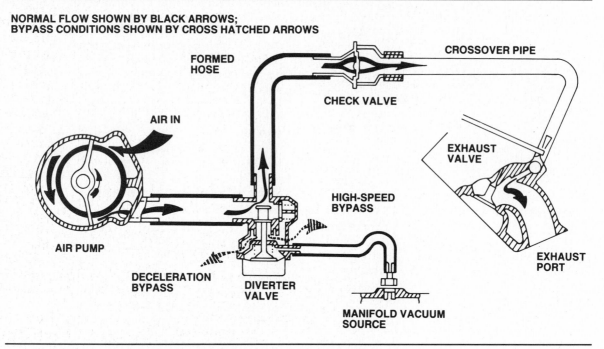

**NORMAL FLOW SHOWN BY BLACK ARROWS;
BYPASS CONDITIONS SHOWN BY CROSS HATCHED ARROWS**

Figure 18-8. Air distribution in a typical air injection system. (AC-Delco)

2. An internal air manifold, figure 18-7, distributes the air to the exhaust port near each exhaust valve through passages cast in the cylinder head or the exhaust manifold. This method is used mainly with larger engines.

The fresh air from the pump passes through the air injector tubes or the manifold to nozzles in the exhaust port. These are made of stainless steel to resist the high exhaust temperatures. This design injects the air in the exhaust at a time when, and at a point where, the gases are the hottest. It also ensures that each cylinder will receive equal air injection for the exhaust. Figure 18-8 shows how air is distributed by a typical air injection system.

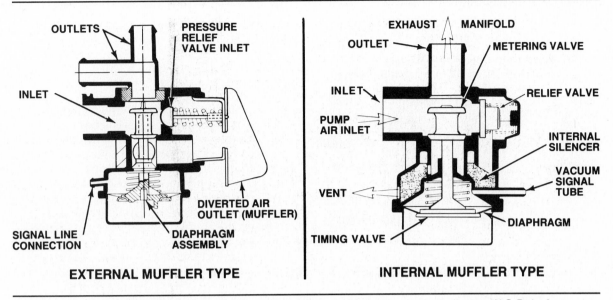

Figure 18-10. Two types of diverter valves. They differ mainly in placement of the air silencer. (AC-Delco)

Antibackfire Valve

During engine deceleration, high intake manifold vacuum enriches the air-fuel mixture. If air is allowed to flow into the exhaust manifold during deceleration, it will combine with excess unburned fuel in the exhaust. The result is engine **backfire** — a rapid combustion of the unburned gases that can destroy a muffler. To prevent engine backfire, the air pump flow must be shut off during deceleration. This is done by a diaphragm-operated antibackfire, or backfire suppression, valve. Two kinds of valves have been used: the gulp valve and the diverter valve. They differ in the direction in which they redirect the airflow.

Gulp valve

Early air injection systems used a **gulp valve**, figure 18-9. When intake manifold vacuum is applied to the valve diaphragm, it causes the air valve to move. This redirects the pump air to the intake manifold to lean out the enriched air-fuel mixture during deceleration.

The gulp valve is connected to the intake manifold by two hoses. The large hose is the air discharge hose. The small hose is the sensing hose that sends manifold vacuum to the gulp valve to operate the diaphragm. There are balance restrictions or bleed holes inside the valve that equalize pressure on both sides of the diaphragm after a few seconds. Even if manifold vacuum is high, the gulp valve only stays open for a few seconds until pressure equalizes.

Any sudden change in vacuum will operate the gulp valve. This is one of the undesirable

features that led to its replacement. For example, the gulp valve will open when the engine starts. This can cause hard starting and a rough idle. Another problem with the gulp valve is that when the throttle is closed at high speeds while manifold vacuum is low, the valve may not open for a few seconds. During this time, an engine backfire can occur. The gulp valve largely has been replaced by the diverter valve on late-model cars.

Diverter valve

The **diverter valve**, figure 18-10, is also known as a dump or bypass valve. Like the gulp valve, the diverter valve uses a diaphragm operated by manifold vacuum to redirect the airflow from the air pump. However, the pump air passes through the diverter valve continuously on its way to the air injection manifold. During deceleration, manifold vacuum operates the

Backfire: The accidental combustion of gases in an engine's intake or exhaust system.

Gulp Valve: A valve used in an air injection system to prevent backfire. During deceleration it redirects air from the air pump to the intake manifold where the air leans out the rich air-fuel mixture.

Diverter Valve: Also called a dump valve. A valve used in an air injection system to prevent backfire. During deceleration it "dumps" the air from the air pump into the atmosphere.

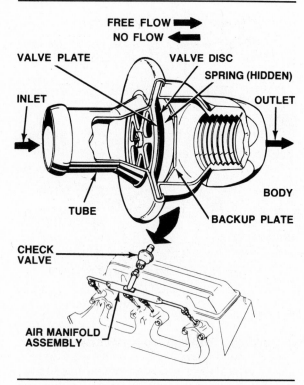

Figure 18-11. The check valve protects the system against reverse flow of exhaust gases. (AC-Delco)

valve diaphragm to divert, or dump, the air directly to the atmosphere, not to the intake manifold.

Some diverter valves vent the air to the engine air cleaner for muffling. Others vent it through a muffler and filter built into the valve. Diverter valves are used with vacuum solenoids, vacuum differential valves, vacuum vent valves, and idle valves to fine tune air injection. Because diverter valves do not affect the air-fuel mixture in the intake manifold, they are more trouble free than the gulp valves.

Diverter valves also may have a pressure relief valve to keep the pump from building up too much pressure in the system. This kind of diverter valve is used with an air pump that does not have a built-in relief valve.

Check Valves

All injection systems use one or more 1-way check valves, figure 18-11, to protect the air pump from harmful reverse flow of exhaust gases from the engine. A check valve contains a spring-type metallic disc or reed that closes the air line under exhaust backpressure. Check valves are located between the air manifold and the diverter or gulp valve. If exhaust pressure is higher than air injection pressure, or if the air

pump fails, the check valve spring closes the valve to prevent the reverse flow of exhaust.

SECOND-GENERATION AIR INJECTION SYSTEMS

Air injection systems used with catalytic converters do the same thing as the basic air injection system just described. They help to oxidize the HC and CO in the exhaust gases by adding fresh air into the exhaust system. These newer, second-generation systems use many of the same components as the noncatalytic systems.

However, adding the catalytic converter changed the air distribution needs of the engine. The role of the air injection system shifted. It is no longer mainly an oxidation device; now it has become an assist device for the converter. Air injection on late-model cars is used to improve converter efficiency and to accelerate catalyst warmup.

Air Injection and Catalytic Converters

Air can be injected near each exhaust port, into the exhaust manifold outlet, into the exhaust pipe ahead (upstream) of the converter, or directly into the converter (downstream). Air flowing into these points mixes with the exhaust gases and continues the oxidation process inside the converter. During the early stages of engine operation, this injected air increases the exhaust temperature. This brings the converter to operating temperature more quickly and increases its efficiency, because catalysts work best at temperatures of about 400° to 500°F (204° to 260°C).

However, too much HC and CO in the exhaust of an engine that is cold (choke closed) or idling for a long time can damage the catalysts or overheat the converter. Switching the injected air from the upstream exhaust ports to a downstream point near or at the converter helps to dilute the HC and CO concentration in the exhaust. Electronic engine control microprocessors monitor the duration of injection as well as engine temperature to prevent catalyst overheating.

Engines with electronic engine control systems also use one or more exhaust oxygen (O_2) sensors. These devices must reach a certain temperature before they start to function. Air injection in the exhaust stream during engine warmup helps the sensor reach operating temperature more rapidly.

With the introduction of NO_x reduction converters (also called dual-bed, TWC, or 3-way

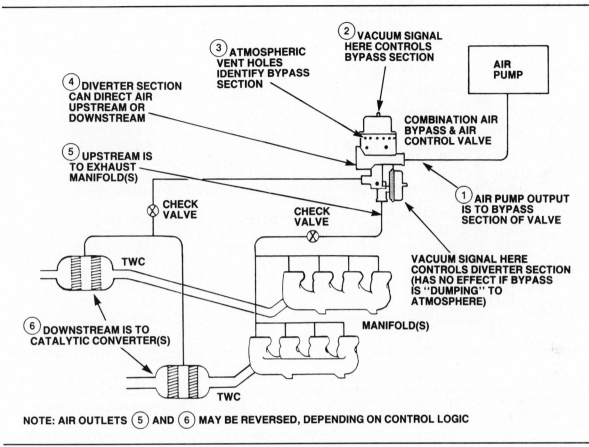

② **VACUUM SIGNAL HERE CONTROLS BYPASS SECTION**

③ **ATMOSPHERIC VENT HOLES IDENTIFY BYPASS SECTION**

④ **DIVERTER SECTION CAN DIRECT AIR UPSTREAM OR DOWNSTREAM**

AIR PUMP

COMBINATION AIR BYPASS & AIR CONTROL VALVE

⑤ **UPSTREAM IS TO EXHAUST MANIFOLD(S)**

① **AIR PUMP OUTPUT IS TO BYPASS SECTION OF VALVE**

⊗ **CHECK VALVE**

CHECK VALVE ⊗

VACUUM SIGNAL HERE CONTROLS DIVERTER SECTION (HAS NO EFFECT IF BYPASS IS "DUMPING" TO ATMOSPHERE)

TWC

MANIFOLD(S)

⑥ **DOWNSTREAM IS TO CATALYTIC CONVERTER(S)**

TWC

NOTE: AIR OUTLETS ⑤ AND ⑥ MAY BE REVERSED, DEPENDING ON CONTROL LOGIC

Figure 18-12. Operation of Ford's managed thermactor air (MTA) system using a combination bypass-diverter valve. (Ford)

converters) in 1977, the problem of air distribution became even more complex. The oxidation process *adds* O_2 to HC and CO, but the reduction process *removes* O_2 from NO_x compounds. Therefore, a reduction converter requires more air in the exhaust to bring it to operating temperature. When it reaches operating temperature, however, it works most efficiently with *less* air in the exhaust. This seeming paradox, along with the problems of catalyst damage and converter overheating, led to the development and use of air switching systems.

Figure 18-12 shows the operation of Ford's managed thermactor air (MTA) system used with a 3-way converter. The system is typical of current air-switching systems and controls airflow in three ways:

1. It dumps pump air to the atmosphere during periods of rich exhaust, such as deceleration, extended high-speed, or high-load operation. This prevents backfiring or catalyst damage and removes the pump load from the engine.

2. It directs pump airflow upstream to the exhaust manifold during cold engine start-up when the converter also is cold.

3. When the engine and converter reach operating temperature, it redirects pump airflow downstream to a point between the reduction and oxidation catalysts.

■ **Don't Oil The Air Pump!**

No air injection system is completely quiet. Usually pump noise increases in pitch as engine speed increases. If the drive belt is removed and the pump shaft turned by hand, it will squeak or chirp. Many who work on their own cars and even some mechanics are not aware that air injection pumps are permanently lubricated, and require no periodic maintenance.

Suppose you pinpoint the air pump as the source of the noise. It would seem that a few squirts of oil would silence it. See those three small holes in the housing? While it is easy to mistake them from oiling points, these are actually vents. Don't oil them. More than a few pumps have failed because someone assumed that taking "good" care of the pump would make it last longer!

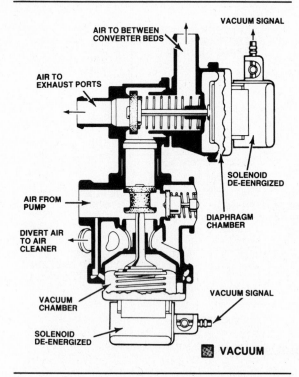

Figure 18-13. GM uses this electrically signaled diverter valve for switching and diverting tasks on air injection systems used with CCC engines.

The downstream point of entry for air injection is usually the mid-bed of the catalyst in a 3-way converter system, figure 18-12. If separate reduction and oxidation converters are used, an air inlet tube connects to the pipe between them. Air switching is necessary because the air injection requirements have changed. Once the system has reached operating temperature, additional air added to the exhaust will cause the O_2 sensor to send false signals to the engine control computer and reduce the efficiency of the NO_x reduction converter.

Air Management Valves

Air switching is done by air management valves. This term is used to describe the large variety of diverter, bypass, and air control or switching valves that have been used by domestic carmakers in their air injection systems since 1975. For example, GM currently uses more than 16 variations of air management valves with its computer command control (CCC) systems. Figure 18-13 shows one such valve used by GM.

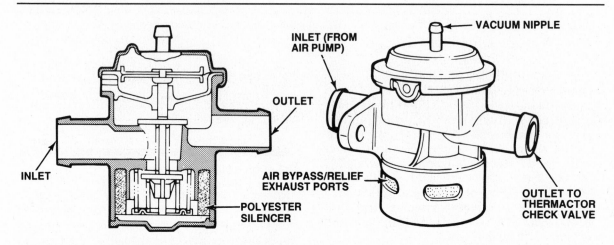

Figure 18-14. A typical normally closed air bypass valve. (Ford)

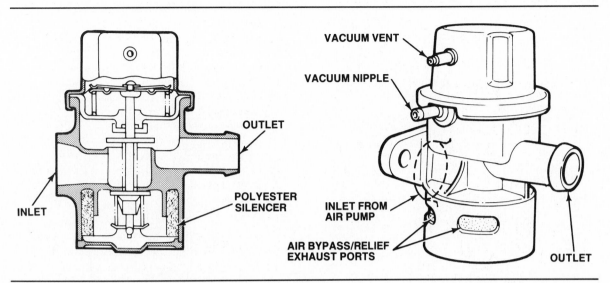

Figure 18-15. A typical normally open air bypass valve. (Ford)

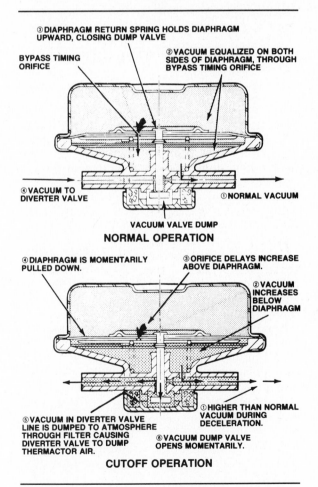

③ DIAPHRAGM RETURN SPRING HOLDS DIAPHRAGM UPWARD, CLOSING DUMP VALVE

BYPASS TIMING ORIFICE

② VACUUM EQUALIZED ON BOTH SIDES OF DIAPHRAGM, THROUGH BYPASS TIMING ORIFICE

④ VACUUM TO DIVERTER VALVE

① NORMAL VACUUM

VACUUM VALVE DUMP

NORMAL OPERATION

④ DIAPHRAGM IS MOMENTARILY PULLED DOWN.

③ ORIFICE DELAYS INCREASE ABOVE DIAPHRAGM.

② VACUUM INCREASES BELOW DIAPHRAGM

⑤ VACUUM IN DIVERTER VALVE LINE IS DUMPED TO ATMOSPHERE THROUGH FILTER CAUSING DIVERTER VALVE TO DUMP THERMACTOR AIR.

① HIGHER THAN NORMAL VACUUM DURING DECELERATION.

⑥ VACUUM DUMP VALVE OPENS MOMENTARILY.

CUTOFF OPERATION

Figure 18-16. This vacuum differential valve dumps the vacuum from the diverter valve whenever vacuum increases suddenly. (Ford)

Ford also uses many types of valves. These valves are divided into two general groups: normally closed or normally open. Normally closed valves, figure 18-14, provide air to the exhaust system during normal engine operation, brief idle periods, and some acceleration conditions. When vacuum is low or nonexistent, pump air is dumped through the valve silencer ports.

Normally open valves with a vacuum vent, figure 18-15, provide a timed air dump during deceleration. They also dump air when the signal port vacuum is less than 1.5 psi (10 kPa) more than the vent port vacuum. This mode protects the catalyst from overheating.

Various vacuum differential valves, vacuum delay valves, and bleed valves are used to control the vacuum to Ford diverter valves. The vacuum differential valve, figure 18-16, is in the vacuum line to the diverter valve. Sudden changes in vacuum cause the differential valve to dump the vacuum that goes to the diverter valve. This causes the diverter valve to cut off the air injection and dump the air pump output to the atmosphere.

Some Ford diverter valves, figure 18-17, have a built-in vacuum differential valve. The top of the valve also has a vacuum vent. When the vent is closed, the valve acts as a normal diverter valve. When the vacuum vent is open, any manifold vacuum above 2 psi (13.5 kPa) causes the diverter valve to dump the air pump output. An air temperature electric switch and an electric solenoid open the vent line at cold temperatures.

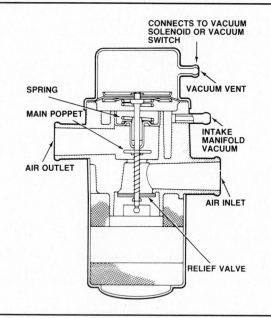

Figure 18-17. This Ford diverter valve has a vacuum vent that allows air to be dumped during cold engine operation. (Ford)

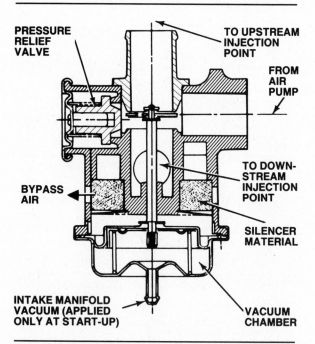

Figure 18-19. Cutaway of Chrysler's air switch/relief valve. (Chrysler)

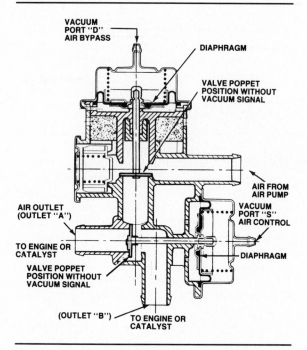

Figure 18-18. Ford's combination air bypass/air control valve handles air switching and diverting tasks. (Ford)

Ford's combination air bypass/air control valve, figure 18-18, and the GM electric divert/electric air switching (EDES) valve, figure 18-13, combine the divert and switching functions in one valve. The divert, or air control, section of the valve protects the converter

during open-throttle operation and high temperatures. The air switching section sends air to the exhaust ports during open-loop operation. When the engine control system moves into the closed-loop mode, the valve switches airflow to a point between the converter beds.

Chrysler uses a combination switch/relief valve, figure 18-19, controlled by a coolant vacuum switch cold open (CVSCO) or a vacuum solenoid. On a cold start, air is injected as close as possible to the exhaust valves. When engine coolant temperature reaches the point where exhaust gas recirculation (EGR) begins, the CVSCO or vacuum solenoid shuts off the vacuum signal to the valve. This causes the valve to send most of the pump air downstream to the catalytic converter. The rest of the pump air continues to reach the exhaust ports by passing through slots in the upstream valve seat.

Regardless of manufacturer or system, most air control valves operate with:

● Manifold vacuum or ported vacuum working on a vacuum diaphragm in the valve
● Output pressure of the air pump working against vacuum or a spring in the vacuum chamber
● Opening or closing of the vacuum supply to the diaphragm by one or more solenoids in the vacuum line.

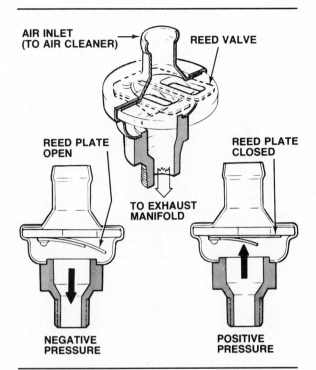

Figure 18-20. A pulse air injection valve opens when negative pressure occurs in the exhaust manifold. (Ford)

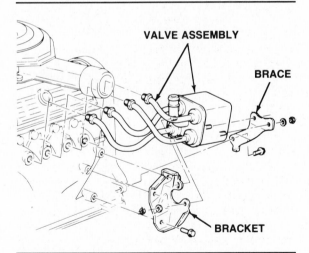

Figure 18-21. GM's pulse air injection system is called PAIR and was first used on the 1975 Cosworth Vega. (Chevrolet)

PULSE AIR INJECTION

The use of pulse air injection has increased on domestic vehicles since 1975. Pulse air systems use a pulse air valve, figure 18-20, instead of an air pump. This device is similar to (but not interchangeable with) the check valve used in an air pump system. The pulse air valve is a

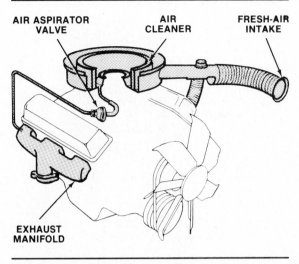

Figure 18-22. Chrysler's air aspirator system has remained unchanged on carbureted models since its introduction in 1977.

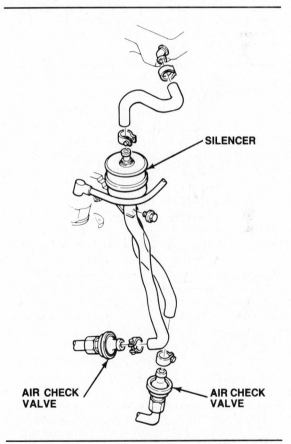

Figure 18-23. Ford's Thermactor II (pulse air) system uses an external silencer. (Ford)

spring-loaded diaphragm or reed valve that is connected to the exhaust system.

Pulse air valves can be connected with tubing to each exhaust port, figure 18-21, or to the exhaust manifold, figure 18-22. Some sys-

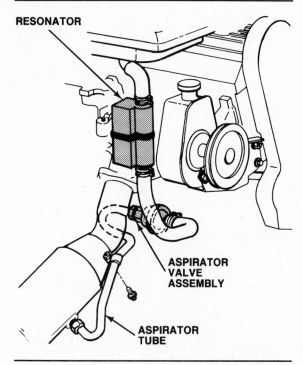

Figure 18-24. Chrysler's turbocharged 4-cylinder engines use a resonator in the line between the air cleaner and the aspirator valve. (Chrysler)

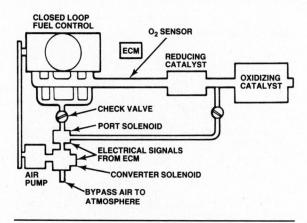

Figure 18-25. A typical General Motors computer-controlled AIR system. (GM)

tems use two pulse air valves. One is connected to the exhaust manifold and the other sends air to the converter. An external silencer, figure 18-23, or resonator, figure 18-24, may be installed between the valve and air cleaner to reduce airflow noise.

Each time an exhaust valve closes, there is a period when manifold pressure drops below atmospheric pressure. During these low pressure (slight vacuum) pulses, the air injection valve opens to admit fresh air to the exhaust, figure 18-20. When exhaust pressure rises above atmospheric pressure, the valve acts as a check valve and closes.

Since the pulse air injection system draws its air from the air cleaner, it eliminates the power-consuming air pump. This makes the system useful with small 4- and 6-cylinder engines with no power to waste. However, it works best at low engine speeds when extra air is needed most by the catalytic converter. At high engine speeds, the vacuum pulses occur too rapidly for the valve to follow, and the internal spring simply keeps the valve closed.

Pulse air systems have other disadvantages. They must be connected upstream in the exhaust system where negative pressure pulses are strong. This means that pulse air cannot be switched downstream for use in the converter. The system works best on vehicles equipped

with only an HC-CO oxidation catalyst, although a few models with 3-way converters use pulse air to provide additional air in the exhaust when the engine is cold. Such systems contain valves in the supply line between the air cleaner and pulse air valve to shut off air supply once the engine warms up.

COMPUTER-CONTROLLED AIR SWITCHING

A recent trend in air injection is to put the control of the system in the engine control microprocessor. Figure 18-25 shows a typical system used by General Motors. Air management in the GM system is provided by a combined air control and air switching valve, figure 18-26. Each function of the valve has a vacuum solenoid controlled by the electronic control module (ECM). When the CCC system is in open-loop, both solenoids are grounded by the ECM. This allows airflow from the pump to pass through the control valve to the air switching valve, which routes it to the exhaust ports. Once the CCC system switches to closed-loop operation, the ECM de-energizes the air switching solenoid. This redirects the airflow from the exhaust ports to a point between the catalysts for as long as the system remains in closed-loop. However, if the ECM recognizes the need to divert air, it de-energizes the air control solenoid, causing the valve to switch airflow to the air cleaner. This may be done under the following circumstances:

- Excessively rich operation
- Deceleration
- High engine rpm operation (when air pressure exceeds the internal relief valve setting)
- The ECM sees a problem or failure in the system and turns on the "CHECK ENGINE" light.

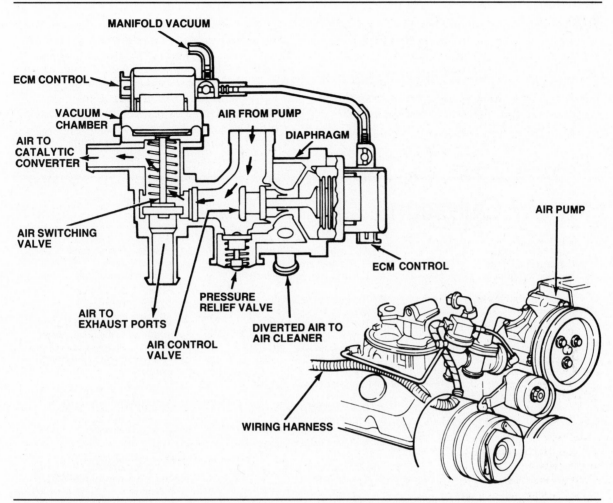

Figure 18-26. The GM computer-controlled AIR system control valve combines the divert and air switching functions in a single housing. Solenoids at both ends of the valve activate the functions on command from the ECM. (GM)

The Ford MTA system, figure 18-12, is similar in operation to the GM system just described, but uses separate solenoids which connect to the combination air bypass/air control valve with vacuum lines.

Pulse air systems also may be computer-controlled. For example, the GM PAIR system contains a shut-off valve to control the air supply to the valve assembly. During a cold engine start, and whenever the CCC system is in the open-loop mode, the ECM energizes the shut-off valve. This opens the fresh air line at a time when the engine needs it most. When the CCC system goes into the closed-loop mode, the ECM de-energizes the shutoff valve, closing off the fresh air line.

SUMMARY

Air injection is one of the oldest methods used to control HC and CO exhaust emissions. The injected air mixes with hot exhaust gas as it leaves the combustion chambers to further oxidize HC and CO emissions. All air injection systems used on domestic cars operate in essentially the same way, regardless of manufacturer.

With the advent of the catalytic converter, the role of the air injection system changed from primary oxidation device to that of an assist device for the converter. Oxidation converters that reduce HC and CO emissions require more air; reduction converters that minimize NO_x emissions require less air. Since these two functions are incompatible with the basic air injection system, air switching or control valves are used with converter applications. These valves send pump air upstream near the exhaust port or manifold when the engine is started, and downstream near or to the converter once the engine reaches normal operating temperature.

The pulse air injection system relies on exhaust pulses instead of an air pump to draw fresh air from the air cleaner into the exhaust. Pulse air systems work best with oxidation converters and are most efficient at low engine speeds. Some engine control systems now use the microprocessor to control the air injection or pulse air systems. This provides a quicker response to changing engine requirements.

Review Questions

Choose the single most correct answer.
Compare your answers with the correct answers on page 414.

1. The American Motors air injection system is called:
   a. Air Injection System
   b. Air guard
   c. Thermactor Air Injection System
   d. Air Injection Reactor

2. The main reason for an air injection system is to:
   a. Oxidize HC and CO exhaust emissions
   b. Reduce NO_x exhaust emissions
   c. Eliminate crankcase emissions
   d. Eliminate evaporative HC emissions

3. Oxidation of HC and CO emissions produces:
   a. HCO and CO_2
   b. H_2CO_3 and CO_2
   c. H_2O and CO_2
   d. All of the above

4. Which of the following is true?
   a. All air pumps can be rebuilt
   b. Two-vane pumps can be rebuilt
   c. No air pumps can be rebuilt
   d. Three-vane pumps can be rebuilt

5. Two-vane pumps have:
   a. An impeller-type fan for filter
   b. An integral wire mesh filter
   c. A hose to the clean side of the air cleaner
   d. A separate air filter

6. Air injection nozzles are made of:
   a. Copper
   b. Stainless steel
   c. Aluminum
   d. Vanadium

7. The two types of air injection backfire suppressor valves are:
   a. The check valve and the gulp valve
   b. The gulp valve and the diverter valve
   c. The diverter valve and the relief valve
   d. The diverter valve and the check valve

8. The Chrysler air switching valve is controlled by:
   a. A vacuum solenoid
   b. A reed valve
   c. A bypass timing orifice
   d. A coolant vacuum switch

9. This illustration shows:
   a. A relief valve
   b. A solenoid operated vacuum valve
   c. A gulp valve
   d. A diverter valve

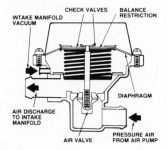

10. Pulsed air injection works best at:
    a. Low speeds
    b. High speeds
    c. Idle
    d. Deceleration

11. Mechanic A says that a gulp valve diverts air to the atmosphere. Mechanic B says that a diverter valve diverts air to the intake manifold.
    Who is right?
    a. A only
    b. B only
    c. Both A and B
    d. Neither A nor B

12. Mechanic A says that air injection helps an exhaust catalyst reach its operating temperature faster. Mechanic B says that a catalyst is most efficient when cooled by air injection.
    Who is right?
    a. A only
    b. B only
    c. Both A and B
    d. Neither A nor B

Chapter

19

Spark Timing Control Systems

Proper timing of the ignition spark can help car engines reduce exhaust emissions and meet U.S. Federal HC and NO_x standards. General Motors started using auxiliary spark timing control on some 1970 cars, with the systems achieving widespread use in 1971 to 1974 cars. Each automaker developed slightly different spark timing controls, according to engine requirements and emission standards for each model year, but the systems and devices all operate on the same principles.

This chapter explains these principles of spark timing control systems, their effect on engine emission levels, and how the specific manufacturers' systems function.

SPARK TIMING, COMBUSTION, AND EMISSION CONTROL

Approximately 3 milliseconds (0.003 second) elapse from the instant the air-fuel mixture ignites until its combustion is complete. Remember that this burn time is a function of *time* and not of piston travel or crankshaft degrees. The ignition spark must occur early enough so that the combustion pressure reaches its maximum just after top dead center, when the piston is beginning its downward power stroke. Combustion should be completed by about 10° after top dead center (atdc). If the spark occurs too soon before top dead center, the rising piston will be opposed by combustion pressure. If the spark occurs too late, the force on the piston will be reduced. In both cases, power will be lost. In extreme cases, the engine could be damaged. Ignition must start at the proper instant for maximum power and efficiency.

As engine speed increases, piston speed increases. If the air-fuel ratio remains relatively constant, the fuel burning time will remain constant. However, at greater engine speed, the piston will travel farther during this burning time. Ignition timing must be changed to ensure that maximum combustion pressure occurs at the proper piston position.

For example, consider an engine, figure 19-1, that requires 0.003 second for the fuel charge to burn and that achieves maximum power if the burning is completed at 10° atdc.

● At an idle speed of 625 rpm, figure 19-1A, the crankshaft rotates about 11 degrees in 0.003 second. Therefore, timing must be set at 1° btdc to allow ample burning time.

● At 1,000 rpm, figure 19-1B, the crankshaft rotates 18 degrees in 0.003 second. Ignition should begin at 8° btdc.

● At 2,000 rpm, figure 19-1C, the crankshaft rotates 36 degrees in 0.003 second. Spark timing must be advanced to 26° btdc.

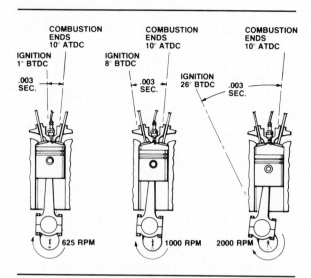

Figure 19-1. An example of ignition timing.

Change in timing is called spark advance, or ignition advance. In this chapter, you'll learn how this relates to emission control.

Spark timing affects combustion temperature. Firing the spark at precisely the right instant, as in the example above, creates the maximum possible amount of heat and pressure, and the maximum possible engine power. Unfortunately, this great heat also creates a large amount of NO_x exhaust emissions.

An efficient combustion process starts with an advanced spark. High-temperature combustion occurs at an early point during the process. However, NO_x emission levels are increased by higher combustion temperatures. Late in the combustion process, the exhaust gases are cooler. They do not heat up the exhaust manifold as much. But this cooler exhaust temperature produces a greater amount of HC emissions.

Spark timing controls keep ignition timing retarded during idle and low-speed operation, when the air-fuel mixture is rich. Retarded timing reduces the peak combustion temperature because ignition occurs when cylinder pressure is lower. This helps to reduce NO_x formation. At the same time, the greatest combustion temperatures occur during the end of combustion. This results in higher exhaust temperatures, which reduce the amount of HC in the exhaust.

SPARK ADVANCE

Two basic factors govern ignition timing: engine speed and load. All changes in timing are related to these two factors:

● Timing increases, or advances, as engine speed increases. It decreases, or retards, as engine speed decreases.

● Timing decreases, or retards, as load increases. It increases, or advances, as load decreases.

Optimum ignition timing under any given combination of these basic factors results in maximum cylinder pressure. This delivers maximum power, with a minimum of exhaust emissions and the best possible fuel economy.

When ignition occurs too early, the combustion pressure slows down the piston. If timing is too far advanced, the increased combustion pressure causes engine knock. When ignition occurs too late, power is lost because the piston is too far down on its power stroke to benefit from the combustion pressure.

In most breaker-point ignition systems, the distributor has two spark advance mechanisms that react to engine operating changes and alter ignition timing:

1. The centrifugal advance changes ignition timing to match engine speed by altering the position of the distributor cam or trigger wheel on the distributor shaft.
2. The vacuum advance changes ignition timing to match engine load by altering the position of the breaker points or the magnetic pickup coil.

These changes in position alter the time, relative to crankshaft position, at which the ignition primary circuit is opened and the spark is delivered to the combustion chamber.

Late-model electronic ignition systems use the computer to control engine timing. The advent of computer controls made separate spark timing control systems obsolete. You will learn more about these systems later in this chapter.

Centrifugal Advance — Speed

The centrifugal, or mechanical, advance mechanism in a breaker-point distributor consists of two weights connected to the distributor drive shaft by two springs, figure 19-2A. The distributor cam, or electronic trigger wheel, and the distributor rotor are mounted on another shaft. The second shaft fits over the drive shaft like a sleeve. Drive shaft motion is transmitted to the second shaft through the centrifugal advance weights. When the weights move, the relative position of the drive shaft and the second shaft is changed.

As engine speed increases, distributor shaft rotational speed increases. The advance weights move outward because of centrifugal force. This outward movement of the weights shifts the second shaft and cam, or trigger wheel, figure 19-2B. As a result, the primary circuit opens earlier in the compression stroke and the spark occurs earlier.

A

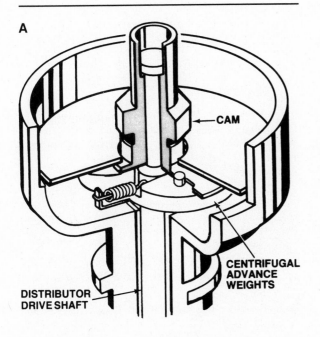

CAM

CENTRIFUGAL
ADVANCE
WEIGHTS

DISTRIBUTOR
DRIVE SHAFT

B

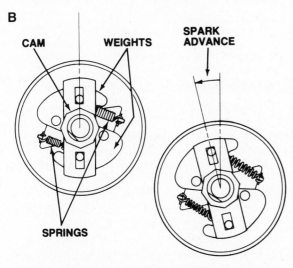

CAM   WEIGHTS   SPARK
ADVANCE

SPRINGS

Figure 19-2. (A) The centrifugal advance weights transfer the distributor drive shaft rotation to the cam and rotor. (B) When the centrifugal advance weights move, the position of the cam and rotor changes.

A control spring connects each advance weight to the distributor drive shaft. These springs are selected to allow the correct amount of weight movement and ignition advance for each engine.

At low engine speeds, spring tension holds the weights in, so that initial timing is maintained. As engine speed increases, centrifugal force overcomes spring tension and the weights move outward. The advance is not a large, rapid change, but rather a slow, gradual shift. Figure 19-3 shows a typical centrifugal advance curve. The advance curve can be changed by

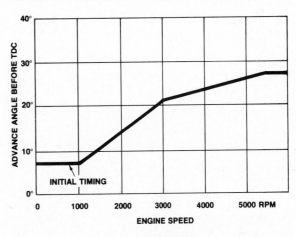

Figure 19-3. A typical centrifugal advance curve.

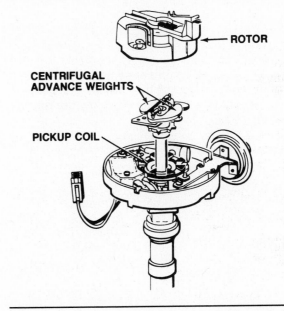

ROTOR

CENTRIFUGAL
ADVANCE WEIGHTS

PICKUP COIL

Figure 19-4. Delco-Remy V-6 and V-8 distributors have the centrifugal advance mechanism above the pickup coil and trigger wheel, as shown, or above the cam and rotor.

changing the tension of the control springs. Remember that centrifugal advance responds to engine speed.

In most distributors, the centrifugal advance mechanism is mounted below the cam and

Spark Timing: A way of controlling exhaust emissions by controlling ignition timing. Vacuum advance is delayed or shut off at low and medium speeds, reducing NO_x and HC emissions.

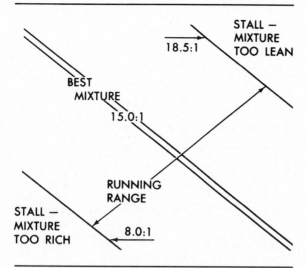

Figure 19-5. Air-fuel ratio limits for a 4-stroke gasoline engine. (Chevrolet)

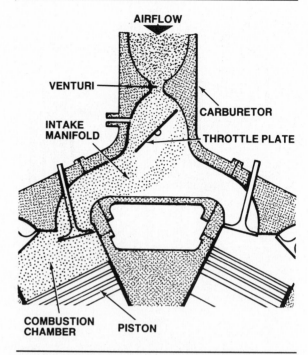

Figure 19-6. Air flows through the carburetor and intake manifold to reach the combustion chambers.

breaker points, figure 19-2A, or below the trigger wheel and pickup coil. Delco-Remy V-6 and V-8 distributors, as well as some Japanese distributors, have the advance mechanism above the cam or trigger wheel, just below the rotor, figure 19-4.

Vacuum Advance — Load

The vacuum advance mechanism allows efficient engine performance within a range of air-fuel ratios. These ratios are important, since there are limits as to how rich or how lean they can be and still remain fully combustible. The air-fuel mixtures with which an engine can operate efficiently range from 8 to 18.5 to 1 by weight, figure 19-5.

These ratios generally are stated as eight parts of air combined with one part of gasoline (8:1), which is the richest mixture that an engine can tolerate and still fire regularly. A ratio of 18.5 parts of air mixed with one part of gasoline (18.5:1) is the leanest mixture that an engine can tolerate without misfiring.

An average air-fuel ratio is about 15:1. This mixture takes about 3 milliseconds to burn. A lean mixture (one with more air and less fuel) requires more time to burn. The ignition timing must be advanced to provide maximum combustion pressure at the correct piston position. A rich mixture (one with more fuel and less air) burns more quickly and emits more exhaust pollutants. Ignition timing should be retarded for efficient combustion and emission control.

Engine Vacuum

The reciprocating engine can be considered as an air pump. As a piston moves downward, air pressure in the cylinder decreases and atmospheric air rushes in to fill the void.

The fuel delivery system uses this air movement to carry fuel to the cylinders. We'll use a carbureted engine to explain what occurs. The air must travel through the carburetor, figure 19-6, to reach the cylinders. Under several operating conditions, air movement caused by the downstroke of a piston is not great enough to draw fuel into the cylinder. The carburetor forces the air to flow through a restriction called a venturi. This speeds the airflow and creates a low pressure (vacuum) area. Fuel is drawn into the airflow by the vacuum and the resulting air-fuel mixture enters the cylinder combustion chambers. The air-fuel ratio changes as the vacuum in the carburetor changes.

The ignition timing also must be changed as the air-fuel ratio changes, so that the mixture has the correct amount of time to burn. The vacuum advance mechanism is connected to a small hole, or port, in the carburetor just above the throttle plate, figure 19-7. This is called ported vacuum. When a vacuum exists at the port, timing is advanced. When no vacuum

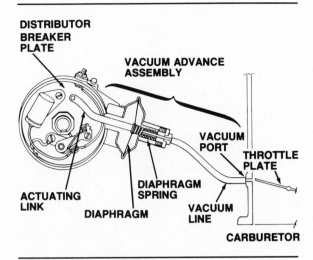

Figure 19-7. The vacuum advance assembly is connected to a port in the carburetor.

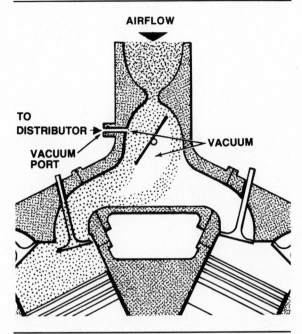

Figure 19-9. When the throttle is partially open, the port is exposed to manifold vacuum.

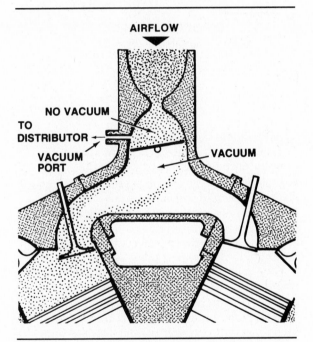

Figure 19-8. When the throttle is closed during idle or deceleration, there is no vacuum at the port.

exists at the port, the timing remains at a basic setting or is affected only by the centrifugal advance. The vacuum advance mechanism will be explained in more detail later; the following paragraphs explain the relationship between carburetor vacuum and the need for advanced timing.

The automobile driver controls engine load and carburetor vacuum through the action of the throttle plate, which is a variable restriction in the carburetor airflow. When the engine is at idle, figure 19-8, the throttle plate is almost closed. Very little air flows through the carburetor to mix with the fuel. With a rich air-fuel mixture, no spark advance is necessary. High vacuum exists in the intake manifold, but there is no vacuum at the port because it is above the closed throttle plate. There is no vacuum-controlled spark advance with a closed throttle. The large amount of vacuum *below* the throttle plate exists because of the small amount of air entering the engine. This is called manifold vacuum.

When the throttle plate is partially open, figure 19-9, more air can flow through the carburetor. The air-fuel mixture becomes lean and requires an advanced spark. Since the port is now exposed to vacuum, ignition timing is advanced.

At medium cruising speeds, the engine operates with a partly open throttle and an air-fuel ratio of 15 or 16. Airflow velocity is high and the vacuum signal at the carburetor port is strong enough to provide enough vacuum advance for these relatively lean air-fuel mixtures.

A lean mixture is not the only factor that affects ignition timing requirements. During part-throttle operation, the cylinders are only partially filled with the air-fuel mixture. Because there is less to compress, the compression pressure is less. A less highly compressed mixture takes longer to burn.

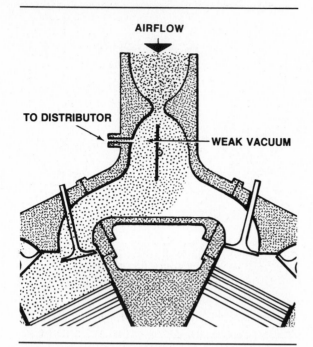

Figure 19-10. When the throttle is fully open, the vacuum at the port is too weak to cause any vacuum advance.

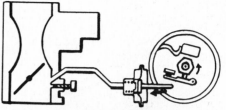

**THROTTLE OPENED
VACUUM PORT   UNCOVERED
VACUUM SPARK   ADVANCE INTRODUCED**

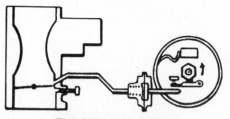

**THROTTLE CLOSED
VACUUM PORT COVERED
NO VACUUM SPARK ADVANCE**

Figure 19-11. Vacuum at the carburetor port causes the position of the breaker plate to shift.

At wide-open throttle, figure 19-10, the power circuit in the carburetor provides a richer mixture than at part-throttle cruising. Also, the cylinders are more completely filled because airflow volume increases. These factors cause a faster burn that requires the ignition timing to slightly retard, or decrease, for proper efficiency. Manifold vacuum and ported vacuum drop at wide-open throttle and become just about equal. Vacuum may drop to as little as 2.5 psi (19.9 kPa), and typically 3.5 psi (23.7 kPa) are required to operate a vacuum advance mechanism.

Vacuum advance also retards when the throttle is opened quickly from idle or a part throttle position. This occurs because airflow velocity lags behind throttle opening. This reduced vacuum does not draw enough fuel through the main metering circuit of the carburetor, so the accelerator pump supplies extra fuel. This momentarily rich mixture does not need as much spark advance for complete combustion. Because the vacuum at the vacuum port of the carburetor is low when the throttle is first opened quickly, the vacuum advance decreases, or retards, to meet the needs of the momentarily rich mixture. Remember that vacuum advance responds to engine load.

Vacuum advance mechanism
The vacuum advance mechanism at the distributor, figure 19-7, consists of the:
● Movable breaker plate on which the points or electronic pickup coil is mounted
● Vacuum assembly, a housing with a flexible diaphragm and a spring
● Actuating link that connects the vacuum diaphragm to the breaker plate
● Tubing to connect the vacuum unit to the vacuum source.

When the diaphragm is pulled toward a vacuum, it pulls the actuating link, figure 19-11. This rotates the breaker plate to change the position of the point rubbing block in relation to the cam. This increases, or advances, the ignition timing. In an electronic system, the pickup is shifted relative to the trigger wheel to advance the timing.

Some older vacuum advance units can be adjusted by changing spring tension against the diaphragm.

Some distributors used by Ford, Volkswagen, and AMC have a dual-diaphragm vacuum unit to retard timing during certain engine operating conditions. In these systems, figure 19-12, one diaphragm acts as we just described to advance ignition timing. The second diaphragm is exposed to intake manifold vacuum (between

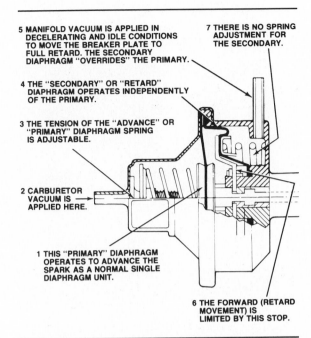

5 MANIFOLD VACUUM IS APPLIED IN DECELERATING AND IDLE CONDITIONS TO MOVE THE BREAKER PLATE TO FULL RETARD. THE SECONDARY DIAPHRAGM "OVERRIDES" THE PRIMARY.

7 THERE IS NO SPRING ADJUSTMENT FOR THE SECONDARY.

4 THE "SECONDARY" OR "RETARD" DIAPHRAGM OPERATES INDEPENDENTLY OF THE PRIMARY.

3 THE TENSION OF THE "ADVANCE" OR "PRIMARY" DIAPHRAGM SPRING IS ADJUSTABLE.

2 CARBURETOR VACUUM IS APPLIED HERE.

1 THIS "PRIMARY" DIAPHRAGM OPERATES TO ADVANCE THE SPARK AS A NORMAL SINGLE DIAPHRAGM UNIT.

6 THE FORWARD (RETARD MOVEMENT) IS LIMITED BY THIS STOP.

Figure 19-12. Operation of a typical dual-diaphragm vacuum advance assembly. (Ford)

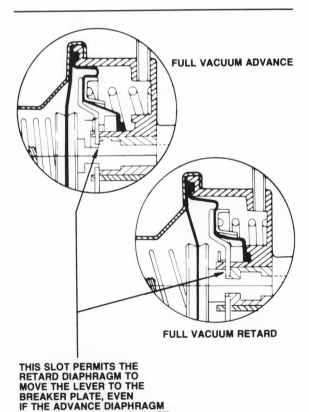

FULL VACUUM ADVANCE

FULL VACUUM RETARD

THIS SLOT PERMITS THE RETARD DIAPHRAGM TO MOVE THE LEVER TO THE BREAKER PLATE, EVEN IF THE ADVANCE DIAPHRAGM IS AT FULL VACUUM ADVANCE.

Figure 19-13. In a dual-diaphragm unit, the retard diaphragm can override the action of the advance diaphragm. (Ford)

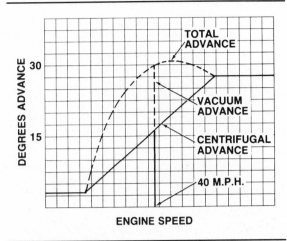

Figure 19-14. This typical advance curve shows that the total advance is the total of vacuum plus centrifugal advance.

the throttle plate and the combustion chamber) which is high during idle and closed-throttle deceleration. This vacuum acts on the second diaphragm to shift the breaker plate in the retard direction.

For exhaust emission control, the retard diaphragm can override the advance diaphragm. This causes timing to retard even when the advance diaphragm is at advanced position, figure 19-13.

Total Ignition Advance

The two types of advance mechanisms work on different parts of the distributor to advance timing. Therefore, their effects add: the total ignition advance equals the sum of the centrifugal advance plus the vacuum advance. This total advance is then added to the initial ignition timing to determine actual ignition timing under any condition. Figure 19-14 shows the advance curves of a typical ignition system.

SPARK TIMING CONTROLS OF THE 1960's

Spark timing control systems were introduced in the 1960's to reduce exhaust emissions. Certain types of pollutants are produced under specific engine operating conditions. Spark timing control systems help reduce HC, and to some extent, NO_x emissions. Retarded timing reduces combustion temperature to help reduce NO_x formation. At the same time, higher temperatures are created toward the end of combustion. This results in higher exhaust temperatures, which reduces the amount of HC in the exhaust.

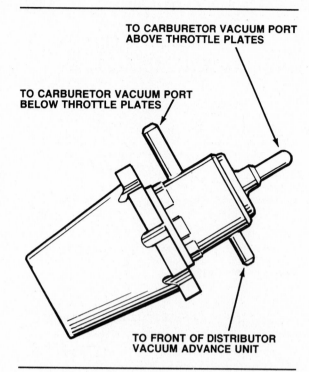

TO CARBURETOR VACUUM PORT
ABOVE THROTTLE PLATES

TO CARBURETOR VACUUM PORT
BELOW THROTTLE PLATES

TO FRONT OF DISTRIBUTOR
VACUUM ADVANCE UNIT

Figure 19-15. A deceleration vacuum advance valve.

Early emission control equipment advanced or retarded ignition timing under certain engine operating conditions, usually during starting, deceleration, and idle.

The deceleration vacuum advance valve, figure 19-15, was used during the late 1960's on Chrysler, Ford, AMC, and Pontiac cars with manual transmissions. In a manual transmission car, the air-fuel mixture becomes extremely rich during deceleration or gear shifting.

During deceleration, this valve momentarily switches the vacuum for the vacuum advance from a low-vacuum source at the carburetor to a high-vacuum source at the manifold, then back to the low-vacuum source. This prevents over-retard during deceleration or gear shifting which could cause high CO emissions in some engines.

While the deceleration vacuum advance valve was effective against CO emissions, it did not limit HC and NO_x emissions. As emission limits for these pollutants became stricter in the early 1970's, use of the device was discontinued and manufacturers developed other devices which worked against all three major pollutants.

SPARK TIMING CONTROLS OF THE 1970'S

These systems were designed primarily to delay vacuum advance at low and intermediate

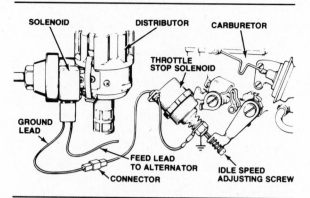

SOLENOID   DISTRIBUTOR   CARBURETOR

THROTTLE
STOP SOLENOID

GROUND
LEAD

FEED LEAD
TO ALTERNATOR   IDLE SPEED
CONNECTOR   ADJUSTING SCREW

Figure 19-16. A distributor retard solenoid.

speeds, and to allow it during high-speed cruising. The most common types of control systems used in the early 1970's were:

- Distributor solenoids
- Vacuum delay valves
- Speed- and transmission-controlled timing.

Distributor Solenoids

A distributor vacuum retard solenoid was used on some 1970-71 Chrysler cars with V-8 engines and automatic transmissions. This electric solenoid is attached to, and controls the action of, the distributor vacuum advance unit, figure 19-16.

The solenoid is energized by contacts mounted on a carburetor throttle stop solenoid. When the throttle closes, the carburetor solenoid touches the idle adjusting screw to complete the ground circuit. The contacts in the carburetor solenoid carry current to the distributor solenoid windings. Because the distributor solenoid plunger is connected to the vacuum diaphragm, solenoid movement shifts the breaker plate in the retard direction.

When engine speed increases, the idle adjusting screw breaks contact with the carburetor solenoid. Current to the distributor solenoid stops and normal vacuum advance is allowed.

Some 1972-73 Chrysler V-8 distributors have a spark timing advance solenoid that promotes better starting by providing a 7½-degree spark advance. The solenoid is mounted in the distributor vacuum unit, figure 19-17. It is activated by power from the starter relay at the same terminal that sends power to the starter solenoid. The solenoid is activated only while the engine is cranking.

The starting advance solenoid is not an emission control device by itself, but it allows lower basic timing settings, which help in emission control while providing advanced timing for quicker starting.

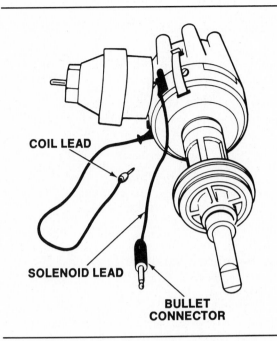

Figure 19-17. The installation of a distributor advance solenoid.

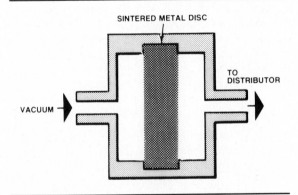

Figure 19-18. A cross-sectional view of a vacuum delay valve containing a sintered metal disc that slows the application of vacuum.

Vacuum Delay Valve

The vacuum delay valve "filters" the carburetor vacuum, making it take longer to get to the distributor vacuum advance unit. Generally, vacuum must be present in the system for 15 to 30 seconds before it affects the advance mechanism.

One method of vacuum delay is used in Ford's spark delay valve (SDV) system, figure 19-18. In this design, vacuum must work its way through a sintered, or sponge-like, metal disc to reach the distributor. Many GM engines also use this type of spark delay valve.

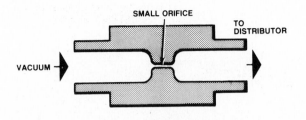

Figure 19-19. A cross-sectional view of a vacuum delay valve using a small orifice to delay the application of vacuum.

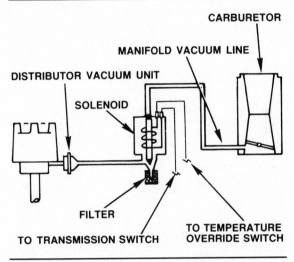

Figure 19-20. A simplified transmission-controlled spark system.

Another method of vacuum delay is used in Chrysler's orifice spark advance control (OSAC) system, figure 19-19. A small orifice is placed in the vacuum line to delay vacuum buildup.

Manufacturers often use vacuum delay valves along with other emission control systems. All valves operate on one of the two principles just described.

Speed- and Transmission-Controlled Timing

These systems prevent any distributor vacuum advance when the vehicle is in a low gear or is traveling slowly. A solenoid controls the application of vacuum to the advance mechanism, figure 19-20. Current through the solenoid is controlled by a switch that reacts to vehicle operating conditions.

A control switch used with a manual transmission reacts to shift lever position. A control switch used with an automatic transmission usually reacts to hydraulic fluid pressure. Both systems prevent any vacuum advance when the car is in a low or intermediate gear.

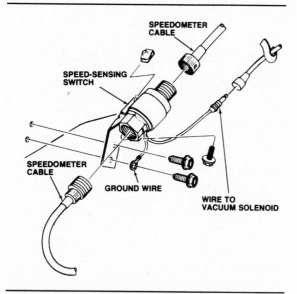

Figure 19-21. Most speed-controlled spark systems use a speed-sensing switch such as this. (Cadillac)

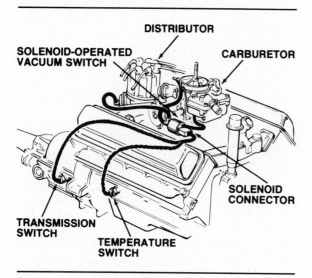

Figure 19-22. A typical General Motors transmission-controlled spark system.

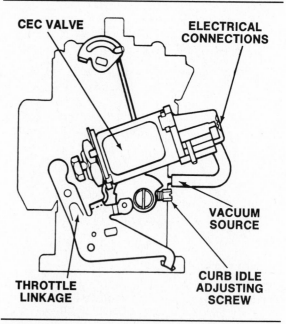

Figure 19-23. The General Motors combination emission control (CEC) valve.

A speed-sensing switch may be connected to the vehicle speedometer cable, figure 19-21. The switch signals an electronic control module when the vehicle speed is below a certain specified level. The module triggers a solenoid that controls engine vacuum at the distributor.

Both vacuum-delay systems and speed- and transmission-controlled systems usually have an engine temperature bypass. This allows normal vacuum advance at high and low engine temperatures. Before March 1973, some systems had an **ambient temperature** override switch. Most of these switches were discontinued at the direction of the Environmental

Protection Agency (EPA). Later temperature override systems sense engine coolant temperature or underhood temperature.

SUMMARY OF CARMAKERS' SYSTEMS

The systems using the principles you just learned are known by many different trade names and are all a little different. The following paragraphs briefly describe the major systems used by domestic manufacturers.

General Motors

Transmission-controlled spark (TCS)
Introduced by GM in 1970, the TCS system was used to reduce the exhaust emissions of many GM cars and light-duty trucks during all but high-gear operation. This was done primarily by a switch in the shift linkage and a solenoid which controlled vacuum to the distributor vacuum advance unit, figure 19-22.

When the vehicle is in any gear except high, the switch energizes the solenoid, cutting off vacuum to the advance mechanism and retarding the timing. When the vehicle is in high gear, the switch deenergizes the solenoid. This restores manifold vacuum to the advance mechanism, providing normal vacuum advance. Transmission control switch and solenoid operation on GM systems varies for different models and years.

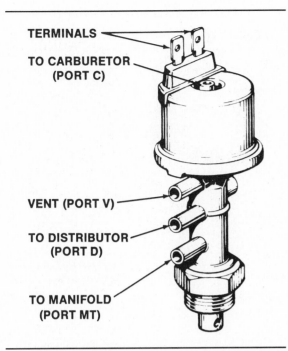

TERMINALS

TO CARBURETOR
(PORT C)

VENT (PORT V)

TO DISTRIBUTOR
(PORT D)

TO MANIFOLD
(PORT MT)

Figure 19-24. A distributor vacuum control switch.

Some vehicles use an additional temperature switch with a time delay relay. The switch acts as a thermal override to allow normal vacuum advance in low gears when the engine coolant was below a preset temperature (various control temperatures were used, depending upon the vehicle). The time delay provides full vacuum advance for about 20 seconds after the ignition is turned on for easier starting and warmup.

Combination emission control (CEC) valve
The CEC valve is part of the TCS system on some 1971-73 GM cars. Mounted on the side of the carburetor, figure 19-23, the CEC valve is a simple solenoid with a vacuum valve at one end and a throttle check rod extending from the other. The valve provides vacuum spark advance control and deceleration throttle position control in high gear.

Distributor vacuum control switch
Some Oldsmobiles used this switch, figure 19-24, along with a normally closed TCS switch. The vacuum control switch does the dual jobs of a TCS solenoid and a thermal vacuum switch. When the engine is at normal operating temperature, it permits vacuum spark advance only with the transmission in high gear. As engine coolant temperature rises, the switch applies full manifold vacuum to the distributor regardless of transmission gear position.

With the transmission in other than high gear, the TCS switch closes to energize the solenoid and seal off the carburetor port, venting distributor vacuum to the atmosphere. In high gear, the deenergized solenoid seals off the vent port, applying carburetor vacuum to the distributor. This switch prevents engine overheating caused by prolonged idling with a retarded spark. The thermal part of the switch seals off the vent port, applying manifold vacuum to the distributor.

A combination thermostatic vacuum switch (TVS) and TCS solenoid used on some Buick engines performs the same function as the distributor vacuum control switch.

Speed control switch (SCS) system
Distributor vacuum advance on 1972 Cadillacs and Pontiacs is controlled by vehicle speed instead of transmission gear. The main parts of

Ambient Temperature: The temperature of the air surrounding a particular device or location.

■ **Are Spark Timing Controls Confusing?**

As you study this chapter, you may be confused by the various spark control systems. If so, here is a summary which may help you understand this form of emission control.

The greatest amount of exhaust emissions are produced at idle and during low-speed operation. *All* spark control systems are designed to reduce these emissions by preventing vacuum advance at idle, and when the car is in the lower gears. This means that a spark control system should work *only* when the engine is at normal operating temperature, and the car is operating at low speeds or in the low gears. When the engine is at normal operating temperature and the car is at cruising speed or in high gear, vacuum advance takes place just as it would without a spark control system.

All spark control systems have some form of temperature control to prevent the system from operating if the engine is too cold or too hot.

Here are three troubleshooting tips:

1. If vacuum advance *is* present when the engine is at idle or in the low gears, then a speed sensor, transmission switch, control module, or solenoid is defective.
2. If vacuum advance is *not* present when the engine is in high gear, then a speed sensor, transmission switch, control module, or solenoid is defective.
3. If vacuum advance is *not* present when the engine is cold or overheated, then a temperature sensor or switch is defective.

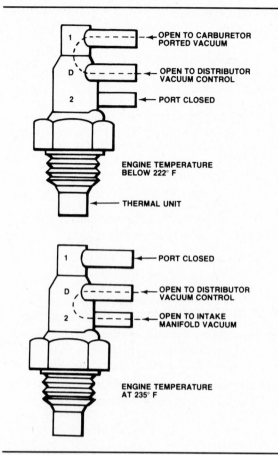

Figure 19-25. The thermostatic vacuum valve (TVV) used by AMC.

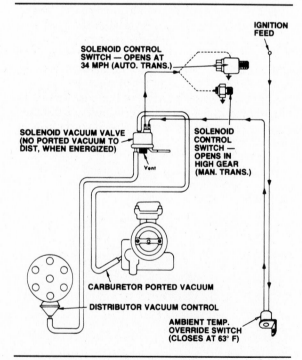

Figure 19-26. The AMC transmission-controlled spark system. (AMC)

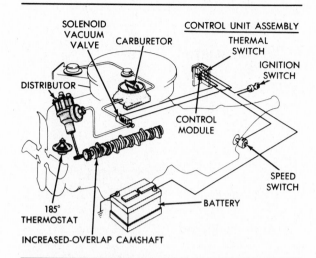

Figure 19-27. Chrysler's NO$_x$ control system. (Chrysler)

the SCS system are a speed sensor connected to the speedometer drive, and a solenoid that prohibits vacuum advance at speeds below 35 mph (56 kph) unless the engine begins to overheat. Vacuum advance also is eliminated at speeds below 25 mph (40 kph) during deceleration.

American Motors

Before 1971, AMC used dual-diaphragm vacuum units and decelerations valves to control exhaust emissions. Some cars have a thermostatic vacuum valve, figure 19-25, that permits normal vacuum advance at low engine temperatures. As engine temperature rises, the valve sends manifold vacuum to the distributor advance unit.

After 1971, many AMC models use a speed- or transmission-controlled spark timing system similar to General Motors' TCS, figure 19-26. Before March 1973, automatic transmission cars used a speed-sensing control switch. On later models, fluid pressure operates the control switch. On manual transmission models, gear lever position activates the switch.

Some AMC cars have an ambient temperature or coolant temperature override switch. When used with a TCS system, vacuum from the carburetor is controlled, but vacuum from the manifold is *always* applied when coolant temperature is low.

Chrysler

NO$_x$ spark control
Chrysler's 1971-72 cars sold in California have a TCS system similar to those of GM and AMC. An ambient temperature switch was used only

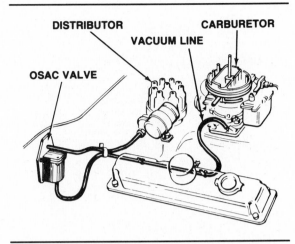

Figure 19-28. The Chrysler orifice spark advance control (OSMC) system.

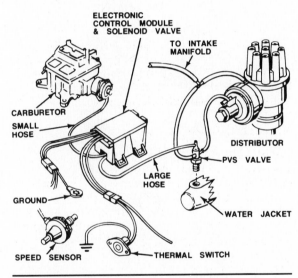

Figure 19-29. Ford's electronic distributor modulator (EDM) system.

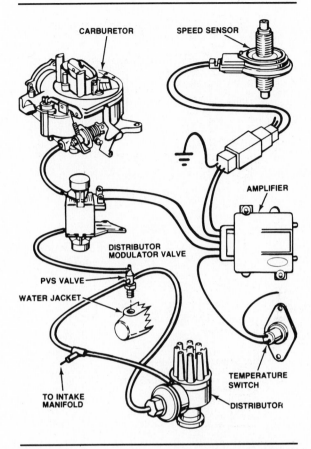

Figure 19-30. The Ford electronic spark control (ESC) system.

on 1971 cars to permit vacuum advance at all times when the temperature was low.

An electric control unit actuates the solenoid vacuum valve on automatic transmission cars, figure 19-27. This is a reversing relay that receives signals from grounding switches: a temperature switch, a speed switch, and a vacuum switch (1971 only). All switches must be open to activate the control unit, energize the solenoid, and cut off the vacuum advance.

Orifice spark advance control (OSAC) system
The OSAC system was introduced on 1973 models. Its main component is the OSAC valve, figure 19-28. This valve contains a 1-way restriction, or orifice, that delays the rate of

vacuum buildup in the vacuum advance unit during acceleration from idle to part throttle. During closed-throttle deceleration or wide-open throttle acceleration, there is no spark retard delay.

OSAC valves with ambient temperature control are mounted on the vehicle firewall; those without the temperature control are attached to the air cleaner cover. This system was phased out in 1975 models and later models when the catalytic converter was introduced.

Ford

Electronic distributor modulator (EDM) system
The EDM system, figure 19-29, was used on some 1970-71 Ford cars to permit vacuum advance only during high-gear operation. Above a selected speed, the speed sensor sends a voltage signal to the electronic control module. The ambient temperature switch overrides the speed sensor signal at low temperatures to permit normal vacuum advance.

A ported vacuum switch (PVS) overrides the module when coolant temperature is high. This

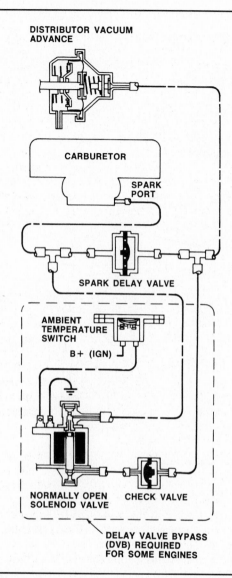

Figure 19-31. Ford's spark delay valve (SDV) system.

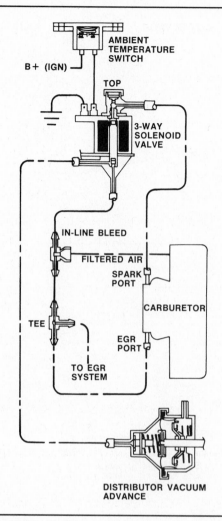

Figure 19-32. Ford's temperature-activated vacuum (TAV) system.

applies intake manifold vacuum to the distributor. The module is mounted under the instrument panel and determines whether or not the solenoid which it contains will be energized. At speeds above 25 mph (40 kph), the solenoid applies carburetor port vacuum to the vacuum advance unit; during deceleration, it denies vacuum to the advance unit.

Electronic spark control (ESC) system
The ESC system used on some 1972 Ford cars is very similar to the EDM system just described. It differs mainly in that the ESC module and the solenoid are separate, figure 19-30, while the earlier EDM module contained the solenoid.

Spark delay valve (SDV) system
This system uses a spark delay valve to restrict the vacuum control applied to the carburetor

during acceleration, figure 19-31. Since the valve is installed between the carburetor and the distributor, vacuum must pass through a sintered metal disc to reach the advance unit. When carburetor vacuum decreases, a check valve in the SDV opens to relieve vacuum from the advance unit.

On engines built before 1973, an ambient temperature switch energizes a solenoid to apply vacuum to the distributor through a delay valve bypass.

Transmission-regulated spark (TRS and TRS + 1) systems
The Ford TRS system is similar to the system used by GM and other carmakers. The TRS + 1 system has two separate vacuum systems controlled electrically by a transmission gear selector switch and an ambient air temperature switch. The TRS part of the system is identical to the previous TRS system; the + 1 function

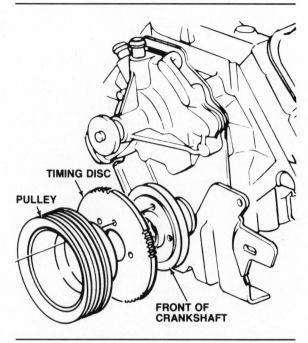

Figure 19-33. Crankshaft position signals can be taken directly from the crankshaft.

provides a choice between the carburetor spark port or the carburetor EGR port as a vacuum source using a 3-way solenoid valve as a switching device. This introduces EGR operation under certain transmission gear and temperature conditions.

Temperature-activated vacuum (TAV) systems
Some Ford cars use the TAV system to match vacuum spark advance to engine requirements by switching between two vacuum sources, figure 19-32. The 3-way vacuum solenoid valve is connected to the distributor and the carburetor spark and EGR ports.

An ambient temperature switch controls the 3-way valve. At high temperature, the EGR vacuum controls spark advance; at low temperature, the spark port vacuum controls advance.

The cold-temperature-activated vacuum (CTAV) system is a variation of TAV. The ambient temperature switch is located in the air cleaner. The temperature required to energize the solenoid is greater than in the TAV system, and a latching relay allows only one cycle of the temperature switch each time the ignition is turned on.

DECLINE AND FALL OF EARLY SPARK TIMING CONTROLS

Many spark timing control systems were discontinued when catalytic converters were introduced in 1975. Converters reduce HC and CO emissions, and exhaust gas recirculation provides more effective control of NO_x emissions.

Emission standards continued to tighten during the mid-1970's. During this period, fuel efficiency requirements also became important. Meeting the challenge of more stringent emission standards and better fuel economy required far more accurate ignition system performance than mechanical devices could deliver. Centrifugal and vacuum advance mechanisms simply could not respond quickly enough to changes in engine operating conditions to provide the necessary accuracy, so major manufacturers turned to computer-controlled ignition systems.

EARLY ELECTRONICALLY CONTROLLED TIMING

As we saw in Chapter 14, the electronic control module in a computer-controlled ignition system receives signals from various system sensors. These signals may include information on coolant temperature, atmospheric pressure and temperature, throttle position (and rate of change of position), and crankshaft position. Integrated circuits (IC's) in the control module are programmed to interpret this information and calculate the proper ignition timing for each individual spark.

Unlike today's sophisticated integrated systems, the early computer-controlled systems worked with the manufacturers' standard solid-state ignition systems. Some changes were made to the standard ignition because it no longer had to control spark timing, but many parts remained the same.

These early computer-controlled ignition systems used by domestic manufacturers can be divided into two types. In one type, rotation of the distributor shaft sends a crankshaft position signal to the control module. The other type receives crankshaft position information from a sensor mounted near the crankshaft, figure 19-33. The sensor reacts to the rotation of a special disc attached to the crankshaft itself.

Signals taken directly from the crankshaft are more accurate than those taken from the distributor shaft. The gears or chain driving the camshaft and the gears driving the distributor shaft have tolerances, or looseness. While small in actual measurements, these tolerances can combine to cause a significant difference between crankshaft position and ignition timing.

The early systems used by the domestic carmakers are called:

- Chrysler Electronic Lean-Burn (ELB)
- GM Microprocessed Sensing and Automatic Regulation (MISAR)
- Ford Electronic Engine Control (EEC).

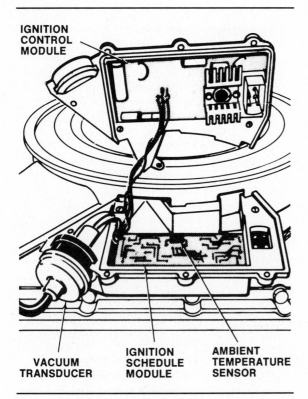

IGNITION
CONTROL
MODULE

VACUUM
TRANSDUCER

IGNITION
SCHEDULE
MODULE

AMBIENT
TEMPERATURE
SENSOR

Figure 19-34. The early Chrysler electronic lean-burn (ELB) computer contains two separate modules. Later models use one module for all functions.

The electronic timing and regulation function of all three systems is similar, but the electronics are fundamentally different. The lean-burn system uses an analog computer, while MISAR and EEC use digital microprocessors.

The practical difference between analog and digital electronics in this case is that a digital computer can *instantly* alter timing from 1 to 65 degrees. An analog computer must calculate through all the points on a theoretical curve to make the same adjustment. (Since an electronic spark advance adjustment takes only a few milliseconds, this fact is not really significant to the driver or technician). Also, a digital system is more flexible and more economical to build than an analog system.

Chrysler

Chrysler's electronic lean-burn (ELB) system was introduced on some 1976 models and used through 1978. The spark control computer is mounted on the air cleaner, figure 19-34. Early models have two printed circuit boards: the ignition schedule module and the ignition control module. Computers used with 1977 5.2L (318-cid) engines and all 1978 systems have only one circuit board to do both jobs.

The ignition schedule module receives signals from various engine sensors and computes them to determine the exact spark timing required. It then directs the ignition module to advance or retard the timing accordingly.

Six or seven sensors, figures 19-35 and 19-36, feed information to the computer.
● The Start pickup coil in the distributor provides a fixed amount of advance during cranking.
● The Run pickup coil supplies a basic timing signal and allows the computer to determine engine speed. The single circuit board computer system uses only one pickup coil to provide all timing signals.
● The coolant temperature sensor on the water pump housing signals the computer when coolant temperature is low.
● The air temperature sensor is located inside the computer. It is a thermistor that provides a varying amount of resistance with changing air temperature.
● The carburetor switch sensor tells the computer if the engine is at idle or off idle.

The other two sensors are transducers, devices that change mechanical movement to an electrical signal. The transducer has a coil and a movable metal core. A small amount of voltage applied to the coil will vary in strength as the core moves within the coil. This varying voltage signal is interpreted by the computer.
● The throttle position transducer has a core connected to the throttle lever. Core movement tells the computer the position and rate of change of position of the throttle plates.
● The vacuum transducer contains a diaphragm that is exposed to engine vacuum. Movement of the diaphragm moves the core and signals the computer of changes in engine vacuum.

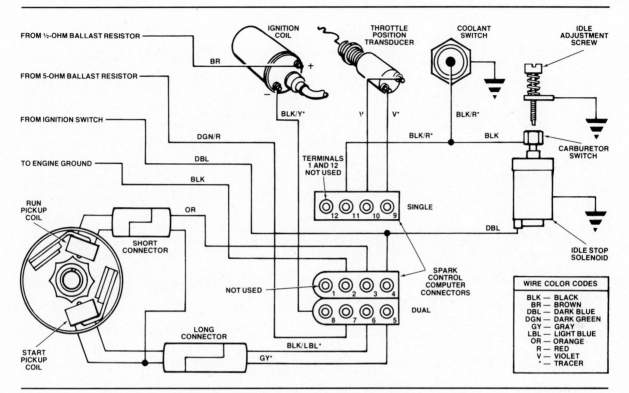

Figure 19-35. 1976-77 (except 5.2L/318-cid) V-8 ELB system wiring diagram.

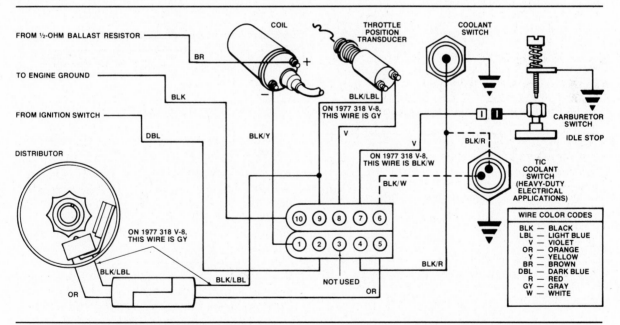

Figure 19-36. 1977 5.2L/318-cid V-8 and all 1978 (except 4-cylinder) ELB system wiring diagram.

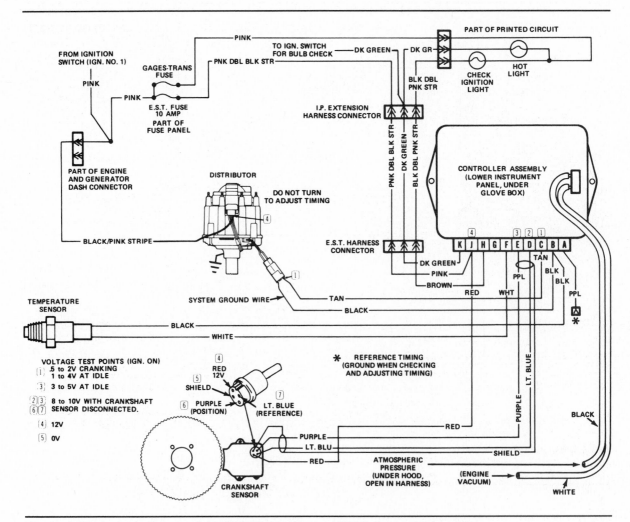

Figure 19-37. A circuit diagram of the 1977 General Motors MISAR system. (Oldsmobile)

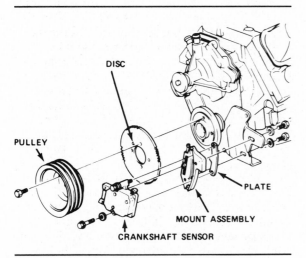

Figure 19-38. The MISAR system takes crankshaft position signals from a special disc on the front of the engine. (Oldsmobile)

General Motors

The MISAR system, figure 19-37, was used on 1977-78 Oldsmobile Toronados. The microprocessor is contained in a control module mounted under the instrument panel. The module monitors signals from these engine sensors:

- Coolant temperature
- Manifold vacuum
- Atmospheric pressure
- Crankshaft speed and position.

The coolant sensor is not a simple on/off switch, but rather a thermistor which varies resistance with changes in temperature.

The vacuum sensor is located in the control module. The sensor is a solid-state unit connected to the intake manifold by a vacuum line. A second line connected to the module is open in the engine compartment to provide an atmospheric pressure signal.

Crankshaft speed and position signals are provided by a rotating disc and a stationary

Figure 19-39. Ford's electronic engine control (EEC-I) system. (Ford)

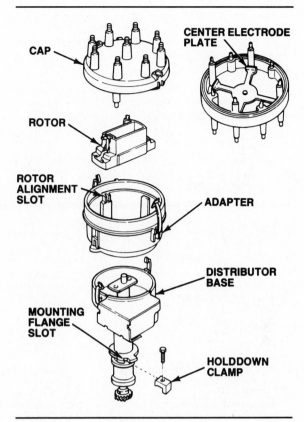

Figure 19-40. Components of Ford's EEC-I distributor. (Ford)

sensor on the front of the engine, figure 19-38. These components replace the distributor pickup coil and trigger wheel. In 1978, the crankshaft speed and position sensor was moved into the distributor.

Ford

Ford introduced its electronic engine control (EEC) system, figure 19-39, on the 1978 Lincoln Versailles. This system controls both spark timing and EGR valve operation. A digital microprocessor installed in the passenger compartment receives signals from various sensors.

An electromagnetic pickup at the flywheel end of the crankshaft provides crankshaft position and speed signals. It also replaces the distributor pickup coil and trigger wheel. Other information is fed into the computer by a throttle position sensor, coolant temperature sensor, barometric pressure sensor, inlet temperature sensor, manifold pressure sensor, and EGR valve position sensor.

The module determines the best spark timing and EGR valve operating mode. A spark timing signal is sent to the Dura-Spark solid-state ignition control module.

The EEC-I distributor, figure 19-40, differs considerably from the standard distributor. Since its only job is to distribute the ignition spark, it contains only a rotor, coil electrode, and spark plug electrodes. The distributor cap and rotor are designed to minimize the chances

of cross-firing, when the spark jumps to the incorrect spark plug electrode.

SUMMARY

Exhaust emissions are greatest during idle and part-throttle. The best way to reduce HC emissions during these periods is to retard the ignition timing. Retarded spark timing creates higher temperatures at the end of the combustion process and allows more time for the mixture to vaporize. This maintains adequate burning of the air-fuel mixture and lowers the HC and CO emissions. Retarded spark timing also reduces peak combustion pressures and combustion chamber temperatures. This reduces NO_x formation.

The two primary factors which determine spark advance are engine speed and load. Most breaker-point distributors have two devices to advance or retard spark timing in response to changing engine operation conditions. The centrifugal advance alters the position of the distributor cam or trigger wheel on the distributor shaft, thereby changing engine timing to match engine speed. The vacuum advance alters the position of the breaker points or magnetic pickup coil with respect to the cam or trigger wheel. It changes timing to match engine load.

Many different spark timing emission control systems have been used. Regardless of their design and operation, these systems all regulate distributor vacuum advance. Vacuum advance generally is allowed only at cold startup, during high gear operation, or if the engine overheats.

Auxiliary spark control systems are not used on all engines. Development of exhaust gas re-circulation systems, catalytic converters, and the expanded use of air injection systems has gradually eliminated the use of transmission-controlled spark and speed-controlled spark systems on most late-model engines.

Electronic spark timing systems appeared because more accurate control was required. They all use an electronic control module to regulate ignition timing for the best combination of emission control, fuel economy, and driveability.

Review Questions

Choose the single most correct answer.
Compare your answers with the correct answers on page 414.

1. Retarded spark timing:
   a. Occurs during idle and low-speed operation
   b. Reduces peak combustion temperature
   c. Reduces NO_x emissions
   d. All of the above

2. The deceleration vacuum advance valve was most effective in limiting which type of emission?
   a. NO_x
   b. HC
   c. CO_2
   d. CO

3. A sintered metal disc is used with:
   a. The CEC system
   b. Speed-controlled spark systems
   c. The SDV system
   d. Transmission-controlled spark systems

4. Speed- and transmission-controlled timing prevent distributor vacuum advance when the car is:
   a. In high gear
   b. In low gear
   c. Travelling at high speed
   d. None of the above

5. The GM TCS system is controlled by a:
   a. Transmission switch
   b. Throttle-vacuum switch
   c. Speedometer switch
   d. Diverter valve

6. The CEC valve:
   a. Holds the throttle open in low gear
   b. Provides vacuum advance during idle
   c. Holds the throttle open during high-gear acceleration
   d. Is deenergized in high gear

7. The distributor vacuum control switch applies full vacuum:
   a. At all times
   b. Only at idle
   c. At increased engine coolant temperature regardless of transmission gear position
   d. Only in low gear under heavy load

8. In the SCS system at low speeds:
   a. The solenoid is deenergized
   b. The speed switch is closed
   c. The speed switch is open
   d. The solenoid is energized and the speed switch is open

9. The OSAC valves on mid-70's Chryslers are located:
   a. Between the engine and the carburetor
   b. Between the engine and the exhaust manifold
   c. In the cooling system
   d. Between the carburetor and the distributor

10. The MISAR system was introduced in the:
    a. Oldsmobile Toronado
    b. AMC Pacer
    c. Lincoln Versailles
    d. Cadillac Seville

11. Which of the following was *not* one of the carmakers' early electronically controlled timing systems?
    a. ELB
    b. TCS
    c. MISAR
    d. EEC

12. Many spark timing control systems were discontinued because:
    a. Catalytic converters reduce HC and CO emissions
    b. EGR systems reduce NO_x emissions
    c. Computer-controlled ignition systems are more accurate than centrifugal and vacuum advance mechanisms
    d. All of the above

13. In Chrysler's electronic lean-burn system, which of the following is a transducer?
    a. The Start pickup coil
    b. The air temperature sensor
    c. The Run pickup sensor
    d. The throttle position sensor

14. On Ford's earliest EEC system, crankshaft speed and position signals are provided by:
    a. The distributor pickup coil
    b. The trigger wheel
    c. An electromagnetic pickup at the flywheel
    d. A rotating disc and stationary sensor on the front of the engine

15. In the TAV system, spark advance is controlled at high temperatures by:
    a. Manifold vacuum
    b. EGR vacuum
    c. The electronic module
    d. A latching relay

16. The early EDM system has:
    a. A speed sensor
    b. An electronic control module
    c. A 3-way solenoid
    d. All of the above

17. AMC cars with automatic transmissions used a speed-sensing control switch:
    a. Before 1971
    b. After 1975
    c. Before 1973
    d. After 1976

20

Exhaust Gas Recirculation

Exhaust gas recirculation (EGR) is an emission control system that reduces the amount of oxides of nitrogen (NO_x) produced during combustion. When NO_x is present in the atmosphere and acted upon by sunlight, it combines with hydrocarbons to form photochemical smog, the prime air pollutant.

In this chapter, you will learn:
- The principles of EGR systems
- How a basic EGR system reduces NO_x formation and prevents engine detonation
- EGR system variations used by domestic automakers.

NO_x FORMATION

Under normal circumstances, nitrogen and oxygen do not combine unless temperatures exceed 2,500°F (1372°C). When ignition timing is correct, maximum heat and pressure are created in an engine's combustion chambers. Whenever combustion chamber temperatures exceed 2,500°F (1371°C), nitrogen and oxygen combine rapidly to form large amounts of NO_x.

Since peak combustion chamber temperature is controlled by ignition timing, the spark timing control systems discussed in Chapter 19 were the first attempts to meet NO_x control requirements. By retarding spark timing slightly, less pressure and heat are produced. This holds combustion chamber temperatures below the level at which NO_x forms rapidly. Spark timing control systems were used to control NO_x formation until 1972, when Federal test procedures to determine NO_x levels were changed. As a result, a more effective way of controlling NO_x was needed.

EGR and How It Works

Although small amounts of NO_x are formed at temperatures below 2,500°F (1371°C), these quantities can be easily controlled. But once combustion chamber temperatures reach 2,500°F (1371°C) or more, NO_x formation increases rapidly.

There are four basic ways to reduce NO_x formation:

1. Enrich the air-fuel mixture. This allows the engine to run cooler but increases HC and CO emissions, as well as reducing fuel economy.
2. Lower the compression ratio. This has been done to allow engines to burn unleaded gasoline, but is a limited means of controlling NO_x. Too low a ratio results in inefficient combustion and increased HC and CO emissions.
3. Retard spark timing slightly. This method was used until 1972, but could not provide

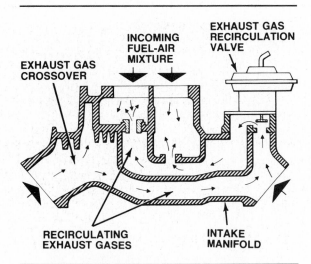

Figure 20-1. Basic exhaust gas recirculation methods.

enough control to meet the more stringent emission standards after that date.

4. Dilute the incoming air-fuel mixture with a small amount of **inert gas** to lower combustion temperatures. Since exhaust gases are relatively inert, a portion of them can be recirculated to dilute the mixture. This is the most efficient method of lowering combustion temperatures and reducing NO_x emissions without affecting engine performance, fuel economy and other exhaust emissions.

Exhaust gas recirculation (EGR) is done by routing small quantities (6 to 10 percent) of exhaust gas from the engine's exhaust ports to the intake manifold, figure 20-1. This exhaust gas dilutes the incoming air-fuel mixture in the cylinder. Since exhaust gas contains no oxygen, the resulting air-fuel-exhaust gas mixture is not as powerful when ignited. This means that it will not create as much heat as an undiluted air-fuel mixture would produce.

How EGR Affects Combustion

Since it does not require much exhaust gas to cool down peak combustion temperatures, recirculation must be held to very low quantities. Even when the EGR valve used to reroute the exhaust gas is wide open, the orifice through which the gas passes is very small.

Because the amount of NO_x produced at low engine speeds is very small, exhaust gas recirculation is neither required nor desirable at idle. It also is undesirable during high-speed driving at wide-open throttle, since it adversely affects efficient operation and good driveability. Maximum recirculation is required only during cruising and acceleration at speeds between

30 and 70 mph, (48 and 113 kph), when NO_x formation is greatest. Engine temperature also is a determining factor in recirculation. When engine temperature is low, NO_x formation is also low, and recirculation is eliminated to produce fast warmup and better driveability.

The dilution of the air-fuel mixture with EGR reduces the energy involved in combustion. This in turn reduces the amount of heat created in the combustion chamber, as well as the peak combustion pressure. The power output of early engines equipped with EGR systems was less than that of the same engine without EGR, giving rise to the idea that exhaust gas recirculation automatically meant a reduction in power output. During the early years of EGR, many drivers were convinced that they could increase the performance of their engine by disconnecting the EGR system, a popular myth that still exists today. For many years now, EGR dilution of the air-fuel charge has been calculated into the fuel system calibration and it is required for maximum performance.

EGR and Engine Detonation

A secondary effect of EGR is that it helps to control engine detonation. Detonation (engine ping) is an explosion or uneven burning of the air-fuel charge which generally takes place near the end of the combustion burn time because of the high pressure and temperature created. Detonation reduces engine power and efficiency. In some situations, detonation can severely damage the engine.

In the days when tetraethyl lead was added to gasoline, detonation was not a great problem, but it has become one for the engines operating on the unleaded gasoline required with catalytic converters. Diluting the air-fuel charge with exhaust gas helps reduce the temperature and pressure that cause detonation.

The use of EGR with electronically-controlled late-model engines permits split-second changes in ignition timing that further minimize the problem of detonation. In essence, this incorporates the timing retard provided by the older spark timing control systems, but more efficiently. You can prove this by simply disconnecting the EGR system on such an engine and listen to the pinging that results. Reconnect the system and the pinging disappears.

SYSTEM COMPONENTS AND OPERATING PRINCIPLES

The devices used to recirculate exhaust gas are described in the following paragraphs. Since

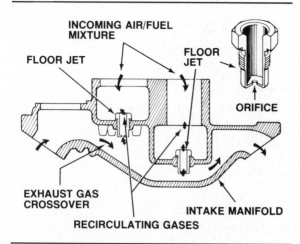

Figure 20-2. Chrysler's floor jet EGR system.

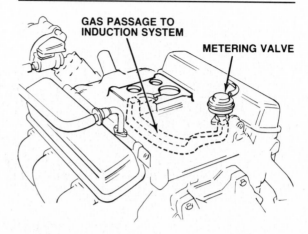

Figure 20-4. EGR metering valve mounted on the intake manifold.

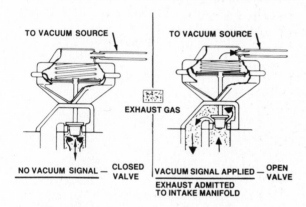

Figure 20-3. Typical single-diaphragm EGR valve. (AC-Delco)

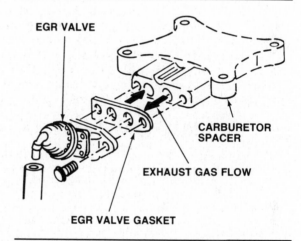

Figure 20-5. This EGR valve is mounted on a spacer installed between the carburetor and the intake manifold.

system designs and controls differ from one manufacturer to another, we will discuss each of the carmakers later in this chapter.

EGR Floor Jets

The floor jet system, figure 20-2, is an historical curiosity used only on some 1972-73 Chrysler-built engines. On V-8 engines, stainless steel jets are threaded into the floor of the intake manifold under the carburetor. With 6-cylinder engines, the jet is in the intake manifold hot spot beneath the carburetor. The jets provide an opening between the exhaust passage and the intake manifold. In this way, manifold vacuum controls how much exhaust is drawn into the intake system through the preset jet orifice. Floor jets are the simplest of all EGR system designs. However, they are also unsatisfactory because they allow exhaust gas to enter the intake manifold at all times. This causes rough engine operation when the engine is idling or when warming up.

EGR Valves

Introduced on 1972 Buicks, the EGR valve is a spring-loaded, vacuum-operated, poppet-type valve, figure 20-3. Modulating valves and tapered-stem valves also are used, and they operate about the same as the poppet type. This valve meters the exhaust gas entering the intake system. The EGR valve is mounted on the intake manifold, figure 20-4, or on a plate under the carburetor, figure 20-5. The valve

Inert Gas: A gas that will not undergo chemical reaction.

Exhaust Gas Recirculation (EGR): A way of reducing NO_x emissions by directing unburned exhaust back through an engine's intake.

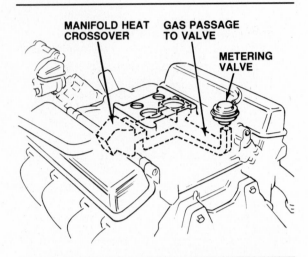

Figure 20-6. Exhaust gas passages to the EGR valve.

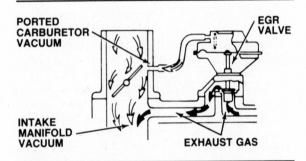

Figure 20-7. This EGR valve is operated by ported vacuum.

may be connected to the intake and exhaust systems by internal passages in the intake manifold, figures 20-4 and 20-6, or in some cases by external steel tubing. The EGR valve is held in the closed position by the spring. The valve is opened by ported or venturi vacuum from the carburetor, depending on the system design.

Ported Vacuum Systems

EGR systems controlled by ported vacuum use a slot-type port in the carburetor throttle body above the throttle plate, figure 20-7. This port is connected to the EGR valve by a vacuum line. With the throttle plate closed, no vacuum is transmitted. As the throttle plate opens, the port is exposed to increasing manifold vacuum. The amount of exhaust gas flow depends on manifold vacuum, throttle position, and exhaust gas backpressure. Because ported vacuum cannot be greater than intake manifold vacuum, recirculation at wide-open throttle is prevented. The valve opens at a point greater than the relatively weak manifold vacuum produced during wide-open throttle operation.

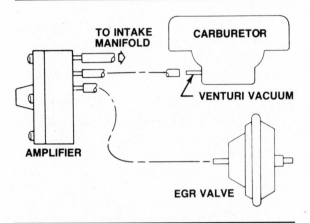

Figure 20-8. EGR systems controlled by venturi vacuum require a vacuum amplifier. (Ford)

Venturi Vacuum Systems

EGR systems controlled by venturi vacuum use a vacuum port at the throat of the carburetor venturi to provide a control vacuum, figure 20-8. Since this control vacuum is very weak, a vacuum amplifier boosts it enough to operate the EGR valve. The amount of exhaust gas flow depends mainly on engine intake airflow. It is also affected by intake vacuum and exhaust gas backpressure.

Recirculation at wide-open throttle is prevented by a relief valve, or dump diaphragm, that compares venturi and manifold vacuum. When the diaphragm senses that the throttle is wide open, the stored vacuum in the amplifier is vented to the atmosphere. This limits the vacuum output reaching the EGR valve to that provided by manifold vacuum. Since the EGR valve opens at a point greater than the relatively weak manifold vacuum, the valve remains closed at wide-open throttle.

Vacuum amplifiers

The vacuum amplifier, figure 20-9, converts the weak venturi control vacuum into one that is strong enough to operate the EGR valve. It does this by storing manifold vacuum in a reservoir inside the amplifier unit. This guarantees enough vacuum, regardless of variations in manifold vacuum. Whenever the venturi is equal to, or greater than, manifold vacuum, a relief valve (dump diaphragm) vents the reservoir, cancelling the output EGR signal.

A vacuum amplifier working perfectly would produce an accurate, repeatable, and precise proportion between venturi air flow and EGR flow. However, vacuum amplifiers are no longer used on late-model EGR systems because storage reservoirs tend to leak in actual use.

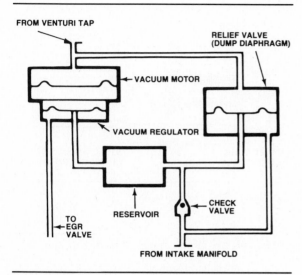

Figure 20-9. A vacuum amplifier is controlled by venturi vacuum and uses manifold vacuum to operate the EGR valve. (Ford)

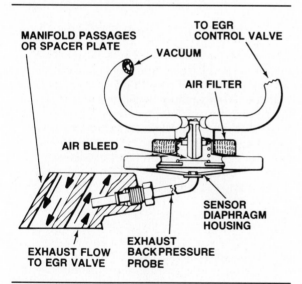

Figure 20-10. Typical EGR backpressure transducer.

Backpressure Transducer

This diaphragm-operated sensor unit, figure 20-10, has a tube that extends into an exhaust gas passage. When high exhaust backpressure is sensed through the tube, the diaphragm closes an air bleed hole in the EGR vacuum line. This provides maximum EGR during acceleration, when backpressure is high. As backpressure drops, a spring moves the sensor diaphragm to reopen the vacuum line bleed. Since this decreases the vacuum at the EGR valve, the amount of exhaust gas recirculated also is reduced.

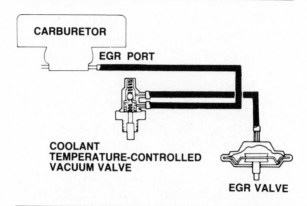

Figure 20-11. The coolant-temperature-controlled vacuum valve cuts off vacuum to the EGR valve when the engine is cold. (Ford)

Modulating Devices

Engineers have developed various methods of modulating, or adjusting, EGR valve operation in relation to engine operating conditions. High and low ambient temperature vacuum modulators weaken the vacuum signal to the EGR valve. A coolant temperature override switch or valve, figure 20-11, may be used to eliminate EGR vacuum below certain engine operating temperatures. Some venturi-vacuum-controlled systems use a time delay solenoid to shut off vacuum to the EGR valve for about half a minute after the ignition is turned on.

A vacuum-bias valve is used to bleed off part of the EGR vacuum signal under high-manifold-vacuum conditions to eliminate **high-speed surge**. A dual-diaphragm EGR valve, figure 20-12, sometimes is used. This uses manifold vacuum to help the valve spring offset the carburetor vacuum under certain cruising conditions.

Closed-Loop Electronic Control

When the EGR system is part of a closed-loop electronic control system, it contains a sensor that informs the microprocessor of how the system is working. It may do this by sending a position signal indicating how far the valve is open, or a pressure signal indicating the EGR gas flow rate. The microprocessor compares the sensor data with its program to determine whether the flow rate should be increased,

High-Speed Surge: A sudden increase in engine speed caused by high manifold vacuum pulling in an excess air-fuel mixture.

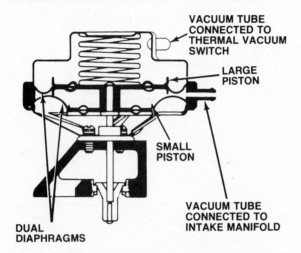

Figure 20-12. The dual-diaphragm EGR valve modulates the exhaust gas flow.

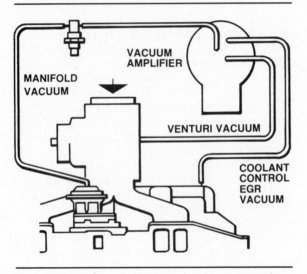

Figure 20-13. Chrysler's external vacuum reservoir.

maintained, or decreased. It then signals a solenoid in the EGR system.

The solenoid operation can be precisely controlled through the pulse width modulation (PWM). This means that the solenoid is continuously cycled on and off a fixed number of times per second. The solenoid is on (energized) for part of each cycle, and off for the rest of that cycle. The percentage of the total cycle time that the solenoid is energized is called its duty cycle. The duty cycle is calculated by the input from various other sensors to the microprocessor and determined by a timed voltage pulse to the solenoid from the microprocessor. The microprocessor varies, or modulates, this pulse width to establish the duty cycle and achieve the desired solenoid output.

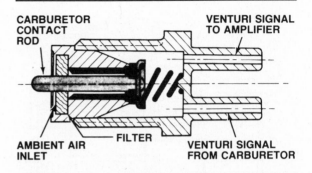

Figure 20-14. Wide-open-throttle EGR dump valve. (Chrysler)

CHRYSLER EGR SYSTEMS

Chrysler used the floor jet system, figure 20-2, described earlier, on some 1972-73 models. The system was supplemented in 1973 with both the ported and venturi vacuum systems and discontinued at the end of the 1973 model year.

Venturi Vacuum System

The venturi vacuum system was changed on some 1975 and later engines to include an external vacuum reservoir mounted on a bracket and attached to the vacuum amplifier, figure 20-13. After the internal vacuum reservoir has been vented, or dumped, the external reservoir supplies manifold vacuum for exhaust recirculation until the internal reservoir can be refilled.

Wide-Open-Throttle Dump Valve

A wide-open-throttle dump valve, figure 20-14, was added to some engines in March 1976. These engines use a delay-dump amplifier, which can hold the EGR valve open too long at wide-open throttle. Mounted on the carburetor, the wide-open throttle EGR dump valve overrides the delay-dump amplifier at wide-open or near wide-open throttle by mechanically bleeding the vacuum to the amplifier. This closes the EGR valve immediately.

Coolant Control Valves

Chrysler has used both a coolant control EGR valve (CCEGR) and a coolant vacuum switch cold closed (CVSCC). These devices are combined as a coolant temperature sensor with a molded vacuum port connector, figure 20-15. The sensor was first added to 1974 systems to block vacuum flow to the EGR valve whenever coolant temperature is less than specified for the particular engine. Some 1974 and later engines use an EGR delay timer. This electric timer mounted on the firewall operates an

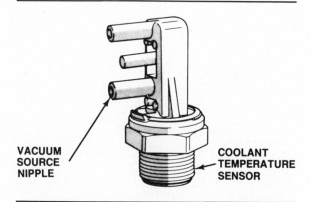

Figure 20-15. The CCEGR and CVSCC valves are combined as a coolant temperature sensor. (Chrysler)

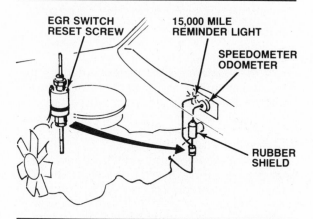

Figure 20-16. Chrysler's EGR maintenance reminder system.

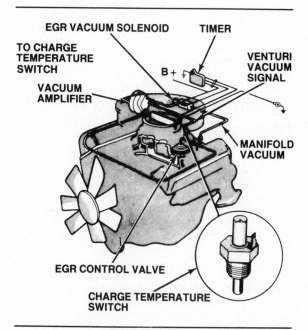

Figure 20-17. Chrysler's EGR system with charge temperature switch and timer. (Chrysler)

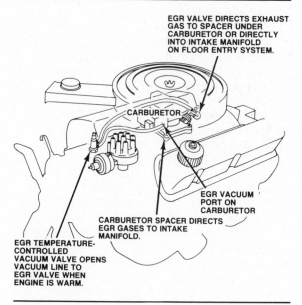

Figure 20-18. Basic Ford EGR system. (Ford)

engine-mounted vacuum solenoid to prevent exhaust gas recirculation for about half a minute after the ignition is turned on.

EGR Maintenance Reminder System

An EGR maintenance reminder system, figure 20-16, was used on 1975 engines only. This mileage counting device signals the driver every 15,000 miles (24,000 kilometers) to have the EGR system inspected. An instrument panel reminder light comes on at 15,000-mile (24,000-kilometer) intervals and remains lighted until the switch attached to the speedometer cable is manually reset.

Charge Temperature Switch

The charge temperature switch, figure 20-17, used on some 6-and 8-cylinder Chrysler engines, measures air-fuel mixture temperature in the intake manifold and prevents EGR flow unless the temperature is above the switch specification. When the intake charge temper-

ature reaches about 60°F (16°C), the charge temperature switch opens, allowing the charge temperature timer (EGR delay timer) to time out. At the end of the timer cycle, vacuum flows to the EGR valve and it opens to pass exhaust gas into the intake manifold.

FORD EGR SYSTEMS

The basic Ford EGR system, figure 20-18, was introduced on all Ford-built cars in 1973, except

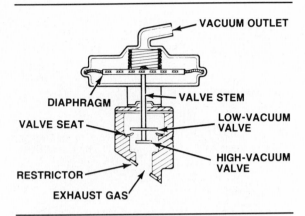

Figure 20-19. Ford modulating EGR valve. (Ford)

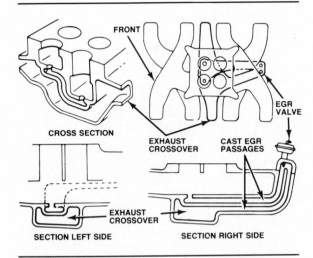

Figure 20-20. Ford 1974 floor entry EGR system. (Ford)

those using the 1,600-cc and 2,000-cc engines. The basic system has three main components: an EGR valve, a temperature-controlled vacuum valve, and a carburetor or throttle body spacer. Some later-model Ford engines have been certified by the Environmental Protection Agency (EPA) for use without an EGR system.

EGR Valve

Three types of EGR valves are used: poppet, modulating, and tapered stem. The poppet valve contains a spring-loaded diaphragm, a valve and valve stem, and a flow restrictor. Ported carburetor vacuum opens the valve at a specific vacuum level and allows exhaust gas to enter the valve. Exhaust gas flow to the combustion chambers is controlled by the flow restrictor in the valve body inlet port.

The modulating valve, figure 20-19, has an extra disc valve on the stem below the main valve, and it operates much like a poppet valve. However, at a specific vacuum level, the lower disc valve restricts exhaust gas flow to the valve chamber. This modulating action improves the driveability of some engines.

The tapered-stem valve operates in the same way. Exhaust gas flow is modulated by the tapered stem as it gradually unseats to permit an increasing gas flow.

The three valve types are not interchangeable between engine types or model years.

Ford uses various ways to deliver exhaust gas to the EGR valve. On many inline engines, exhaust gas is drawn from the exhaust manifold through an external stainless steel tube. Most V-6 and V-8 engines use a passage in the exhaust crossover of the intake manifold. This routes the gas through the carburetor or throttle body spacer to the EGR valve. The EGR

valve then meters the gas back through a separate passage in the spacer to the carburetor primary venturis or throttle body, where it is mixed with the air-fuel mixture.

Some 1974 engines use a floor-entry system, figure 20-20. The intake manifold has two runners cast in the floor under the intake runners. One connects the exhaust crossover to the EGR valve. The other connects the valve with two holes in the manifold floor directly under the carburetor primary venturis. This allows exhaust gas to mix with the air-fuel mixture before entering the combustion chambers. This is a valve-controlled system and is not the same as Chrysler's floor jet EGR system, which does not use a valve.

Temperature-Controlled Vacuum Valve

A temperature-controlled vacuum valve in the engine cooling system cuts off vacuum to the EGR valve when the engine is cold, figure 20-18. Ford calls this valve a ported vacuum switch (PVS). It works the same as American Motors' coolant temperature switch (CTO) and Chrysler's coolant-controlled EGR switches (CCEGR or CVSCC). The valve may be installed in a heater hose or in the intake manifold water jacket. Different temperature settings are used for different Ford engines. The valves are color coded for identification.

Spacer Plate

Ford's spacer plate contains separate inlet and outlet passages to the EGR valve. The valve is mounted at one end of the plate with a gasket. In some engines, an EGR cooler is installed between the valve and plate to improve EGR flow

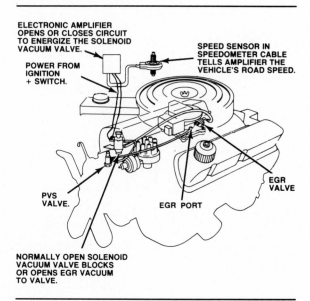

Figure 20-21. Ford high-speed EGR modulator subsystem. (Ford)

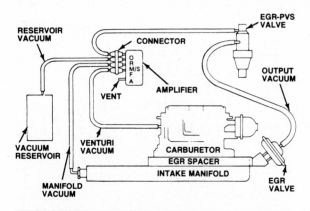

Figure 20-22. Ford venturi vacuum control subsystem with single-connector amplifier. (Ford)

and valve durability. The spacer is installed between the carburetor or throttle body and the intake manifold with top and bottom gaskets to prevent leaks. Spacer plates are not interchangeable between engine types.

System Modifications

Basic EGR system operation is modified by the following subsystems.

High-speed EGR modulator subsystem
Some Ford V-8 engines have this subsystem, figure 20-21, to improve durability at speeds above 64 mph (103 kph). This subsystem shuts off carburetor vacuum to the EGR valve to prevent recirculation. A speed sensor driven by the speedometer cable signals an electronic module. When road speed exceeds 64 mph (103 kph), the module closes the open solenoid valve. This, in turn, closes the EGR vacuum port and vents the vacuum outside the valve. As road speed drops below 64 mph (103 kph), the module deenergizes the solenoid valve. This closes the vent and opens the EGR vacuum line.

Venturi vacuum control
Some 1974 6-cylinder engines use venturi vacuum rather than ported vacuum. A vacuum amplifier, figure 20-22, is connected between the EGR valve and intake manifold vacuum. When air flow through the venturi is high enough, venturi vacuum opens the amplifier. This allows manifold vacuum to open the EGR valve.

The EGR valve remains closed at idle, since there is no venturi vacuum produced at idle. Whenever venturi vacuum is equal to or greater than manifold vacuum (such as under wide-open throttle), the vacuum amplifier dumps the output vacuum to the EGR valve, causing the valve to close. Some of the amplifiers contain an output bias. This is a bleed designed for quicker EGR valve opening.

■ Engine Modifications Can't Do The Whole Job

Since the automobile was discovered to be the biggest source of air pollution, many changes in engine design have been made to "clean it up". These engine modifications, such as EGR systems, have reduced exhaust emissions, but it is impossible to eliminate the major cause of HC emissions simply by changing the engine design. This is because the major cause of HC emissions is the effect of the "quench area" on combustion.

The quench area is the inner surface of the combustion chamber. When the ignition flame front passes through the combustion chamber, it burns the fuel charge as it goes until the quench area is reached. This is a thin layer between .002" and .010" (.05-mm and .25-mm) thick at the edge of the combustion chamber. When the flame front reaches the quench area, it is snuffed out because the quench area is so close to the cylinder head water jacket that the temperature there is too low for combustion to continue. Consequently, hydrocarbons within the quench area do not burn. They are ejected from the cylinder on every exhaust stroke along with the exhaust gases formed by combustion and enter the atmosphere as pollutants.

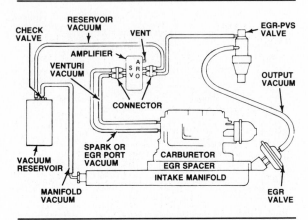

Figure 20-23. Ford venturi vacuum control subsystem with dual-connector amplifier. (Ford)

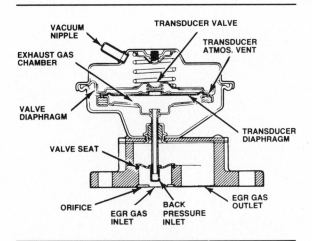

Figure 20-25. This Ford EGR valve has a built-in exhaust backpressure transducer. (Ford)

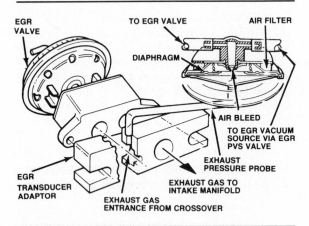

Figure 20-24. EGR backpressure transducer (sensor). (Ford)

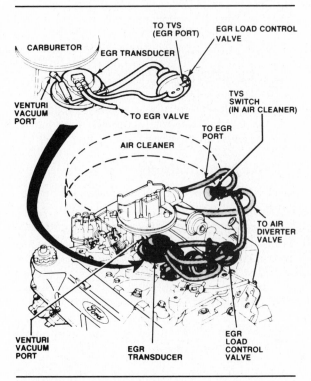

Figure 20-26. Ford air temperature vacuum switch and EGR load control valve installation on a 6.6L/400-cid V-8. (Ford)

A single-connector amplifier with all ports on one side is used with 1974-75 systems, figure 20-22. A dual-connector amplifier with ports on both sides is used with 1975 and later systems, figure 20-23.

Backpressure transducer
Late 1975 and most later engines use a backpressure sensor, or transducer, figure 20-24. This is connected between the EGR valve and the intake manifold. It modulates exhaust gas flow by varying the vacuum to the EGR valve, according to exhaust backpressure.

On some 1977 models, the backpressure transducer is within the EGR valve, figure 20-25. This combination unit has an internal exhaust gas chamber and a transducer diaphragm to sense exhaust backpressure through a hollow valve stem. This combination EGR valve-transducer constantly meters exhaust flow according to exhaust backpressure.

Many late-model 4-cylinder Ford engines use a remote backpressure variable transducer (BVT) to modulate EGR vacuum relative to two different sources of backpressure.

Temperature control
Some 1976 and later V-8 engines use an air-temperature vacuum switch (TVS) mounted in the air cleaner, figure 20-26, instead ofl the coolant PVS. The TVS also controls the vacuum

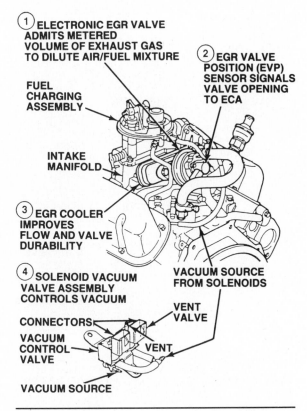

Figure 20-27. Ford closed-loop electronic control EGR system with solenoid vacuum valve control. (Ford)

flow to the diverter valve in the Thermactor air injection system. Since the switch is closed at temperatures below about 60°F (16°C), exhaust gas does not recirculate on a cold engine.

Computer Control and Pulse Width Modulation

Ford uses two slightly different types of control for its EGR systems that are part of the electronic engine control (EEC) system. One type uses solenoid vacuum valve control, figure 20-27. The other type uses an electronic vacuum regulator, figure 20-28. The sensor in each system sends a position signal which tells the microprocessor how far the valve is open.

In each case, the EEC microprocessor compares the sensor data with its program to determine whether the flow rate should be increased, maintained, or decreased. It then signals the solenoid vacuum valve assembly, figure 20-27, or the electronic vacuum regulator, figure 20-28, to maintain or change the vacuum on the EGR diaphragm as required. The EGR valve used in both systems is essentially a ported vacuum

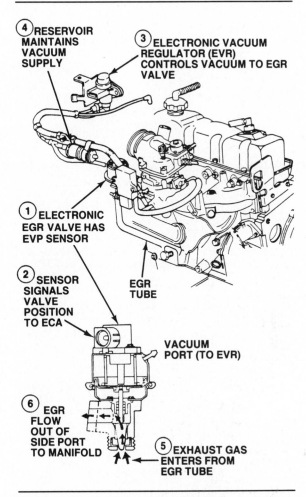

Figure 20-28. Ford closed-loop electronic control EGR system with solenoid vacuum valve control. (Ford)

valve with a position sensor (potentiometer) attached to the top of the EGR valve stem, figure 20-28.

In a system using the solenoid assembly, figure 20-29, the two solenoids "dither", that is, they open and close rapidly to modulate the valve opening. The vacuum valve solenoid provides vacuum to the EGR valve when energized. The vent valve solenoid vents the EGR valve to the atmosphere when deenergized. The vacuum valve solenoid contains an inlet port restrictor to reduce its flow rate compared to the vent valve solenoid. This allows the vent valve solenoid to vent vacuum flow immediately if the vacuum valve solenoid sticks open.

The electronic vacuum regulator, figure 20-30, is a spring-loaded solenoid that responds to the duty cycle established by the EEC microprocessor by applying, trapping, or bleeding off vacuum to the EGR valve as required.

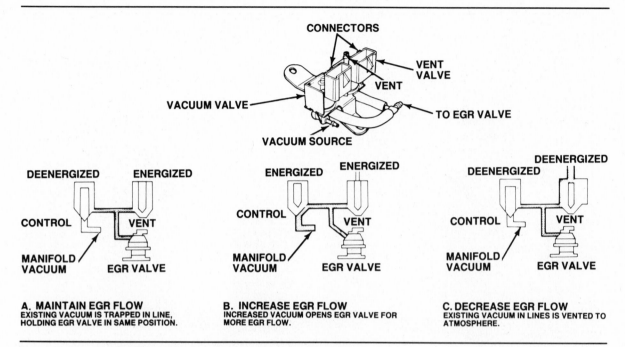

Figure 20-29. Ford's EGR solenoid vacuum valve assembly. (Ford)

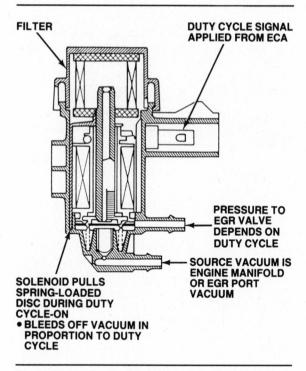

Figure 20-30. Ford's electronic vacuum regulator. (Ford)

GENERAL MOTORS EGR SYSTEMS

Exhaust gas recirculation was introduced by Buick on 1972 models with manual transmissions and all 1972 vehicles sold in California.

The EGR system used on a General Motors vehicle is one of four types:
- Ported vacuum
- Positive backpressure
- Negative backpressure
- Pulse width modulation (PWM).

System Components

You must accurately identify the type of system you are working on because test procedures differ according to the design. Most systems include a coolant temperature override switch or a thermal vacuum switch (TVS) to prevent EGR operation when the engine is cold.

The ported vacuum EGR valve is a conventional design that meters exhaust gas into the intake manifold during part-throttle operation. Some of these systems have a remote-mounted backpressure transducer that allows the EGR valve to supply additional flow under load when ported vacuum is low. The transducer is installed in the vacuum line to the EGR valve and contains an air bleed that is normally open. When exhaust system backpressure to the transducer exceeds a preset level, the air bleed is sealed off and full vacuum is applied to the EGR valve diaphragm.

Single-diaphragm EGR valve

The single diaphragm valve, figure 20-3, contains a single spring-loaded diaphragm connected to the valve by a shaft. As the throttle

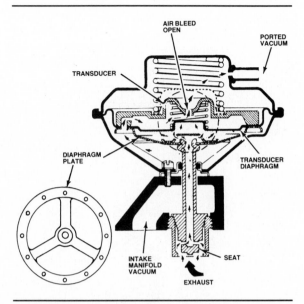

Figure 20-31. General Motors' positive backpressure EGR valve has a built-in transducer. (Buick)

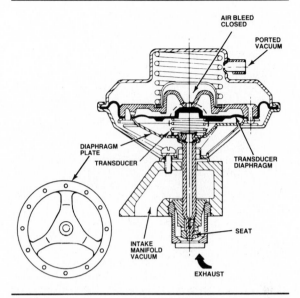

Figure 20-32. General Motors' negative backpressure EGR valve. (AC-Delco)

valve opens, vacuum is applied to the diaphragm from a carburetor EGR port. The vacuum pulls the diaphragm up to open the valve and permit EGR flow. Variations in the strength of the vacuum signal control the quantity of exhaust gas that is recirculated.

Dual-diaphragm EGR valve

The dual-diaphragm EGR valve, figure 20-12, was an early attempt to modulate EGR flow. It contains two diaphragms with differing effective areas and two vacuum ports. The upper diaphragm receives a ported signal from the carburetor; the lower diaphragm is rigidly attached to the shaft, forming a vacuum chamber connected to manifold vacuum. High manifold vacuum (during cruising) pulls the valve partially closed, decreasing EGR flow. Low manifold vacuum (acceleration) allows the ported vacuum signal to open the valve and increase EGR flow.

Positive backpressure EGR valve

Before the introduction of the positive backpressure EGR valve, a separate backpressure transducer or backpressure valve (BPV) was used with single-diaphragm EGR valves to regulate the vacuum signal according to engine load.

Building the transducer into the valve resulted in the positive backpressure EGR valve, figure 20-31, which first appeared in 1977. This valve uses both engine vacuum and exhaust backpressure to control EGR flow. Exhaust system backpressure travels up the inside of the

EGR valve stem and moves a diaphragm to seal off the air bleed when pressure exceeds a preset level. It provides a greater amount of recirculation during heavy engine loads than the single-diaphragm valve. EGR systems using this valve regulate the timed vacuum to the EGR valve according to exhaust backpressure level.

Negative backpressure EGR valve

This valve, figure 20-32, was introduced in 1979 for use on engines that have relatively little exhaust backpressure. In a negative-backpressure system, the transducer air bleed is normally closed. As ported vacuum opens the EGR valve, a negative pressure signal from the vacuum in the intake manifold is buffered by the exhaust system pressure and travels up the inside of the EGR valve stem to the backside of the transducer diaphragm. If the pressure signal is low enough (high vacuum), it will open the air bleed and reduce the amount of EGR. This modulating process goes on constantly and has the advantage of supplying a consistent percentage of EGR under most operating conditions.

Other EGR system controls

EGR flow is blocked on many GM engines when the engine is cold. You can easily determine if a temperature control device is used, because it will be found in the EGR vacuum line. If no temperature control is used, the EGR vacuum line will run directly between the valve and the carburetor EGR port. EGR systems

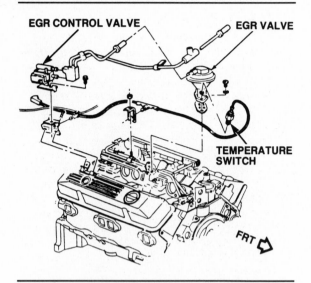

Figure 20-33. The GM PWM EGR system used on 5.0L and 5.7L engines. (GM)

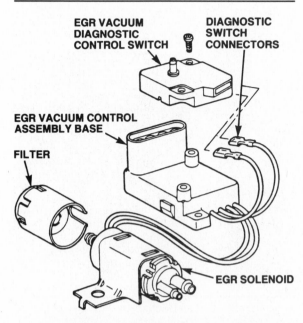

Figure 20-34. GM PWM systems use a diagnostic vacuum switch and serviceable filter on the EGR solenoid. (GM)

with temperature control may use one of three different methods to control the EGR flow:

1. A coolant temperature sensor located in the thermostat housing, intake manifold, or engine coolant passage
2. A temperature-sensitive vacuum switch in the intake manifold to sense intake mixture temperature
3. A temperature-sensitive valve near the thermostat housing or strapped to the engine to sense radiated heat from the engine.

Some engines may use a combination of these methods. For example, the system may be designed to keep EGR off until both the coolant temperature and mixture temperature or radiated heat reach specified values.

In addition to these controls, GM EGR systems may use a delay valve or restrictor in the EGR vacuum line to delay valve closing under specified conditions.

EGR with Computer Command Control (CCC)

The first GM EGR systems to be computer-controlled were on Cadillacs with electronic fuel injection. When the engine reached a specified temperature, the electronic control module turned on an electric solenoid. This passed vacuum to the EGR valve and allowed exhaust gas flow. This same concept has been further refined with current PWM systems which first appeared on 1983 models.

Pulse-width-modulated EGR systems

The typical pulse-width-modulated system, figure 20-33, uses a PWM solenoid and works much the same as the Ford system previously described. In this design, the computer operates the solenoid continuously at a fixed frequency of 32 Hz. This is similar to the operation of a carburetor mixture control solenoid but at a faster rate. To regulate the amount of vacuum applied to the EGR valve, the computer varies the solenoid duty cycle (modulates the pulse width), the ratio of on-time to off-time. The duty cycle is calculated by the ECM according to data from various sensors which measure factors such as engine coolant temperature, throttle position, mass airflow, manifold air temperature, engine rpm, and transmission gear.

For most driving conditions, the duty cycle ranges from 10 to 90 percent. A low duty cycle provides less vacuum to the EGR valve while a high duty cycle provides more. The solenoid operates at zero duty cycle only when the transmission or transaxle is in Park or Neutral.

The EGR solenoid may be a normally open or normally closed version, depending on the system. A vent filter installed on the solenoid, figure 20-34, should be replaced every 30,000 miles (48,000 kilometers).

Figure 20-35. The integrated electronic EGR valve was introduced on 1987 GM Generation II 2.8L engines.

A diagnostic switch senses exhaust gas temperature as it enters the intake manifold through the EGR valve. If a vacuum circuit failure occurs, the diagnostic switch will set a code in the ECM memory and turn on the instrument panel "CHECK ENGINE" or "SERVICE ENGINE SOON" light. The diagnostic switch can take different forms, as shown in figures 20-33 and 20-34.

There are two basic variations of the PWM system; one was developed by Chevrolet and the other by Buick, but both are used in many different GM models. Both also use throttle position and manifold vacuum as their basic regulating inputs. The Chevrolet system came out first and has a thermal sensor in the base of the EGR valve to delay vacuum application until engine temperature reaches 175°F (79°C). The Chevrolet system does not respond to vacuum changes as rapidly as does the Buick system, and is not fast enough to signal a loss of EGR on turbocharged engines.

The Buick system is more sophisticated and includes a current-regulating module that maintains constant control voltage to the solenoid for more accurate EGR regulation. The Buick system also includes a vacuum switch that monitors the amount of vacuum applied to the EGR valve; if the vacuum level is outside the programmed limits, the switch will set a trouble code in the ECM memory.

An integrated electronic EGR valve, figure 20-35, was introduced on some 1987 GM engines. The top of the EGR valve contains a vacuum regulator and pintle position sensor sealed inside a nonremovable plastic cover, figure 20-36. This combines all of the control devices in one assembly. The internal solenoid is normally open, causing the vacuum signal to be vented when EGR is not desired. A voltage regulator converts the ECM signal and regulates the current to the solenoid as required to establish the proper pulse width modulation. The pintle position sensor provides a voltage output to the ECM which increases as the duty cycle increases, allowing the ECM to monitor valve operation. A serviceable filter on the side of the assembly (arrow, figure 20-35) provides clean air to the vacuum regulator.

Some GM V-6 engines use an aspirated EGR system, figure 20-37. This delivers EGR flow under low engine vacuum conditions for detonation control. The **aspirator** is a combined regulator and venturi mounted in the air cleaner. The aspirator control valve is installed between the aspirator pressure outlet and the carburetor manifold vacuum port. If engine manifold vacuum drops under the aspirator control valve calibration, the valve opens to pass air through the aspirator venturi and create a supplemental vacuum. Once manifold vacuum returns above the aspirator control valve calibration, the valve closes to prevent airflow through the aspirator venturi.

Since the air pump flow goes to the exhaust ports in the open-loop mode (and would not flow through the aspirator), an orifice in the AIR control valve acts as a partial bypass, allowing the aspirator to function in open-loop if required. When the engine is in closed-loop mode, the AIR control valve sends pump airflow through the aspirator and into the air cleaner.

AMERICAN MOTORS EGR SYSTEMS

The basic AMC EGR system, as introduced on 1973 models, uses a vacuum-operated EGR valve. When the carburetor throttle is opened

Aspirator: A regulator and venturi installed in the air cleaner to provide EGR flow under low engine vacuum conditions to control detonation.

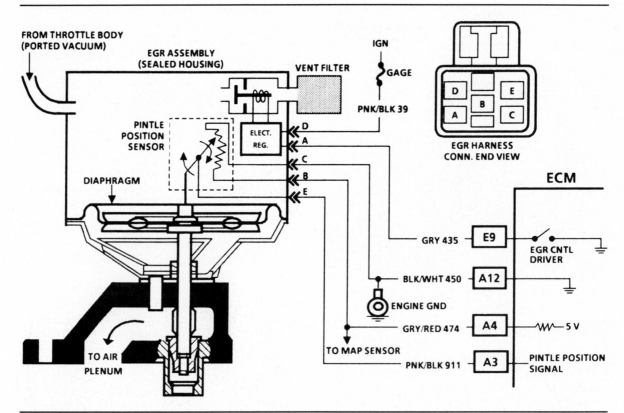

Figure 20-36. The integrated electronic EGR valve contains a built-in vacuum regulator, solenoid, and pintle position sensor. (Oldsmobile)

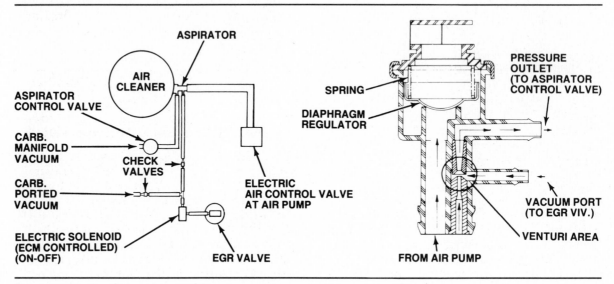

Figure 20-37. The GM aspirated EGR system used on some V-6 engines. (GM)

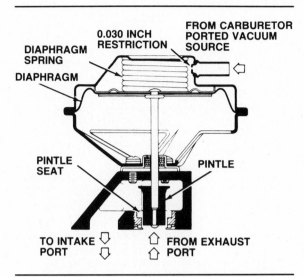

Figure 20-38. This American Motors EGR valve is typical of most single-diaphragm EGR valves. (AMC)

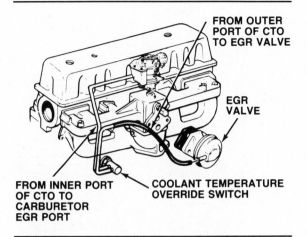

Figure 20-40. Basic AMC 6-cylinder EGR installation. (AMC)

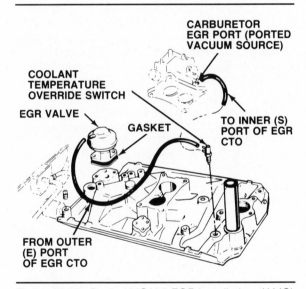

Figure 20-39. Basic AMC V-8 EGR installation. (AMC)

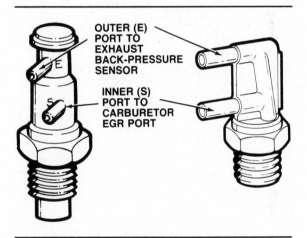

Figure 20-41. American Motors' 2-port EGR coolant temperature override (CTO) switch. (AMC)

Coolant Temperature Override Switch

This EGR coolant temperature override (CTO) switch prevents normal EGR valve operation when engine coolant temperature is below 115°F (46°C). The 3-port switch originally used has been replaced by a 2-port switch, figure 20-41, on 1974 and later models. At coolant temperatures above 115°F or 160°F (46°C or 71°C) depending on the model, vacuum is permitted to reach the EGR valve.

beyond the idle position, ported vacuum is applied to the normally closed EGR valve diaphragm, figure 20-38. Moving upward against coil spring pressure, the diaphragm opens the **pintle valve**. This valve permits exhaust gas to be drawn into the engine intake from the manifold crossover exhaust passages on V-8 engines, figure 20-39, or from below the carburetor heat riser on 6-cylinder engines, figure 20-40. The EGR valve is closed during idle and deceleration to prevent a rough idle. California cars use an exhaust backpressure transducer, figure 20-10.

American Motors also uses the following override and modulator devices to control EGR system operation.

Pintle Valve: A valve shaped much like a hinge pin. In an EGR valve, the pintle is attached to a normally closed diaphragm. When ported vacuum is applied, the pintle rises from its seat and allows exhaust gas to be drawn into the engine's intake system.

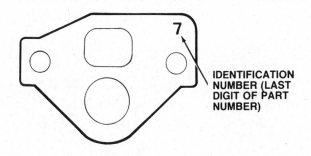

Figure 20-42. American Motors' EGR restrictor plate. (AMC)

Low- and High-Temperature Vacuum Signal Modulators

Used only on 1973 models, these modulators are connected to the EGR vacuum control line. When air temperature is under 60°F (16°C) the low-temperature modulator weakens the vacuum to the EGR valve. This decreases the amount of exhaust gas flow. At temperatures above 115°F (46°C), the high-temperature modulator restores vacuum to the EGR valve and increases exhaust gas flow.

Exhaust Backpressure Sensor

Used only on California cars during 1973-75, the backpressure sensor, figure 20-10, was extended to all 1976 and later models. The sensor permits EGR flow only when the engine is at normal operating temperature and exhaust backpressure is high. If backpressure is not high enough, and the EGR-CTO switch does not open, the EGR vacuum signal is vented to the atmosphere.

Restrictor Plate

A stainless steel restrictor plate, figure 20-42, is used on some 1974-75 California models, and on all 1976-77 cars. Located between the intake manifold and spacer, the plate is adjusted for each engine and exhaust system. It limits EGR flow rate and improves driveability.

SUMMARY

Combustion chamber temperatures that exceed 2,500°F (1371°C) cause nitrogen and oxygen to combine rapidly, forming large quantities of NO_x, a prime air pollutant. To reduce NO_x as much as possible, automakers use exhaust gas recirculation (EGR) systems to meter small quantities of exhaust gas into the incoming air-fuel mixture. This dilutes the fuel charge and results in lower combustion chamber temperatures.

Each automaker uses some variation of the basic system to achieve exhaust gas recirculation. Various temperature sensors, flow valves, and other control devices in the EGR system modulate the flow of exhaust gas to maintain driveability while reducing NO_x emissions.

With the advent of more sophisticated engine control systems, EGR system control has become a function of the engine control computer. The computer determines when and how much EGR flow is required and activates a pulse-wide-modulated (PWM) solenoid. The solenoid duty cycle established by the computer determines the flow rate and quantity.

Review Questions

Choose the single most correct answer.
Compare your answers with the correct answers on page 414.

1. Photochemical smog is a result of:
   a. Sunlight + NO_x + HC
   b. Sunlight + NO_x + CO_2
   c. Sunlight + CO + HC
   d. Sunlight + NO_x + CO

2. NO_x forms in an engine under:
   a. High pressure and low temperature
   b. Low pressure and low temperature
   c. High temperature and high pressure
   d. All of the above

3. Which of the following is a relatively inert gas?
   a. NO_x
   b. HC
   c. Exhaust gases
   d. Air

4. EGR is most desirable at:
   a. Speeds of 30-70 mph (48-113 kph)
   b. Idle speeds
   c. High speeds
   d. Low engine temperature

5. Which of the following is *not* part of an exhaust gas recirculation system?
   a. Chrysler's floor jets
   b. Buick's EGR valve
   c. Ported vacuum systems
   d. Slow-idle solenoid

6. Which is *not* true of EGR valves?
   a. They operate on venturi vacuum systems
   b. They operate on ported vacuum
   c. They may be mounted on the intake manifold
   d. They operate at wide-open throttle

7. Which is *not* part of the AMC EGR system?
   a. Low- and high-temperature vacuum signal modulators
   b. Floor jets
   c. Restrictor plate
   d. CTO switch

8. A Chrysler EGR delay timer prevents exhaust gas recirculation:
   a. During wide-open throttle
   b. At idle
   c. For 30 seconds after ignition
   d. At medium speeds

9. The Ford temperature-controlled vacuum valve is called:
   a. CTO
   b. CCEGR
   c. PVS
   d. OSAC

10. The high-speed EGR modulator subsystem shuts off the EGR valve above:
    a. 24 mph (39 kph)
    b. 34 mph (55 kph)
    c. 64 mph (103 kph)
    d. 84 mph (135 kph)

11. The Ford TVS, in addition to controlling EGR vacuum, controls:
    a. The temperature control override
    b. The ported vacuum switch
    c. The diverter valve vacuum
    d. All of the above

12. The combined transducer and EGR valve was introduced in which Buick model year?
    a. 1975
    b. 1976
    c. 1977
    d. 1978

13. Which is *not* one of the basic ways to reduce NO_x formation?
    a. Enrich the air-fuel mixture
    b. Lower the compression ratio
    c. Advance spark timing slightly
    d. Dilute the incoming air-fuel mixture with a small amount of inert gas

14. In a closed-loop electronic control system, a sensor informs the microprocessor of how the system is functioning by sending a:
    a. Position signal
    b. Pressure signal
    c. Either a or b
    d. Neither a nor b

15. Which is *not* a type of EGR valve in use?
    a. Poppet
    b. Modulating
    c. Tapered stem
    d. Pintle

16. General Motors' computer-controlled EGR systems use:
    a. A solenoid vacuum valve
    b. An electronic vacuum regulator
    c. An integrated electronic EGR valve
    d. A ported vacuum valve with a position sensor

17. Which is *not* a temperature control method used to control EGR flow?
    a. A coolant temperature sensor
    b. A temperature-sensitive vac-uum switch
    c. A radiant temperature-sensitive valve
    d. A temperature delay valve or restrictor

18. In a GM computer-controlled EGR system, exhaust gas temperature is sensed by:
    a. A diagnostic switch
    b. A radiant temperature-switch
    c. A vane air flow meter
    d. A manifold absolute pressure sensor

19. An aspirated EGR system helps control detonation by delivering EGR flow under:
    a. Low engine vacuum
    b. High engine vacuum
    c. Ported vacuum
    d. Venturi vacuum

Chapter

21

Catalytic Converters

To meet the strict exhaust emission limits of the late 1970's, automakers turned to the catalytic converter. This device is installed in the exhaust system between the exhaust manifold and the muffler, and generally is positioned under the passenger compartment. Its location is important, since as much of the exhaust heat as possible must be retained for effective operation.

From the outside, the **catalytic converter** looks like a small muffler or resonator. It contains no moving parts. It simply forms a chamber in the exhaust system through which the exhaust gas passes. Inside the converter, the exhaust flows through a honeycomb monolith or pellet-type catalyst material, which turns the exhaust pollutants into harmless byproducts of combustion.

In this chapter, you will learn:

- How the catalytic converter works
- The difference between oxidation and reduction converters
- How to make sure that the converter does its job efficiently.

REDUCING EMISSIONS

One way of lowering hydrocarbons (HC) and carbon monoxide (CO) is to increase the combustion temperature, which causes more complete burning. As combustion temperatures rise, however, so does the formation of the third major pollutant, oxides of nitrogen, or NO_x. A second method of turning exhaust gas into nonpolluting materials is the use of the catalytic converter, figure 21-1. By passing the exhaust gas through a **catalyst** in the presence of oxygen, the HC and CO compounds unite with the oxygen, resulting in two harmless byproducts of the catalytic reaction: water vapor (H_2O) and carbon dioxide (CO_2).

Oxidation Reaction

A catalyst is a substance that starts or increases a chemical reaction, while remaining unchanged by that reaction. Since it only encourages rather than takes part in the reaction, the catalyst is never used up.

To change HC and CO into harmless materials, the catalytic elements (platinum and palladium) start an **oxidation**, or burning, reaction in the catalytic converter. Oxidation is the addition of oxygen to an element or compound. If there is not already enough oxygen in the exhaust, an air pump or aspirator valve supplies extra air. The oxidation mixes the HC and CO with oxygen to form H_2O and CO_2.

Considerable heat is generated by the oxidation process. The heat of the catalyst will

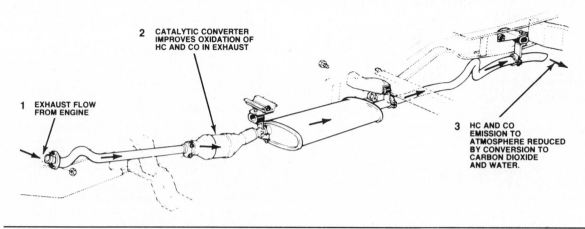

Figure 21-1. Typical catalytic converter installation. (Ford)

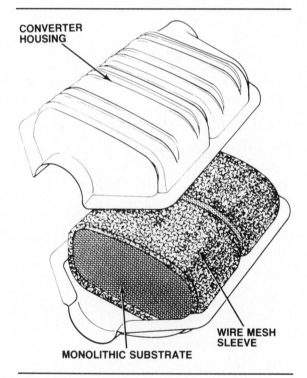

Figure 21-2. Typical catalytic converter with a monolithic substrate.

range from 900°F to 1,600°F (482° to 871°C), and the exhaust gas at the outlet end of the converter will be 50° to 200°F (28° to 111°C) higher than at the inlet end. By 1978, automakers were able to reduce the outside, or "skin", temperatures of converters by as much as 300°F (167°C). The use of smaller engines and changes in engine timing were responsible for this temperature reduction. In spite of the intense heat, however, oxidation does not generate the flame and radiant heat associated with a simple burning reaction.

Reduction Reaction

The catalytic oxidation reaction just described has *no* effect on oxides of nitrogen. NO_x control requires a separate reaction, called **reduction**, rather than oxidation. Reduction is the opposite of oxidation, or the chemical removal of oxygen from a material. The reduction reaction changes NO_x to harmless nitrogen (N_2) and CO_2 by chemically promoting the switch of oxygen from the NO_x to the CO compound. The elements rhodium and platinum are used as reduction catalysts.

CATALYTIC CONVERTER CONSTRUCTION

There are two main types of catalytic converter designs: conventional oxidation converters (COC), and oxidation-reduction converters.

Catalytic Converter: A device installed in an exhaust system that converts pollutants to harmless byproducts through a catalytic chemical reaction.

Catalyst: A substance that causes a chemical reaction, without being changed by the reaction.

Oxidation: The combining of an element with oxygen in a chemical process that often produces extreme heat as a byproduct.

Reduction: A chemical process in which oxygen is taken away from a compound.

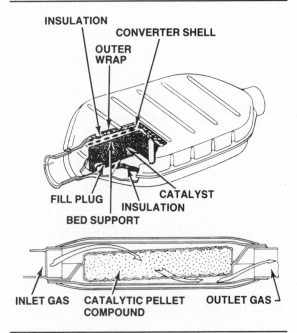

Figure 21-3. Typical catalytic converter with a pellet-type substrate.

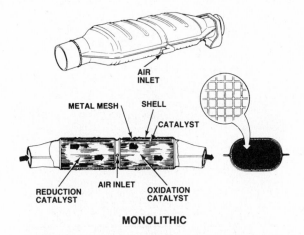

Figure 21-4. Three-way monolithic catalytic converter construction. (GM)

Both consist of two stamped metal pieces welded together to form a round or oval shell. The round type contains a ceramic honeycomb **monolith**, figure 21-2; the oval type contains pellets, figure 21-3. The outer housing is made of aluminum or stainless steel because it must be able to withstand the high temperatures associated with oxidation.

The pellet-type, figure 21-3, creates a fair amount of exhaust restriction, but is less expensive to manufacture, and the pellets in some models can be replaced if they become contaminated. Monolithic converters, figure 21-2, have a honeycomb structure that is much less restrictive, but more expensive to manufacture. This type of converter can only be serviced as a unit.

Conventional Oxidation Converter (COC)

The catalytic element generally used in a conventional oxidation converter (COC) is platinum, or a mixture of platinum and palladium. These two **noble metals** best meet the requirements of an effective catalyst: durability, operating temperature, and chemical activity. The catalyst is deposited on an aluminum oxide or ceramic **substrate** through which the exhaust gas flows. This substrate must provide a catalyst support which can withstand high temperatures.

Monolithic and pellet substrates

Two forms of ceramic substrate material are used in oxidation converters: tiny pellets or a honeycomb monolith. The substrate can either be a laminated (sandwiched) design or an **extruded** design. The catalyst element is deposited on the surface of the substrate material. Both kinds of substrate material — monolith or pellets — provide several thousand square yards or meters of catalyst surface area over which the exhaust gases flow.

In converters using a monolith substrate material, figure 21-2, a diffuser inside the converter shell allows a uniform flow of exhaust gases over the entire area of the substrate. If a diffuser were not used, the gases would tend to flow only through a central portion of the substrate. In converters using a pellet substrate, figure 21-3, the gas flows over the top and down through the substrate layers.

Monolith substrate is a ceramic material which can break easily when subjected to shock or severe jolts. To prevent damage to the core, it is placed inside a stainless steel mesh which acts as a cushion. This also protects the core from thermal shock caused by temperature extremes, and keeps it properly positioned during final assembly of the converter shell.

Reduction Converter

A typical reduction converter uses two substrates of different sizes coated with platinum or rhodium. In a monolith design, these substrates are surrounded by the protective stainless steel mesh and are encased in a stainless steel shell, figure 21-2. In a 3-way converter (described in the next section), the reduction catalyst is located at the front of the converter

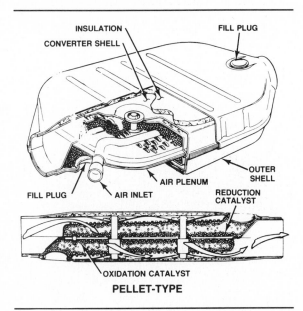

Figure 21-5. Three-way pellet-type catalytic converter construction. (GM)

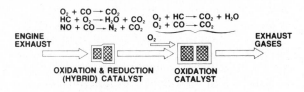

Figure 21-6. Catalytic oxidation and reduction reactions. (Chrysler)

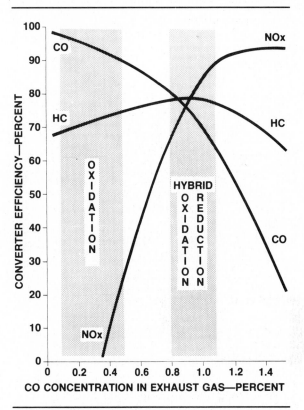

Figure 21-7. Catalyst operating characteristics. (Chrysler)

housing (monolith design), figure 21-4, or at the top of the housing (pellet design), figure 21-5.

Two-Stage, Three-Way Converter (TWC)

Because the oxidation and the reduction reactions oppose each other, both cannot occur at the same time and in the same place. An oxidation and a reduction catalyst can be combined in the same converter, but a second oxidation catalyst is then required for complete emission control. An oxidation-reduction catalyst is often called a 2-stage, 3-way converter. It also is called a TWC (because it works on all three major pollutants), or a hybrid catalyst. A hybrid catalyst and a second oxidation catalyst can be installed in opposite ends of the same converter housing, or in two separate converters, figure 21-6.

The 3-way converter works best to reduce NO_x when the CO level in the exhaust is between 0.8 and 1.5 percent. As the CO level increases or decreases from that percentage, the 3-way converter's efficiency decreases, figure 21-7. Exhaust gas first passes through the hybrid catalyst, where the NO_x is changed to N_2 and CO_2. It then goes through the oxidation catalyst, where the HC and CO are changed to H_2O and CO_2.

Dual-bed monolithic and pellet converters
Catalytic converters using both an oxidation and a reduction catalyst in the same housing are called dual-bed, two-stage or 3-way catalytic converters. Like oxidation converters, the

Monolith: A large block. In a catalytic converter, the monolith is made like a honeycomb to provide several thousand square yards or meters of catalyst surface area.

Noble Metals: Metals, such as platinum and palladium, that resist oxidation.

Substrate: The layer, or honeycomb, of aluminum oxide upon which the catalyst (platinum or palladium) in a catalytic converter is deposited.

Extruded: Shaped by forcing through a die.

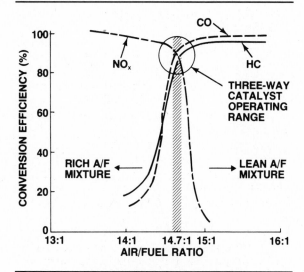

Figure 21-8. Three-way catalyst will only work properly in a narrow air-fuel ratio range. (GM)

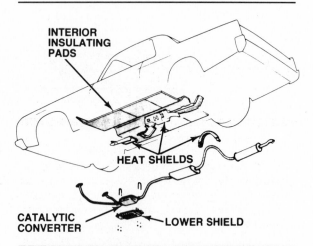

Figure 21-9. Monolithic converters require heat shielding. (Chrysler)

oxidation-reduction converters may use either a monolith or pellet substrate. A 3-way converter using a monolith substrate has the reduction catalyst located at the front of the converter housing, figure 21-4. A 3-way converter with a pellet substrate has one placed above the other, with an air plenum separating the two catalysts, figure 21-5. The reduction catalyst is positioned on top of the plenum. In both designs, the oxygen produced during the reduction catalyst's reaction is used in the oxidation process.

Conversion with a 14.7 air-fuel ratio
Oxidation and reduction converters work most efficiently when the air-fuel mixture is maintained at the stoichiometric ratio of 14.7, figure

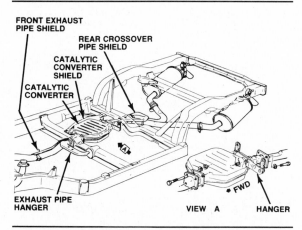

Figure 21-10. Some pellet-type converters may also use heat shields. (Chevrolet)

21-8. The most complete combustion of air and fuel occur at this ratio, resulting in the least amount of harmful pollutants. HC and CO emissions are high at ratios richer than 14.7; NO_x emissions are greatest with ratios leaner than 14.7.

As we learned in Part Four of this manual, the 3-way converter system is used with an air-fuel management feedback system or electronic engine controls. To help maintain the ratio at the ideal, or 14.7, ratio, the 3-way converter system uses an exhaust oxygen (O_2) sensor. This sensor measures the amount of oxygen in the exhaust gas and sends a voltage signal to the engine control microprocessor. The microprocessor then controls the carburetor or fuel injection system to keep the ratio as close as possible to 14.7.

Heat Shields

Monolith converters transmit a considerable amount of heat through their housing. In most installations, heat shields, figure 21-9, are used to protect the passenger compartment, the automatic transmission and its cooler lines, and other parts of the chassis from excessive heat. Pellet-type converters have an insulated housing that does not transmit as much heat as the monolithic type, but some installations also may use heat shields, figure 21-10.

Size, shape, and placement of these shields vary according to the carmaker and model. The underbody floor pan above the converter is heavily shielded and more heat insulation is placed under the floor mats.

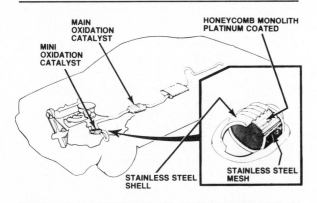

Figure 21-11. Chrysler miniconverter installation. (Chrysler)

FIRST-GENERATION OXIDATION CONVERTERS

The first catalytic converters used only an oxidation catalyst and worked on HC and CO emissions only. These oxidation converters were used from 1975 through 1978 on domestic cars.

AMC and GM cars first used pellet converters made by the AC Division of General Motors. The pellet catalyst can be removed from the converter housing and replaced with fresh catalyst. Chrysler vehicles used a monolithic converter, as did Ford cars. Some 1977 and later Ford vehicles have converters with two monolithic substrates. The Ford converter requires a secondary air source to keep enough oxygen in the exhaust for complete oxidation of the HC. This is provided by the Thermactor air injection system.

Preconverters or Mini-Converters

Some Ford and Chrysler vehicles have used a small oxidation converter connected directly to the exhaust manifold outlet, figure 21-11. This converter contains a small catalyst surface area close to the engine which heats up rapidly to start the HC and CO oxidation process more quickly during cold engine warm-up when the main converter has not yet reached its operating temperature. For this reason, they are often called "light-off" converters, or LOC. The oxidation reaction started in the preconverter is completed by the larger main converter located under the passenger compartment.

SECOND-GENERATION OXIDATION-REDUCTION CONVERTERS

The first 3-way converters were used by Volvo and Saab on a limited number of cars sold in California during 1977. The 2.3-litre, 4-cylinder engine used in 1978 in California Ford vehicles has a closed-loop feedback emission control system. The feedback is provided by the O_2 sensor. The front half of the hybrid converter has an oxidation-reduction catalyst that controls HC, CO, and NO_x. The rear half has only a platinum catalyst to further oxidize HC and CO.

Some of GM's 1978 California cars also have a closed-loop feedback emission control system. It works about the same as the Ford system. This was the first production use of a closed-loop system on a V-type engine. Since 1980, almost every carmaker has used the 3-way design.

Air Switching

We have already seen that the NO_x reduction catalyst works best with less oxygen and a richer HC and CO content in the exhaust gas,

---

■ **Catalytic Converter Odors**

Although catalytic converters control HC and CO emissions, they also produce other undesirable emissions in small quantities. For example, gasoline has a little bit of sulfur in it. This reacts with the water vapor inside a converter to produce hydrogen sulfide. This toxic byproduct has the distinct odor of rotten eggs. The smell is usually most noticeable while the engine is warming up, or during deceleration.

When the odor is very strong at normal operating temperatures, it may mean that the engine is out-of-tune and is running too rich. In many such cases, the carburetor idle mixture screws are not set right, although the problem also may be caused by a high fuel level or some other carburetor problem. Since the odor does not necessarily mean an incorrect mixture adjustment, the carburetor should not be adjusted simply to get rid of the odor. Changing brands of gasoline may help control the odor in some cases, since the amount of sulfur present in gasoline varies from one brand to another.

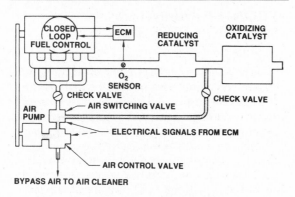

CLOSED LOOP OPERATION

Figure 21-12. Pump air is injected downstream between the reduction and oxidation catalysts when the engine is warm. (AC-Delco)

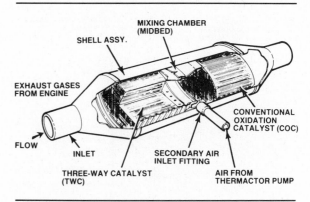

Figure 21-13. Three-way converters contain both catalysts in a single housing with an air inlet between the two catalysts. (Ford)

while the HC and CO oxidation catalyst works best with extra oxygen in the exhaust. For this reason, the air switching function in late-model air injection systems (Chapter 18) routes pump air from the exhaust ports during engine warmup to a downstream point in the exhaust once the engine is at operating temperature.

Pump air must enter the exhaust system at a point between the reduction catalyst and the oxidation catalyst, figure 21-12. Some systems use separate reduction and oxidation converters connected by an air inlet pipe. Most late-model installations, however, house both catalysts in a single converter with an air inlet between the two catalysts, figure 21-13.

CONVERTER OPERATION PRECAUTIONS

Since converters have no moving parts, they require no periodic service. Under Federal law, catalyst effectiveness is warranted for 50,000 miles (80,000 kilometers) or 5 years. However, a catalyst will eventually wear out. When it does, the entire converter (monolith) or catalyst (pellet) must be replaced to maintain effective emission control.

However, there are several ways in which a catalytic converter can be damaged or destroyed. Using the wrong fuel, bad air-fuel mixtures, excessive combustion heat — all of these can damage a converter.

Fuel Requirements

Cars with catalytic converters must use unleaded fuel. The lead and phosphorous additives used in leaded gasoline will coat the catalyst and destroy its efficiency. When leaded fuel

is used, the lead will plate the catalyst and form a coating that prevents the exhaust gases from reaching the catalyst. A change to unleaded gasoline will allow the converter to regain some of its efficiency, depending on how long leaded fuel has been used. However, the only known method of testing a converter's efficiency is with a 4-gas analyzer.

To prevent the use of leaded fuel in cars with catalytic converters, the unleaded gasoline pump nozzles in service stations and the fuel tank filler neck on cars are smaller in diameter than those using leaded fuel. In addition, cars which require unleaded fuel have labels reading "Unleaded Gasoline Only" next to the gas tank filler and on the instrument panel in the driver's compartment. A U.S. Federal law makes it illegal to put leaded fuel in converter-equipped cars.

Fuel system cleaning additives should not be used in the fuel tank or inducted into the engine through the carburetor or fuel injection system on converter-equipped cars, because they may harm the catalyst.

Engine Condition

Excessively high temperatures can reduce converter life and destroy the catalyst core. Converters operate best at internal temperatures up to 1,500°F (816°C). At higher temperatures, the catalyst will start to break up or melt. Although temperature-protection systems are used on some vehicles, they work only under certain conditions. Proper engine maintenance is required to prevent overheating the converter.

When an engine is in poor condition, or needs a tune-up, the exhaust gas fed to the converter contains too much raw fuel. This causes the converter to become a catalytic furnace. If more than two spark plugs misfire at the same time, raw fuel is pumped directly into the catalytic converter. This misfiring over a long time causes internal converter heat to rise rapidly.

Excessive temperatures also may be caused by improper use of an engine in good operating condition. Long idling periods are one of the worst running conditions, since more heat is developed when an engine runs at idle for long periods than when driving at normal highway speeds. Idle periods longer than 10 minutes should be avoided. It is much better to shut off the engine and restart it when required.

Other Precautions

To avoid excessive catalyst temperatures and the possibility of fuel vapors reaching the converter, follow these rules:

1. Do not attempt to start the engine on compression by pushing the vehicle. Use jumper cables instead.
2. Do not crank an engine for more than 40 seconds when it is flooded or firing intermittently.
3. Do not turn off the ignition switch when the car is in motion.
4. Fix problems immediately such as engine dieseling, heavy surging, repeated stalling or backfiring, and other indications of below-normal performance.
5. As much as possible, don't disconnect spark plugs to test the ignition. If an oscilloscope is not available, do not run the engine for more than 30 seconds with the plug wire off or shorted.

SUMMARY

Catalytic converters were first used in 1975. Catalytic converters in the exhaust system promote a chemical reaction to change polluting byproducts of combustion into harmless water and carbon dioxide. Two types of converters are used: oxidation and reduction. The oxidation type removes HC and CO emissions from the exhaust gases; reduction converters remove NO_x. Late-model engines with feedback fuel systems or electronic engine controls use a 3-way converter which combines the oxidation and reduction function.

Early emission controls relied upon spark timing and air injection systems to control HC and CO. When the converter came into use, the role of the air injection system changed from that of a prime oxidation device to an assist device that improved converter efficiency. Cars equipped with an oxidation converter (COC) inject the air into the manifold outlet or exhaust pipe where it helps the catalytic reaction and heats the converter. Cars equipped with a reduction converter (TWC) require that the air be switched to a point downstream once the engine is warm.

Converters use a monolithic or pellet-type substrate coated with platinum, palladium, or rhodium to provide a surface on which the oxidation or reduction reactions occur. These catalysts can be damaged by too much heat, the use of leaded fuel, or too much unburned fuel reaching their surface. The only known way to determine the efficiency of a given converter is with a 4-gas analyzer.

Some early converters have a temperature protection system to prevent damage from overheating. These were abandoned on most 1976 and later vehicles.

Most monolithic converter installations use a series of heat shields to protect chassis components from damage by excessive heat. Pellet-type converters have insulated housings and do not generally require heat shielding.

Converters do not require periodic service during the 50,000-mile (80,000-kilometer) life-span required by U.S. Federal regulations.

Review Questions
Choose the single most correct answer.
Compare your answers with the correct answers on page 414.

1. Catalytic converters:
   a. Increase the HC content in exhaust emissions
   b. Neither add nor remove the oxygen from exhaust emissions
   c. Improve oxidation of HC and NO
   d. Improve oxidation of HC and CO

2. The catalyst material in an oxidation catalytic converter is:
   a. Platinum or palladium
   b. Aluminum oxide
   c. Stainless steel
   d. Ceramic lead oxide

3. A catalyst:
   a. Slows a chemical reaction
   b. Heats a chemical reaction
   c. Increases, but is not consumed by, a chemical action
   d. Combines with the chemicals in the reaction

4. A reduction reaction:
   a. Adds oxygen to a compound
   b. Removes oxygen from a compound
   c. Reduces HC in exhaust gases
   d. Removes H_2O from exhaust gases

5. The outer shell of the oxidation catalyst is made of:
   a. Aluminum oxide pellets
   b. Platinum or palladium
   c. A honeycomb monolith
   d. Aluminum or stainless steel

6. A catalyst is deposited on:
   a. A stainless steel shell
   b. A ceramic substrate
   c. A stainless steel mesh
   d. Any of the above

7. Cars with catalytic converters are required to use:
   a. Premium unleaded fuel
   b. Unleaded fuel
   c. Low-lead fuel
   d. Any of the above

8. Which of the following will damage the converter?
   a. Push starting the engine
   b. Engine misfiring
   c. Long idling periods
   d. All of the above

9. Which of the following is *not* true for cars with catalytic converters?
   a. Engines should not be cranked for more than 60 seconds.
   b. Spark plugs should always be disconnected to test ignition
   c. Engine dieseling, surging, and stalling should be fixed immediately
   d. Ignition should not be turned off while car is moving

10. The reduction converter must be located:
    a. Between the engine and the oxidation converter
    b. Between the oxidation converter and the muffler
    c. After the muffler
    d. Any of the above

11. Hybrid converters were first used on domestic cars in:
    a. 1975
    b. 1976
    c. 1978
    d. 1980

NIASE Technician Certification Sample Test

This sample test is similar in format to the series of tests given by the National Institute for Automotive Service Excellence (ASE). Each of these exams covers one of eight areas of automobile repair and service. The tests are given every fall and spring throughout the United States.

For a technician to earn certification in a particular field, the technician must pass the appropriate test and have at least two years of "hands-on" experience (or a combination of work and auto technician training). Successfully passing all eight tests certifies the technician as a Master Automobile Technician.

In the following sample test, some of the questions were povided by ASE. All of the questions, however, follow the form of the national exams. Learning to take this kind of test will help you if you plan to apply for certification later in your career.

For more information, write to:

National Institute for
    AUTOMOTIVE SERVICE EXCELLENCE
1920 Association Drive, Suite 400
Reston VA 22091-1502

1. The three major air pollutants emitted by motor vehicles are:

   a. Sulfates, particulates, carbon monoxide
   b. Sulfates, carbon monoxide, nitrous oxide
   c. Carbon monoxide, oxides of nitrogen, hydrocarbons
   d. Hydrocarbons, carbon dioxide, nitrous oxide

2. Which should be true of a vehicle's stoichiometric air-fuel ratio?

   a. HC is low; CO, O_2, and CO_2 are high
   b. HC and CO are low; O_2 and CO_2 are high
   c. HC, CO, and O_2 are low; CO_2 is high
   d. HC, CO, O_2, and CO_2 are low

3. While the engine is running, a technician pulls the PCV valve out of the valve cover and puts his thumb over the valve opening. There are no changes in engine operation. Technician A says the PCV valve could be stuck in the open position. Technician B says the hose between the intake manifold (carburetor base) and the PCV valve could be plugged.

   Who is right?
   a. A only
   b. B only
   c. Both A and B
   d. Neither A nor B

4. A car with an air pump emission control system backfires when decelerating. Which of these should the technician check?

   a. The operation of the exhaust manifold check valve
   b. The output pressure of the air pump
   c. The operation of the diverter or gulp valve
   d. The air manifolds for leakage

5. A choke delay valve on a carburetor is the same as a:

   a. Distributor vacuum delay valve
   b. Dual-diaphragm vacuum break
   c. Thermal vacuum valve
   d. Cold-temperature vacuum valve

6. Which of the following is true of an EGR system?

   a. Most early-model systems are regulated by a back-pressure transducer
   b. Most late-model systems are regulated by a back-pressure transducer
   c. Even a poorly maintained EGR system has little effect on driveability
   d. EGR systems have numerous parts scheduled for regular service

7. Pressure and volume tests of a mechanical fuel pump are both below specs.
   Technician A says that an air leak in the fuel line between the tank and the pump could be the cause.
   Technician B says a plugged fuel tank pick-up could be the cause.
   Who is right?
   a. A only
   b. B only
   c. Both A and B
   d. Neither A nor B

8. Air injection systems used with electronic engine controls may have:

   a. Air switching valves
   b. Gulp valves
   c. Diverter valves
   d. All of the above

9. With a firing order of 1-5-3-6-2-4, number 1 cylinder is at tdc with both valves closed when which cylinder is at tdc with the exhaust valve fully open?
   a. 3
   b. 4
   c. 5
   d. 6

10. Centrifugal advance weights in an electronic distributor move the:
    a. Transducer
    b. Baseplate (breaker plate)
    c. Stator
    d. Trigger wheel

11. A compression test has been made on a 6-cylinder engine. Cylinders 3 and 4 have readings of 10 psi. All other cylinders have readings between 130 and 135 psi.
    Technician A says this could be caused by a blown head gasket.
    Technician B says it could be caused by wrong valve timing.
    Who is right?
    a. A only
    b. B only
    c. Both A and B
    d. Neither A nor B

12. Which of these is least likely to cause a car to hesitate (stumble) when the gas pedal is depressed quickly?
    a. Retarded ignition timing
    b. Low carburetor float level
    c. Leaking carburetor accelerator pump check valve
    d. Leaking carburetor power valve

13. The burn time in a typical spark-ignition, gasoline engine is approximately:
    a. 2 milliseconds
    b. 3 milliseconds
    c. 6 milliseconds
    d. 30 milliseconds

14. The number of degrees that a crankshaft rotates between power strokes of individual cylinders is the:
    a. Firing order
    b. Ignition interval
    c. Ignition timing
    d. Timing offset

15. Valve lash is being adjusted on an engine with solid lifters. Technician A says too much valve lash can cause poor engine performance.
    Technician B says not enough valve lash can cause valve burning.
    Who is right?
    a. A only
    b. B only
    c. Both A and B
    d. Neither A nor B

16. Generally a fuel pump should deliver a fuel volume of:
    a. One pint in one minute at 500 rpm
    b. One quart in one minute at 500 rpm
    c. One gallon in one minute at 500 rpm
    d. One quart in five minutes at 500 rpm

17. A cold-start injector is a:
    a. Regulated airflow injector
    b. Mechanical, continuous injector
    c. Solenoid-operated auxiliary injector
    d. Detonation-control injector

18. Which statement is true of a cylinder power balance test?
    a. An engine with electronic idle speed control should be tested at idle
    b. Exhaust emissions should noticeably increase when you cut out one cylinder
    c. Both a and b
    d. Neither a nor b

19. A leaking fuel pump diaphragm may cause:
    a. Backfiring
    b. Crushed fuel tank
    c. Fuel leaks at the filler tube
    d. Diluted oil

20. Most of the sensors used in fuel injection systems are:
    a. Resistors
    b. Transistors
    c. Potentiometers
    d. All of the above

21. When the engine is off, the hot air passage remains open and the cold air passage remains closed in a vacuum-operated air cleaner. This could be caused by:
    a. High ambient air temperature
    b. A broken vacuum motor spring
    c. Low ambient air temperature
    d. The heat control valve stuck closed

22. Which of the following is not a carburetor float adjustment?
    a. Float toe
    b. Float reach
    c. Float drop
    d. Float level

23. A choke adjustment specification that calls for the choke to be set on "index" means:
    a. The choke rod or link must be bent to a specified length
    b. The vacuum break link must be bent until the U-bend is closed
    c. The index mark on the choke cover must be aligned with a similar mark on the housing
    d. The choke link must be placed in the index hole in the choke lever

24. If the rpm gain measured during artificial enrichment tests is less than specified:
    a. The air-fuel ratio is too lean
    b. The engine is overheated
    c. The air-fuel ratio is too rich
    d. The air injection system is not working correctly

25. An engine idles roughly and stalls on light acceleration. When the vacuum line is disconnected from the EGR valve, the problem disappears. This probably means:
    a. The EGR valve is stuck closed
    b. The EGR valve is getting a weak vacuum signal
    c. The EGR valve diaphragm spring is broken
    d. The EGR valve is incorrectly opening at idle

26. High HC exhaust emissions are often due to:
    a. An overheated engine
    b. A restricted air cleaner
    c. Ignition system problems
    d. An inoperative EGR system

27. During inspection of an air injection system, a brittle, burned hose is found on the upstream side of a check valve. This may be due to:
    a. The hose being positioned too close to the engine
    b. The air pump developing excessive air pressure
    c. The injection nozzles being plugged and allowing back-pressure to build up in the system
    d. The check valve allowing exhaust gas to leak back into the air injection system

28. Which of the following should be done before testing engine compression?
    a. Be sure battery is fully charged and engine cranks freely
    b. Be sure engine is a normal temperature and throttle is held open
    c. Disconnect or ground the ignition system
    d. All of the above

29. A cylinder power balance test is performed while the engine is running by:
    a. Shorting the ignition of each cylinder and noting rpm drop
    b. Shorting the ignition of each cylinder and noting rpm gain
    c. Running the ignition for each cylinder in an open circuit condition and noting available secondary voltage
    d. Opening a vacuum port at the carburetor or intake manifold and noting rpm gain

30. Which of the following must be done before carburetor idle speed and mixture adjustment?
    a. Mechanical valve lifter adjustment
    b. Manifold heat control valve service
    c. Complete ignition system service
    d. All of the above

31. On a car with a throttle stop solenoid, normal slow idle is usually adjusted with the solenoid:
    a. Deenergized and retracted
    b. Energized and extended
    c. Removed from the carburetor
    d. None of the above

32. Idle speed adjustment on a carburetor with an idle air bypass circuit is usually made by turning a large screw that:
    a. Varies the throttle valve position
    b. Regulates fuel flow through the idle circuit
    c. Varies airflow through the bypass circuit
    d. Opens an air-fuel passage in the secondary idle circuit

33. Adjusting idle mixture by turning the mixture screws to obtain the smoothest idle and then turning them in leaner until the speed drops a specified amount is called the:
    a. Lean best idle method
    b. Lean drop method
    c. One-quarter turn rich method
    d. None of the above

34. The exhaust manifold heat riser is stuck in the open position.
    Technician A says this can cause poor gas mileage.
    Technician B says this can cause the intake manifold vacuum to be lower than normal.
    Who is right?
    a. A only
    b. B only
    c. Both A and B
    d. Neither A nor B

35. Technician A says the automatic choke is opened by manifold vacuum.
    Technician B says the automatic choke is closed by spring force.
    Who is right?
    a. A only
    b. B only
    c. Both A and B
    d. Neither A nor B

36. The valve shown below controls ignition timing:

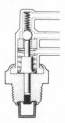

    a. Advance during deceleration
    b. Retard during high-speed driving
    c. Advance during high-temperature idle
    d. Retard during high-temperature idle

37. A late-model engine computer typically cannot recognize that:
    a. A signal is not being furnished
    b. A signal is improbable
    c. A signal is out of limits for too long
    d. A device is worn or in need of replacement

38. Technician A says low-pressure fuel injection systems operate at 20 to 25 psi.
    Technician B says high-pressure fuel injection systems operate at 50 psi or more.
    Who is right?
    a. A only
    b. B only
    c. Both A and B
    d. Neither A nor B

39. Hardened steel caps or plugs installed over the idle mixture screws have been used on most carburetors since:
    a. 1971
    b. 1976
    c. 1980
    d. 1983

40. Liquid fuel in the EEC canister generally indicates a defective:
    a. Head gasket
    b. Vacuum-operated purge valve
    c. Engine control module
    d. Liquid-vapor separator

Glossary
of Technical
Terms

Actuator: An electrical or mechanical device that receives an output signal from a computer and does something in response to that signal.

Adaptive Memory: A feature of computer memory that allows the microprocessor to adjust its memory for computing open-loop operation, based on changes in engine operation.

Adsorption: A chemical action by which liquids or vapors are gathered on the surface of a material. In a vapor storage canister, fuel vapors are attached (adsorbed) to the surface of charcoal granules.

After Top Dead Center: The position of a piston after it has passed top dead center. Abbreviated: atdc. Usually expressed in degrees, such as 5° atdc.

Air Charge Temperature (ACT) Sensor: A thermistor used to measure intake air temperature or air-fuel mixture temperature.

Air Injection: A way of reducing exhaust emissions by injecting air into each of the exhaust ports of an engine. It mixes with the hot exhaust and oxidizes the HC and CO to form H_2O and CO_2.

Air-Fuel Ratio: The ratio of air to gasoline in the air-fuel mixture which enters an engine.

Ambient Temperature: The temperature of the air surrounding a particular device or location.

Ampere: The unit for measuring the rate of electrical current flow.

Analog: A voltage signal or processing action that is continuously variable relative to the operation being measured or controlled.

Analog-to-Digital (AD): An electronic conversion process for changing analog voltage signals to digital voltage signals.

Aneroid Bellows: An accordion-shaped membrane exposed to barometric (atmospheric) pressure. Pressure changes cause the bellows to flex. On a carburetor, this flexing actuates a valve or metering rods to change fuel flow.

Antiknock Value: The characteristic of gasoline that helps prevent detonation or "knocking".

Antioxidant Inhibitors: A gasoline additive used to prevent oxidation and the formation of gum.

Armature: The movable part in a relay. The revolving part in a generator or motor.

Aspirator: A regulator and venturi installed in the air cleaner to provide EGR flow under low engine vacuum conditions to control detonation.

Atmospheric Pressure: The pressure caused by the weight of the earth's atmosphere. At sea level, this pressure is 14.7 psi (101 kPa) at 32°F (0°C).

Atomization: Breaking down into small particles or a fine mist.

Backfire: The accidental combustion of gases in an engine's intake or exhaust system.

Backpressure: The resistance, caused by turbulence and friction, that is created as a gas or liquid is forced through a passage.

Baffle: A plate or obstruction that restricts the flow of air or liquids. The baffle in a fuel tank keeps the fuel from sloshing as the car moves.

Before Top Dead Center: The position of a piston as it nears top dead center. Abbreviated: btdc. Usually expressed in degrees, such as 5° btdc.

Bifurcated: Separated into two parts. A bifurcated exhaust manifold has four primary runners that converge into two secondary runners; these converge into a single outlet into the exhaust system.

Bimetal Temperature Sensor: A device made of two strips of metal welded together. When heated, one side will expand more than the other, causing it to bend.

Binary: A mathematical system consisting of only two digits (0 and 1) which allows a digital computer to read and process input voltage signals.

Block-Learn: The long-term effects of integrator corrections. As such, block-learn complements adaptive memory. If, for example, the computer continually must overcompensate fuel metering to maintain the stoichiometric ratio, it "learns" the correction and adapts its memory to make the correction factor part of its basic program.

Blowby: The leakage of combustion gases and unburned fuel past an engine's piston rings.

Blowoff Valve: A spring-loaded valve that opens when boost pressure overcomes the spring tension.

Boost: A measure of the amount of air pressurization, above-atmospheric, that a supercharger can deliver.

Bore: The diameter of an engine cylinder.

Bottom Dead Center: The exact bottom of a piston stroke. Abbreviated: bdc.

Camshaft Overlap: The period of crankshaft rotation in degrees during which both the intake and exhaust valves are open.

Carbon Monoxide (CO): An odorless, colorless, tasteless poisonous gas. A major pollutant given off by an internal combustion engine.

Carburetor Icing: A condition that is the result of the rapid vaporization of fuel entering a carburetor; the temperature drops enough to freeze the water particles in the airflow.

Catalyst: A substance that causes a chemical reaction, without being changed by the reaction.

Catalytic Converter: A device installed in an exhaust system that converts pollutants to harmless byproducts through a catalytic chemical reaction.

Catalytic Cracking: An oil refining process which uses a catalyst to break down (crack) the larger components of the crude oil. The gasoline produced usually has a lower sulfur content than gasoline produced by thermal cracking.

Central Processing Unit (CPU): The processing and calculating portion of a computer.

Check Valve: A valve that permits flow in only one direction.

Circuit: A circle or unbroken path through which an electric current can flow.

Clearance Volume: The volume of a combustion chamber when the piston is at top dead center.

Closed-Loop: An operational mode in which the engine control microprocessor reads and responds to feedback signals from the EGO sensor, adjusting system operation accordingly.

Combination Valve: A valve on the fuel tanks of some Ford cars that allows fuel vapors to escape to the vapor storage canister, relieves fuel tank pressure, and lets fresh air into the tank as fuel is withdrawn. Similar to a liquid-vapor separator valve.

Compression Ratio: The total volume of an engine cylinder divided by its clearance volume.

Conductor: A material that allows easy flow of electricity.

Continuous Injection System: A fuel injection system in which fuel is injected constantly whenever the engine is running. Bosch K-Jetronic and CIS systems are typical examples.

Conventional Theory of Current Flow: The current flow theory which says electricity flows from positive to negative. Also called positive current flow theory.

Detonation: Also called knocking or spark knock. An unwanted explosion of an air-fuel mixture caused by high heat and compression.

Dieseling: A condition in which extreme heat in an engine's combustion chamber continues to ignite fuel after the ignition has been turned off.

Digital: A 2-level voltage signal or processing function that is either on/off or high/low.

Digital-to-Analog (DA): An electronic conversion process for changing digital voltage signals to analog voltage signals.

Displacement: A measurement of the volume of air displaced by a piston as it moves from bottom to top of its stroke. Engine displacement is the piston displacement multiplied by the number of pistons in an engine.

Diverter Valve: Also called a dump valve. A valve used in an air injection system to prevent backfire. During deceleration it "dumps" the air from the air pump into the atmosphere.

Duty Cycle: The percentage of total time in one complete on-off cycle during which a solenoid is energized.

Dynamometer: A device used to measure mechanical power, such as the power of an engine.

Eccentric: Off center. A shaft lobe which has a center different from that of the shaft.

Electron Theory of Current Flow: The current flow theory which says electricity flows from negative to positive.

Electronic Fuel Injection (EFI): A computer-controlled fuel injection system which gives precise mixture control and almost instant response to all operating conditions at all speed ranges.

Engine Mapping: Vehicle operation simulation procedure used to tailor the onboard computer program to a specific engine/powertrain combination. This program is stored in a PROM or calibration assembly.

Ethanol: Ethyl alcohol distilled from grain or sugar cane.

Evaporative Emission Control (EEC): A way of controlling HC emissions by collecting fuel vapors from the fuel tank and carburetor fuel bowl vents and directing them through an engine's intake system.

Exhaust Gas Recirculation (EGR): A way of reducing NO_x emissions by directing unburned exhaust back through an engine's intake.

Extruded: Shaped by forcing through a die.

Fast-Burn Combustion Chamber: A compact combustion chamber with a centrally located spark plug. The chamber is designed to shorten the combustion period by reducing the distance of flame front travel.

Firing Order: The order in which combustion occurs in the cylinders of an engine.

Flat Spot: The brief hesitation or stumble of an engine caused by a momentary overly lean air-fuel mixture due to the sudden opening of the throttle.

Float Valve: A valve that is controlled by a hollow ball floating in a liquid, such as in the fuel bowl of a carburetor.

Flooding: A condition caused by heat expanding the fuel in a fuel line. The fuel pushes the carburetor inlet needle valve open and fills up the fuel bowl even when more fuel is not needed. Also, the presence of too much fuel in the intake manifold.

Four-Stroke Engine: The Otto cycle engine. An engine in which a piston must complete four strokes to make up one operating cycle. The strokes are: intake, compression, power, and exhaust.

G-Lader: A type of supercharger pump which compresses air by squeezing it through an internal spiral, then forcing it through ports into the engine.

Galvanic Battery: A direct current voltage source, generated by the chemical action of an electrolyte.

Gasohol: A blend of ethanol and unleaded gasoline, usually at a one to nine ratio.

Gulp Valve: A valve used in an air injection system to prevent backfire. During deceleration it redirects air from the air pump to the intake manifold where the air leans out the rich air-fuel mixture.

Headers: Exhaust manifolds on high-performance engines that reduce backpressure by using larger passages with gentle curves.

High-Speed Surge: A sudden increase in engine speed caused by high manifold vacuum pulling in an excess air-fuel mixture.

High-Swirl Combustion Chamber: A combustion chamber in which the intake valve is shrouded or masked to direct the incoming air-fuel charge and create turbulence that will circulate the mixture more evenly and rapidly.

Hydrocarbon (HC): A chemical compound made up of hydrogen and carbon. A major pollutant given off by an internal combustion engine. Gasoline, itself, is a hydrocarbon compound.

Ignition Interval (Firing Interval): The number of degrees of crankshaft rotation between ignition sparks.

Impeller: A rotor or rotor blade used to force a gas or liquid in a certain direction under pressure.

Inert Gas: A gas that will not undergo chemical reaction.

Injection Pump: A pump used on diesel engines to deliver fuel under high pressure at precisely timed intervals to the fuel injectors.

Input Conditioning: The process of amplifying or converting a voltage signal into a form usable by the computer's central processing unit.

Integrator: The ability of the computer to make short-term — minute-by-minute — corrections in fuel metering.

Intercooler: An air-to-air or air-to-liquid heat exchanger used to lower the temperature of the air-fuel mixture by removing heat from the intake air charge.

Keep-Alive Memory (KAM): A form of long-term RAM used mostly with adaptive strategies. Requires a separate power supply circuit to maintain voltage when the ignition is off.

Linear: In a straight line.

Liquid-Vapor Separator Valve: A valve in some EEC fuel systems that separates liquid fuel from fuel vapors.

Lobes: The rounded protrusions on a camshaft that force, and govern, the opening of the intake and exhaust valves.

Manifold Vacuum: Low pressure in an engine's intake manifold, located below the carburetor throttle.

Methanol: Methyl alcohol distilled from wood or made from natural gas.

Micron: A unit of length equal to one millionth of a meter, one one thousandth of a millimeter.

Monolith: A large block. In a catalytic converter, the monolith is made like a honeycomb to provide several thousand square yards or meters of catalyst surface area.

Multipoint (Port) Injection: A fuel injection system in which individual injectors are installed in the intake manifold at a point close to the intake valve. Air passing through the manifold mixes with the injector spray just as the intake valve opens.

Negative Temperature Coefficient (NTC) Resistor: A thermistor whose resistance decreases as the temperature increases.

Noble Metals: Metals, such as platinum and palladium, that resist oxidation.

Normally Aspirated: An engine that uses normal vacuum to draw in its air-fuel mixture. Not supercharged.

Octane Rating: The measurement of the antiknock value of a gasoline.

Ohm: The unit for measuring electrical resistance.

Oil Galleries: Passages in the block and head that carry oil under pressure to various parts of the engine.

Open-Loop: An operational mode in which the engine control microprocessor adjusts the system to function according to predetermined instructions and does not respond to feedback signals from the EGO sensor.

Orifice: A small opening in a tube, pipe, or valve.

Oxidation: The combining of an element with oxygen in a chemical process that often produces extreme heat as a byproduct.

Oxides of Nitrogen (NO$_x$): Chemical compounds of nitrogen given off by an internal combustion engine. They combine with hydrocarbons to produce smog.

Parallel Circuit: A circuit with more than one path for the current to follow.

Particulates: Liquid or solid particles such as lead and carbon that are given off by an internal combustion engine as pollution.

Percolation: The bubbling and expansion of a liquid. Similar to boiling.

Photochemical Smog: A combination of pollutants which, when acted upon by sunlight, forms chemical compounds that are harmful to human, animal, and plant life.

Piezoelectric: Voltage caused by physical pressure applied to the faces of certain crystals.

Piezoresistive: A sensor whose resistance varies in relation to pressure or force applied to it. A piezoresistive sensor receives a constant reference voltage and returns a variable signal in relation to its varying resistance.

Pintle Valve: A valve shaped much like a hinge pin. In an EGR valve, the pintle is attached to a normally closed diaphragm. When ported vacuum is applied, the pintle rises from its seat and allows exhaust gas to be drawn into the engine's intake system.

Plenum: A chamber that stabilizes the air-fuel mixture and allows it to rise to a pressure slightly above atmospheric pressure.

Poppet Valve: A valve that plugs and unplugs its opening by axial motion.

Ported Vacuum: The low-pressure area (vacuum) just above the throttle in a carburetor.

Positive Crankcase Ventilation (PCV): A way of controlling engine emissions by directing crankcase vapors (blowby) back through an engine's intake system.

Positive Temperature Coefficient (PTC) Resistor: A thermistor whose resistance decreases as the temperature increases.

Pre-Ignition: An engine condition in which the air-fuel mixture ignites prematurely due to excessive combustion chamber temperature.

Pressure Differential: A difference in pressure between two points.

Pressure Drop: A reduction of pressure between two points.

Programmable Read-Only Memory (PROM): An integrated circuit chip installed in a computer which contains appropriate operating instructions and database information for a particular application.

Pulsating: To expand and contract rhythmically.

Purge Valve: A vacuum-operated valve used to draw fuel vapors from a vapor storage canister.

Random-Access Memory (RAM): Temporary short-term or long-term computer memory that can be read and changed, but is lost whenever power is shut off to the computer.

Read-Only Memory (ROM): The permanent part of a computer's memory storage function. ROM can be read but not changed, and is retained when power is shut off to the computer.

Reciprocating Engine: Also called piston engine. An engine in which the pistons move up and down or back and forth, as a result of combustion of an air-fuel mixture in the top of the piston cylinder.

Reduction: A chemical process in which oxygen is taken away from a compound.

Reed Valve: A one-way check valve. A reed, or flap, opens to admit a fluid or gas under pressure from one direction, while closing to deny movement from the opposite direction.

Reference Voltage: A constant voltage signal (below battery voltage) applied to a sensor by the computer. The sensor alters the voltage according to engine operating conditions and returns it as a variable input signal to the computer, which adjusts system operation accordingly.

Resistance: Opposition to electrical current flow.

Runners: The passages or branches of an intake manifold that connect the manifold's plenum chamber to the engine's inlet ports.

Scavenging: A slight suction caused by a vacuum drop through a well designed header system. Scavenging helps pull exhaust gases out of an engine cylinder.

Series Circuit: A circuit with only one path for the current to follow.

Series-Parallel Circuit: A circuit in which some loads are wired in series and some are wired in parallel.

Siamesed: Joined together. A siamesed port on an intake manifold is a single port that supplies the air-fuel mixture to two cylinders.

Sintered: Welded together without using heat, forming a porous material such as the metal disc used in some vacuum delay valves.

Siphoning: The flowing of a liquid as a result of a pressure differential, without the aid of a mechanical pump.

Spark Timing: A way of controlling exhaust emissions by controlling ignition timing. Vacuum advance is delayed or shut off at low and medium speeds, reducing NO_x and HC emissions.

Stoichiometric Ratio: An ideal air-fuel mixture for combustion in which all oxygen and all fuel will be completely burned.

Stratified Charge Engine: An engine that uses 2-stage combustion: first is combustion of a rich air-fuel mixture in a precombustion chamber, then combustion of a lean air-fuel mixture occurs in the main combustion chamber.

Stroke: One complete top-to-bottom or bottom-to-top movement of an engine piston.

Substrate: The layer, or honeycomb, of aluminum oxide upon which the catalyst (platinum or palladium) in a catalytic converter is deposited.

Sulfur Oxides: Chemical compounds given off by processing and burning gasoline and other fossil fuels. As they decompose, they combine with water to form sulfuric acid.

Supercharging: Use of an air pump to deliver an air-fuel mixture to the engine cylinders at a pressure greater than atmospheric pressure.

Temperature Inversion: A weather pattern in which a layer or "lid" of warm air keeps the cooler air beneath it from rising.

Tetraethyl Lead: A gasoline additive used to help prevent detonation.

Thermal Cracking: A common oil refining process which uses heat to break down (crack) the larger components of the crude oil. The gasoline which is produced usually has a higher sulfur content than gasoline produced by catalytic cracking.

Thermistor (Thermal Resistor): A resistor especially built to change its resistance as the temperature changes.

Thermostatic: Referring to a device that automatically responds to temperature changes in order to activate a switch.

Throttle Body Injection (TBI): A fuel injection system in which one or two injectors are installed in a carburetor-like throttle body mounted on a conventional intake manifold. Fuel is sprayed at a constant pressure above the throttle plate to mix with the incoming air charge.

Top Dead Center: The exact top of a piston's stroke. Also a specification used when tuning an engine. Abbreviated: tdc.

Transducer: A device that changes one form of energy into another.

Turbo Lag: The time interval required for a turbocharger to overcome inertia and spin up to speed.

Turbocharger: A supercharging device that uses exhaust gases to turn a turbine that forces extra air-fuel mixture into the cylinders.

Two-Stroke Engine: An engine in which a piston makes only two strokes to complete one operating cycle.

Vacuum Lock: A stoppage of fuel flow caused by insufficient air intake to the fuel tank.

Vacuum: A pressure less than atmospheric pressure.

Vapor Lock: A condition in which bubbles are formed in a car's fuel system when the fuel gets hot enough to boil. Flow is stopped or restricted as a result.

Vaporization: Changing a liquid, such as gasoline, into a gas (vapor).

Venturi Vacuum: Low pressure in the venturi of a carburetor, caused by fast airflow through the venturi.

Venturi: A restriction in an airflow, such as in a carburetor, that speeds the airflow and creates a vacuum.

Volatility: The ease with which a liquid changes from a liquid to a gas or vapor.

Volt: The unit for measuring the amount of electrical force.

Voltage Drop: The amount of voltage required to move current through a load.

Voltage: The electromotive force that causes current flow. The potential difference in electrical force between two points when one is negatively charged and the other is positively charged.

Volumetric Efficiency: The comparison of the *actual* volume of air-fuel mixture drawn into an engine to the *theoretical maximum* volume that could be drawn in. Written as a percentage.

Wastegate: A diaphragm-actuated bypass valve used to limit turbocharger boost pressure by limiting the speed of the exhaust turbine.

Water Injection: A method of lowering the air-fuel mixture temperature by injecting a fine spray of water which evaporates as it cools the intake charge.

Water Jackets: Passages in the head and block that allow coolant to circulate throughout the engine.

Index

Answers to Review and NIASE Questions

Chapter 1: Introduction to Fuel Systems and Emission Controls
1-C 2-C 3-D 4-A 5-C 6-D 7-B
8-D 9-C 10-B 11-C 12-B 13-D
14-A

Chapter 2: Engine Operating Principles
1-C 2-B 3-A 4-D 5-B 6-D 7-D
8-D 9-D 10-D 11-A 12-C 13-D
14-B 15-A 16-B 17-C 18-C 19-D

Chapter 3: Engine Air-Fuel Requirements
1-A 2-C 3-D 4-A 5-C 6-B 7-D
8-B 9-D 10-B 11-B 12-D 13-C
14-C 15-B 16-D 17-A

Chapter 4: Fuel Tanks, Lines, and Evaporative Emission Controls
1-C 2-B 3-A 4-C 5-D 6-C 7-A
8-D 9-C 10-D 11-B 12-B 13-A
14-C 15-B 16-B 17-D

Chapter 5: Fuel Pumps and Filters
1-C 2-B 3-C 4-B 5-D 6-A 7-D
8-C 9-B 10-A 11-A 12-D 13-C
14-A

Chapter 6: Air Cleaners and Filters
1-D 2-B 3-B 4-C 5-A 6-C 7-A
8-C 9-D 10-C 11-B 12-D

Chapter 7: Intake and Exhaust Manifolds
1-C 2-D 3-D 4-D 5-C 6-B 7-B
8-A 9-B 10-C 11-B 12-D 13-?
14-?

Chapter 8: Basic Carburetion
1-C 2-B 3-A 4-D 5-C 6-A 7-B
8-D 9-C 10-C 11-B 12-A 13-C
14-D 15-A 16-B 17-C 18-D 19-A
20-C 21-A 22-B 23-A 24-B 25-C
26-C 27-A 28-B 29-B 30-A

Chapter 13: Electronic Fuel Metering Control
1-B 2-C 3-A 4-C 5-D 6-B 7-C
8-A 9-D 10-A 11-C 12-D 13-A
14-C 15-B 16-A 17-C 18-D 19-C
20-C

Chapter 14: Electronic Engine Control Systems
1-A 2-C 3-C 4-A 5-D 6-C 7-C
8-A 9-B 10-A 11-C 12-D 13-A
14-B 15-C 16-B

Chapter 15: Gasoline Fuel Injection Systems
1-A 2-C 3-A 4-C 5-C 6-A 7-B
8-B 9-A 10-D 11-B 12-D 13-A
14-A 15-D

Chapter 16: Supercharging and Turbochargers
1-C 2-A 3-B 4-D 5-C 6-B 7-C
8-A 9-D 10-D 11-B 12-A 13-B
14-C

Chapter 17: Positive Crankcase Ventilation
1-D 2-B 3-D 4-C 5-A 6-C 7-D
8-D 9-D 10-A 11-D 12-B 13-C
14-D 15-B

Chapter 18: Air Injection
1-B 2-A 3-C 4-D 5-A 6-B 7-B
8-D 9-C 10-A 11-D 12-C

Chapter 19: Spark Timing Control Systems
1-D 2-D 3-C 4-B 5-A 6-C 7-C
8-B 9-D 10-A 11-B 12-D 13-D
14-C 15-B 16-D 17-C

Chapter 20: Exhaust Gas Recirculation
1-A 2-C 3-C 4-A 5-D 6-D 7-B
8-C 9-C 10-C 11-C 12-C 13-C
14-C 15-D 16-C 17-D 18-A 19-A

Chapter 21: Catalytic Converters
1-D 2-A 3-C 4-B 5-D 6-B 7-B
8-D 9-B 10-D 11-C

NIASE Technician Certification Sample Test
1-C 2-C 3-B 4-C 5-A 6-B 7-C
8-D 9-D 10-D 11-A 12-D 13-B
14-B 15-C 16-B 17-C 18-C 19-D
20-A 21-B 22-B 23-C 24-C 25-D
26-C 27-D 28-D 29-A 30-D 31-B
32-C 33-B 34-A 35-A 36-C 37-D
38-B 39-C 40-D